Advances in Functional Rubber and Elastomer Composites

Advances in Functional Rubber and Elastomer Composites

Editor

Md Najib Alam

Basel • Beijing • Wuhan • Barcelona • Belgrade • Novi Sad • Cluj • Manchester

Editor
Md Najib Alam
School of Mechanical Engineering
Yeungnam University
Gyeongsan
Korea, South

Editorial Office
MDPI AG
Grosspeteranlage 5
4052 Basel, Switzerland

This is a reprint of articles from the Special Issue published online in the open access journal *Polymers* (ISSN 2073-4360) (available at: https://www.mdpi.com/journal/polymers/special_issues/Adv_Funct_Rubber_Elastomer_Compos).

For citation purposes, cite each article independently as indicated on the article page online and as indicated below:

Lastname, A.A.; Lastname, B.B. Article Title. *Journal Name* **Year**, *Volume Number*, Page Range.

ISBN 978-3-7258-1638-5 (Hbk)
ISBN 978-3-7258-1637-8 (PDF)
doi.org/10.3390/books978-3-7258-1637-8

© 2024 by the authors. Articles in this book are Open Access and distributed under the Creative Commons Attribution (CC BY) license. The book as a whole is distributed by MDPI under the terms and conditions of the Creative Commons Attribution-NonCommercial-NoDerivs (CC BY-NC-ND) license.

Contents

About the Editor vii

Preface ix

Md Najib Alam
Advances in Functional Rubber and Elastomer Composites
Reprinted from: *Polymers* **2023**, *16*, 1726, doi:10.3390/polym16121726 1

Bettina Fasolt, Fabio Beco Albuquerque, Jonas Hubertus, Günter Schultes, Herbert Shea and Stefan Seelecke
Electrode Impact on the Electrical Breakdown of Dielectric Elastomer Thin Films
Reprinted from: *Polymers* **2023**, *15*, 4071, doi:10.3390/polym15204071 5

Ming Liu, Iikpoemugh Elo Imiete, Mariapaola Staropoli, Pascal Steiner, Benoît Duez, Damien Lenoble, et al.
Hydrophobized MFC as Reinforcing Additive in Industrial Silica/SBR Tire Tread Compound
Reprinted from: *Polymers* **2023**, *15*, 3937, doi:10.3390/polym15193937 23

Zhaoyang Wang, Yankai Lin, Zhanxu Li, Yumeng Yang, Jun Lin and Shaojian He
Effect of Fluorosilicone Rubber on Mechanical Properties, Dielectric Breakdown Strength and Hydrophobicity of Methyl Vinyl Silicone Rubber
Reprinted from: *Polymers* **2023**, *15*, 3448, doi:10.3390/polym15163448 41

Federica Magaletti, Fatima Margani, Alessandro Monti, Roshanak Dezyani, Gea Prioglio, Ulrich Giese, et al.
Adducts of Carbon Black with a Biosourced Janus Molecule for Elastomeric Composites with Lower Dissipation of Energy
Reprinted from: *Polymers* **2023**, *15*, 3120, doi:10.3390/polym15143120 53

Sanghoon Song, Haeun Choi, Junhwan Jeong, Seongyoon Kim, Myeonghee Kwon, Minji Kim, et al.
Optimized End Functionality of Silane-Terminated Liquid Butadiene Rubber for Silica-Filled Rubber Compounds
Reprinted from: *Polymers* **2023**, *15*, 2583, doi:10.3390/polym15122583 71

Daria Slobodinyuk, Alexey Slobodinyuk, Vladimir Strelnikov and Dmitriy Kiselkov
Simple and Efficient Synthesis of Oligoetherdiamines: Hardeners of Epoxyurethane Oligomers for Obtaining Coatings with Shape Memory Effect
Reprinted from: *Polymers* **2023**, *15*, 2450, doi:10.3390/polym15112450 90

Ahmed A. Bakhsh
Optimization of Polyolefin-Bonded Hydroxyapatite Graphite for Sustainable Industrial Applications
Reprinted from: *Polymers* **2023**, *15*, 1505, doi:10.3390/polym15061505 107

Muhammad Yasar Razzaq, Joamin Gonzalez-Gutierrez, Muhammad Farhan, Rohan Das, David Ruch, Stephan Westermann and Daniel F. Schmidt
4D Printing of Electroactive Triple-Shape Composites
Reprinted from: *Polymers* **2023**, *15*, 832, doi:10.3390/polym15040832 127

Saedah R. Al-Mhyawi, Nader Abdel-Hamed Abdel-Tawab and Rasha M. El Nashar
Synthesis and Characterization of Orange Peel Modified Hydrogels as Efficient Adsorbents for Methylene Blue (MB)
Reprinted from: *Polymers* **2023**, *15*, 277, doi:10.3390/polym15020277 143

Jae Kap Jung, Ji Hun Lee, Sang Koo Jeon, Un Bong Baek, Si Hyeon Lee, Chang Hoon Lee and Won Jin Moon
H_2 Uptake and Diffusion Characteristics in Sulfur-Crosslinked Ethylene Propylene Diene Monomer Polymer Composites with Carbon Black and Silica Fillers after High-Pressure Hydrogen Exposure Reaching 90 MPa
Reprinted from: *Polymers* **2023**, *15*, 162, doi:10.3390/polym15010162 . . . . **163**

Quoc-Viet Do, Takumitsu Kida, Masayuki Yamaguchi, Kensuke Washizu, Takayuki Nagase and Toshio Tada
Anomalous Strain Recovery after Stress Removal of Graded Rubber
Reprinted from: *Polymers* **2022**, *14*, 5477, doi:10.3390/polym14245477 . . . . **180**

Fahad Alhashmi Alamer and Ghadah A. Almalki
Fabrication of Conductive Fabrics Based on SWCNTs, MWCNTs and Graphene and Their Applications: A Review
Reprinted from: *Polymers* **2022**, *14*, 5376, doi:10.3390/polym14245376 . . . . **189**

Md Najib Alam, Vineet Kumar and Sang-Shin Park
Advances in Rubber Compounds Using ZnO and MgO as Co-Cure Activators
Reprinted from: *Polymers* **2022**, *14*, 5289, doi:10.3390/polym14235289 . . . . **208**

Vineet Kumar, Md Najib Alam, Sang-Shin Park and Dong-Joo Lee
New Insight into Rubber Composites Based on Graphene Nanoplatelets, Electrolyte Iron Particles, and Their Hybrid for Stretchable Magnetic Materials
Reprinted from: *Polymers* **2022**, *14*, 4826, doi:10.3390/polym14224826 . . . . **223**

Yang-Sook Jung, Sunhee Lee, Jaehyeung Park and Eun-Joo Shin
One-Shot Synthesis of Thermoplastic Polyurethane Based on Bio-Polyol (Polytrimethylene Ether Glycol) and Characterization of Micro-Phase Separation
Reprinted from: *Polymers* **2022**, *14*, 4269, doi:10.3390/polym14204269 . . . . **241**

Vineet Kumar, Siraj Azam, Md. Najib Alam, Won-Beom Hong and Sang-Shin Park
Novel Rubber Composites Based on Copper Particles, Multi-Wall Carbon Nanotubes and Their Hybrid for Stretchable Devices
Reprinted from: *Polymers* **2022**, *14*, 3744, doi:10.3390/polym14183744 . . . . **263**

Chunhao Yang, Wuning Ma, Zhendong Zhang and Jianlin Zhong
Low-Velocity Impact Behavior of Sandwich Plates with FG-CNTRC Face Sheets and Negative Poisson's Ratio Auxetic Honeycombs Core
Reprinted from: *Polymers* **2022**, *14*, 2938, doi:10.3390/polym14142938 . . . . **281**

About the Editor

Md Najib Alam

Dr. Md Najib Alam obtained his B.Sc. and M.Sc. degrees from the University of Kalyani, India in 2006 and 2008, respectively. He completed his Ph.D. at the same university in 2014, where he addressed numerous challenges related to rubber vulcanization chemistry. Following his Ph.D., Dr. Alam pursued postdoctoral research at Chulalongkorn University, Thailand, from 2015 to 2019. From 2019 to 2022, he served as a research fellow at the School of Mechanical Engineering, Yeungnam University, Republic of Korea. In 2022, he transitioned to the role of assistant professor at the same institution. Throughout his research career, Dr. Alam has published over 50 original research articles in internationally recognized, peer-reviewed journals, including the *Journal of Industrial and Engineering Chemistry, Composites Science and Technology, and Composites Part B: Engineering*. Currently, he is serving as a guest editor of various Special Issues in MDPI's *Polymers*. His research focuses on rubber vulcanization, novel rubber composites, stretchable sensors, and smart materials.

Preface

Rubbers or elastomers are essential polymer materials that play a crucial role in modern civilization. Although they are rarely used in their pure forms, their properties can be significantly enhanced through the incorporation of various foreign materials. These additives are broadly categorized into curatives and fillers. Reinforcing fillers greatly improve the mechanical properties and durability of rubber composites, while curatives help to form three-dimensional network structures, providing chemical and physical stability to the processed rubber.

With continuous advancements in the science and technology of rubber materials, rubber composites are now being utilized in cutting-edge technologies such as stretchable electronics, sensing devices, dielectric energy harvesting, smart damping systems, and membrane technology. However, the increasing demand for rubber products poses challenges for both industry and the environment. Current fillers, derived from fossil fuels, are becoming increasingly expensive, and some curatives and fillers cause ecological pollution. Therefore, exploring alternative materials is crucial. In cases where alternatives are not feasible, certain modifications can reduce the use of foreign materials while maintaining similar properties for industrial applications.

This reprint provides readers with insights into various techniques used to fabricate rubber composites and their advanced applications. It offers valuable concepts on the mechanisms of rubber-filler interactions and their role in enhancing properties. Additionally, this reprint discusses potential alternative fillers that could be beneficial for environmental remediation or economic advantages. Each chapter thoroughly covers the background of the innovations, presents results and discussions, and concludes with important findings. I am confident that readers will gain many innovative ideas that will be beneficial for their research or academic pursuits.

Md Najib Alam
Editor

Editorial

Advances in Functional Rubber and Elastomer Composites

Md Najib Alam

School of Mechanical Engineering, Yeungnam University, 280, Daehak-ro, Gyeongsan 38541, Republic of Korea; mdnajib.alam3@gmail.com

1. Introduction

Two crucial innovations—mastication and vulcanization—have revolutionized the use of rubber in our daily lives [1,2]. Initially, natural rubber was utilized to manufacture shoes, waterproof jackets, toys, and other items; however, the discovery of vulcanization in the mid-19th century has significantly transformed its applications. For example, solid rubber strips started to be used in wheels, greatly enhancing riding comfort in vehicles. With continuous advancements in science and manufacturing technologies, modern tires are now easier to produce, as well as being more durable, safer for driving, and more fuel-efficient.

At the beginning of the 20th century, the demand for rubber surged, but its supply was limited, prompting the search for new sources [3]. In 1909, a team led by Fritz Hofmann at Bayer laboratory successfully created the first synthetic rubber from isoprene monomers. In 1931, another synthetic rubber, neoprene, was synthesized, and a few years later, in 1935, a series of synthetic rubbers known as Buna rubbers were developed through the copolymerization of two monomers. Subsequently, many types of synthetic rubbers have been produced, featuring a wide range of chemical, thermal, and oil-resistant properties [3]. These special features enabled their use in applications such as oil fields and machinery.

Recently, elastomeric materials have found applications in advanced electronic devices; artificial intelligence; and robotics, medical, and health monitoring technologies [4–6]. For these applications, rubber or elastomers must possess special characteristics. For instance, elastomers can be made electrically or magnetically sensitive by incorporating electrically conductive or magnetic fillers [7–9]. These materials can then be used in smart sensing devices. Additionally, rubber composites with specific fillers exhibit excellent electromagnetic wave absorption properties. Some rubber composites can even function as transducers, converting mechanical energy into electrical power, which can be utilized for generating green energy [10,11].

Fillers are immensely important for making rubber composites for various applications. However, some fillers exhibit low performance abilities due to their surface characteristics, while others are expensive or pose human and environmental toxicity risks. In such cases, alternative or modified fillers can be beneficial [12,13]. Additionally, modifying rubber itself can enhance composite properties. By adopting these strategies, rubber composites have the ability to become more advanced, practical, cost-effective, and sustainable for our societies and industries.

2. Overview of Published Articles

Fasolt et al. [14] found that electrodes impact the breakdown voltage in dielectric elastomer actuators, which can be improved using stiffer, silicone rubber-based electrodes. Furthermore, Liu et al. [15] used hydrophobized, bio-based, microfibrillated cellulose as a reinforcing filler in silica/styrene butadiene rubber (SBR) tire tread compounds, observing a notable increase in mechanical properties compared to the sole use of silica/SBR compounds. Meanwhile, Wang et al. [16] enhanced the mechanical, dielectric, and hydrophobic properties of commonly used Methyl Vinyl Silicone Rubber by blending it with Fluorosilicone Rubber. Carbon black and silica remain dominant fillers in the rubber industry;

Citation: Alam, M.N. Advances in Functional Rubber and Elastomer Composites. *Polymers* **2024**, *16*, 1726. https://doi.org/10.3390/polym16121726

Received: 23 May 2024
Revised: 10 June 2024
Accepted: 11 June 2024
Published: 17 June 2024

Copyright: © 2024 by the author. Licensee MDPI, Basel, Switzerland. This article is an open access article distributed under the terms and conditions of the Creative Commons Attribution (CC BY) license (https://creativecommons.org/licenses/by/4.0/).

Magaletti et al. [17] utilized modified carbon black with a bio-sourced Janus molecule to reinforce rubber, significantly reducing energy dissipation compared to conventional silica or carbon black-reinforced rubber compounds. Moreover, Song et al. [18] improved silica filler dispersion in rubber by using liquid styrene butadiene rubber end-functionalized with a silane coupling agent, resulting in significant improvements in rolling resistance, snow traction, and abrasion resistance compared to the traditionally used treated distillate aromatic extract (TDAE) oil as a filler dispersant. Slobodinyuk et al. [19] synthesized new shape memory polymers for self-healing coatings using oligomers with terminal epoxy groups from oligotetramethylene oxide dioles; they developed a high-yield synthesis method for oligoetherdiamines (94%), involving acrylic acid and aminoethylpiperazine, which enhanced the thermal and mechanical properties of urethane polymers, achieving over 95% shape fixity and over 94% shape recovery. Ahmed A. Bakhsh [20] explored the mechanical and thermal properties of polyolefin–hydroxyapatite nanocomposites using HDPE and LDPE matrices, finding significant enhancements with minor decreases at the 40% level of hydroxyapatite loading. Razzaq et al. [21] reported the 4D printing of electro-active, triple-shape composites made from polyester urethane (PEU), polylactic acid (PLA), and multiwall carbon nanotubes (MWCNTs). These composites, suitable for fused filament fabrication, demonstrated the triple-shape effect through resistive heating, offering potential applications in space, robotics, and actuation technologies. Additionally, Al-Mhyawi et al. [22] developed an adsorbent hydrogel using acrylic acid and orange peel via free radical polymerization to remove methylene blue (MB) from water; optimized and characterized using SEM and BET analysis, the hydrogel showed an 84% adsorption in 10 min and proved to be reusable for up to ten times, demonstrating an efficient and eco-friendly method for water treatment. Jung et al. [23] studied the effects of carbon black (CB) and silica fillers on H_2 permeation in sulfur-crosslinked ethylene propylene diene monomer (EPDM) polymers. CB-blended EPDMs exhibited dual sorption, while neat and silica-blended EPDMs followed Henry's law. CB-filled EPDMs reduced H_2 diffusivity as a result of an increased tortuosity, suggesting its potential use as a sealant material for H_2 refueling stations. Do et al. [24] investigated the mechanical responses of graded styrene–butadiene rubber (SBR) with varying crosslink densities compared to homogenously vulcanized SBR; graded SBR showed a good elasticity and a significant warpage after stress removal, indicating a prolonged shrinking stress on the high-crosslink surface, enhancing crack resistance and slow strain recovery. Alam et al. [25] explored using MgO as a co-activator to reduce conventional ZnO levels in rubber vulcanization. A 3:2 MgO:ZnO weight ratio significantly shortened the curing times and enhanced mechanical properties, providing a safer, high-performance alternative for industrial applications. Kumar et al. [26] developed stretchable magnetic composites using silicone rubber mixed with graphene nanoplatelets (GNPs) and electrolyte–iron particles (EIPs). These composites, cured at room temperature, exhibited enhanced mechanical and magnetic properties. GNPs provided high stiffness and stretchability, while hybrids of GNPs and EIPs showed an improved mechanical performance and magnetic sensitivity, which is ideal for soft robotics. Furthermore, Jung et al. [27] synthesized a series of bio-based thermoplastic polyurethanes (TPUs) using bio-based polyether polyol and 1,4-butanediol (BDO), with aromatic (4,4-methylene diphenyl diisocyanate: MDI) and aliphatic (bis(4-isocyanatocyclohexyl) methane: H_{12}MDI) isocyanates. Various micro-phase structures were identified and matched with specific TPU samples, including H-BDO-2.0, M-BDO-2.0, H-BDO-2.5, and M-BDO-3.0. In another study, Kumar et al. [28] developed stretchable devices using silicone rubber composites with multi-wall carbon nanotubes (MWCNTs) and copper particles. The hybrid composites showed an optimal stiffness and stretchability, generating ~6 V with a cycle durability of over 0.4 million, making them suitable for flexible electronics and piezoelectric energy-harvesting applications. Yang et al. [29] examined the low-velocity impact response of sandwich plates with functionally graded carbon nanotube-reinforced composite (FG-CNTRC) face sheets and a Ti-6Al-4V auxetic honeycomb core. Using first-order shear deformation theory and Hamilton's principle, they analyzed the impact response, considering various stacking sequences, CNT volume fractions, and impact conditions. In a review paper, Alamer et al. [30] discussed different carbon-based conductive materials on fabrics, notably carbon nanotubes and graphene, highlighting their superior properties and pivotal roles in electronic device applications across various fields

3. Summary and Future Outlook

High-level filler dispersion is crucial for optimizing the properties of rubber composites. In this Special Issue, numerous researchers have developed various techniques, both physical and chemical, to enhance filler dispersion. Despite these advancements, there remains a significant potential for improvement, particularly for functional elastomer composites that require specialized elastomers and fillers for specific applications. Researchers have created functional rubber composites, such as electroactive and magnetoactive rubber composites, which are promising for advanced engineering applications in robotics and sensing. Others have developed structural composites that are useful in separation, purification, and structural technologies. Curing studies of these functional composites are also essential for ensuring their reliability and to address environmental safety concerns, highlighting the need for future research in this area. The potential for advancements in the properties of functional rubber and elastomeric materials is boundless. Therefore, continued research in this special field is imperative for future progress.

Funding: This research received no external funding.

Acknowledgments: The author thanks all the contributors and reviewers for their valuable contributions and support from the section editors of this Special Issue.

Conflicts of Interest: The author declares no conflicts of interest.

References

1. Guise-Richardson, C. Redefining Vulcanization: Charles Goodyear, patents, and industrial control, 1834–1865. *Technol. Cult.* **2010**, *51*, 357–387. [CrossRef]
2. Princi, E. Basics on Rubber. In *Rubber: Science and Technology*; De Gruyter: Berlin, Germany; Boston, MA, USA, 2019; pp. 1–16. [CrossRef]
3. Wikipedia Contributors. Synthetic Rubber. In *Wikipedia, The Free Encyclopedia*; Wikimedia Foundation: San Francisco, CA, USA, 2024; Available online: https://en.wikipedia.org/w/index.php?title=Synthetic_rubber&oldid=1217827503 (accessed on 18 May 2024).
4. Souri, H.; Banerjee, H.; Jusufi, A.; Radacsi, N.; Stokes, A.A.; Park, I.; Sitti, M.; Amjadi, M. Wearable and stretchable strain sensors: Materials, sensing mechanisms, and applications. *Adv. Intell. Syst.* **2020**, *2*, 2000039. [CrossRef]
5. Stern, U. Electronic skin: From flexibility to a sense of touch. *Nature* **2021**, *591*, 685.
6. Yuan, Y.; Liu, B.; Li, H.; Li, M.; Song, Y.; Wang, R.; Wang, T.; Zhang, H. Flexible wearable sensors in medical monitoring. *Biosensors* **2022**, *12*, 1069. [CrossRef] [PubMed]
7. Park, M.; Park, J.; Jeong, U. Design of conductive composite elastomers for stretchable electronics. *Nano Today* **2014**, *9*, 244–260. [CrossRef]
8. Alam, M.N.; Choi, J. Highly reinforced magneto-sensitive natural-rubber nanocomposite using iron oxide/multilayer graphene as hybrid filler. *Compos. Commun.* **2022**, *32*, 101169. [CrossRef]
9. Alam, M.N.; Kumar, V.; Lee, D.J.; Choi, J. Synergistically toughened silicone rubber nanocomposites using carbon nanotubes and molybdenum disulfide for stretchable strain sensors. *Compos. Part B Eng.* **2023**, *259*, 110759. [CrossRef]
10. Alam, M.N.; Kumar, V.; Jung, H.S.; Park, S.S. Fabrication of High-Performance Natural Rubber Composites with Enhanced Filler–Rubber Interactions by Stearic Acid-Modified Diatomaceous Earth and Carbon Nanotubes for Mechanical and Energy Harvesting Applications. *Polymers* **2023**, *15*, 3612. [CrossRef]
11. Gołąbek, J.; Strankowski, M. A Review of Recent Advances in Human-Motion Energy Harvesting Nanogenerators, Self-Powering Smart Sensors and Self-Charging Electronics. *Sensors* **2024**, *24*, 1069. [CrossRef]
12. Abdul Salim, Z.A.S.; Hassan, A.; Ismail, H. A review on hybrid fillers in rubber composites. *Polym. Plast. Technol. Eng.* **2018**, *57*, 523–539. [CrossRef]
13. Chang, B.P.; Gupta, A.; Muthuraj, R.; Mekonnen, T.H. Bioresourced fillers for rubber composite sustainability: Current development and future opportunities. *Green Chem.* **2021**, *23*, 5337–5378. [CrossRef]
14. Fasolt, B.; Albuquerque, F.B.; Hubertus, J.; Schultes, G.; Shea, H.; Seelecke, S. Electrode Impact on the Electrical Breakdown of Dielectric Elastomer Thin Films. *Polymers* **2023**, *15*, 4071. [CrossRef] [PubMed]
15. Liu, M.; Imiete, I.E.; Staropoli, M.; Steiner, P.; Duez, B.; Lenoble, D.; Scolan, E.; Thomann, J.S. Hydrophobized MFC as Reinforcing Additive in Industrial Silica/SBR Tire Tread Compound. *Polymers* **2023**, *15*, 3937. [CrossRef] [PubMed]

16. Wang, Z.; Lin, Y.; Li, Z.; Yang, Y.; Lin, J.; He, S. Effect of Fluorosilicone Rubber on Mechanical Properties, Dielectric Breakdown Strength and Hydrophobicity of Methyl Vinyl Silicone Rubber. *Polymers* **2023**, *15*, 3448. [CrossRef] [PubMed]
17. Magaletti, F.; Margani, F.; Monti, A.; Dezyani, R.; Prioglio, G.; Giese, U.; Barbera, V.; Galimberti, M.S. Adducts of Carbon Black with a Biosourced Janus Molecule for Elastomeric Composites with Lower Dissipation of Energy. *Polymers* **2023**, *15*, 3120. [CrossRef] [PubMed]
18. Song, S.; Choi, H.; Jeong, J.; Kim, S.; Kwon, M.; Kim, M.; Kim, D.; Jeon, H.; Paik, H.J.; Chung, S.; et al. Optimized End Functionality of Silane-Terminated Liquid Butadiene Rubber for Silica-Filled Rubber Compounds. *Polymers* **2023**, *15*, 2583. [CrossRef] [PubMed]
19. Slobodinyuk, D.; Slobodinyuk, A.; Strelnikov, V.; Kiselkov, D. Simple and Efficient Synthesis of Oligoetherdiamines: Hardeners of Epoxyurethane Oligomers for Obtaining Coatings with Shape Memory Effect. *Polymers* **2023**, *15*, 2450. [CrossRef] [PubMed]
20. Bakhsh, A.A. Optimization of Polyolefin-Bonded Hydroxyapatite Graphite for Sustainable Industrial Applications. *Polymers* **2023**, *15*, 1505. [CrossRef] [PubMed]
21. Razzaq, M.Y.; Gonzalez-Gutierrez, J.; Farhan, M.; Das, R.; Ruch, D.; Westermann, S.; Schmidt, D.F. 4D Printing of Electroactive Triple-Shape Composites. *Polymers* **2023**, *15*, 832. [CrossRef]
22. Al-Mhyawi, S.R.; Abdel-Tawab, N.A.H.; El Nashar, R.M. Synthesis and Characterization of Orange Peel Modified Hydrogels as Efficient Adsorbents for Methylene Blue (MB). *Polymers* **2023**, *15*, 277. [CrossRef]
23. Jung, J.K.; Lee, J.H.; Jeon, S.K.; Baek, U.B.; Lee, S.H.; Lee, C.H.; Moon, W.J. H_2 Uptake and Diffusion Characteristics in Sulfur-Crosslinked Ethylene Propylene Diene Monomer Polymer Composites with Carbon Black and Silica Fillers after High-Pressure Hydrogen Exposure Reaching 90 MPa. *Polymers* **2022**, *15*, 162. [CrossRef] [PubMed]
24. Do, Q.V.; Kida, T.; Yamaguchi, M.; Washizu, K.; Nagase, T.; Tada, T. Anomalous Strain Recovery after Stress Removal of Graded Rubber. *Polymers* **2022**, *14*, 5477. [CrossRef] [PubMed]
25. Alam, M.N.; Kumar, V.; Park, S.S. Advances in Rubber Compounds Using ZnO and MgO as Co-cure Activators. *Polymers* **2022**, *14*, 5289. [CrossRef] [PubMed]
26. Kumar, V.; Alam, M.N.; Park, S.S.; Lee, D.J. New Insight into Rubber Composites Based on Graphene Nanoplatelets, Electrolyte Iron Particles, and Their Hybrid for Stretchable Magnetic Materials. *Polymers* **2022**, *14*, 4826. [CrossRef] [PubMed]
27. Jung, Y.S.; Lee, S.; Park, J.; Shin, E.J. One-Shot Synthesis of Thermoplastic Polyurethane Based on Bio-Polyol (Polytrimethylene Ether Glycol) and Characterization of Micro-Phase Separation. *Polymers* **2022**, *14*, 4269. [CrossRef] [PubMed]
28. Kumar, V.; Azam, S.; Alam, M.N.; Hong, W.B.; Park, S.S. Novel Rubber Composites Based on Copper Particles, Multi-Wall Carbon Nanotubes and Their Hybrid for Stretchable Devices. *Polymers* **2022**, *14*, 3744. [CrossRef] [PubMed]
29. Yang, C.; Ma, W.; Zhang, Z.; Zhong, J. Low-Velocity Impact Behavior of Sandwich Plates with FG-CNTRC Face Sheets and Negative Poisson's Ratio Auxetic Honeycombs Core. *Polymers* **2022**, *14*, 2938. [CrossRef]
30. Alhashmi Alamer, F.; Almalki, G.A. Fabrication of Conductive Fabrics Based on SWCNTs, MWCNTs and Graphene and Their Applications: A Review. *Polymers* **2022**, *14*, 5376. [CrossRef]

Disclaimer/Publisher's Note: The statements, opinions and data contained in all publications are solely those of the individual author(s) and contributor(s) and not of MDPI and/or the editor(s). MDPI and/or the editor(s) disclaim responsibility for any injury to people or property resulting from any ideas, methods, instructions or products referred to in the content.

Article

Electrode Impact on the Electrical Breakdown of Dielectric Elastomer Thin Films

Bettina Fasolt [1,2,*], Fabio Beco Albuquerque [3], Jonas Hubertus [4], Günter Schultes [4], Herbert Shea [3] and Stefan Seelecke [2]

1. Intelligent Material Systems Lab, Center for Mechatronics and Automation Technology, ZeMA gGmbH, DE-66121 Saarbrücken, Germany
2. Intelligent Material Systems Lab, Department of Systems Engineering, Department of Materials Science and Engineering, Saarland University, DE-66121 Saarbrücken, Germany; stefan.seelecke@imsl.uni-saarland.de
3. LMTS Soft Transducers Laboratory, EPFL Ecole Polytechnique Fédérale de Lausanne, CH-2002 Neuchâtel, Switzerland; herbert.shea@epfl.ch (H.S.)
4. Sensors and Thin Film Group, University of Applied Sciences, DE-66117 Saarbrücken, Germany; jonas.hubertus@htwsaar.de (J.H.); guenter.schultes@htwsaar.de (G.S.)
* Correspondence: b.fasolt@zema.de

Abstract: Dielectric Elastomer Actuators (DEAs) enable the realization of energy-efficient and compact actuator systems. DEAs operate at the kilovolt range with typically microampere-level currents and hence minimize thermal losses in comparison to low voltage/high current actuators such as shape memory alloys or solenoids. The main limiting factor for reaching high energy density in high voltage applications is dielectric breakdown. In previous investigations on silicone-based thin films, we reported that not only do environmental conditions and film parameters such as pre-stretch play an important role but that electrode composition also has a significant impact on the breakdown behavior. In this paper, we present a comprehensive study of electrical breakdown on thin silicone films coated with electrodes manufactured by five different methods: screen printing, inkjet printing, pad printing, gold sputtering, and nickel sputtering. For each method, breakdown was studied under environmental conditions ranging from 1 °C to 80 °C and 10% to 90% relative humidity. The effect of different manufacturing methods was analyzed as was the influence of parameters such as solvents, silicone content, and the particle processing method. The breakdown field increases with increasing temperature and decreases with increasing humidity for all electrode types. The stiffer metal electrodes have a higher breakdown field than the carbon-based electrodes, for which particle size also plays a large role.

Keywords: dielectric breakdown test; electrode manufacturing methods; influence electrodes; silicone films; carbon black; environmental conditions

Citation: Fasolt, B.; Albuquerque, F.B.; Hubertus, J.; Schultes, G.; Shea, H.; Seelecke, S. Electrode Impact on the Electrical Breakdown of Dielectric Elastomer Thin Films. *Polymers* 2023, *15*, 4071. https://doi.org/10.3390/polym15204071

Academic Editor: Md Najib Alam

Received: 23 August 2023
Revised: 26 September 2023
Accepted: 4 October 2023
Published: 12 October 2023

Copyright: © 2023 by the authors. Licensee MDPI, Basel, Switzerland. This article is an open access article distributed under the terms and conditions of the Creative Commons Attribution (CC BY) license (https://creativecommons.org/licenses/by/4.0/).

1. Introduction

The utilization of DEAs as electromechanical transducers is attractive for a number of different applications such as valves [1,2], pumps [3], and switches [4,5], as well as in haptic devices [6,7], wearables [8,9], and soft robotics applications [10–14]. DEAs are lightweight and offer silent, energy-efficient actuation without the use of rare earth materials. Additionally, the actuators' self-sensing properties enable smart applications without the need for external sensors.

A standard DEA typically consists of a compliant dielectric elastomer membrane sandwiched between two stretchable electrodes [15,16]. When a voltage is applied to the electrodes, electrostatic forces lead to a reduction of the membrane thickness along with a simultaneous lateral expansion, thus resulting in voltage-controlled motion.

An important limiting factor for high voltage applications is dielectric breakdown. The breakdown field strength E_{BD} represents the maximum value of the electrical field which can be withstood by the membrane. The Maxwell pressure scales as E_BD^2 while the elastic energy density scales as E_BD^4 [17] It is thus essential to be able to operate DEAs at high electric fields and to understand on which parameters the maximal breakdown field depends. Environmental conditions such as temperature and humidity, as well as the pre-stretch of the membrane, and the dielectric material itself are only a selection of possible parameters known to have an influence on the breakdown voltage [18–24]. In a previous breakdown study by Fasolt et al. conducted on pure silicone film and film with screen-printed carbon black (CB) electrodes, it was discovered that in addition to the above parameters the electrode also had a significant impact on the breakdown, lowering the breakdown field by up to 20% [25]. To validate the influence of electrodes on the breakdown behavior, the results were compared with a study conducted by Albuquerque and Shea [16], which tested silicone film with sputtered gold electrodes under various environmental conditions. The results of the breakdown fields varied significantly but because the film thickness, pre-stretch, and electrode material were also different, a direct comparison was not possible. Other published studies about breakdown behavior of silicone thin film used yet different electrode materials: Förster-Zügel et al. used graphite powder and a shadow mask [26], Stoyanov et al. sprayed carbon nanotubes [27], Jiang et al. applied conductive carbon grease [28], Albuquerque and Shea applied CB electrodes by pad printing [29], and Zakaria et al. sputtered silver electrodes [30]. In order to establish a framework for transferability of measured results between different studies, the current paper systematically investigates the effect of different electrode materials and electrode manufacturing methods over a wide temperature and humidity range.

Specifically, this study provides a comprehensive breakdown investigation conducted on electrodes applied by four different manufacturing methods, using the same test setup and environmental conditions, ranging from 1 °C to 80 °C and 10% to 90% relative humidity. Three research groups collaborated on this project and provided different electrodes manufactured with application methods used in their labs: screen printing, inkjet printing, pad printing, gold sputtering, and nickel sputtering. The electrodes were applied on the same 20 µm-thick silicone dielectric material, Wacker Elastosil 2030/20 µm, and the same bi-axial pre-stretch of the film was used for all samples.

Each manufacturing method has its unique application scope, shown in Table 1. Manufacturing methods such as spraying, spin coating, blade casting, or 3D printing are also possible but were not included in this study [31–33]. Sputtered metal electrodes have a high conductivity and nanometer-scale thicknesses but usually lose conductivity when stretched. Hubertus et al. [34–36] describe a method where electrodes are sputtered on a pre-stretched film and subsequently released so that they exhibit a strongly wrinkled configuration, enabling subsequent stretching within the pre-stretch range and even above. Carbon black electrodes are attractive low-cost materials and can be applied by high-throughput and scalable processes such as screen printing, pad printing, or inkjet printing. These electrodes can remain conductive even at large deformations and hence are a widely used material for soft actuator and sensor applications. Other printable electrode materials such as carbon nanotubes, silver, and graphenes are not included in the study.

The breakdown behavior of the different manufacturing methods is systematically analyzed in this paper and possible breakdown-affecting parameters are discussed. As a reference, experiments were also conducted on samples without applied electrodes. The investigations conducted in this paper are divided into three main sections. The focus of the first group is the influence of environmental conditions such as temperature and humidity on the breakdown behavior. The results are shown for all different types of electrodes. Sections 3.1–3.3 focus on the differences in breakdown behavior for the different types of electrodes. Influencing parameters such as the mechanical effect of the manufacturing method and stiffness (pull-in effect) are discussed. Sections 3.4–3.6 examine possible influencing parameters for the carbon black-based electrodes such as solvents, carbon black processing (mixing and milling), and silicone content. The experiments of Section 3.3,

Sections 3.4–3.6 were conducted under standard environmental conditions (20 °C/55% rel. humidity).

Table 1. Application range of electrode manufacturing methods for DEAs used in this study and associated advantages/disadvantages.

	Advantage	Disadvantage
Screen-printing	• Very fast process, ready up-scaling to mass production • Printing of small and large areas • High acceptable range of ink viscosity (500–10,000 mPas)	• New screen for each design → not ideal for prototyping • Material waste for prototyping due to large minimum amount of ink needed for first print • Can only print on flat or rounded surfaces • Mechanical impact of screen on DE film
Pad-printing	• Printing on irregular shaped surfaces possible • Fast process • Medium range of acceptable ink viscosity (1500–2000 mPas)	• Stencil necessary → not ideal for prototyping • Cannot print on large areas • Mechanical impact of soft pad on DE film
Inkjet-printing	• No screen or stencil necessary • Ideal for prototyping • Contactless process—no mechanical impact	• Very low range of acceptable ink viscosity (10–20 mPas) • Ink needs high solvent content • Clogging of nozzle requires frequent cleaning procedures • Slow process, poorly suited for mass production
Sputtering	• Nanometer-thick high-conductivity electrodes • Microscale actuator designs possible with subsequent laser ablation • Ideal for micro-structures incl. connections using laser ablation	• Slow and complex process for laboratory sputtering systems, involving a vacuum step • Pre-stretch of film required for sputtering and subsequently releasing to avoid cracks → DE operation preferable within pre-stretched range • High investment cost for mass production

2. Experimental Setup and Procedure

2.1. Test Setup

The breakdown tests were conducted using a custom-built automated electrical breakdown test setup. A detailed description of the development and design is given in [37] During the tests, two gold-plated electrodes, subsequently denoted measurement electrodes, with a diameter of 6.3 mm, one moveable on a pivoting arm and one fixed, make contact with the silicone film and voltage is applied at a rate of 0.5 kV/s until breakdown. The breakdown voltage is defined as the voltage when the current flow through the material reached a value of 150 µA. A LabVIEW test software automatically stops the voltage application and records the breakdown value when the admissible current flow is detected by the HypotMAX 7710 Dielectric Withstand Tester (Associated Research, Lake Forest, IL, USA). The movable top electrode is flat, and the bottom electrode has a convex shape with a radius of curvature of 26 mm. The geometry configuration flat top and flat bottom was also investigated for the same film and pre-stretch, but only a negligible difference in the breakdown field was detected. Therefore, the flat/convex shape, also compatible with standardization suggestions from Carpi et al. [38], was chosen for the measurement electrodes, as it also features minimal membrane interactions during spot positioning. After breakdown, the electrodes separate, and the tester automatically moves to the next position. The test setup was placed in a climate chamber Vötsch CLIMEEVENT C/600/40/3 to be able to conduct all breakdown tests in a controlled environment. The tester design allows for consecutive testing without the need to open the climate chamber. Figure 1 shows a schematic illustration of the breakdown tester and the steps carried out for each measurement point.

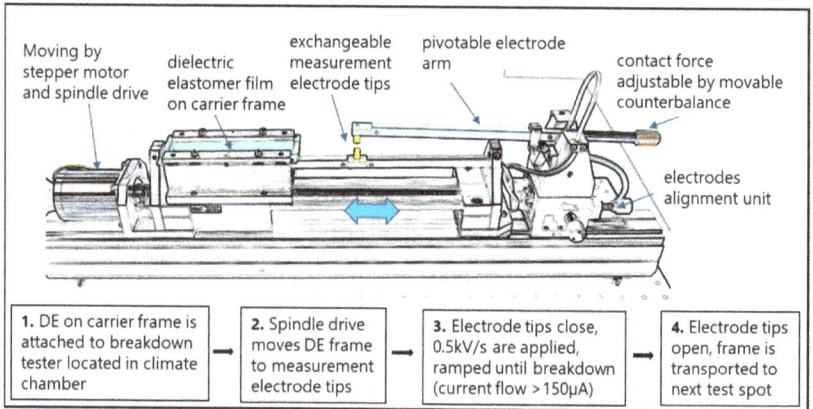

Figure 1. Schematic sketch (top) and diagram (bottom) of the breakdown tester and measurement steps.

The tests were conducted on pure film and film with applied electrodes. To compare these results, all manufactured electrodes had the same diameter as the measurement electrodes. Two different test frames were used. The frames for the screen-printed, nickel-sputtered, and pure film were prepared at the iMSL lab on a metal frame with eleven measurement points, and the gold-sputtered, pad-printed, and inkjet-printed electrodes were prepared at the EPFL lab on plastic frames with three measurement points each. Figure 2 shows a picture of the automated tester with the measurement electrodes and an example of the screen-printed and gold-sputtered electrode samples located in the tester as well as the placement of the tester in the climate chamber.

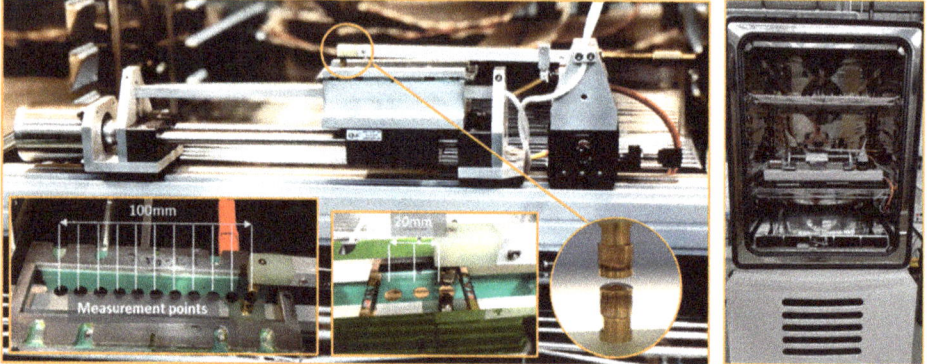

Figure 2. Left: Automated breakdown tester with measurement electrodes (convex/flat design for voltage application) and example of screen-printed and gold-sputtered electrodes. Right: Test setup in climate chamber CLIMEEVENT C/600/40/3, Fa. Vötsch.

2.2. Materials and Sample Preparation

Wacker Elastosil 2030/20 μm, pre-stretched bi-axially by $\lambda_1 = \lambda_2 = 1.3$, is used as a dielectric film for all samples. The average sample thickness after stretching is 11.8 μm. The samples without electrodes, with screen-printed electrodes, and with nickel-sputtered electrodes are pre-stretched using an automated stretcher, consisting of two separately controlled stepper motors and vacuum clamps for the fixation of the DE film. Screen-printed samples are first transferred onto special printer frames to allow for the printing of two designs at the same time and are subsequently transferred onto the test frames. The nickel-sputtered samples and the samples used for tests without electrodes are directly transferred to the test frames used in the breakdown tester. This procedure is described in more detail in [37]. The gold-sputtered, inkjet-printed, and pad-printed samples are pre-stretched using a circular stretcher and transferred to PMMA frames coated with a pressure-sensitive adhesive (Adhesives Research ArClear). The thickness is controlled using a white light transmission interferometer, as described in [39].

2.2.1. Gold-Sputtered Samples

The gold electrodes ($\approx$20 nm thick) are applied over a mask using a Jeol JFC-1200 gold coater, Jeol USA, Inc., Peabody, MA, USA (Argon, 8 Pa, 140 s at 20 mA), as described in [40]. The coating is carried out on both sides, leading to circular 6.3 mm diameter electrodes.

2.2.2. Nickel-Sputtered Samples

Ten nm-thick nickel electrodes are deposited by a DC magnetron sputter process in a laboratory vacuum chamber. At the beginning of the process, the samples, covered with a shadow mask, are inserted into the vacuum chamber and placed on a movable sample holder. The pumping process is started until a background pressure of less than 1×10^{-5} mbar is reached. Prior to sputtering, three pump–purge cycles are executed, where Argon is let into the chamber up to a pressure of 1×10^{-1} mbar and pumped out again. A constant gas flow of 15 sccm Argon in combination with an appropriate position of a downstream throttle then leads to a constant sputter pressure of 1.5×10^{-3} mbar. The sputtering process is started by first pre-sputtering the magnetron target for 1.5 min, while the sample is still located outside the influence of the target. After that, the sample is transferred under the target and coated with a 10 nm-thick nickel thin film. A total of 300 W is applied to the target and held constant during the whole process. With a target-to-substrate distance of 4.5 cm, a 10 nm-thick nickel thin film is manufactured in 5 s. The geometry of the deposited thin film is realized by using a shadow mask, lasering 11 circles with a diameter of 6.3 mm and a distance of 10 mm between the centers of the circles out of a 50 µm-thick metal foil.

2.2.3. Screen-Printed Samples

A screen-printing process requires electrode materials for printing and a screen provided with the desired sample design [41]. The electrode material is prepared using a mixture of 83 wt.% solvent (50% Coats screen VD60, 50% Wacker Belsil DM 1 Plus), 3.8 wt.% carbon black, and 13.2 wt.% PDMS. The solvent is added to achieve the viscosity necessary for screen printing. The preparation of the electrode material is carried out in multiple steps. First, the carbon black and solvents are blended in a planetary mixer. This mixture is subsequently ground in a three-roll mill and, after adding PDMS, again blended in the planetary mixer. A screen with two sets of eleven dots, each with a diameter of 6.3 mm, is prepared, and the electrode material is screen printed onto the pre-stretched film, achieving a thickness of 3 µm. After heat curing at 150 °C for 10 min, the electrodes are screen printed onto the other side of the film and heat cured for another 60 min at 150 °C. The film is then transferred onto two test frames consisting of 11 electrode dots designed for testing in a breakdown tester. After curing, the solvents are evaporated, and the mixture consists of 25 wt.% carbon black and 75 wt.% silicone.

2.2.4. Pad-Printed Samples

The carbon black-PDMS composite electrode is a 4 ± 1 µm-thick pad-printed electrode comprising 0.8 g of carbon black (Ketjenblack EC-300J, Nouryon, Amsterdam, Netherlands) dispersed in 8 g of silicone elastomer (Silbione LSR 4305, Elkem, Oslo, Norway) of A:B ratio

1:1 and 32 g of a mixture of 50% isopropanol and 50% iso-octan, prepared following Rosset et al. [39] using a planetary mixer. After pad printing, the samples with the pad-printed electrodes are cured at 80 °C for 1 h. The pad-printing process is repeated on the opposite membrane side, leading to 6.3 mm diameter electrodes.

2.2.5. Inkjet-Printed Samples

The inkjet-printed electrodes (3 ± 1 μm thick) contain carbon black (Ketjenblack EC-300J), a dispersant (Wacker Belsil SPG 128 VP, WACKER Chemie AG, Munic, Germany), and a solvent (DOWSIL OS2, Dow Chemical Company, Midland, MI, USA) and were applied following the process described in Schlatter et al. [42]. Circular electrodes of 6.3 mm were directly printed on the membrane.

2.3. Test Procedure

For characterization under well-defined environmental conditions, the breakdown tester is placed in a climate chamber. The samples are tested at six different environmental conditions, covering a wide range of possible conditions from low via medium to high temperatures with low, medium, and high humidities, Table 2. The difference in water content between low and high humidity at 1 °C is only 4 g/m^3, which is outside of the adjustable range of the climate chamber.

Table 2. Temperatures and relative humidities of test conditions.

	Temperature and % RH	
1 °C—undefined % RH	20 °C 10% RH	80 °C 10% RH
	20 °C 55% RH—defined as standard environmental condition	
	20 °C 90% RH	80 °C 90% RH

The respective conditions are adjusted, and when stable, an additional 30 min remain before the test samples are placed into the chamber. One test sample is immediately fixed onto the movable station of the breakdown tester and five more samples are placed on a storage rack located above the tester. After 60 min, thermal equilibrium is reached in the test samples, and the tests are conducted on all test points consecutively (eleven electrodes on the large frames and three on the small frames). The tested frame is removed and replaced with a sample from the storage rack, which will then be re-stocked with samples from outside the climate chamber so that the minimum dwell time for each sample is 1 h. When conditions are stable again after replacement, another 10 min remain before the next test is conducted. This procedure is repeated until results for 15 measurement points per condition are available.

3. Results

The motivation of this study was to understand how different electrode materials and their manufacturing methods influence dielectric breakdown behavior in thin silicone films. First, the influence of six different environmental conditions is discussed in Section 3.1. Then, Sections 3.2 and 3.3 discuss the mechanical effects associated with the different electrode materials. This includes the mechanical impact due to the manufacturing method as well as the different stiffnesses of the metal and non-metal electrodes. These experiments are conducted at defined standard condition (20 °C/55% RH). In Sections 3.4–3.6, the focus is on the carbon black-based electrodes, which are of particular relevance as they are not only used for lab breakdown measurements but also in applications with dielectric elastomer actuator devices. Here, parameters important for manufacturing, i.e., solvents, carbon black processing, and silicone content, and their impact on the breakdown behavior are systematically studied.

Results from the breakdown measurements performed for each test case typically lead to plots such as the one shown in Figure 3, displaying breakdown results for all electrode types at standard conditions. Each test point is consecutively tested by connecting the setup electrodes of the breakdown tester to the manufactured top and bottom electrode and applying a voltage. The voltage is ramped with 500 V/s until breakdown, which is defined by a current flow of >150 µA through the dielectric material. The breakdown voltage is recorded for each breakdown point in the tested sequence until 15 points are tested. The results in Figure 3 indicate significant differences between the breakdown voltages for metal-sputtered electrodes and carbon black (CB) electrodes, showing an average of more than 400 volts (>20%) higher for the sputtered electrodes than the CB electrodes. The samples without electrodes are tested as a reference and are consistently higher than for all the samples with applied electrodes.

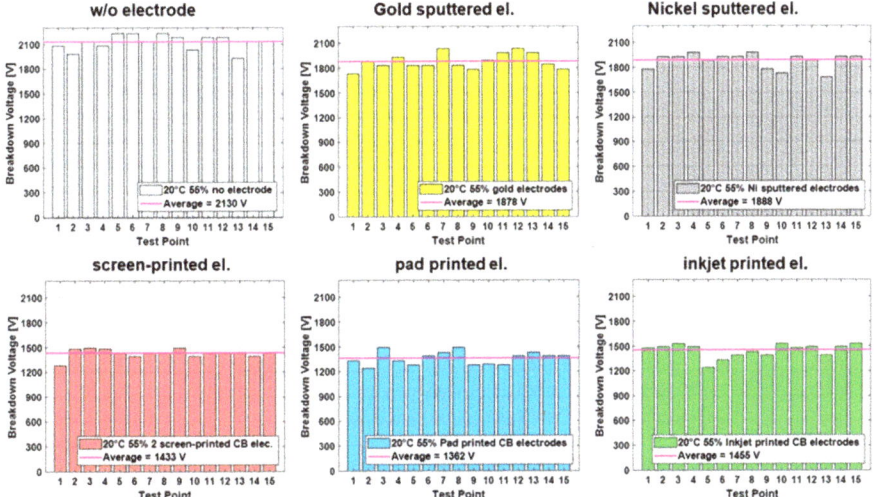

Figure 3. Breakdown voltage of Elastosil 2030/20 at 20 °C 55% RH for all electrode types, with 15 measurement points per sample.

However, to compare the different conditions and electrodes, a bar graph illustration is impractical. Therefore, for the remaining figures in this paper, a boxplot design is used, which allows for a compact comparison and interpretation of the different cases. For example, the breakdown of single spots (depicted as red cross) due to imperfections in the film are easily identified through outliers, while additionally the margin of the breakdown values is visible by the width of the box, including the value for the median.

Furthermore, the breakdown field, defined as the breakdown voltage divided by the initial film thickness, rather than the breakdown voltage, is introduced, because it allows for a better comparison with published measurements conducted on samples with different membrane thicknesses. The small thickness change when the voltage is applied is not taken into account for the calculation of the electric field.

Figure 4 gives a comprehensive overview of the test results for each electrode type under all of the environmental conditions. The experiments were conducted at low, ambient, and high temperatures: 1 °C, 20 °C, and 80 °C. The ambient and high temperatures are tested each at low relative humidity (10% RH) and high relative humidity (90% RH). The difference in water content between the low and high humidity at 1 °C is only 4 g/m^3, which is outside of the range adjustable in the climate chamber. Therefore, the results in Figure 4 for 1 °C are shown with an unspecified relative humidity.

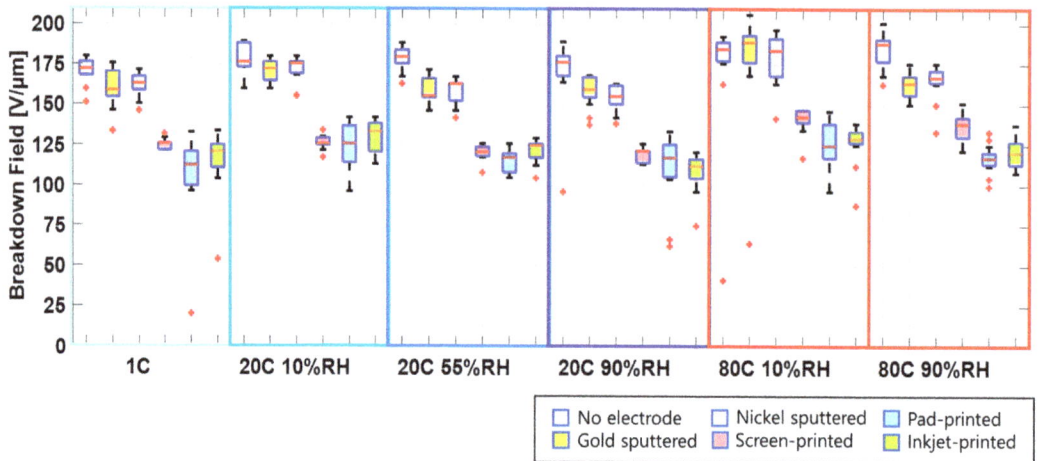

Figure 4. Breakdown field of Elastosil 2030/20 at different environmental conditions for pure silicone film (no electrode) and for film with electrodes applied using different manufacturing methods.

Depending on the environmental conditions and the manufacturing method of the electrodes, breakdown fields ranging from 100 V/µm to 200 V/µm are measured. It is important to point out that the samples are pre-stretched bi-axially by 30%, and therefore the breakdown field is higher than what is expected for un-stretched samples. The influence of pre-stretching on the breakdown field is reported in different studies [25,43–45].

From Figure 4, one sees that the breakdown field for films without electrodes as well as films with sputtered electrodes is significantly higher than for samples with carbon black electrodes, for all environmental conditions.

3.1. Influence of Temperature and Humidity on Breakdown Behavior

Even though breakdown fields vary between electrode materials and deposition methods, two trends are apparent for all electrode types: the breakdown field increases with an increasing temperature and decreases with increasing humidity. This is shown in Figure 5, where two temperatures at low and high humidity are displayed for a metal and a carbon black electrode. These two observations will be discussed in the following sections.

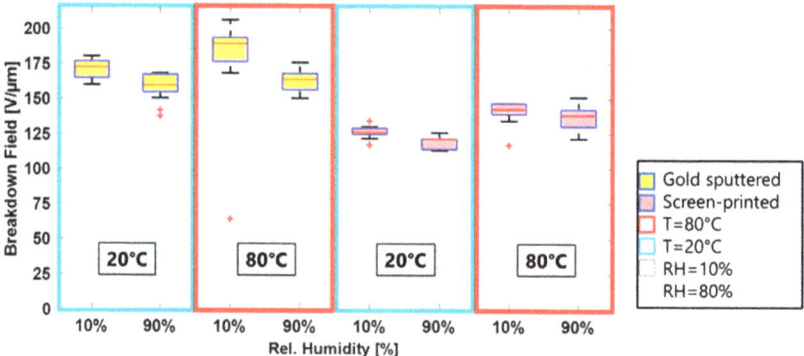

Figure 5. Influence of temperature (red and blue box) and relative humidity (light blue background) on the breakdown field of silicone film Elastosil 2030/20 for gold electrodes and screen-printed carbon black electrodes.

3.1.1. Increase of Breakdown Field with Increasing Temperature

The exact composition of Elastosil 2030 is not known to the authors, but it is assumed that it mainly consists of polydimethlylsiloxan (PDMS). PDMS is a silicone polymer with a long backbone of alternating units of silicon and oxygen (Si-O) and two side chains of methyl on each silicon. The silicone molecule is helical, and intermolecular forces are low, hence easy rotational movements along the backbone are possible. The tetrahedral structure of the silicon in the chains prevents tight packing, thus the free volume in PDMS is high. Additionally, the methyl groups can rotate freely around the backbone [46,47], allowing for even more flexibility. How this composition is affected by temperature change can be explained using the free energy F of the system, defined in Equation (1), with U: Internal Energy, T: temperature, and S: the entropy [48].

$$F = U - TS \qquad (1)$$

The most stable condition is when the free energy is minimized. Silicone elastomers behave nearly entropy elastic above the glass transition temperature ($-126\ °C$ for Elastosil 2030), and therefore the contribution of the internal energy to the free energy can be neglected in our tested temperature range. From Equation (1), an increase in temperature and entropy will be more energetically favorable. An increasing temperature causes the PDMS molecules to move and rotate more freely and increase in entropy. This state has two consequences for the breakdown behavior. One is a stiffening effect and hence increase in Young's modulus [49]. The other one is a prolonged breakdown through the membrane due to the high activity of the chains, which was observed by Du et al. in their investigation about the treeing characteristics at different temperatures [50].

A different influence of the temperature on the breakdown behavior for different electrode materials was not observed.

3.1.2. Decrease of Breakdown Field with Increasing Relative Humidity

With increasing humidity at a constant temperature, a reduction in the breakdown field was observed for all electrodes. The high water vapor permeability of 3000 g/m^2/24 h for Elastosil 2030/20 µm, [51], a result of the high free volume, allows for the water vapor to rapidly diffuse into and through the membrane. The free volume is a function of the temperature, and when the temperature increases, the free volume and the permeability increase as well [52]. Furthermore, the polymer chains become more mobile, and the diffusion of the water molecules is favored. The mobility of the water vapor molecules significantly depends on the temperature as well. Thus, higher temperatures mean higher gas mobility which subsequently results in a higher gas diffusivity and permeability [53].

Depending on the temperature and humidity, the membrane can absorb 0.1 to 0.25 wt.% of water [54,55]. The absorbed water molecules in the membrane lower the dielectric properties of the material because the water introduces additional charge carriers and thus breakdown takes place at lower breakdown fields.

Figure 5 shows the breakdown fields for metal (gold) electrodes and carbon black (screen-printed) electrodes for two temperatures, at low and high humidity. The lower breakdown field for higher humidities is clearly visible for both electrodes. However, while the median difference (calculated by subtracting the medial breakdown at 90% RH from 10% RH and divided by the 10% RH value times 100) for the CB electrodes is around 3% for both temperatures, it is significantly higher, 13.5%, for the gold electrodes at 80 $°C$ and 90% RH.

3.2. Effect of Electrode Manufacturing-Induced Mechanical Actions on Breakdown Behavior

Section 3.1 illustrated the influence of the environmental conditions on the breakdown field and reported how the breakdown field is different for different electrode types for a given environmental condition. This difference can either be due to processes during manufacturing or due to specific electrode properties and their influence on the film. This section explores whether the manufacturing process is responsible for a mechanical change of the film. Figure 6 shows a breakdown plot for all electrode types, combining data from all temperatures and humidities in one box for each electrode type (90 data points per electrode).

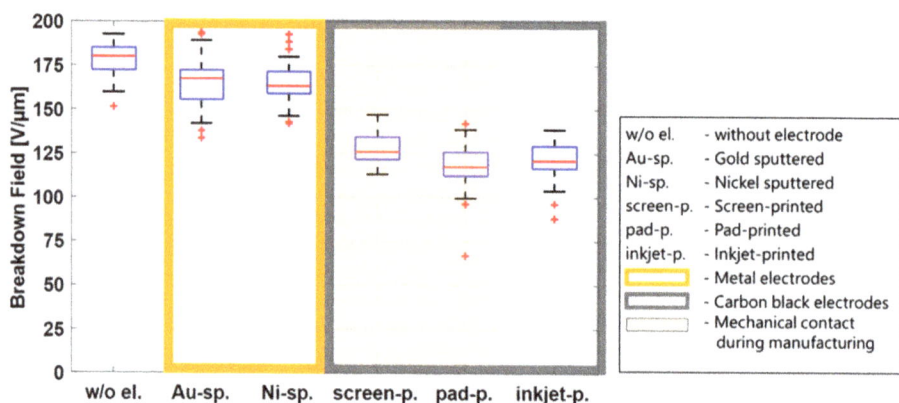

Figure 6. Summarized breakdown field results from all temperature and humidity tests conducted in Figure 4 subdivided into samples without electrodes, metal electrodes, and CB electrodes.

The sputtering process and the inkjet-printing process are non-contact manufacturing methods. During screen printing and pad printing, direct mechanical contact with the film is necessary. In the screen-printing process, the mesh touches the film, and a squeegee applies the electrode material through the mesh. In the pad-printing process, the pad, which transfers the electrode material onto the film, stamps the electrode material directly onto the film. Figure 6 shows higher breakdown fields for the sputtered electrodes, but the results for the inkjet-printing electrodes are in the same order of magnitude as for the contact processes screen printing and inkjet printing, indicating that the mechanical impact is not a decisive parameter that influences the breakdown behavior. Because a breakdown field reduction is observed whenever electrodes are applied—compared to the reference samples without electrodes—other parameters responsible for this phenomenon are discussed in the following sections.

3.3. Influence of Electrode Stiffness

Figure 6 shows that the breakdown field of the samples with metal electrodes (gold and nickel) is significantly higher than that of the samples with CB electrodes, regardless of temperature or humidity. Therefore, the following investigations of possible parameters influencing the breakdown behavior of the film will only be discussed under standard environmental conditions (20 °C and 55% RH).

An important difference between metal and CB electrodes is the stiffness of the material. When a voltage is applied, the Maxwell stress induces a thinning of the membrane and subsequently an increase in the electric field. In an ideal silicone film, where incompressibility is assumed, the thinning of the membrane will result in an area extension of the film.

Two parameters generate an electric field increase: the thinning of the dielectric membrane due to the Maxwell stress and the voltage increase due to the experimental procedure. While soft CB electrodes allow for both of these mechanisms to occur, gold electrodes are extremely stiff and hence impede the area extension. Because of the membrane's incompressibility, this subsequently strongly suppresses the thinning of the material under the application of a voltage. In this way, the electric field in the gold electrode case only increases due to the voltage increase and is thus lower than the electric field, resulting from the same voltage with soft CB electrodes. To confirm this theory, two limiting cases are investigated. One case is a very soft electrode, consisting only of finely milled CB powder. The other is a rigid electrode. Here, the external measurement electrodes are used, which adhere to the pure film when voltage is applied, thus restricting expansion. For this case, the highest breakdown field is expected. The CB powder adds nearly no stiffness to the film; thus, the lowest breakdown field is expected. Figure 7 shows the breakdown field for the four discussed configurations. The results confirm the theory that with an increasing rigidity of the electrode, the breakdown field will also increase.

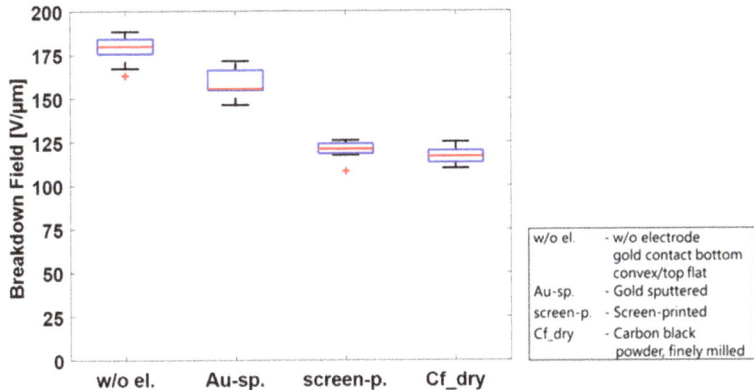

Figure 7. Comparison of electrodes with different stiffnesses (external-measurement electrodes, sputtered gold electrodes, screen-printed CB/PDMS electrodes, and finely milled CB powder) and their influence on the breakdown field at 20 °C/55% RH.

Another possible mechanism for the higher breakdown voltage in the case with only measurement electrodes might be the fact that the contact surface is reduced in comparison to sputtered electrodes due to surface roughness and potential micro-sized air bubbles.

The previous sections explained the factors that are mainly responsible for the different breakdown behavior of metal and carbon black electrodes. The following sections will focus on the material parameters of carbon black electrodes and their possible influence on the breakdown behavior. Metal electrodes are not included in these sections.

3.4. Carbon Black Electrodes: Influence of Solvents with and without Carbon Black

During the manufacturing process of carbon black electrodes via screen printing, pad printing, and inkjet printing, solvents are necessary to provide the electrode material with the viscosity required for the respective process. Even though the solvents are fully evaporated when the electrodes are cured, an influence on the breakdown during the manufacturing is possible. This section will investigate a possible influence of solvents on the breakdown behavior of the DE membrane under standard environmental conditions.

First, the breakdown effect of the three solvent mixtures used for screen printing, S1 (50% Belsil/50% VD60), pad printing, S2 (50% Iso-octane/50% Isopropanol), and inkjet printing, S3 (OS2), was studied. These tests were conducted on twelve test frames, four for each mixture, pre-stretched with Elastosil 2030/20 µm, identical to the frames prepared in the sections above. A total of 0.2 mL of each solvent mixture per test frame was applied by a syringe along the length of the frame and immediately distributed over an area of 110 mm × 20 mm using a spatula. Immediately after application, mixtures S1 and S3 led to significant swelling of the membrane, whereas less swelling but the formation of micro bubbles on the surface of the film was observed using mixture S2. To determine if these phenomena influence the breakdown behavior and also if they are dwell-time dependent, two samples of each mixture were heat cured immediately and two samples were first stored for 24 h before heat curing at 150 °C for one hour. The samples were then tested in the breakdown tester, and the results are shown in Figure 8. For comparison, the results of untreated samples are also included in this figure. No significant differences in the breakdown fields are observed between solvent mixtures, regardless of whether the solvents remained on the sample before curing or were heat cured immediately. Compared to the breakdown field of the untreated samples, only a slightly lower breakdown field (~10 V/μm between medians) is visible. This indicates that the swelling of the membrane after application and heat curing is only temporary and has no major impact on the membrane. The solvent by itself is therefore not the sole parameter responsible for the different breakdown behavior of sputtered metal electrodes and carbon black electrodes, but it may have an influence when carbon black is added to the mixture.

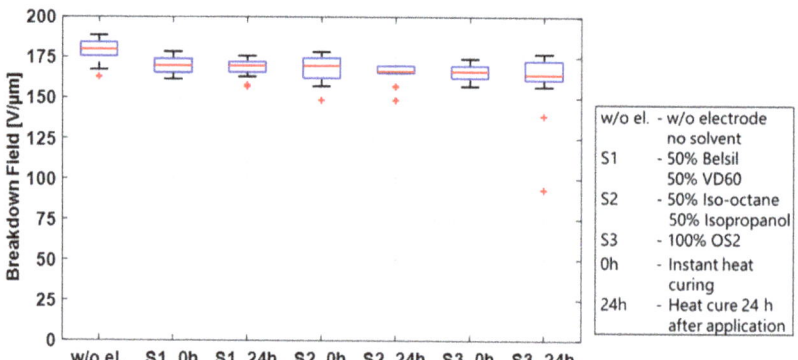

Figure 8. Influence of different solvents and their dwell time before heat curing at 150 °C on the breakdown field of Elastosil 2030/20 μm.

To investigate the influence of the solvent mixtures in the presence of carbon black, 0.5 g of CB was added to 6 g of each solvent mixture and processed in a planetary mixer. A planetary mixer was chosen because the different volatilities of the solvents would result in different evaporation rates when processed in a three-roll mill. The CB/solvent mixtures were manually applied using a stencil and a spatula. As in the test conducted above, two frames of each mixture were immediately heat cured, and two frames were stored 24 h before heat curing at 150 °C for 1 h. A picture of the electrode dots on the film as well as the breakdown results are shown in Figure 9. The addition of CB to the solvents has a significant impact on the breakdown field, regardless of the solvent mixture or dwell time, lowering the breakdown field by 40%. This phenomenon can be explained by two different mechanisms. On the one hand, the soft electrode layer covers the entire surface and thus detects all irregularities and changes in the thickness of the film, which are specified by the data sheet to $+/-5\%$. The breakdown field will therefore always be determined by the thinnest spot of the film. Second, as discussed above, when solvents are applied to the membrane, swelling of the membrane or the forming of small bubbles on the surface are observed. When CB particles are present, it allows the particles to embed into the membrane, effectively thinning the dielectric layer. A calculation of the thickness reduction of the dielectric layer based on the difference in the breakdown field of the pure film (median 180 V/μm) and the film with the CB/solvent electrodes (median 110 V/μm) results in a reduction of 4.5 μm, or a 2.25 μm layer of embedded CB particles on each side. Even though the primary aggregate size of conductive carbon blacks is in the nanometer range, conglomerates > 150 nm up to micrometer structures form when solvents are added [56,57].

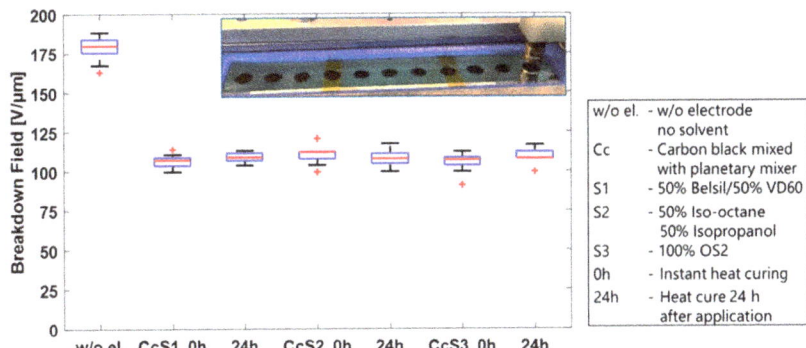

Figure 9. Influence of different solvents mixed with CB powder (without milling), dwelling time before heat curing on the breakdown behavior, and a picture with electrodes applied by hand using a stencil.

The results of Figure 9 are now compared to the results at the standard environmental conditions of Figure 4 to determine whether the lower breakdown field of the manufactured CB electrodes is solely a function of the solvent/CB mixtures, or if additional material parameters influence the breakdown behavior as well. Therefore, each solvent mixture is compared to the respective manufacturing method using this solvent. The results are shown in Figure 10, where the CB/solvent and manufacturing method for each solvent are combined in one box for better comparability. It is clearly visible that the breakdown field of all three manufactured electrodes is considerably higher than that of the CB/solvent mixture, indicating that not only do carbon black and solvents have an impact on the breakdown behavior but also that other manufacturing or material parameters are relevant as well. Two additional parameters possibly influencing the breakdown are (i) the fineness of the carbon black particles used in the process and (ii) the amount of PDMS added to the electrode material. Both will be investigated in the next sections.

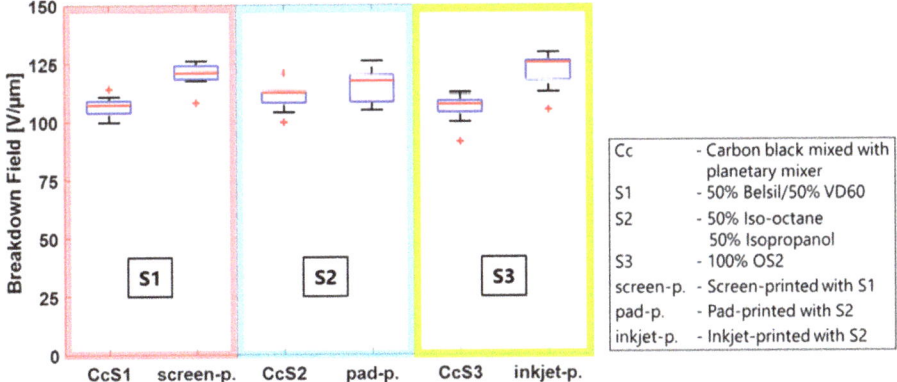

Figure 10. Influence of three solvent mixtures in CB powder and processed in their respective manufacturing method on the breakdown behavior.

3.5. Influence of Carbon Black Processing

Section 3.4 discussed the importance of investigating a possible impact of the fineness of the CB particles in the applied electrodes on the breakdown behavior. Because CB particles agglomerate when blended with most liquids, a grinding process is necessary to break down the agglomerates to smaller sizes. The grinding process was different in the three manufacturing methods. Though the process is explained in Section 2.1 and in the references, a short comparison is necessary for a better understanding: the CB/solvent electrodes used in Section 3.4 were mixed using a planetary mixer; the CB for pad printing was also mixed using a planetary mixer, but steel balls were added to increase grinding. The CB in the screen-printing electrodes was first mixed in a planetary mixer and subsequently milled in a three-roll mill. Lastly, the CB for the inkjet formulation was first ground in a three-roll mill and subsequently sonicated in an ultrasound bath, with a 10 min waiting period for the larger particles to settle down before decanting. The fineness of the CB mixtures increases from CB/solvent to inkjet formulation in the order above.

The experiments were conducted on electrodes consisting of carbon black, and the solvent with the lowest vapor pressure, S1, was processed with the planetary mixer as an example of a coarser blend and a three-roll mill as an example for a finer blend. To investigate the influence on the membrane when no swelling from the solvents occurs, the experiments were additionally conducted with a CB/distilled water mixture. Both sample sets were prepared in the same way and heat cured at 150 °C for 1 h.

The results in Figure 11 show a higher breakdown field when the carbon black is more finely milled than in the coarser samples. While this effect is only small in the electrodes with solvents, it is considerably more pronounced in the electrodes without solvents. A possible explanation could be that due to the swelling of the membrane, when in contact with solvents, CB particles can be embedded in the surface layer of the PDMS membrane, and the difference between very fine and coarser particles will not be as pronounced. Without solvents and no swelling, however, the particles will stay on the surface of the membrane. Larger particles could damage the film more when an electric field is applied, probably due to the sharp edges and higher imprint. If the CB powder is very fine, the damaging effect on the film is likely reduced and a higher breakdown field than for the fine CB particles and solvents is achieved. The presence as well as the processing of the carbon black is an important parameter to influence the breakdown behavior of a DE.

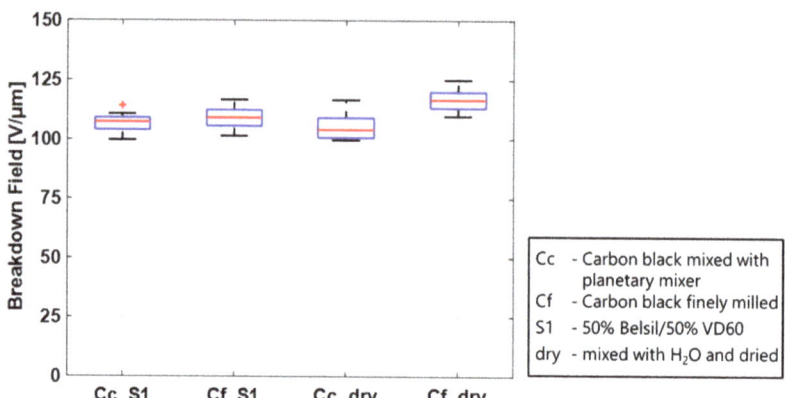

Figure 11. Influence of carbon black particle fineness on the breakdown behavior of Elastosil 2030/20. Particles were either mixed in a three-roll mill for a very fine homogeneous distribution or mixed using a planetary mixer, obtaining a slightly coarser structure.

3.6. Influence of Silicone Content (PDMS)

The influence of solvent and carbon black has been investigated, and the last important material parameter to study is the silicone content. The silicone content in the electrode mixtures varies depending on the manufacturing method. The highest PDMS content—90 wt.% after curing—is used in pad printing. The screen-printed electrodes consist of 75 wt.% PDMS, and the inkjet electrode material is not mixed with PDMS at all but with a silicone polyglucoside dispersant (~60 wt.% after solvent removal).

The study in this section was conducted on mixtures containing four different PDMS concentrations—0 wt.%, 45 wt.%, 75 wt.%, and 90 wt.%—after curing (without solvents). A stock mixture of CB and PDMS was prepared using a planetary mixer and was subsequently milled in a three-roll mill. The required test concentrations were then blended with PDMS using the planetary mixer. The electrode dots containing PDMS were applied using the screen-printing method. The electrodes without silicone were not processable in a screen printer and were applied as described above using a stencil. All the samples were cured at 150 °C for one hour. The results are depicted in Figure 12 and show a significant impact of the PDMS content in the electrodes on the breakdown behavior.

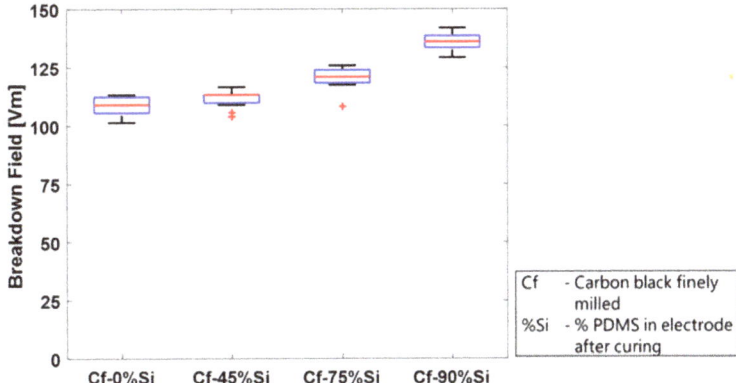

Figure 12. Influence of silicone content in the electrode matrix on the breakdown behavior of Elastosil 2030/20 at 20 °C/55% rel. humidity in wt.%. All electrodes except Cf-0%Si are applied by the screen-printing method.

The breakdown field increases with an increasing PDMS content, from a median of 109 V/μm without silicone to 136 V/μm in the mixture with 90 wt.% silicone. This could be explained by the fact that the higher the silicone content, the more the particles are embedded in the silicone structure, and damage due to sharp particles is reduced. Furthermore, less CB particles are directly located on the surface but are enclosed in the dielectric matrix. This proposed explanation is in agreement with the results of Section 3.5, where finely milled CB dust present on the top of the film and not embedded in the silicone also leads to lower breakdown fields. It should be mentioned, however, that this is only an observation regarding the breakdown field. A higher PDMS/carbon black ratio has not only the advantage of the higher breakdown field and better attachment to the film but also the disadvantage of a lower electrical resistance, as discussed in detail by Willian et al. [58].

4. Conclusions

This paper presented results on the dielectric breakdown of silicone-based electroactive polymer actuators to give a comprehensive understanding about breakdown-influencing parameters. A particular focus was on the effects of electrode composition and electrode manufacturing. These results will allow for a comparison and interpretation of results from different published studies, using these results to better design given applications. As previous work [19] indicated a significant impact of the presence of electrodes on the PDMS membrane, it is important to understand which parameters influence the electric breakdown behavior. This paper focused on a systematic study of different electrodes including gold- and nickel-sputtered electrodes as well as carbon black-based electrodes applied by widely used manufacturing methods such as screen printing, pad printing, or inkjet printing. In addition to a systematic comparison of the different materials, the present work also studied the effect of environmental parameters, such as temperature and humidity, on the breakdown behavior.

We found that adding electrodes lowers the breakdown field compared to films without electrodes. The effect, however, is low for sputtered metal electrodes but significant when carbon black electrodes are applied, reducing the breakdown field by up to 30%. Possible parameters responsible are identified as the mechanical impact during manufacturing, the stiffness of the electrode, and material parameters. Material parameters include the used solvent, carbon black processing, and the silicone content. They were investigated and discussed in detail.

The mechanical impact due to direct contact with the membrane during screen printing and pad printing has no effect on the breakdown behavior compared to inkjet printing, where no direct contact is present. The influence of the electrode stiffness is clearly visible, showing that breakdown tests using stiff metal electrodes yielded a higher breakdown field.

Further investigations should focus on material parameters used for carbon black electrode manufacturing, because these electrodes are common for actuator and sensor applications and are applied by fast and scalable processes. Different manufacturing methods require different solvents, thus the solvent mixtures used during the three manufacturing methods were investigated separately. The data indicate only a minor influence of the solvents on the breakdown field, regardless of the dwelling time on the film before curing. When carbon black is added to the solvents, however, the breakdown field is reduced by up to 40%. This effect is slightly improved when the fineness of the carbon black particles is increased, e.g., through processing in a three-roll mill. Another important parameter influencing the breakdown behavior is the silicone content in the electrode material. The breakdown field is significantly increased when silicone is added to the electrode material and increases with increasing silicone content.

The study on the influence of various environmental conditions was carried out for all types of electrodes with temperatures between 1 °C and 80 °C and humidities ranging from 10%RH to 90%RH. An increasing breakdown behavior with an increasing temperature and decreasing humidity was observed for all electrode types.

Author Contributions: Conceptualization, B.F. and S.S.; methodology, B.F., S.S., H.S. and G.S.; software, B.F.; validation, B.F.; formal analysis, B.F., F.B.A. and J.H.; investigation, B.F.; resources, B.F., F.B.A. and J.H.; writing—original draft preparation, B.F.; writing—review and editing, B.F., H.S., F.B.A., J.H., G.S. and S.S.; visualization, B.F.; supervision, S.S.; project administration, B.F. All authors have read and agreed to the published version of the manuscript.

Funding: This research received no external funding.

Institutional Review Board Statement: Not applicable.

Data Availability Statement: The data presented in this study are available on request from the corresponding author.

Acknowledgments: We gratefully acknowledge the support from WACKER Chemie AG, who supplied the Elastosil 2030 film and other chemicals necessary for the tests.

Conflicts of Interest: Author Bettina Fasolt was employed by the company Intelligent Material Systems Lab, Center for Mechatronics and Automation Technology, ZeMA gGmbH, Saarbrücken, Germany. All authors declare that the research was conducted in the absence of any commercial or financial relationships that could be construed as a potential conflict of interest.

References

1. Hill, M.; Rizzello, G.; Seelecke, S. Development and experimental characterization of a pneumatic valve actuated by a dielectric elastomer membrane. *Smart Mater. Struct.* **2017**, *26*, 085023. [CrossRef]
2. Giousouf, M.; Kovacs, G. Dielectric elastomer actuators used for pneumatic valve technology. *Smart Mater. Struct.* **2013**, *22*, 104010. [CrossRef]
3. Cao, C.; Gao, X.; Conn, A.T. A Magnetically Coupled Dielectric Elastomer Pump for Soft Robotics. *Adv. Mater. Technol.* **2019**, *4*. [CrossRef]
4. Ciarella, L.; Richter, A.; Henke, E.F.M. Digital electronics using dielectric elastomer structures as transistors. *Appl. Phys. Lett.* **2021**, *119*, 261901. [CrossRef]

5. Pniak, L.; Almanza, M.; Civet, Y.; Perriard, Y. Ultrahigh-Voltage Switch for Bidirectional DC-DC Converter Driving Dielectric Elastomer Actuator. *IEEE Trans Power Electron.* 2020, *35*, 13172–13181. [CrossRef]
6. Gratz-Kelly, S.; Krüger, T.; Rizzello, G.; Seelecke, S.; Moretti, G. An audio-tactile interface based on dielectric elastomer actuators. *Smart Mater. Struct.* 2023, *32*, 034005. [CrossRef]
7. Ji, X.; Liu, X.; Cacucciolo, V.; Civet, Y.; El Haitami, A.; Cantin, S.; Perriard, Y.; Shea, H. Untethered Feel-Through Haptics Using 18-μm Thick Dielectric Elastomer Actuators. *Adv. Funct. Mater.* 2021, *31*, 2006639. [CrossRef]
8. Zhao, H.; Hussain, A.M.; Israr, A.; Vogt, D.M.; Duduta, M.; Clarke, D.R.; Wood, R.J. A Wearable Soft Haptic Communicator Based on Dielectric Elastomer Actuators. *Soft Robot.* 2020, *7*, 451–461. [CrossRef]
9. Bolzmacher, C.; Biggs, J.; Srinivasan, M. Flexible dielectric elastomer actuators for wearable human-machine interfaces. In Proceedings of the Smart Structures and Materials 2006: Electroactive Polymer Actuators and Devices (EAPAD), SPIE, San Diego, CA, USA, 26 February–2 March 2006; Volume 6168, p. 616804. [CrossRef]
10. Henke, E.F.M.; Schlatter, S.; Anderson, I.A. Soft Dielectric Elastomer Oscillators Driving Bioinspired Robots. *Soft Robot.* 2017, *4*, 353–366. [CrossRef]
11. Li, W.B.; Zhang, W.M.; Zou, H.X.; Peng, Z.K.; Meng, G. Multisegment annular dielectric elastomer actuators for soft robots. *Smart Mater. Struct.* 2018, *27*. [CrossRef]
12. Guo, Y.; Liu, L.; Liu, Y.; Leng, J. Review of Dielectric Elastomer Actuators and Their Applications in Soft Robots. *Adv. Intell. Syst.* 2021, *3*, 2000282. [CrossRef]
13. Mahmoudinezhad, M.H.; Anderson, I.; Rosset, S. A Skin-Like Soft Compression Sensor for Robotic Applications. *Soft Robot.* 2023, *10*, 687–700. [CrossRef] [PubMed]
14. Ji, X.; Liu, X.; Cacucciolo, V.; Imboden, M.; Civet, Y.; El Haitami, A.; Cantin, S.; Perriard, Y.; Shea, H. An Autonomous Untethered Fast Soft Robotic Insect Driven by Low-Voltage Dielectric Elastomer Actuators. *Sci. Robot.* 2019, *4*, eaaz6451. [CrossRef] [PubMed]
15. Hajiesmaili, E.; Clarke, D.R. Dielectric elastomer actuators. *J. Appl. Phys.* 2021, *129*, 151102. [CrossRef]
16. Pelrine, R.; Kornbluh, R.; Pei, Q.; Joseph, J. High-Speed Electrically Actuated Elastomers with Strain Greater Than 100%. *Science* 2000, *287*, 836–839. [CrossRef]
17. Duduta, M.; Hajiesmaili, E.; Zhao, H.; Wood, R.J.; Clarke, D.R. Realizing the potential of dielectric elastomer artificial muscles. *Proc. Natl. Acad. Sci. USA* 2019, *116*, 2476–2481. [CrossRef]
18. Haddad, G.; Wong, K.L.; Gupta, R.K. Dielectric Breakdown Characteristics of HTV Silicone Rubber under Multiple Stress Conditions. In Proceedings of the 2014 International Symposium on Electrical Insulating Materials, Niigata, Japan, 1–5 June 2014.
19. Gatti, D.; Haus, H.; Matysek, M.; Frohnapfel, B.; Tropea, C.; Schlaak, H.F. The dielectric breakdown limit of silicone dielectric elastomer actuators. *Appl. Phys. Lett.* 2014, *104*, 052905. [CrossRef]
20. Yu, L.; Vudayagiri, S.; Jensen, L.A.; Skov, A.L. Temperature dependence of dielectric breakdown of silicone-based dielectric elastomers. *Int. J. Smart Nano Mater.* 2020, *11*, 129–146. [CrossRef]
21. Albuquerque, F.B.; Shea, H. Influence of humidity, temperature and prestretch on the dielectric breakdown strength of silicone elastomer membranes for DEAs. *Smart Mater. Struct.* 2020, *29*, 105024. [CrossRef]
22. Gerratt, A.P.; Bergbreiter, S. Dielectric breakdown of PDMS thin films. *J. Micromech. Microeng.* 2013, *23*. [CrossRef]
23. Taine, E.; Andritsch, T.; Saeedi, I.A.; Morshuis, P.H.F. Size effect and electrical ageing of PDMS dielectric elastomer with competing failure modes. *Smart Mater. Struct.* 2023, *32*, 105021. [CrossRef]
24. Prakash Prabhakar, O.; Sahu, R.K. Effects of Soft and Hard Fillers on Electromechanical Properties and Performance of Polydimethylsiloxane Elastomer for Actuator Applications. 2023. Available online: https://doi.org/10.21203/rs.3.rs-2559565/v2 (accessed on 26 September 2023).
25. Fasolt, B.; Welsch, F.; Jank, M.; Seelecke, S. Effect of actuation parameters and environment on the breakdown voltage of silicone dielectric elastomer films. *Smart Mater. Struct.* 2019, *28*, 094002. [CrossRef]
26. Förster-Zügel, F.; Solano-Arana, S.; Klug, F.; Schlaak, H.F. Dielectric breakdown strength measurements with silicone-based single-layer dielectric elastomer transducers. *Smart Mater. Struct.* 2019, *28*, 075042. [CrossRef]
27. Stoyanov, H.; Brochu, P.; Niu, X.; Lai, C.; Yun, S.; Pei, Q. Long lifetime, fault-tolerant freestanding actuators based on a silicone dielectric elastomer and self-clearing carbon nanotube compliant electrodes. *RSC Adv.* 2013, *3*, 2272–2278. [CrossRef]
28. Jiang, L.; Betts, A.; Kennedy, D.; Jerrams, S. Investigation into the electromechanical properties of dielectric elastomers subjected to pre-stressing. *Mater. Sci. Eng. C* 2015, *49*, 754–760. [CrossRef]
29. Albuquerque, F.B.; Shea, H. Influence of electric field, temperature, humidity, elastomer material, and encapsulation on the lifetime of dielectric elastomer actuators (DEAs) under DC actuation. *Smart Mater. Struct.* 2021, *30*, 125022. [CrossRef]
30. Zakaria, S.; Morshuis, P.H.F.; Benslimane, M.Y.; Yu, L.; Skov, A.L. The electrical breakdown strength of pre-stretched elastomers, with and without sample volume conservation. *Smart Mater. Struct.* 2015, *24*, 055009. [CrossRef]
31. Chortos, A.; Hajiesmaili, E.; Morales, J.; Clarke, D.R.; Lewis, J.A. 3D Printing of Interdigitated Dielectric Elastomer Actuators. *Adv. Funct. Mater.* 2020, *30*, 1907375. [CrossRef]
32. Cohen, A.J.; Kollosche, M.; Yuen, M.C.; Lee, D.Y.; Clarke, D.R.; Wood, R.J. Batch-Sprayed and Stamp-Transferred Electrodes: A New Paradigm for Scalable Fabrication of Multilayer Dielectric Elastomer Actuators. *Adv. Funct. Mater.* 2022, *32*, 2205391. [CrossRef]
33. Krüger, T.S.; Çabuk, O.; Maas, J. Manufacturing Process for Multilayer Dielectric Elastomer Transducers Based on Sheet-to-Sheet Lamination and Contactless Electrode Application. *Actuators* 2023, *12*, 95. [CrossRef]

34. Hubertus, J.; Fasolt, B.; Linnebach, P.; Seelecke, S.; Schultes, G. Electromechanical evaluation of sub-micron NiCr-carbon thin films as highly conductive and compliant electrodes for dielectric elastomers. *Sens. Actuators A Phys.* **2020**, *315*, 112243. [CrossRef]
35. Hubertus, J.; Croce, S.; Neu, J.; Seelecke, S.; Rizzello, G.; Schultes, G. Laser Structuring of Thin Metal Films of Compliant Electrodes on Dielectric Elastomers. *Adv. Funct. Mater.* **2023**, *33*, 2214176. [CrossRef]
36. Hubertus, J.; Neu, J.; Croce, S.; Rizzello, G.; Seelecke, S.; Schultes, G. Nanoscale Nickel-Based Thin Films as Highly Conductive Electrodes for Dielectric Elastomer Applications with Extremely High Stretchability up to 200%. *ACS Appl. Mater. Interfaces* **2021**, *13*, 39894–39904. [CrossRef]
37. Welsch, F.; Fasolt, B.; Seelecke, S. Dielectric breakdown test setup for dielectric elastomers: Design and validation. In Proceedings of the SPIE Smart Structures and Materials + Nondestructive Evaluation and Health Monitoring, Denver, CO, USA, 5–6 March 2018; Volume 10594, p. 43. [CrossRef]
38. Carpi, F.; Anderson, I.; Bauer, S.; Frediani, G.; Gallone, G.; Gei, M.; Graaf, C.; Jean-Mistral, C.; Kaal, W.; Kofod, G.; et al. Standards for dielectric elastomer transducers. *Smart Mater. Struct.* **2015**, *24*, 105025. [CrossRef]
39. Rosset, S.; Araromi, O.A.; Schlatter, S.; Shea, H.R. Fabrication process of silicone-based dielectric elastomer actuators. *J. Vis. Exp.* **2016**, *108*, e53423. [CrossRef]
40. Albuquerque, F.B. *Lifetime of Dielectric Elastomer Actuators under DC Electric Fields*; EPFL: Lausanne, Switzerland, 2022. [CrossRef]
41. Fasolt, B.; Hodgins, M.; Rizzello, G.; Seelecke, S. Effect of screen printing parameters on sensor and actuator performance of dielectric elastomer (DE) membranes. *Sens. Actuators A Phys.* **2017**, *265*, 10–19. [CrossRef]
42. Schlatter, S.; Grasso, G.; Rosset, S.; Shea, H. Inkjet Printing of Complex Soft Machines with Densely Integrated Electrostatic Actuators. *Adv. Intell. Syst.* **2020**, *2*, 2000136. [CrossRef]
43. Iannarelli, A.; Niasar, M.G.; Ross, R. The effects of static pre-stretching on the short and long-term reliability of dielectric elastomer actuators. *Smart Mater. Struct.* **2019**, *28*, 125014. [CrossRef]
44. Zakaria, S.; Yu, L.; Kofod, G.; Skov, A.L. The influence of static pre-stretching on the mechanical ageing of filled silicone rubbers for dielectric elastomer applications. *Mater. Today Commun.* **2015**, *4*, 204–213. [CrossRef]
45. Kofod, G. Dielectric Elastomer Actuators. Ph.D. Thesis, Technical University of Denmark, Lyngby, Denmark, 2001.
46. Mazurek, P.; Vudayagiri, S.; Skov, A.L. How to tailor flexible silicone elastomers with mechanical integrity: A tutorial review. *Chem. Soc. Rev.* **2019**, *48*, 1448–1464. [CrossRef]
47. Shin-Etsu Silicone Chemical Co., Ltd. Characteristic Properties of Silicone Rubber Compounds. Available online: https://www.shinetsusilicone-global.com/catalog/pdf/rubber_e.pdf (accessed on 23 June 2023).
48. Polmanteer, K.E. Current Perspectives on Silicone Rubber Technology. *Rubber Chem. Technol.* **1981**, *54*, 1051–1080. [CrossRef]
49. Rey, T.; Chagnon, G.; Le Cam, J.B.; Favier, D. Influence of the temperature on the mechanical behaviour of filled and unfilled silicone rubbers. *Polym. Test.* **2013**, *32*, 492–501. [CrossRef]
50. Du, B.X.; Ma, Z.L.; Gao, Y.; Han, T. Effect of Ambient Temperature on Electrical Treeing Characteristics in Silicone Rubber. *Trans. Dielectr. Electr. Insul.* **2011**, *18*, 401–407. [CrossRef]
51. Wacker Chemical Corporation. Datenblätter Elastosil Film 2030 250/20. 2016. Available online: https://www.wacker.com (accessed on 10 March 2021).
52. Chang, K.S.; Chung, Y.C.; Yang, T.H.; Lue, S.J.; Tung, K.L.; Lin, Y.F. Free volume and alcohol transport properties of PDMS membranes: Insights of nano-structure and interfacial affinity from molecular modeling. *J. Memb. Sci.* **2012**, *417–418*, 119–130. [CrossRef]
53. Zhang, H.; Cloud, A. The Permeability characteristics of silicone Rubber. In Proceedings of the Sampe Fall Technical Conference, Global Advances in Material and Process Engineering, Dallas, TX, USA, 6–9 November 2006.
54. Gong, B.; Tu, Y.; Zhou, Y.; Li, R.; Zhang, F.; Xu, Z.; Liang, D. Moisture Absorption Characteristics of Silicone Rubber and Its Effect on Dielectric Properties. In Proceedings of the 2013 Annual Report Conference on Electrical Insulation and Dielectric Phenomena, Chenzhen, China, 20–23 October 2013.
55. Hillborgl, H.; Gedde, U.W. Hydrophobicity Changes in Silicone Rubbers. *IEEE Trans. Dielectr. Electr. Insul.* **1999**, *6*, 703–717. [CrossRef]
56. Neffati, R.; Brokken-Zijp, J.M.C. Structure and porosity of conductive carbon blacks. *Mater. Chem. Phys.* **2021**, *260*, 124177. [CrossRef]
57. Ketjenblack®. Available online: https://www.fuelcellstore.com/spec-sheets/ketjenblack-ec-300j-fact-sheet.pdf (accessed on 19 July 2023).
58. Willian, T.P.; Fasolt, B.; Motzki, P.; Rizzello, G.; Seelecke, S. Effects of Electrode Materials and Compositions on the Resistance Behavior of Dielectric Elastomer Transducers. *Polymers* **2023**, *15*, 310. [CrossRef]

Disclaimer/Publisher's Note: The statements, opinions and data contained in all publications are solely those of the individual author(s) and contributor(s) and not of MDPI and/or the editor(s). MDPI and/or the editor(s) disclaim responsibility for any injury to people or property resulting from any ideas, methods, instructions or products referred to in the content.

Article

Hydrophobized MFC as Reinforcing Additive in Industrial Silica/SBR Tire Tread Compound

Ming Liu [1], Iikpoemugh Elo Imiete [1], Mariapaola Staropoli [1], Pascal Steiner [2], Benoît Duez [2], Damien Lenoble [1], Emmanuel Scolan [1] and Jean-Sébastien Thomann [1,*]

[1] Material Research and Technology Department (MRT), Luxembourg Institute of Science and Technology (LIST), 41 Rue du Brill, L-4422 Belvaux, Luxembourg; mingliu_0603@yahoo.com (M.L.)
[2] Goodyear Innovation Center Luxembourg (GIC*L), Avenue Gordon Smith, L-7750 Colmar-Berg, Luxembourg
* Correspondence: jean-sebastien.thomann@list.lu

Abstract: Silica is used as reinforcing filler in the tire industry. Owing to the intensive process of silica production and its high density, substitution with lightweight bio-based micro fibrillated cellulose (MFC) is expected to provide lightweight, sustainable, and highly reinforced tire composite. MFC was modified with oleoyl chloride, and the degree of substitution (DS) was maintained between 0.2 and 0.9. Subsequently, the morphology and crystallinity of the modified MFC were studied and found to be significantly dependent on the DS. The advantages associated with the use of the modified MFC in synergy with silica for the reinforcement of styrene butadiene rubber (SBR) nanocomposite was investigated in comparison with silica/SBR compound. The structural changes occasioned by the DS values influenced the processability, curing kinetics, modulus-rolling resistance tradeoff, and tensile properties of the resultant rubber compounds. We found that the compound made with modified MFC at a DS of 0.67 (MFC16) resulted to the highest reinforcement, with a 350% increase in storage modulus, 180% increase in Young's modulus, and 15% increase in tensile strength compared to the referenced silica-filled compounds. Our studies show that MFC in combination with silica can be used to reinforce SBR compound for tire tread applications.

Keywords: micro fibrillated cellulose; surface functionalization; silica reinforcement; nanocomposite; elastomer

1. Introduction

There is a huge drive to replace petrochemical-based materials with renewable resources in various applications (e.g., packaging, glass-fiber-reinforced composites, tire compounds, etc.). For example, the tire industry utilizes vast amounts of non-renewable fillers for reinforcement, such as carbon black and silica. Carbon black is derived from fossil fuel, and silica is derived from minerals. Apart from their non-renewable nature, silica has an undesirably high density, and its production is both time and energyconsuming [1–3]. Cellulose and cellulose derivatives such as micro fibrillated cellulose (MFC) constitute abundant, inexpensive, lightweight, and renewable fillers [4,5] compared to carbon black and silica. The potential use of cellulosic derivatives as a reinforcing filler has been demonstrated in several publications [6–8]. The drawback of cellulose is its hydrophilic character, which results to poor dispersion and weak interfacial adhesion when incorporated in hydrophobic polymer for composite fabrication. Therefore, modification of cellulosic fibers is necessary to avoid these drawbacks for improved performance of cellulose-based composites.

The use of modified cellulose fibers in combination with the currently used fillers could be a robust approach targeted at incorporating MFC and gradually phasing out non-renewable fillers. This could be achieved by the addition or partial substitution of silica and carbon black with MFC, which may lead to a more sustainable design of innovative compounds useful for the tire industry. In this respect, our group has demonstrated recently that a combination of fillers within styrene butadiene rubber (SBR)-based nanocomposites results in novel mechanical reinforcement behavior [9]. Therefore, the use of bio-sourced micro fibrillated fibers such as MFCs in combination with silica in rubber could possibly enhance the synergy previously observed between silica and anisotropic structures.

Cellulose fibers and other cellulose derivatives have been combined with silica and carbon black in rubber matrices with very promising results [10–13]. These studies showed that the dispersion and mechanical properties would need to be improved when the matrix is highly hydrophobic, such as SBR. The high hydrophobicity of SBR would therefore require an efficiently modified MFC to fit as a reinforcing agent. The modification process of MFC often leads to loss of crucial fiber properties [14] which are needed for reinforcement.

A limited number of studies have been dedicated to filled rubber compounds prepared by melt mixing MFCs into elastomers. One study was reported in 2018 [15] using a melt-mixing process with chemically modified freeze-dried MFCs in SBR. The freeze-dried MFCs were esterified with palmitoyl chloride or 3,3′dithiopropionic acid chloride using a gas-phase protocol and subsequently incorporated into rubber elastomers using an internal mixer. This result demonstrated that hydrophobized MFCs could be successfully incorporated into rubber elastomers using current industrially applied melt-mixing processes. The properties of the resulting MFC composite were promising. However, the freeze-drying processing before and after the surface functionalization of MFCs cannot be easily upscaled for large-scale industrial production. Also, the storage modulus offers the possibility of improvement.

Water removal is required for MFC modification and is crucial, as several modification reactions are sensitive to moisture. Optimizing the process of water removal by avoiding freeze or oven drying is important to reduce cost and preserve the fibrillar morphology, which is desirable for high levels of reinforcement. Alternatively, there are reports of hydrophobic modification of cellulose derivatives that are not sensitive to water [16,17]. However, they are governed by reaction equilibria, the use of excess reactants, and low reaction yield, which negatively impact on reproducibility. These processes would be difficult to upscale, with reproducibility and cost as the main drawbacks. These studies, at best, present the potentials of MFCs and do not outline key aspects that can translate laboratory results into industrial compounds. One of the disconnects between laboratory experiments and industrial manufacturing is the effective replication of processing condition. We considered these drawbacks in our studies and went a step further to study the processability of the fabricated compounds, which is crucial for upscaling.

In this study, we developed a scalable process using modified MFCs to substitute different amounts of silica in MFC/Silica/SBR nanocomposites. The MFCs were modified with oleoyl chloride at various degrees of substitution (DS) to hydrophobize the MFC surface and facilitate dispersion in the SBR matrix. The same role can be assigned to TDAE oil, which facilitates the incorporation of filler into polymeric matrices. Therefore, we modulated the amount of processing oil in the recipe to compensate for a higher DS while maintaining a filler level needed to meet the percolation threshold. MFC modification started with solvent-exchanging MFCs slurry from water to DMAc using a rotary evaporator. Thereafter, the MFCs were functionalized and studied with FTIR, XRD, SEM, and TGA. Finally, the modified MFCs were melt-mixed with nanosilica and SBR in a Brabender internal mixer. The properties of the resulting composites were studied to understand the effect of the DS, TDAE oil, and filler amount on the mechanical properties and processability of the resultant elastomeric compounds.

2. Materials and Methods

2.1. Materials

Micro fibrillated cellulose (MFC) slurry (Exilva F01V) with solid content of about 12 wt.% was purchased from Borregaard, Sarpsborg, Norway. Dimethylacetamide (DMAc) and tetrahydrofuran (THF) were purchased from Carl Roth (Karlsruhe, Germany). Oleoyl chloride (>89%) and anhydrous pyridine were purchased from Sigma-Aldrich (St. Louis,

MO, USA), while polybutadiene (PB), styrene butadiene rubber (SBR), treated distillate aromatic extract (TDAE oil), zinc oxide, stearic acid, nano silica (200 nm), Bis(triethoxysilylpropyl) disulfide (TESPD), N-(1,3-dimethylbutyl)-N′-phenyl-p-phenylenediamine (6PPD), sulfur, 2-mercaptobenzothiazole (MBT), diphenyl guanidine (DPG), and N-cyclohexyl-2-benzothiazylsulfenamide (CBS) were provided by Goodyear Tire and Rubber Company, Luxembourg.

2.2. MFC Modification

The esterification of MFCs with oleoyl chloride was carried out in DMAc solution after solvent exchanging the MFC slurry from water to DMAc using a rotary evaporator. The reaction shown in Figure 1 is an addition/elimination reaction wherein the fatty acid derivative is grafted on the hydroxyl groups of MFCs, with the formation of pyridinium chloride as a byproduct. The process used about 15 g MFCs slurry dispersed in 150 g DMAc with a high shear homogenizer for 5 min to form a homogeneous MFC slurry. The MFC slurry was then concentrated using a rotary evaporator under a vacuum of 50 mbar and a temperature of 80 °C until 80–90 g water had been evaporated. Subsequently, an equal amount of DMAc (80–90 g) was poured into the concentrated MFC slurry, and the mixture was further homogenized with a homogenizer for 1 min at 15,000 rpm. The new MFC slurry was concentrated using a rotary evaporator at an elevated vacuum of 20 mbar at 90 °C until half of the initial mass had been lost. The final MFC in DMAc solution for esterification was obtained by further homogenization in DMAc with an adjusted concentration of 1% dry matter content of MFC. The wight of MFC was determined gravimetrically after drying the solution under 100 mbar at 105 °C for 24 h.

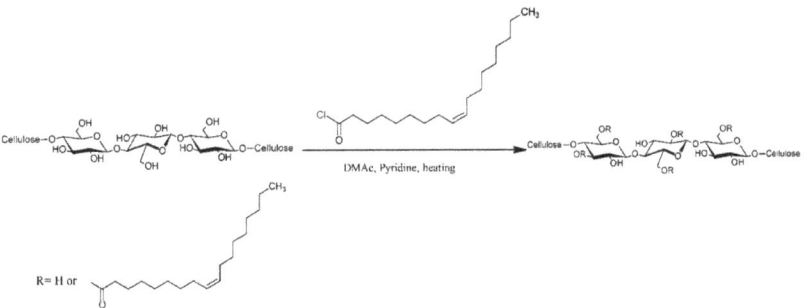

Figure 1. Schematic drawing for the esterification reaction between cellulose and oleoyl chloride.

After the solvent exchange, the esterification of MFC proceeded by adding a certain amount of pyridine, MFC/DMAc slurry, and oleoyl chloride in a reaction chamber under a nitrogen atmosphere and stirred at 90 °C and 1000 rpm for 24 h using a Carousel Tornado overhead stirring system. The molar ratio of the oleoyl chloride/OH group of MFCs was between 0.3 and 0.9, and the molar ratio of oleoyl chloride/pyridine was 1.0. After esterification, the modified MFCs were washed three times with THF using a centrifuge. The degree of substitution (DS) of modified MFCs was determined based on the gained mass of the modified MFCs and the initial mass of MFCs before esterification [18]. By varying the molar ratio of the oleoyl chloride/OH group of MFCs, modified MFCs with low DS (LDS) of 0.24, medium DS1 (MDS1) of 0.49, medium DS2 (MDS2) of 0.67, and high DS (HDS) of 0.91 were obtained.

2.3. Composite Preparation

The compounding of rubber composites was performed using a HAAKE PolyLab OS internal mixer (Thermo Scientific) with a mixing chamber volume of 85 cm^3 according to the recipes shown in Table 1 and the formulation shown in Table 2. The Applied compounding parameters are described in Figure 2.

Table 1. The general composition of the rubber compounds.

Stage	Composition	Amount (phr)
NP1	Polystyrene butadiene	80
	Polybutadiene	20
	TDAE oil	3.75–25
	Zinc oxide	0.5
	Stearic acid	3
	Silica	50–80
	MFCs	10
NP2	6PPD	2.5
	TESPD silane	8
	Silica	15
PR	Zinc oxide	2
	Sulfur	1.1
	MBT	0.3
	DPG	3.2
	CBS	2.3

Table 2. Rubber compounds and their compositions.

	Type of MFC	Degree of Substitution (DS)	Amount (phr)		
			Silica	MFCs	TDAE
Control	/	/	80	0	25
MFC14	LDS	0.24	70	10	21
MFC15	MDS1	0.49	70	10	17
MFC16	MDS2	0.67	70	10	14
MFC17	HDS	0.91	70	10	10
MFC18	HDS	0.91	60	10	10
MFC19	HDS	0.91	60	10	3.75
MFC20	HDS	0.91	50	10	10

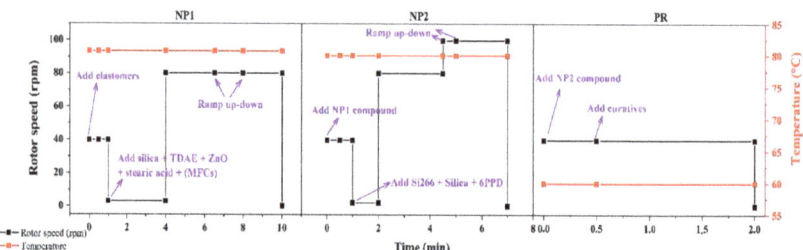

Figure 2. Applied compounding parameters, including rotor speed and temperature of the mixing chamber in the NP1, NP2, and PR stages.

The compound-filling factor was set at 0.75, corresponding to a volume of 63.75 cm^3. The temperature of the mixer was set and maintained at 80 °C for the NP1 and NP2 stages and 60 °C for the PR stage. Between mixing stages, the compounds were further homogenized by being passed through a two-roll mill 6 times with a roller rotation speed of 32 rpm, then through another roller at a speed of 24 rpm with a gap of 2 mm. A total of 8 batches of compounding were performed with different amounts of silica, chemically modified MFCs, and TDAE processing oil (shown in Table 2). An amount of 50 g of the processed green compounds was vulcanized under hot pressing at 150 °C and 170 bar for 30 min.

2.4. Characterizations

2.4.1. Chemical Composition Analysis

Chemically modified MFCs were ground with a microfine grinder (IKA, MF 10.1; IKA®-Werke GmbH, Breisgau, Germany) through a 1 mm screen. Ground samples were hydrolyzed using a two-step sulfuric acid process [19]. After acid hydrolysis, the hydrolysate was collected for monosaccharide analysis, and Klason lignin content (acid-insoluble residues) was gravimetrically determined. The chemical composition of the hydrolysate was analyzed by high-performance anion-exchange chromatography with pulsed amperometric detection (HPAEC-PAD) [20].

2.4.2. Attenuated Total Reflectance–Fourier Transform Infrared Spectroscopy (ATR-FTIR)

ATR-FTIR analyses of unmodified and chemically modified MFCs were carried out using a Thermo Scientific Nicolet iS50 FT-IR spectrometer. For the analysis, unmodified and chemically modified MFC films were applied to a diamond cell, and the transmission spectra between 400 and 4000 cm^{-1} were measured at room temperature.

2.4.3. X-Ray Diffraction (XRD)

X-ray diffraction was performed on unmodified and modified MFCs using a Bruker AXS X-ray diffractometer equipped with a filtered Cu Kα radiation source (λ = 0.1542 nm) at an operating voltage and current of 45 kV and 40 mA, respectively, using a 2D detector. The sample crystallinity index (CI, %) was calculated from the XRD spectra using the Segal method based the height of the 200 peak (I_{200}, 2θ = 22.7°) and amorphous peak at 2θ = 18° (I_{AM}) between the 200 and 110 peaks (Equation (1)) [21]. I_{200} represents the sum of crystalline and amorphous material, while I_{AM} represents amorphous material only.

$$CI(\%) = \frac{I_{200} - I_{AM}}{I_{200}} \quad (1)$$

2.4.4. Morphological Characterization

Unmodified and modified MFCs were first deposited on a STEM grid of a carbon film deposited on 400 mesh Cu. The morphology of MFCs was subsequently observed using a focused ion beam (FIB) scanning transmission electron microscope (STEM) under a transmission mode operated at 30 kV.

2.4.5. Thermogravimetric Analysis (TGA)

Dynamic thermogravimetric measurements were performed using a Discovery TGA TA instrument (New Castle, DE, USA). Temperature programs for dynamic tests were run from room temperature to 700 °C at a heating rate of 10 °C/min. The tests were carried out under a nitrogen atmosphere (25 mL/min) and in an air atmosphere (25 mL/min).

2.4.6. Moving Die Rheometer (MDR)

The curing behavior of the prepared compounds was evaluated by MDR with an MDR 2000 rheometer (Alpha Technologies, Bellingham, WA, USA) at a frequency of 1.667 Hz, a strain of 0.5 degrees, and a temperature of 160 °C for 60 min. Samples were 43 mm in diameter and 2 mm thick.

2.4.7. Dynamic Mechanical Analysis (DMA)

Dynamic mechanical testing was carried out on a GABO Eplexor DMA. The temperature dependence of the viscoelastic properties were measured from −80 °C to 80 °C at a frequency of 10 Hz, dynamic strain amplitude of 0.5%, and static strain of 1%. The temperature was increased in steps of 1 °C, and the sample was thermally equilibrated before testing at each temperature. A 150 N load cell was used to perform measurements on cured rubber with a rectangular specimen geometry of 6.35 mm × 37 mm × 2 mm. The phase angle (δ) and E^*, which is the magnitude of the complex modulus, were directly determined by testing. Then, the storage modulus (E'), loss modulus (E''), and tan delta (tanδ) were calculated according to the following formula.

$$E' = E^* \times \cos\delta; E'' = E^* \times \sin\delta; \text{ and } \tan\delta = E''/E' \quad (2)$$

2.4.8. Tensile Test

Tensile specimens were cut from the 2 mm thick cured rubber sheets in the mill direction using a DIN 53504-S2 (22) cutting die with a gauge length of 50 mm and a width of 4 mm. The tensile testing of each specimen was performed using an Instron Model 5864 Electro-Mechanical Test Instrument with a 1 kN load cell (Instron Corp., 2525–806 1 kN; Norwood, MA, USA). Each specimen was extended at a crosshead rate of 200 mm/min until the break. The tensile measurements were conducted by testing 3 specimens for each sample under ambient conditions; standard deviations of the results are presented in the relevant section.

3. Results and Discussion

3.1. ATR-FTIR Analysis

Oleic modified MFC (OL-MFC) was synthesized by reacting oleoyl chloride on the hydroxyl groups of MFCs in pyridine (Figure 1). An excess amount of oleoyl chloride was applied to achieve OL-MFCs with various degrees of substitution (DS = 0.24–0.91). The success of the esterification modification of MFCs was confirmed by FTIR spectroscopy (Figure 3). Compared with the FTIR spectra of unmodified MFCs, the characteristics of grafted ester pendant groups indicated by a carbonyl C=O stretching vibration at 1740 cm^{-1}, antisymmetric C-O-C stretching at 1230 cm^{-1} [22], and alkenyl C=C stretching at 3010 cm^{-1} were observed for all modified MFCs, confirming the successful modification of MFCs with oleoyl chloride. With the increase in DS, an increase in the intensity of the carbonyl C=O peaks, the peaks at 3010 cm^{-1} assigned to C=C stretching, and the peaks at 2928 and 2849 cm^{-1} assigned to C-H stretching vibrations was observed. The increased intensities confirmed the more pendant groups of oleoyl chloride were grafted onto the MFC backbone. Furthermore, the low intensity of O-H at 3400 cm^{-1} demonstrates that large OH groups on the modified MFCs were replaced with the hydrophobic aliphatic chain of oleoyl chloride.

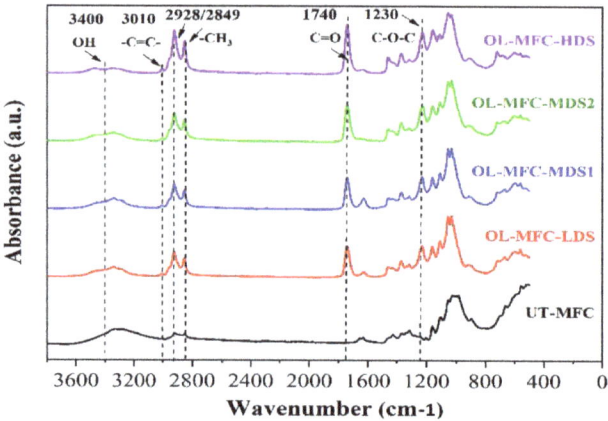

Figure 3. FTIR spectra of untreated MFC and chemically modified MFC with different degrees of substitution (DS).

3.2. XRD Analysis

The unmodified MFCs exhibited very high cellulose content of up to 95% as indicated by glucan content (Table 3) and low hemicellulose content of about 3% (indicated by xylan and mannan) and low Klason lignin of 0.3%. As a result, a high crystallinity index (CI) of up to 75% for the cellulose-rich MFCs was measured (Figure 4), which is consistent with what has been reported for unbleached pulps and MFCs produced from unbleached pulps [23–25]. The XRD patterns shown in Figure 4 indicate the transformation of the cellulose crystal structure with an increase in DS after chemical modification. As can be seen from the spectra, the unmodified and OL-MFC-LDS samples exhibit a similar and typical cellulose Iβ crystalline structure with characteristic peaks at 2θ = 14.9°, 16.7°, 20.6°, 22.7°, and 34.4° for the $1\bar{1}0$, 110, 021, 200, and 004 diffraction planes, respectively [21,26].

Table 3. The chemical compositions of unmodified MFCs used in this study.

		Amount (%)			
Arabinan	Galactan	Glucan	Xylan	Mannan	Klason Lignin
n.d.	n.d.	94.5 (2.8)	2.1 (0.2)	1.2 (0.1)	0.3 (0.2)

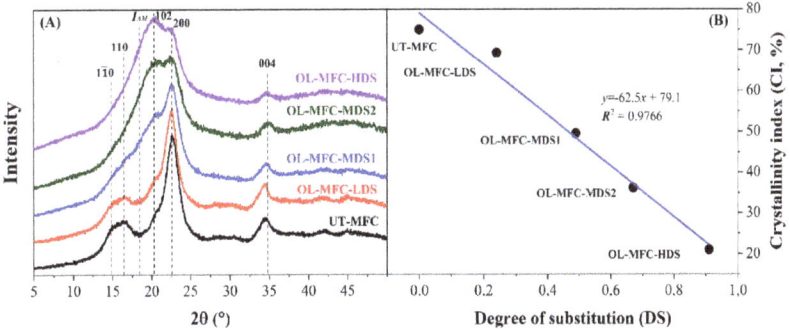

Figure 4. XRD spectra of unmodified MFCs and chemically modified MFCs with different degrees of substitutions (**A**) and crystalline index (CI) values vs. degree of substitutions (**B**).

With a further increase in DS, two diffraction planes ($1\bar{1}0$ and 110) almost disappeared in the XRD spectra of the modified MFCs, and the intensity of the peaks corresponding to the 200 and 004 diffraction planes became weak (Figure 4). In contrast, the peak at 2θ = 20.6° corresponding to the 200 diffraction plane increased with increased DS values, becoming the most intensive peak in the XRD patterns of OL-MFC HDS.

In addition, the intensity of the peak at 2θ = 18°, which was attributed to the contribution of amorphous components of the materials, increased consistently in parallel with that of the peak at 2θ = 20.6°. These changes indicate that the crystalline structure of cellulose was greatly altered when DS was above 0.24. The alteration of the crystalline properties of MFCs due to high grafting should be avoided, as crystallinity contributes to the mechanical properties of MFCs and MFC composites. Similar results have been observed for both heterogeneously [27] and homogeneously [26] modified cellulose fibers. However, the changes in the intensity at 2θ = 14.9° and 16.7° are not consistent with the study reported by Almasi et al. (2015), where the intensity of the two peaks was found to remain constant during the esterification of freeze-dried MFCs with oleic acid at comparable DS values [18]. This is presumably due to the differences in the conditions applied for the modification.

The unmodified MFCs had the highest CI of up to 75%, while a progressive decrease in the CI alongside an increase in the DS values of the chemically modified MFCs was observed. The CI of OL-MFC-LDS was 69%, decreasing to 49% for OL-MFC-MDS1, 36% for OL-MFC-MDS2, and 21% for OL-MFC-HDS (Figure 4). Interestingly, the CI showed a negative linear correlation with the DS of modified MFCs, with a slope of −62.5. This further confirms the damage of the cellulose crystallinity due to the introduction of fatty acid hydrocarbon chains into the cellulose polycrystalline domains.

3.3. STEM Morphology

Chemical modification of MFCs starts with the easily accessible OH groups. Subsequently, it proceeds to the amorphous regions of the cellulose at the initial stage, even with a low dosage of modifier (e.g., OL-MFC-LDS and OL-MFC-MDS1). This results in modified MFCs with improved swelling capacity in the solvents, alongside a preserved three-dimensional network morphology (images B and C in Figure 5). Large nanofibrils were observed on the modified MFCs due to aggregations upon drying (images B and C vs. image A in Figure 5). The increased swelling capacity of the MFCs in the solvent allows for the diffusion of reagents and modifiers deep into the amorphous spaces of the fibrils, followed by esterification with the internally available hydroxyl groups. This contributes to the disruption of hydrogen bonding between fibrils or cellulose chains [28]. Consequently, individual fibrils could be separated from the bundles of fibrils, and cellulose chains could probably be detached from the surface fibrils with extended modification under a high dosage of chemical reagents (i.e., OL-MFC-MDS2) (image D in Figure 5).

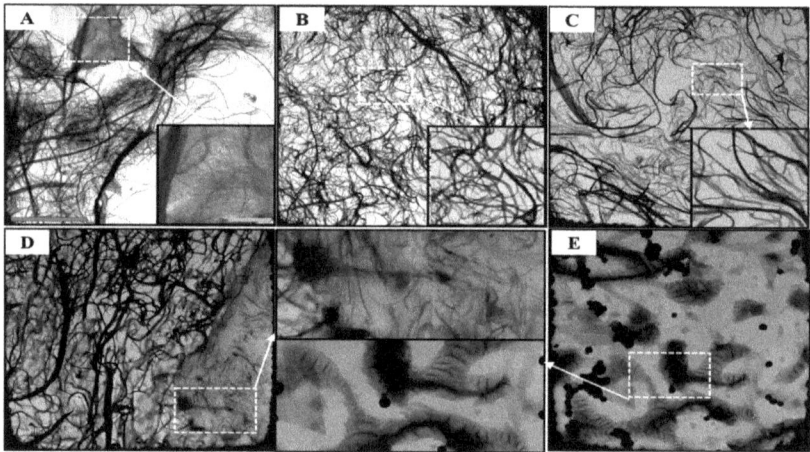

Figure 5. STEM images of untreated MFC and chemically modified MFC with different degrees of substitution ((**A**)—UT-MFC; (**B**)—OL-MFC-LDS; (**C**)—OL-MFC-MDS1; (**D**)—OL-MFC-MDS2; (**E**)—OL-MFC-HDS) (red scale bar in image A represents 10 µm, and images (**B–E**) have the same scales; the blue scale bar in image A represents 2 µm, and all magnified images have the same scale).

When the degrees of substitution increased for OL-MFC-MDS2, the disruption in the OH interaction could not destroy the network structure of the nanofibrils. With a further increase in the DS value to 0.91 (i.e., OL-MFC-HDS), the network structure and the micro-fibrillar integrity were entirely lost (image E in Figure 5). There is the possibility that the disruption of the hydrogen bond network of MFCs at high grafting could degrade the fibrils and engineer a new type of interaction. A possible Van der Waals type of interaction could convolute the fibrillar morphology into a new structure, as shown in Figure 5E. At higher magnification, it appears that the agglomerates are most likely comprised of the degraded and aggregated fiber and fibril fragments. Degradation of MFCs due to surface modification was previously reported [29]. A series of chemically modified MFCs with varied DS values were reported, with convincing evidence that showed a clear evolution in the morphology of MFCs with extended surface modification. Overall, the evolution of the morphology of MFCs shown in Figure 5 shows the progressive degradation of nanofibrils and cellulose chains because of the continuous introduction of fatty acid side chains into both amorphous and polycrystalline domains of cellulose. These results are consistent with the changes in the XRD spectra patterns shown in Figure 4.

3.4. Thermal Stability Analysis

Studies on the thermal stability of the chemically modified MFCs under nitrogen and air atmospheres are presented in Figure 6. The unmodified MFCs featured a substantial decomposition around 260–380 °C, representing the degradation of the cellulose backbone, with a minor weight loss below 150 °C due to the loss of volatiles and moisture [26,30,31]. The MFCs comprise up to 95% cellulose, with little hemicellulose and lignin (Table 3) and negligible weight loss at temperatures ranging from 150 to 260 °C and above 380 °C [32–34] under nitrogen and air atmospheres.

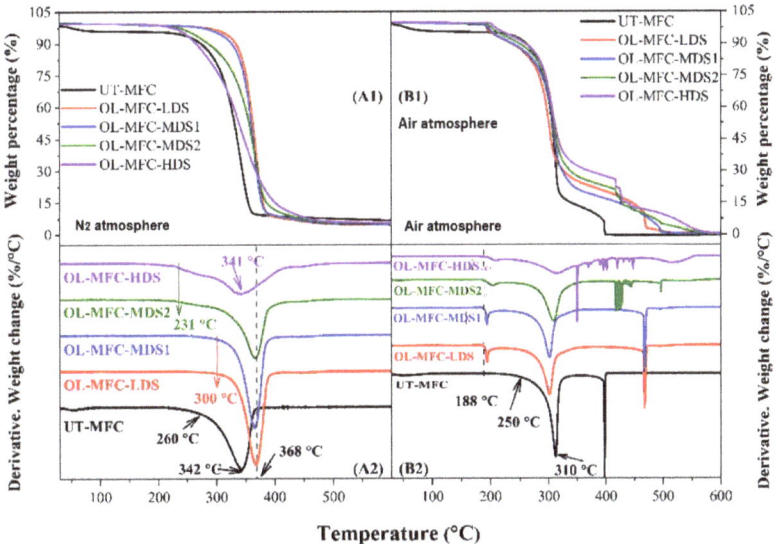

Figure 6. Thermogravimetric analysis (TGA) of unmodified MFCs and chemically modified MFCs in a N_2 atmosphere (**A1,A2**) and air atmosphere (**B1,B2**). (**A1,B1**) Thermogravimetric (TG) curves; (**A2,B2**) derivative thermogravimetric (DTG) curves.

Upon chemical modification, distinct changes in the thermogravimetric analysis (TGA) and the derivative thermogravimetric (DTG) curves were observed when compared to the unmodified MFCs. Compared to the unmodified MFCs, the weight loss remained stable below 150 °C, irrespective of the test condition (i.e., air or N_2), which may be a result of the hydrophobic character of the grafted oleoyl groups. A similar finding was reported for cellulose laurate esters [26]. In a N_2 atmosphere (graphs A1 and A2 in Figure 6), similar degradation behavior above 150 °C was observed for all the chemically modified MFCs compared with that of the unmodified MFCs.

At low DS values up to 0.49, the onset decomposition temperature shifted to a higher temperature from about 260 °C to 300 °C, implying an increase in thermal stability after the esterification reaction. This improvement in thermal stability suggests a rearrangement on the cellulose backbone to form a new ordered structure because of the long-chain fatty acid groups [26,35]. However, at very high grafting, the impact on the thermal stability became pronounced. The high level of grafting possibly led to the degradation of the fibers (STEM images, Figure 5) because of the large disruption of the MFC microstructure. This likely exposed the cellulose structure to easy thermal degradation. Although the polycrystalline structure of cellulose was partially altered or damaged, the results are in line with previous studies [26,35–37].

On the contrary, the samples with higher DS values (i.e., 0.67 and 0.91) started to decompose at a lower temperature of about 230 °C, which is 30 °C lower than that of the unmodified MFCs. The morphological changes—mainly the rearrangement of cellulose chains and their counterparts, as seen in Figure 5—are more likely to be responsible for the decreased onset decomposition temperature. Different TGA and DTG patterns were recorded in the air atmosphere due to the oxidation reactions of the attached fatty acid groups. All the highly modified MFCs remained stable up to 188 °C, which is about 70 °C lower compared to the 260 °C for the unmodified MFCs, and began to decompose thereafter. The decrease in the initial decomposition temperature is presumably due to the low stability of double bonds in the attached fatty acid groups [38]. Before the onset of decomposition, a slight gain of mass was observed because of the uptake of oxygen at the beginning of the oxidation of the unsaturated bonds [38,39]. The decomposition rate was almost constant from 200 to 260 °C and became substantially higher from 260 to 350 °C owing to the progressive oxidation of the remaining alkyl chains attached to the cellulose. The last decomposition stage was recorded between 350 and 580 °C, where the total oxidation of the carbonaceous residues formed in the former stage occurred [38] in a dynamic oxidation atmosphere for all samples.

3.5. Compounding and Rubber Compound Properties
3.5.1. Processability Analysis

The processability of modified MFCs was studied from the maximum temperature (T_m), maximum rotor torque (T_{qm}), and work done during each stage of compounding (Table 4). The degree of substitution and the formulations of the modified MFCs had a considerable influence on the processability of the modified MFCs. Generally, the maximum temperature, the maximum rotor torque, and the work done in all the mixing stages (NP1, NP2, and PR) increased relative to the increase in the DS. The T_m and T_{qm} in NP2 and the total work done for compounding of the MFC14 sample with OL-MFC-LDS were 139.6 °C, 45.6 Nm, and 257.2 KJ, respectively. With an increase in DS for OL-MFC-MDS2 (MFC16), the values increased to 144.4 °C, 47.5 Nm, and 289.9 KJ, respectively, corresponding to 3%, 4%, and 13% compared to MFC14. As the DS further increased in sample MFC17, those values increased by 5%, 24%, and 20%, respectively, compared to MFC14 (Figure 7).

Table 4. A summary of compounding parameters during each mixing stage for all rubber compounds.

Sample	NP1		NP2		PR		Total Work
	Tm (°C)	W (KJ)	Tm (°C)	W (KJ)	Tm (°C)	W (KJ)	W (KJ)
Control	147.9	133.8	138.1	117.5	77.6	14.6	265.9
MFC14	141.1	128	139.6	112.9	79	16.3	257.2
MFC15	141.4	142.2	144.2	126.6	80.8	17.7	286.5
MFC16	141.5	143.9	144.4	126.7	80.5	19.3	289.9
MFC17	145.7	152.2	146.8	138.6	81.7	16.8	307.6
MFC18	140.3	136.5	143.4	124	77.4	15.9	276.4
MFC19	143	152.5	147.7	141.7	83.6	19.8	314
MFC20	135.3	117.8	139.1	110.5	78.1	16.4	244.7

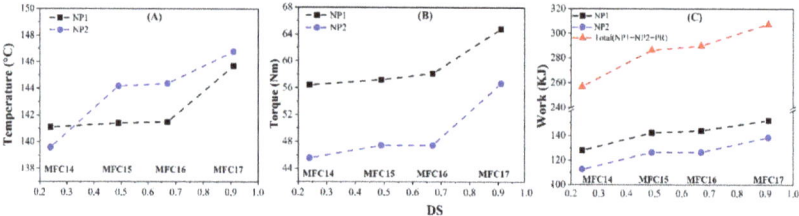

Figure 7. The changes in the maximum mixing temperature (A), the maximum rotor torque (B), and work done (C) in different mixing stages with increased DS of chemically modified MFCs (from MFC14 to MFC15, MFC16, and MFC17).

The compounding process demonstrates the exothermic heat of mixing involving strong interactions between molecules or nano- or micro-sized particles, probably along with chemical reactions. The increase in the compounding parameters alongside the increase in the DS values of MFCs attests to the improved interactions between the modified MFCs and other materials in the mix. The increases in the mixing temperature, rotor torque, and work done with increased DS values from MFC14 to MFC15 (Figure 7) are primarily due to the enhanced surface hydrophobicity of MFCs, leading to improved elastomers/modified MFCs and modified MFC/silanized silica interactions. With further increases in the DS for MDS2, the increase in those compounding parameters plateaued (Figure 7), showing that the hydrophobicity of MFCs reached the highest achievable level without losing the fibrillar network (Figure 5). However, another sharp increase was observed in those compounding parameters from MFC16 to MFC17 (HDS). This is assumed to be related to both the enhanced hydrophobicity of MFCs and the significant alteration of the morphologies. As shown in Figure 5, the supposed delamination of the surface fibers of cellulose chains and the rearrangement of the cellulose microstructure occurred due to the enhanced hydrophobicity.

Furthermore, impacts of the silica dosage and the total amount of processing oil on the compounding parameters were noticed (Table 4 and Figure 8). With the removal of 10 phr silica, decreases in the maximum mixing temperature, the maximum rotor torque, and work done in the mixing stage of NP1 and NP2 were observed due to the decreasing viscosity of the mixtures occasioned by the low filler volume, which could also explain the drop in the mixing temperature [40].

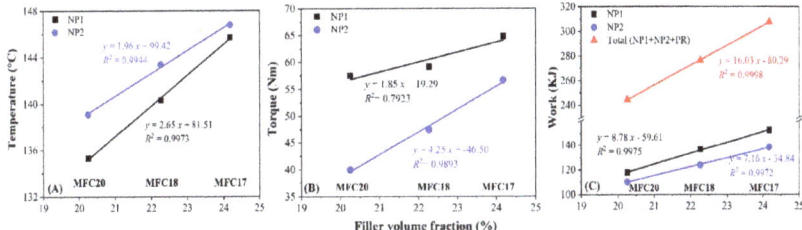

Figure 8. Changes in the maximum mixing temperature (**A**), maximum rotor torque (**B**), and work done during compounding (**C**) with increased total filler volume fraction (from compound MFC17, to MFC18 and MFC20).

3.5.2. MDR Analysis

The cure characteristics of the compounds were studied using a moving die rheometer (MDR), as presented in Table 5 and Figure 9. The maximum torque (T_{max}) represents the achieved crosslink density and the degree of reinforcement of the filler in the matrix. The minimum torque (T_{min}) alludes to the viscosity of the green compounds and the interactions between the fillers and the matrix. It can be observed that the DS of the modified MFCs impacted the maximum and minimum torque, which increased with increasing DS and filler volume fraction. This can be attributed to the possible contributions of the double bonds of the aliphatic chains to the vulcanization reaction. Thereafter, the T_{min} and T_{max} torque plateaued when DS values were above 0.67 (i.e., OL-MFC-MDS2). The progressive increase in the maximum and minimum torque with the increase in filler volume fraction is simply due to the increased crosslink contributions of the increase in filler volume.

Table 5. Vulcanization characteristics of each rubber compound measured by a moving die rheometer (MDR).

Sample	Min Torque (T_{min}) (Nm)	Max Torque (T_{max}) (Nm)	25% Cure T_{25} (min)	Optimum Cure T_{90} (min)
Control	2.9	23.5	4.6	20.5
MFC14	2.9	22.3	6.1	22.5
MFC15	3.7	27.3	5.5	22.3
MFC16	4.0	28.2	5.1	22.5
MFC17	4.2	29.0	6.0	24.3
MFC18	3.3	25.5	5.3	23.0
MFC19	4.5	29.6	4.7	22.0
MFC20	2.9	23.5	4.6	20.5

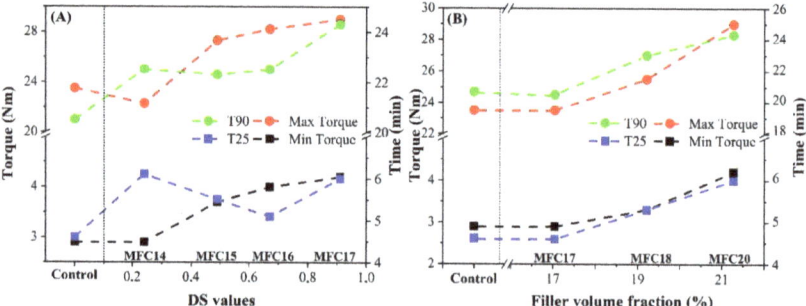

Figure 9. Changes in the maximum torque, minimum torque, T_{25}, and T_{90} with increased DS values of the modified MFCs (**A**) and with increased filler volume fraction (**B**). (T_{25}, time required for 25% cure development; T_{90}, time required for 90% cure development).

The changes in the cure kinetics at 25% cure (T_{25}) and optimum cure time (T_{90}) were observed to be partly influenced by the DS and filler volume fraction, as compounds with modified MFC showed a distinct behavior relative to the control. The t_{25} maintained a duration of 4.6–6.1 min relative to the control, regardless of the degree of substitution. It was also observed that the optimum cure time of the compounds with modified MFCs reached 24.3 min compared with the control (20.5 min). This can likely be attributed to the low reactivity of the fatty acid modifier with other crosslinking agents (e.g., TESPT silane and sulfur), especially when a high dosage of the modifier was incorporated (i.e., for high DS of MFCs). The contributions from the aliphatic double bonds to the crosslinking may have also further delayed the curing time, requiring more time to utilize the double bond in the crosslinking process. Furthermore, the progressive increase in t_{25} and t_{90} as filler loading increased is presumably due to the increased in surface area because of the high filler loadings [40,41]. Overall, the highest optimum cure t_{90} was achieved at about 24 min. Therefore, it would be advantageous to maintain a cure window not exceeding 30 min at the same curing temperature (i.e., 150 °C). This would result in good compound performance and avoid overcuring with possible reversions.

3.5.3. Rubber Process Analysis (RPA) and Dynamic Mechanical Analysis (DMA)

The dynamic mechanical behavior of the fabricated MFC/silica/SBR rubber compounds was studied and compared with that of a silica/SBR compound as a control. Figure 10 shows the normalized tangent delta (TD) alongside the storage modulus (G′) at 1% and 10% strain. An ideal compound is required to have an acceptably high level of stiffness (G′) and low tangent delta (TD). These properties change during strain-imposed deformation because of the filler network breakdown. It can be observed that compounds with higher stiffness (G′) tend to have a higher TD and vice-versa. The tradeoff between stiffness and damping at 1% strain presented MFC19 as a compound with interesting behavior. A further increase in the strain to 10%, indicated MFC18 and MFC19 as having a better tradeoff compared to the control. These changes along the progressive low strain demonstrate the extent of filler network breakdown and recovery described as the Payne effect [42,43]. At low strain up to 10%, good stiffness and TD were retained for MFC18 and 19 compared to the control. This is an indication that the filler network of these compounds was resilient and was not subject to considerable damage. A low Payne effect such as that seen in these compounds could be useful for tire tread applications because the tread is subjected to dynamic deformations and requires compounds with low energy dissipation resulting from filler network breakdown. These properties were achieved by modulating the DS and compound recipe for a better tradeoff. The two compounds were made with 20 phr less silica compared to the reference. Eventually, a reasonable level of stiffness was maintained while keeping an acceptable level of TD. The stiffness could be further enhanced at low filler loading by reducing the TDAE oil. These results demonstrate that a percolation threshold needed for reinforcement can be achieve at low filler loading by substituting 10 phr of modified MFC with 20 phr of silica.

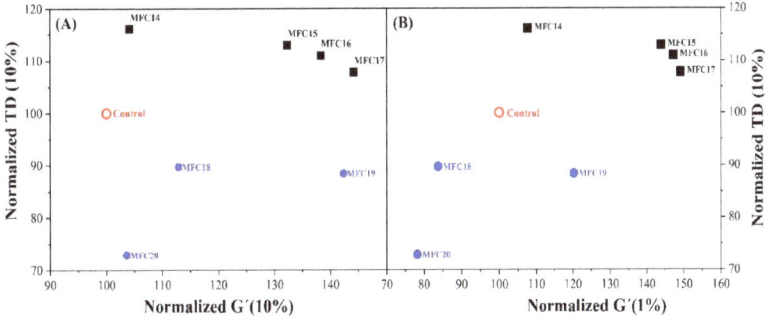

Figure 10. Normalized TD (10%) versus G' (10%) (**A**) and G' (1%) (**B**) for silica/chemically modified MFC hybrid rubber composites measured using an RPA.

The high level of stiffness observed in some compounds (MFC15, 16, and 17) offers new opportunities for applications of MFC-reinforced polymer in aspects of tires not requiring high damping properties.

The behavior of the compounds was further studied at different temperatures. Some of the properties studied include the tangent delta at 60 °C, which is used as a rolling resistance indicator [44], and the tan delta at 0 °C, which is used as an indicator for wet traction [45], as well as the modulus of the compounds at 30 °C. The influences of DS of MFCs and filler loadings on the modulus, rolling resistance, and wet traction of the compound are shown in Table 6. Above a DS of 0.67 (MFC16), a decrease in the modulus was observed. This reduction in the modulus could possibly be attributed to the loss of the reinforcing fibrillar properties of the MFC due to grafting. A progressive loss of crystallinity and compromised morphology were observed. These crucial fiber properties are important determinants of the properties of the final compound, especially the modulus. Given the comparable values of the tan delta of the modified MFC and control, an appreciable decrease in T_g of the compounds made with MFCs provides greater flexibility for applications. The low T_g values of the MFC compounds are probably due to the low T_g of the fatty acid modifier compared to that of the TDAE processing oil [46].

The improved modulus compared to the control also confirms the high reinforcing efficiency of the MFCs in the elastomeric matrix. Alongside the appropriate degree of substitution (not compromising the fiber properties), the modulus can be enhanced by good interfacial adhesion [47,48].

Table 6. Dynamic mechanical analysis (DMA) results for the referenced rubber compounds and the rubber compounds with chemically modified MFC fibers.

Sample	T_g (°C)	TD (0 °C)	TD (60 °C)	E' (MPa, 30 °C)
Control	−21.8	0.306	0.081	12.6
MFC14	−24.0	0.258	0.095	20.5
MFC15	−24.2	0.228	0.107	36.2
MFC16	−24.8	0.205	0.111	56.3
MFC17	−23.6	0.241	0.097	29.7
MFC18	−23.1	0.249	0.084	21.5
MFC19	−23.1	0.236	0.087	28.8
MFC20	−23.3	0.237	0.080	17.6

3.5.4. Tensile Properties

The stress–strain behaviour of all investigated samples is shown in Figure 11 and tensile properties shown in Table 7. Three typical tensile behaviors were observed (graph A in Figure 11). The first behavior (type I) of the referenced silica-filled SBR/PBD compound is characterized by strain-dependent modulus indicated by an upturned curve, leading to increased slope as the strain increases, which could be due to the limited chain extensibility [43]. The second behavior (type II) is linearly elastic, corresponding to a straight stress–strain curve. The last behavior (type III) includes some plastic flow, as indicated by a slightly bent curve at low deformations (1–50%), followed by linearly elastic behavior until failure.

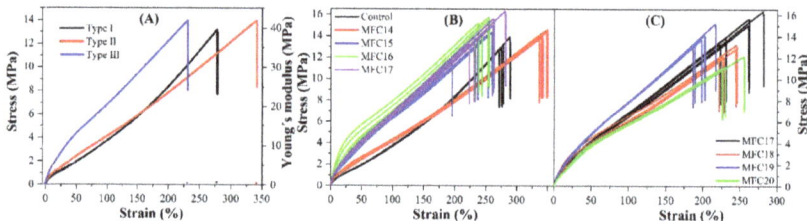

Figure 11. Tensile curves of (**A**) dominant tensile behavior of MFC/silica/SBR and silica/SBR compounds; (**B**) tensile evolution of the control, MFC14, 15, 16, and 17; and (**C**) tensile curves of MFC17, 18, 19, and 20.

Table 7. Tensile properties of rubber compounds with chemically modified MFC fibers. Values are shown as mean (standard deviation).

Sample	Tensile Strength (MPa)	Young's Modulus (MPa)	Strain (%)
Control	13.1 (0.5)	3.8 (0.0)	277.1 (6.9)
MFC14	14.2 (0.4)	3.4 (0.1)	343.2 (5.7)
MFC15	13.4 (1.2)	4.7 (0.3)	245.2 (13.3)
MFC16	15.0 (0.7)	4.7 (0.1)	243.9 (10.0)
MFC17	14.8 (1.2)	5.1 (0.1)	251.8 (24.4)
MFC18	12.4 (0.6)	4.2 (0.1)	233.3 (11.0)
MFC19	13.8 (0.8)	5.8 (0.1)	198.5 (12.0)
MFC20	11.5 (0.6)	3.7 (0.1)	238.4 (15.6)

The compound with the lowest DS (i.e., MFC14) exhibited type II behavior. However, an increase in the DS of MFCs resulted to compounds showing some plastic deformation at low strains (1–50%, type III behavior). The plastic deformation at low strains became the most pronounced for the compounds incorporated with MFC16. This could be due to the evolution of plasticized domains on the fibers from the hydrophobization. Ordinarily, the weight of the fatty acid chain can promote delamination of the surface fibers and make them function as plasticizers instead as reinforcing agents. This, in turn, could lead to the slippage of the hydrophobic cellulose chains on the surface against the relative hydrophilic internal cellulose chains towards the direction of the external load [48]. As the DS value further increased to 0.91, the resulting compounds (MFC17, 18, 19, and 20) exhibited decreased plastic deformation. This could be a result of the reduced sizes and rearrangement of cellulose chains (Figure 5), with enhanced interactions keeping the chains from sliding towards each other.

Apart from the changes in the stress–strain behavior, a significant increase in Young's modulus of the compounds was observed at low strains (<50%) with increased DS (graph B in Figure 11). The highest values of the Youngs's modulus at 25% was determined to be 8.1 MPa for the MFC16 compounds, which is 1.8 times higher than that of the referenced silica-filled compounds. There was also a rise in the modulus at higher strains increased DS. For example, Young's modulus at 100% strain increased gradually with increased DS values and reached about 5.0 MPa when the DS was above 0.49 (i.e., MFC16). This in line with previous observations [49–54], further confirming the high reinforcing efficiency of MFCs in rubber composites. In addition, a high tensile strength of 15 MPa was achieved with the addition of modified cellulose fibers compared with the reference compound made of silica (13 MPa). The strain at break was higher for the MFC14 compound, while it started to decrease as the DS further increased, resulting in more brittle compounds. The results, in general, are consistent with previously reported findings that the incorporation of cellulose fibers can result in a significant improvement in the modulus and, to a lesser extent, in the tensile strength.

4. Conclusions

The chemical modification of MFCs with oleoyl chloride provided hydrophobic domains needed for good dispersion in an SBR matrix and good filler/polymer interactions. MFCs with a DS of 0.2–0.9 were successfully synthesized in pyridine and compounded with silica and SBR. The compounds were fabricated by substituting 10–30 phr silica with 10 phr of modified MFCs. As the DS values increased, significant impacts on the morphology and crystallinity of modified MFCs were observed. These changes were found to further affect the processability of the compounds, as well as the curing behavior, modulus-rolling resistance tradeoff, and tensile properties of the resulting rubber compounds. Overall, the highest reinforcement was achieved for the MFC compound with a DS value of 0.67 (i.e., MFC16). This compound had a high surface hydrophobicity and also retained the fibrillar network structure. The most important drawback in the compound properties was the wear abrasion properties (results not presented), which require improvement. However, this study demonstrates that a small amount of MFCs can be used to replace a large amount of silica, resulting in improvements in reinforcement and mechanical properties. The improved properties achieved with the incorporation of modified MFCs opens potential applications for the use of sustainable bio-based materials to produce tire compounds and lightweight composites.

Author Contributions: Conceptualization, M.L., E.S. and J.-S.T.; methodology, M.L. and P.S.; validation, M.L., P.S. and E.S.; formal analysis, M.L, P.S., I.E.I. and M.S.; writing—original draft preparation, M.L., M.S. and I.E.I.; review and editing, D.L., B.D., M.S. and J.-S.T.; supervision, J.-S.T., B.D. and E.S. All authors have read and agreed to the published version of the manuscript.

Funding: This research was funded by the National Research Found of Luxembourg (FNR) (grant number IPBG16/11514551/TireMat-Tech). The APC was funded by FNR.

Institutional Review Board Statement: Not applicable.

Data Availability Statement: Data is contained within the article.

Conflicts of Interest: The authors declare that there is no conflict of interest.

References

1. Affandi, S.; Setyawan, H.; Winardi, S.; Purwanto, A.; Balgis, R. A facile method for production of high-purity silica xerogels from bagasse ash. *Adv. Powder Technol.* **2009**, *20*, 468–472. [CrossRef]
2. Fernandes, I.J.; Santos, R.V.; Santos ECA dos Rocha, T.L.A.C.; Domingues Junior, N.S.; Moraes, C.A.M. Replacement of Commercial Silica by Rice Husk Ash in Epoxy Composites: A Comparative Analysis. *Mater. Res.* **2018**, *21*, 1–10. [CrossRef]
3. Götze, J.; Möckel, R. (Eds.) *Quartz: Deposits, Mineralogy and Analytics*; Springer: Berlin/Heidelberg, Germany, 2012. [CrossRef]
4. Li, T.; Chen, C.; Brozena, A.H.; Zhu, J.Y.; Xu, L.; Driemeier, C.; Dai, J.; Rojas, O.J.; Isogai, A.; Wågberg, L.; et al. Developing fibrillated cellulose as a sustainable technological material. *Nature* **2021**, *590*, 47–56. [CrossRef] [PubMed]
5. Ferreira, E.S.; Rezende, C.A.; Cranston, E.D. Fundamentals of cellulose lightweight materials: Bio-based assemblies with tailored properties. *Green Chem.* **2021**, *23*, 3542–3568. [CrossRef]
6. Roy, K.; Pongwisuthiruchte, A.; Chandra Debnath, S.; Potiyaraj, P. Application of cellulose as green filler for the development of sustainable rubber technology. *Curr. Res. Green Sustain. Chem.* **2021**, *4*, 100140. [CrossRef]
7. Sirviö, J.A.; Visanko, M.; Heiskanen, J.P.; Liimatainen, H. UV-absorbing cellulose nanocrystals as functional reinforcing fillers in polymer nanocomposite films. *J. Mater. Chem. A* **2016**, *4*, 6368–6375. [CrossRef]
8. He, M.; Zhou, J.; Zhang, H.; Luo, Z.; Yao, J. Microcrystalline cellulose as reactive reinforcing fillers for epoxidized soybean oil polymer composites. *J. Appl. Polym. Sci.* **2015**, *132*, 42488. [CrossRef]
9. Staropoli, M.; Rogé, V.; Moretto, E.; Didierjean, J.; Michel, M.; Duez, B.; Steiner, P.; Thielen, G.; Lenoble, D.; Thomann, J.S. Hybrid Silica-Based Fillers in Nanocomposites: Influence of Isotropic/Isotropic and Isotropic/Anisotropic Fillers on Mechanical Properties of Styrene-Butadiene (SBR)-Based Rubber. *Polymers* **2021**, *13*, 2413. [CrossRef]
10. Bai, W.; Li, K. Partial replacement of silica with microcrystalline cellulose in rubber composites. *Compos. Part A Appl. Sci. Manuf.* **2009**, *40*, 1597–1605. [CrossRef]
11. Xu, S.H.; Gu, J.; Luo, Y.F.; Jia, D.M. Effects of partial replacement of silica with surface modified nanocrystalline cellulose on properties of natural rubber nanocomposites. *Express Polym. Lett.* **2012**, *6*, 14–25. [CrossRef]
12. Kazemi, H.; Mighri, F.; Park, K.W.; Frikha, S.; Rodrigue, D. Effect of Cellulose Fiber Surface Treatment to Replace Carbon Black in Natural Rubber Hybrid Composites. *Rubber Chem. Technol.* **2022**, *95*, 128–146. [CrossRef]
13. Lopattananon, N.; Jitkalong, D.; Seadan, M. Hybridized reinforcement of natural rubber with silane-modified short cellulose fibers and silica. *J. Appl. Polym. Sci.* **2011**, *120*, 3242–3254. [CrossRef]
14. Li, K.; Mcgrady, D.; Zhao, X.; Ker, D.; Tekinalp, H.; He, X.; Qu, J.; Aytug, T.; Cakmak, E.; Phipps, J.; et al. Surface-modified and oven-dried microfibrillated cellulose reinforced biocomposites: Cellulose network enabled high performance. *Carbohydr. Polym.* **2021**, *256*, 117525. [CrossRef] [PubMed]
15. Fumagalli, M.; Berriot, J.; de Gaudemaris, B.; Veyland, A.; Putaux, J.-L.; Molina-Boisseau, S.; Heux, L. Rubber materials from elastomers and nanocellulose powders: Filler dispersion and mechanical reinforcement. *Soft Matter* **2018**, *14*, 2638–2648. [CrossRef] [PubMed]
16. Hu, Z.; Berry, R.M.; Pelton, R.; Cranston, E.D. One-Pot Water-Based Hydrophobic Surface Modification of Cellulose Nanocrystals Using Plant Polyphenols. *ACS Sustain. Chem. Eng.* **2017**, *5*, 5018–5026. [CrossRef]
17. Dhuiège, B.; Pecastaings, G.; Sèbe, G. Sustainable Approach for the Direct Functionalization of Cellulose Nanocrystals Dispersed in Water by Transesterification of Vinyl Acetate. *ACS Sustain. Chem. Eng.* **2019**, *7*, 187–196. [CrossRef]
18. Almasi, H.; Ghanbarzadeh, B.; Dehghannia, J.; Pirsa, S.; Zandi, M. Heterogeneous modification of softwoods cellulose nanofibers with oleic acid: Effect of reaction time and oleic acid concentration. *Fibers. Polym.* **2015**, *16*, 1715–1722. [CrossRef]
19. Sluiter, A.; Hames, B.; Ruiz, R.O.; Scarlata, C.; Sluiter, J.; Templeton, D.; Crocker, D.L.A.P. Determination of Structural Carbohydrates and Lignin in Biomass. *Biomass Anal Technol. Team Lab Anal Proced.* **2004**, 1–14.
20. Liu, M.; Fernando, D.; Daniel, G.; Madsen, B.; Meyer, A.S.; Ale, M.T.; Thygesen, A. Effect of harvest time and field retting duration on the chemical composition, morphology and mechanical properties of hemp fibers. *Ind. Crops. Prod.* **2015**, *69*, 29–39. [CrossRef]
21. Thygesen, A.; Oddershede, J.; Lilholt, H.; Thomsen, A.B.; Ståhl, K. On the determination of crystallinity and cellulose content in plant fibres. *Cellulose* **2005**, *12*, 563–576. [CrossRef]
22. DIN 53504; Testing of Rubber—Determination of Tensile Strength at Break, Tensile Stress at Yield, Elongation at Break and Stress Values in a Tensile Test. Deutsches Institut für Normung: Berlin, German, 2009.
23. Rostami, J.; Mathew, A.P.; Edlund, U. Zwitterionic Acetylated Cellulose Nanofibrils. *Molecules* **2019**, *24*, 3147. [CrossRef] [PubMed]
24. Priya, S.; Khan, G.; Uddin, M.; Haque, M.; Islam, M.; Abdullah-Al-Mamun, M.; Gafur, M.A.; Alam, M.S. Characterization of Micro-fibrillated Cellulose Produced from Sawmill Wastage: Crystallinity and Thermal Properties. *Am. Chem. Sci. J.* **2015**, *9*, 1–8. [CrossRef]
25. Daicho, K.; Saito, T.; Fujisawa, S.; Isogai, A. The Crystallinity of Nanocellulose: Dispersion-Induced Disordering of the Grain Boundary in Biologically Structured Cellulose. *ACS Appl. Nano Mater.* **2018**, *1*, 5774–5785. [CrossRef]
26. Lengowski, E.C.; Muñiz GIB de Andrade AS de Simon, L.C.; Nisgoski, S. Morphological, Physical and Thermal Characterization of Microfibrillated CelluloseE. *Rev. Árvore* **2018**, *42*, e420113. [CrossRef]

27. Wen, X.; Wang, H.; Wei, Y.; Wang, X.; Liu, C. Preparation and characterization of cellulose laurate ester by catalyzed transesterification. *Carbohydr. Polym.* **2017**, *168*, 247–254. [CrossRef]
28. Freire, C.S.R.; Silvestre, A.J.D.; Neto, C.P.; Belgacem, M.N.; Gandini, A. Controlled heterogeneous modification of cellulose fibers with fatty acids: Effect of reaction conditions on the extent of esterification and fiber properties. *J. Appl. Polym. Sci.* **2006**, *100*, 1093–1102. [CrossRef]
29. Sheltami, R.M.; Kargarzadeh, H.; Abdullah, I. Effects of Silane Surface Treatment of Cellulose Nanocrystals on the Tensile Properties of Cellulose-Polyvinyl chloride Nanocomposite. *Sains Malays.* **2015**, *44*, 801–810. [CrossRef]
30. Andresen, M.; Johansson, L.-S.; Tanem, B.S.; Stenius, P. Properties and characterization of hydrophobized microfibrillated cellulose. *Cellulose* **2006**, *13*, 665–677. [CrossRef]
31. Zhao, H.; Yan, H.; Liu, M.; Zhang, C.; Qin, S. Pyrolytic characteristics and kinetics of the marine green tide macroalgae, Enteromorpha prolifera. *Chinese J. Oceanol. Limnol.* **2011**, *29*, 996–1001. [CrossRef]
32. Liu, M.; Baum, A.; Odermatt, J.; Berger, J.; Yu, L.; Zeuner, B.; Thygesen, A.; Holck, J.; Meyer, A.S. Oxidation of lignin in hemp fibres by laccase: Effects on mechanical properties of hemp fibres and unidirectional fibre/epoxy composites. *Compos. Part A Appl. Sci. Manuf.* **2017**, *95*, 377–387. [CrossRef]
33. Zhao, H.; Yan, H.X.; Zhang, M.M.; Liu, M.; Qin, S. Pyrolysis Characteristics and Kinetics of Enteromorpha Clathrata Biomass: A Potential Way of Converting Ecological Crisis "Green Tide" Bioresource to Bioenergy. *Adv. Mater. Res.* **2010**, *113–116*, 170–175. [CrossRef]
34. Müller-Hagedorn, M.; Bockhorn, H.; Krebs, L.; Müller, U. A comparative kinetic study on the pyrolysis of three different wood species. *J. Anal. Appl. Pyrolysis* **2003**, *68–69*, 231–249. [CrossRef]
35. Francisco-Fernández, M.; Tarrío-Saavedra, J.; Naya, S.; López-Beceiro, J.; Artiaga, R. Classification of wood using differential thermogravimetric analysis. *J. Therm. Anal. Calorim.* **2015**, *120*, 541–551. [CrossRef]
36. Huang, F.-Y. Thermal Properties and Thermal Degradation of Cellulose Tri-Stearate (CTs). *Polymers* **2012**, *4*, 1012–1024. [CrossRef]
37. Cao, X.; Peng, X.; Zhong, L.; Sun, S.; Yang, D.; Zhang, X.; Sun, R. A novel transesterification system to rapidly synthesize cellulose aliphatic esters. *Cellulose* **2014**, *21*, 581–594. [CrossRef]
38. Labafzadeh, S.R.; Kavakka, J.S.; Sievänen, K.; Asikkala, J.; Kilpeläinen, I. Reactive dissolution of cellulose and pulp through acylation in pyridine. *Cellulose* **2012**, *19*, 1295–1304. [CrossRef]
39. Raba, D.N.; Chambre, D.R.; Copolovici, D.-M.; Moldovan, C.; Copolovici, L.O. The influence of high-temperature heating on composition and thermo-oxidative stability of the oil extracted from Arabica coffee beans. *PLoS ONE* **2018**, *13*, e0200314. [CrossRef]
40. Faria, E.A.D.; Leles, M.I.G.; Ionashiro, M.; Zuppa, T.D.O.; Antoniosi Filho, N.R. Estudo da estabilidade térmica de óleos e gorduras vegetais por TG/DTG e DTA. *Eclética Química* **2002**, *27*, 47–56. [CrossRef]
41. Klie, B.; Teich, S.; Haberstroh, E.; Giese, U. Newmethod for evaluating rubber mixing Quality by means of alternative representation of the Fingerprint chart. *KGK Kautsch. Gummi Kunstst.* **2015**, *68*, 31–38.
42. Chigondo, F.; Shoko, P.; Nyamunda, B.; Guyo, U.; Moyo, M. Maize stalk as reinforcement in natural rubber composites. *Int. J. Sci. Technol. Res.* **2013**, *2*, 263–271.
43. Ramier, J.; Gauthier, C.; Chazeau, L.; Stelandre, L.; Guy, L. Payne effect in silica-filled styrene–butadiene rubber: Influence of surface treatment. *J. Polym. Sci. Part B Polym. Phys.* **2007**, *45*, 286–298. [CrossRef]
44. Bokobza, L.; Gauliard, V.; Ladouce, L. Silica Reinforcement of StyreneButadiene Rubbers. *Elastomers and Plastics.* **2001**, *54*, 177–180.
45. Cichomski, E.M.; Dierkes, W.K.; Tolpekina, T.V.; Schultz, S. Influence of physical and chemical polymer-filler bonds on the wet-traction performance indicators for passenger car tire tread materials. *KGK Kautsch. Gummi Kunstst.* **2014**, *67*, 50–57.
46. Kawahara, S. Controlling Performance of Filled Rubbers. In *Encyclopedia of Polymeric Nanomaterials*; Springer: Berlin/Heidelberg, Germany, 2015; pp. 453–460. [CrossRef]
47. Kuta, A.; Hrdlicka, Z.; Voldanova, J.; Brejcha, J.; Pokorny, J.; Plitz, J. Dynamic Mechanical Properties of Rubbers with Standard Oils and Oils with Low Content of Polycyclic Aromatic Hydrocarbons. *Test Meas.* **2010**, *63*, 120–122.
48. Liu, M.; Thygesen, A.; Summerscales, J.; Meyer, A.S. Targeted pre-treatment of hemp bast fibres for optimal performance in biocomposite materials: A review. *Ind. Crops. Prod.* **2017**, *108*, 660–683. [CrossRef]
49. Liu, M.; Meyer, A.S.; Fernando, D.; Silva, D.A.S.; Daniel, G.; Thygesen, A. Effect of pectin and hemicellulose removal from hemp fibres on the mechanical properties of unidirectional hemp/epoxy composites. *Compos. Part A Appl. Sci. Manuf.* **2016**, *90*, 724–735. [CrossRef]
50. Abraham, E.; Thomas, M.S.; John, C.; Pothen, L.A.; Shoseyov, O.; Thomas, S. Green nanocomposites of natural rubber/nanocellulose: Membrane transport, rheological and thermal degradation characterisations. *Ind. Crops. Prod.* **2013**, *51*, 415–424. [CrossRef]
51. Jarnthong, M.; Wang, F.; Wei, X.Y.; Wang, R.; Li, J.H. Preparation and Properties of Biocomposite Based on Natural Rubber and Bagasse Nanocellulose. *MATEC Web Conf.* **2015**, *26*, 01005. [CrossRef]

52. Kato, H.; Nakatsubo, F.; Abe, K.; Yano, H. Crosslinking via sulfur vulcanization of natural rubber and cellulose nanofibers incorporating unsaturated fatty acids. *RSC Adv.* **2015**, *5*, 29814–29819. [CrossRef]
53. Abraham, E.; Deepa, B.; Pothan, L.A.; John, M.; Narine, S.S.; Thomas, S.; Anandjiwala, R. Physicomechanical properties of nanocomposites based on cellulose nanofibre and natural rubber latex. *Cellulose* **2013**, *20*, 417–427. [CrossRef]
54. Thomas, M.G.; Abraham, E.; Jyotishkumar, P.; Maria, H.J.; Pothen, L.A.; Thomas, S. Nanocelluloses from jute fibers and their nanocomposites with natural rubber: Preparation and characterization. *Int. J. Biol. Macromol.* **2015**, *81*, 768–777. [CrossRef] [PubMed]

Disclaimer/Publisher's Note: The statements, opinions and data contained in all publications are solely those of the individual author(s) and contributor(s) and not of MDPI and/or the editor(s). MDPI and/or the editor(s) disclaim responsibility for any injury to people or property resulting from any ideas, methods, instructions or products referred to in the content.

Article

Effect of Fluorosilicone Rubber on Mechanical Properties, Dielectric Breakdown Strength and Hydrophobicity of Methyl Vinyl Silicone Rubber

Zhaoyang Wang [1], Yankai Lin [2], Zhanxu Li [1], Yumeng Yang [1], Jun Lin [1,*] and Shaojian He [1,*]

[1] State Key Laboratory of Alternate Electrical Power System with Renewable Energy Sources, North China Electric Power University, Beijing 102206, China; wang94269264@163.com (Z.W.); zhanxuli0614@163.com (Z.L.); yumengyang84@163.com (Y.Y.)
[2] Jiangmen Power Dispatching Center of Jiangmen Power Grid, Jiangmen 529000, China; kyoexii@sina.com
* Correspondence: jun.lin@ncepu.edu.cn (J.L.); heshaojian@ncepu.edu.cn (S.H.)

Abstract: Silicone rubber (SIR) is used in high-voltage insulators because of its insulation, and excellent hydrophobicity is very important in harsh outdoor environments. To enhance the hydrophobicity and low-temperature resistance of silicone rubber, methyl vinyl silicone rubber and fluorosilicone rubber (FSIR) blend composites with different ratios were prepared. The samples were characterized and analyzed using scanning electron microscopy, tensile testing, dynamic mechanical analysis and static contact angle testing. The results showed that after blending, SIR and FSIR were well compatible. FSIR had higher elastic modulus and reduced the tensile strength to some extent in SIR/FSIR composites. The addition of a small amount of FSIR made its crystallization temperature decrease from −30 to −45 °C, meaning that the low-temperature resistance was significantly improved. The breakdown strength of SIR/FSIR composites can still be maintained at a high level when a small amount of FSIR is added. The contact angle of the composites increased from 108.9 to 115.8° with the increase in FSIR content, indicating the enhanced hydrophobicity. When the samples were immersed in water for 96 h, the hydrophobicity migration phenomenon occurred. The static contact angle of the samples with less FSIR content had a weaker decreasing trend, which illustrated that the hydrophobicity was maintained at a high level.

Keywords: silicone rubber; fluorosilicone rubber; mechanical properties; breakdown strength; hydrophobicity

1. Introduction

Hydrophobicity materials are widely used in all aspects of industry and life, such as high-voltage insulators, hull surface coatings, oil pipeline inner walls and so on. An insulator is an important and indispensable part of power transmission and transformation lines, and the operation condition is directly related to the stability and safety of the power grid [1,2]. The silicone rubber composite insulator is the latest insulator. Compared to the traditional porcelain insulator and glass insulator, the silicone rubber composite insulator has the advantages of light weight, favorable durability, excellent hydrophobicity, good resistance to dirt flash and being easy to manufacture and maintain [3–8]. Silicone rubber composite insulators are mainly composed of high-temperature vulcanized silicone rubber (HTV) composite umbrella skirt, glass fiber-reinforced epoxy resin core rod and end fittings [9]. Due to the Si–O bond of silicone rubber and its inorganic properties, silicone rubber is superior to ordinary organic rubbers in terms of heat resistance, chemical stability, electrical insulating, abrasion resistance and weatherability [10]. The excellent hydrophobicity and hydrophobic recovery properties of silicone rubber are key factors in its use as a high-voltage outdoor insulation material. As an insulating material used outdoors for a long time, the silicone rubber umbrella skirt will gradually age under the long-term influence of humidity, surface discharge, ultraviolet rays, temperature, smoke and other factors [11,12]. The aging of the material will make the surface of the insulator hydrophobicity deteriorate, resulting in the occurrence of a leakage current and flashover phenomenon, meaning that the hydrophobicity is an important index used to examine the performance of the insulator [13,14].

To enhance the hydrophobicity of silicone rubber, researchers have used various approaches, including surface modifications, such as plasma jet treatment [15,16], spraying [17], laser irradiation [18], or adding fillers to build up nanostructures. Mendoza et al. [19] conducted a comparative assessment of hydrophilic and hydrophobic ZnO nanoparticles and their methods of deposition on the surface hydrophobicity of silicone rubber (PDMS) and glass substrates. An accurate method was proposed to determine the contact angle hysteresis. Nazir et al. [20] added milled glass fires and graphene nanoplatelets as fillers in silicone rubber and found that the composites have excellent fire retardancy and better mechanical strength and hydrophobicity while retaining the required electrical breakdown strength. Zhu [21] believes that the mechanism through which corona discharge weakens the hydrophobicity of silicone rubber is the particles generated by the discharge continuously impacting the SR surface, on which the hydrophilic hydroxyl group is replaced by a polar hydrophobic methyl group. Khan et al. [22] prepared room temperature vulcanized silicone rubber composites, and after 9000-hour aging tests, it was found that the samples using silica as fillers had better hydrophobicity, and the samples with aluminum trihydrate as filler had higher dielectric breakdown strengths. Sheng et al. [23] added a glycerol layer onto the surface of the silicone rubber, and the contact angle of the silicone rubber could be improved by 19.9% via irradiation treatment with glycerol. Du et al. [24] fluoridated the silicone rubber using fluorine gas to obtain the sample with a contact angle of 116°. Surface modification often requires high costs and is a complicated procedure. And these material improvement methods can make silicone rubber significantly improved in a certain aspect, but do not comprehensively consider the purpose of using it at low temperatures. Therefore, it is recommended to use a fluorosilicone rubber (FSIR) and methyl vinyl silicone rubber (SIR) blend to take into account the three aspects of hydrophobicity, insulation and low-temperature resistance.

FSIR has methyl, vinyl and trifluoropropyl side chains, which improve the oil and solvent resistance of the rubber due to the electronic effect and the good shielding effect of the C-F bond on the C=C bond [25,26]. It also has a wide operating temperature range of −60–200 °C and can be operated in cold environments [27]. Sun et al. [28] found that after ultraviolet aging for 2000 h, the FSIR insulators had a larger contact angle. Wei et al. [29] found that the resistance and breakdown properties of phenyl silicone rubber (SiR) were better than those of vinyl SiR and fluoro-SiR, and fluoro-SiR has a higher dielectric constant than the vinyl SiR and phenyl SiR. Polymer blending allows composites to combine the characteristics of both materials. Metivier et al. [30] found that silicone/fluorosilicon mixtures are compatible by adding surface hydrophilic silica particles, and fumed hydrophilic silica can reduce the size of the fluorosilicon phase to 500 nm. Khanra et al. [31] added modified silica in different ratios to fluoroelastomer and silicone rubber blends, which exhibited good compatibility and improved mechanical properties.

The good hydrophobicity and low-temperature resistance of FSIR are important factors conducive to the operation of insulators in harsh environments. Considering the many advantages of FSIR, in this work, FSIR and SIR were used to prepare composites with different ratios. The expectation is that the SIR/FSIR composites can be used in high-voltage insulators in a harsh environment. Scanning electron microscopy (SEM) tests, tensile tests, dynamic mechanical property tests and contact angle tests were carried out to analyze the effects of different ratios on the micro-morphology, mechanical properties, crosslinking density, crystallization temperature and hydrophobicity of the composites. It is proved that the addition of FSIR can improve the material's hydrophobicity and enhance its low-temperature resistance.

2. Experimental

2.1. Materials

The vinyl content of SIR (XHG-110) was 0.08%, and its molecular weight was 670,000; the material was produced by Zhejiang Wynca Chemical Group Co., Jiande, Zhejiang, China. The vinyl content of FSIR (MFVQ 1402) was 0.35%, and its molecular weight was 730,000; the material was produced by Shenzhen Oufut Rubber Products Co., Ltd., Shenzhen, Guangdong, China. Silica adopted HDK®V15 from Wacker, Germany, and its density was 2.2 g·cm^{-3}, its specific surface area was 130–170 m^2·g^{-1}, and its purity was 99.8%. Hydroxy silicone oil (HSO), aluminum hydroxide (ATH), vinyltrimethoxysilane (VTMS), ferric oxide (Fe$_2$O$_3$), 2,5-dimethyl-2,5-di(tert-butylperoxy) hexane (DBPH) and other reagents were commercially available. Table 1 shows the experimental formulations of the composites.

Table 1. Experimental formulations of SIR/FSIR composites (phr).

Ingredients	100/0	95/5	90/10	80/20	70/30	0/100
SIR	100	95	90	80	70	0
FSIR	0	5	10	20	30	100
Silica	30	30	30	30	30	30
ATH	100	100	100	100	100	100
Fe$_2$O$_3$	4	4	4	4	4	4
HSO	5	5	5	5	5	5
VTMS	2	2	2	2	2	2
DBPH	0.5	0.5	0.5	0.5	0.5	0.5

2.2. Preparation

Firstly, SIR was mixed with the reagents of reinforcing agent silica, thermal conductivity enhanced agent ATH, silane coupling agent VTMS, structural control agent HSO, coloring agent Fe$_2$O$_3$ and vulcanizing agent DBPH in an open two-roll mill. The fillers were added in sequence, and the SIR was taken out after uniform mixing. Secondly, FSIR was prepared following the same procedure. Then, SIR and FSIR were mixed in different ratios in the opening machine to obtain different ratios of SIR/FSIR compounds. Finally, after the corresponding optimum curing time(t_{90}) was measured using a vulcanizing instrument (MDR-2000, Shanghai Dejie Machine Equipment Co., Ltd., Shanghai, China), vulcanization was carried out using a flatbed vulcanizing machine (XLB-0350, Zhejiang Huzhou Dongtang Machinery Co., Ltd., Huzhou, Zhejiang, China) at 170 °C.

2.3. Characterization and Measurements

A cold field emission scanning electron microscope (SU8020, Hitachi, Tokyo, Japan) was used to observe the tensile fracture surface of the samples.

The mechanical properties of the composites were tested using a universal material tester (GT-TC2000, Gotech Testing Machines Inc., Taizhong, Taiwan, China), which had a tensile speed of 500 mm·min^{-1}.

The crosslinking density of the composites was tested via the equilibrium swelling method [32]. The mass of the samples before and after immersion were defined as M_1 and M_2 (g), which were obtained by immersing the vulcanized rubber in toluene at room temperature for 72 h. The volume rate of swelling of the rubber, v_2, was calculated according to Equations (1)–(3).

$$v_2 = \frac{v_1}{(v_1 + v_{sol})} \quad (1)$$

$$v_{sol} = \frac{(M_2 - M_1)}{\rho_{sol}} \quad (2)$$

$$v_1 = \frac{M_3}{\rho} \quad (3)$$

where v_{sol} (cm^3·mol^{-1}) is the volume of solvent absorbed by the rubber after swelling, v_1 (cm^3·mol^{-1}) is the volume of rubber, ρ_{sol} (g·mL^{-1}) is the density of solvent, ρ (g·cm^{-3}) is the density of rubber and M_3 (g) is the mass of rubber. The crosslinking density, v_e ($\times 10^{-4}$ mol·cm^{-3}), of the composites was calculated using Equation (4).

$$v_e = \frac{\ln(1 - v_e) + v_2 + \chi \cdot v_2^2}{2v \cdot v_2^{\frac{1}{3}}} \quad (4)$$

where χ is the silicone rubber–toluene interaction parameter (0.465) [33], and v (cm^3·mol^{-1}) is the molar volume of solvent.

The dynamic mechanical analysis (DMA) of the silicone rubber composites were tested using DMA 242 (NETZSCH, Free State of Bavaria, Germany). The tests were carried out at a frequency of 10 Hz, an amplitude of 0.5% and temperature conditions of −180 to 25 °C.

The breakdown strength tests used a voltage breakdown tester manufactured by Beijing Huaji Instrument Co., Beijing, China. The test was performed at an AC voltage of 50 Hz. The insulating properties of the composites were analyzed using the Weibull classical failure model [34]. Equation (5) represents the failure distribution function.

$$F(x) = 1 - e^{-(\frac{x}{\alpha})^\beta} \quad (5)$$

where x is the breakdown strength, α is the scale parameter, and β is the shape parameter. To facilitate the calculation, the above formula can be converted to logarithmic form, as shown in Equation (6).

$$\lg[-\ln(1-F)] = \beta(\lg\alpha - \lg x) \quad (6)$$

In addition, the failure distribution function can be calculated using Equation (7).

$$F(x) = \frac{i - 0.5}{n + 0.25} \quad (7)$$

where i is the number of measurements derived by arranging x in ascending order, and n is the total number of tests for each sample—in this work, $n = 12$. The relevant parameters α and β of the Weibull distribution function were calculated using the least squares method for Equation (6) to derive the Weibull failure model.

The static contact angle (SCA) and hydrophobic migration of the composites were measured using a contact angle meter manufactured by Shanghai Zhongchen Technical Equipment Co., Shanghai, China. Before the test took place, the samples were sequentially wiped with ethanol and ultrapure water and left for 24 h. The SCA test was carried out after the natural evaporation of the water on the surface. A droplet of water with a volume of about 20 µL was dropped on the surface of the sample using a microsyringe. After taking pictures to record the shape of the droplet, five points on the boundary were selected, and the coordinates were recorded. SCA was obtained via fitting. The final result was the average value of five measurements that were taken for each group. For the hydrophobic migration test, the samples were immersed in ultrapure water for 96 h, after which stage the SCA test was performed.

3. Results and Discussion

3.1. Mechanical Properties

SEM testing can be used to observe the dispersion of fillers in the matrix. To observe the microscopic topography of the prepared sample, the cross-section after fracture in the tensile test was observed, and the compatibility of the fillers with rubber was analyzed to carry out the blending scheme. Figure 1a shows micron-sized ATH, which can be seen to have a layered structure. ATH has good adsorption and dispersion, which can improve the thermal conductivity of rubber, enhance the anti-aging effect and strengthen the vulcanization process. However, the addition of FSIR will affect the compatibility of

ATH and the rubber matrix, and the tensile section of the sample was analyzed via SEM. Figure 1b shows the SIR composite. The ATH, which is exposed in the outer layer of the silicone rubber matrix, cannot be observed. Figure 1f shows the fracture in the FSIR composites. The section is smoother and flatter than that of the SIR composites, indicating that its compatibility is better. In Figure 1c–e, unwrapped ATH and tiny cracks can be observed in SEM. Significant bulges and depressions due to material agglomeration can be observed compared to pure FSIR or pure SIR composites. The scheme in which SIR/FSIR is 70/30 shows better compatibility, and the reason for this phenomenon is that the increase in the FSIR content promotes the fusion of ATH and rubber. Therefore, for SIR/FSIR, the interfacial interaction between the matrix and the filler is weaker, resulting in the mechanical properties of the materials being lower than those of the SIR composite materials.

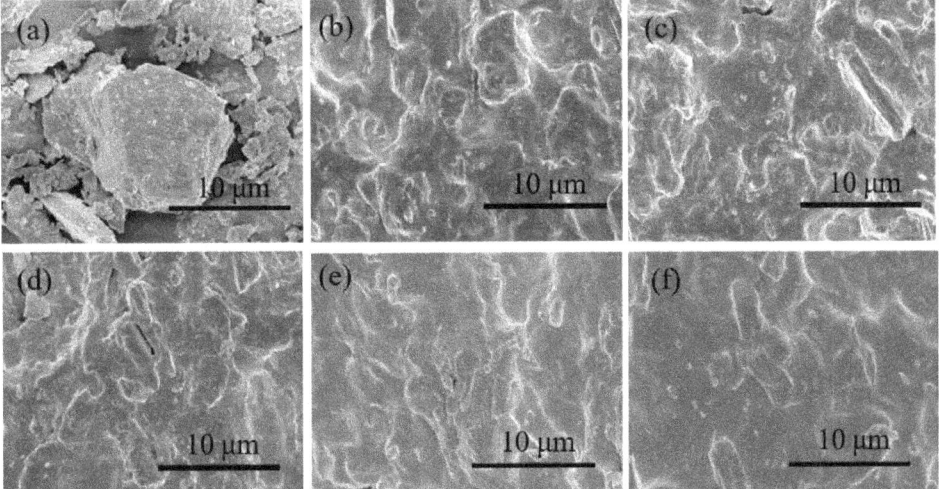

Figure 1. SEM microstructure of tensile sections of the SIR/FSIR composites: (**a**) ATH; (**b**) 100/0; (**c**) 90/10; (**d**) 80/20; (**e**) 70/30; (**f**) 0/100.

Figure 2 shows the stress–strain curves of the SIR/FSIR composites. The elongation at break undergoes very little change with the increase in FSIR contents, which were all around 205%. Table 2 lists the mechanical properties of the SIR/FSIR composites. The tensile strength of the SIR composite is up to 7.3 MPa, while the tensile strength decreases to 5.9 MPa when the content of FSIR is increased to 30 phr. In JB/T 10945-2010, the silicone rubber applied to composite insulators requires the tensile strength to be higher than 3 MPa. Therefore, even if the tensile strengths of the blend composites are reduced via the addition of FSIR, it still meets the industry standard for silicone rubber insulators. The elongation at break of the FSIR composite is only 138%, and the tensile strength is 4.5 MPa, which are much lower than those of the SIR/FSIR composites. When the stress is less than 1.5 MPa, the modulus of elasticity of the composites tends to increase with the increase in the FSIR content. This outcome occurs due to the fact that the FSIR side chain contains a small amount of trifluoropropyl, and the presence of fluorine atoms makes the molecule more polar. This process results in larger intermolecular forces and reduced molecular chain motility. In addition, for SIR/FSIR composites, the hardness of the composites increases, and the stress at 100% strain remains stable, which is about 3.6 MPa with the increase in the FSIR content. Since the percentage of SIR is higher than that of FSIR in the four composites analyzed in this work, the properties of the samples are more similar to those of SIR.

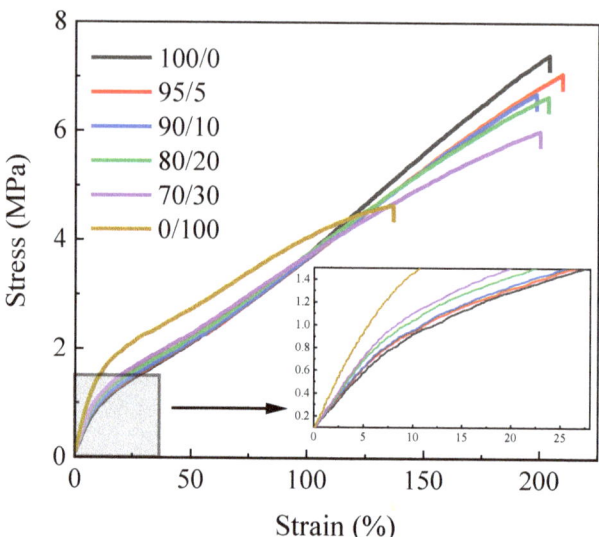

Figure 2. Stress–strain curves of SIR/FSIR composites.

Table 2. Mechanical properties of SIR/FSIR composites.

Properties	100/0	95/5	90/10	80/20	70/30	0/100
Shore A hardness	70	71	70	72	74	77
Tensile strength (MPa)	7.3	6.9	6.6	6.5	5.9	4.5
Elongation at break (%)	204	210	199	204	200	138
Stress at 100% strain (MPa)	3.7	3.6	3.5	3.7	3.6	4.0
Elasticity modulus (MPa)	13.3	14.1	14.1	15.0	16.3	21.1

3.2. Crosslinking Density

After the blending rubber is vulcanized, a cross-linked network is formed inside of the structure. When the crosslinking density increases, it represents the weakening of the motility of the macromolecular chain. Therefore, this test can provide a theoretical analysis from a microscopic perspective for tensile testing. The crosslinking density of the SIR/FSIR composites is given in Figure 3. It is found that the crosslinking density of FSIR composite is 1.55 times that of SIR composite. For the SIR/FSIR composites, the crosslinking density gradually increases with the increase in FSIR content. When the proportion of FSIR increased from 5 to 20 phr, the crosslinking density increased from 2.63×10^{-4} to 3.34×10^{-4} mol·cm^{-3}. The increased crosslinking density indicates that the connection between the macromolecular chains is more compact, and the flexibility of the composite material is reduced and the elastic modulus will increase. In Table 1, the elasticity modulus of FSIR composite is 7.8 MPa higher than that of SIR composites. The macromolecular chain of FSIR contains a small amount of trifluoropropyl, and the electron-absorbing effect of fluorine atoms is stronger and, thus, the free radical reactions are more likely to occur with the vulcanizing agent. As a result, compared to the SIR composite, the SIR/FSIR composites with higher FSIR contents show higher crosslinking densities. However, the crosslinking density of the scheme with an SIR/FSIR ratio of 70/30 is higher than that of the pure FSIR composites. The reason for this outcome is that silica is more likely to interact with non-polar molecules to form crosslinking networks. Therefore, there is an optimal value between the proportions of SIR composites and FSIR composites that maximizes the crosslinking density.

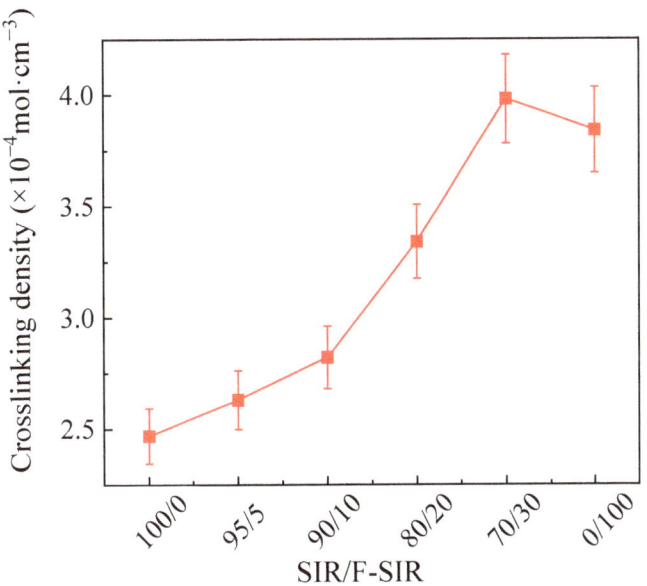

Figure 3. Crosslinking density of SIR/FSIR composites.

3.3. Dynamic Mechanical Properties

Figure 4a shows the storage modulus–temperature curve of the SIR/FSIR composites. The storage modulus of pure SIR is higher than that of pure FSIR under all of the test temperatures. Figure 4b shows the tan delta–temperature curve of SIR/FSIR composites obtained from DMA. The glass-transition temperatures are about −130 °C for all of the blending schemes. When the SIR/FSIR composites are in a glassy state, the storage modulus of the composites gradually decreases as the proportion of FSIR increases.

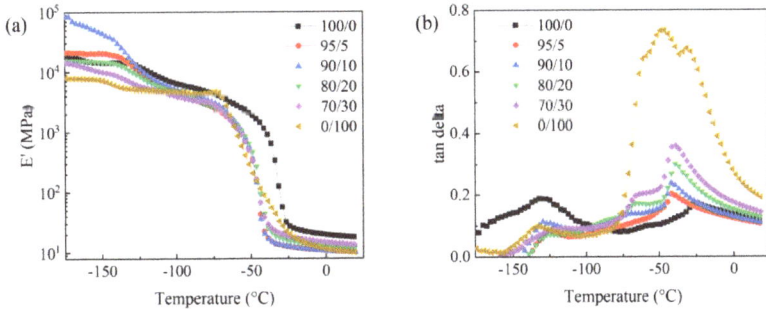

Figure 4. DMA of SIR/FSIR composites: (**a**) storage modulus–temperature curve; (**b**) tan delta–temperature curve.

It can be seen that there are two damping peaks in FSIR, in which the temperature of the crystallization peak is about −50 °C, as shown in Figure 4b. The temperature of the crystallization peak of SIR is about −30 °C. Therefore, FSIR has a superior low-temperature resistance. When SIR/FSIR is 95/5, there is only one crystallization peak, indicating that the compatibility of the two matrices is better at this time. When SIR/FSIR is 70/30, two crystalline peaks are observed, which means that two crystalline phases have been included in the composites. The crystallization peak of FSIR is obviously larger than that of SIR, which occurs because the increase in trifluoropropyl makes the internal polarity of the material larger. The reduction in the crystallization temperature can extend the operating temperature range of the composite. The crystallization temperatures of the larger peaks are concentrated at around −45 °C, indicating that the low-temperature resistance of the composites is significantly improved by the addition of FSIR.

3.4. Breakdown Strength

As an insulator umbrella sleeve material, its insulation performance is one of the most important indicators. The breakdown strength of the composites was investigated, and the Weibull distribution was obtained. Figure 5 shows the Weibull distribution of SIR/FSIR composites. It can be seen that the SIR composite material has excellent insulation properties. The presence of trifluoropropyl makes the molecule polar, which makes the breakdown field strength of the sample smaller and easier to breakdown. SIR/FSIR composites contain two crystalline phases, resulting in more structural defects and reduced insulation properties. Therefore, based on the characterization data, it can also be confirmed that the breakdown strength of the material decreases as the proportion of FSIR in the composites increases. When the content of FSIR is lower, the compatibility between SIR and FSIR is greater, and the decrease in breakdown strength is smaller. Therefore, for the SIR/FSIR composite with 5 phr FSIR, the breakdown strength is comparable to that of SIR composite. However, as the blending ratio of FSIR continues to increase, the breakdown strength continues to decrease. It can be seen that adding excess FSIR will lead to a decrease in the insulation performance of the SIR/FSIR composite.

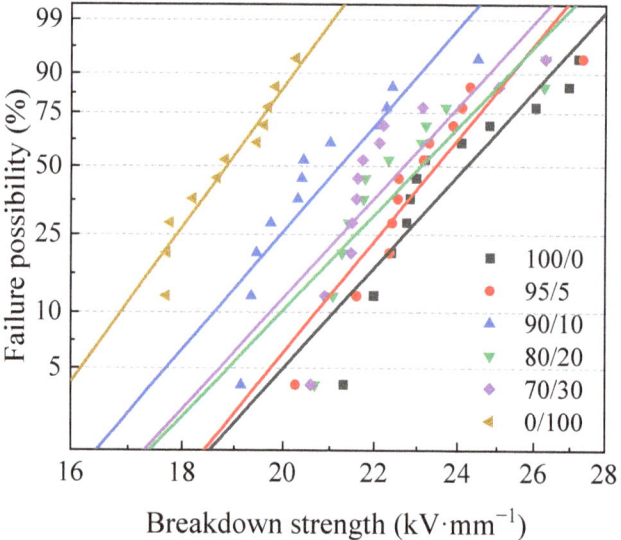

Figure 5. Weibull distributions of SIR/FSIR composites.

3.5. Static Contact Angle

The hydrophobicity of composites is characterized by contact angle testing. When the SCA is larger, the sample surface has better hydrophobicity. Sun et al. [28] tested the contact angle of the two insulator materials of SIR composites and FSIR composites, which were 112.4° and 117.8°. Figure 6 shows the SCA of SIR/FSIR composites. The SCAs are 108.9° and 115.5° for SIR composite and FSIR composite, respectively. The difference in the hydrophobicity of this paper and the work of Sun may be due to experimental formulations and test methods. FSIR has a lower surface free energy, meaning that its hydrophobicity is stronger than that of SIR. In addition, for the SIR/SIR composites, the SCA gradually increases with the increase in the FSIR content. This finding indicates that the higher the content of FSIR, the greater the hydrophobicity of the composites. When the proportion of FSIR is increased to 5 phr, the SCA can be significantly increased by 3.9°. With the SIR/FSIR increasing to 90/10 and 80/20, the SCA is about 113.6° and shows a smooth trend. For the scheme of 70/30, the SCA is slightly higher than the FSIR composites. This observation is the same as the change law of crosslinking density, meaning that the intertwining of molecular chains inside of the material to form a dense crosslinking network is conducive to improving hydrophobicity.

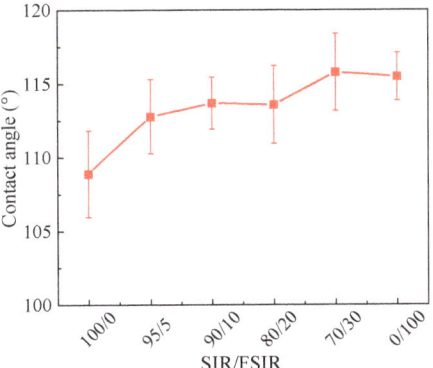

Figure 6. Static contact angles of SIR/FSIR composites.

When the sample is served in a humid environment for a long time, its hydrophobicity changes. Therefore, hydrophobic migration was analyzed. Figure 7 shows the variation in SCA for SIR/FSIR composites measured after 96 h of immersion in water. After 96 h of immersion, the SCA of FSIR composite decreases more significantly than that of SIR composite. For the SIR/FSIR composites, the decreasing trend of SCA is more significant for the increased proportion of FSIR. When SIR/FSIR is 95/5, the SCA decreases by 0.5° after 96 h immersion, while the SCA decreases by 3.6° when SIR/FSIR is 70/30. This phenomenon indicates that the FSIR material has good surface hydrophobicity, but the hydrophobicity migration occurred after internal water immersion. In addition, the SCA of the blended material with FSIR composites after immersion is higher than that of the un-immersed SIR composites. For example, the smallest SCA measured after immersion of all SIR/FSIR blending composites was 112° for the scheme 90/10, but it is still larger than the un-immersed SIR composite. Based on this result, we can further illustrate the effectiveness of FSIR composite in improving hydrophobicity. Insulators made of SIR/FSIR can have better hydrophobic stability in humid environments.

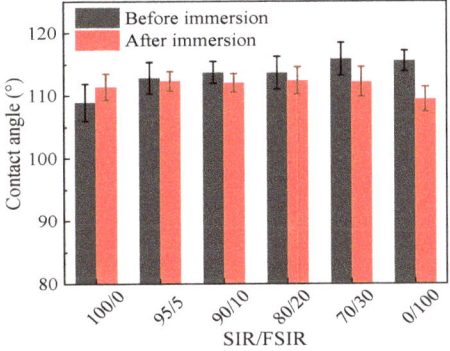

Figure 7. Migration of hydrophobicity in SIR/FSIR composites.

4. Conclusions

In this paper, SIR/FSIR composites with different proportions of matrix materials were prepared and found to have good compatibility. Firstly, the tensile strength decreased from 7.3 to 5.9 MPa when the content of FSIR was increased to 30 phr. The elongation at break and the stress at 100% strain did not significantly change. When the stress was less than 1.5 MPa, the modulus of elasticity of the composites tended to increase with the increase in FSIR content. Secondly, the low-temperature resistance of the composites can be significantly improved when the proportion of FSIR in the composites is relatively small. Compared to SIR composite materials, the crystallization temperature of the composites is reduced from -30 to $-45\ °C$. Thirdly, the breakdown strength test reveals that when the FSIR content is increased, it leads to a decrease in the overall insulating properties of the composites. Finally, the SCA increases as the FSIR content increases for blending composites, and the material hydrophobicity increases. When the samples were immersed for 96 h, the hydrophobicity migration phenomenon occurred. As the proportion of FSIR increased, the SCA decreased more significantly, which indicated that hydrophobicity weakened.

This research provides a reference point for the material formulation of SIR high-voltage insulators. The huge demand for composite insulators has been shown in power transportation, especially in cold weather or harsh environments. Insulators composed of SIR/FSIR composites exhibit the advantage of having wider temperature application range. In addition, owing to the increased hydrophobicity, the insulators using SIR/FSIR composites as an umbrella sleeve material are expected to efficiently avoid flashover, which could prolong the service life of the insulator and reduce the probability of the need to replace the disabled insulator.

Author Contributions: Data curation, Z.W., Y.L. and Z.L.; Formal analysis, Z.W.; Funding acquisition, S.H.; Investigation, Z.W., Y.L. and Z.L.; Methodology, S.H.; Project administration, J.L. and S.H.; Resources, J.L.; Supervision, J.L.; Visualization, Y.Y.; Writing—original draft, Z.W.; Writing—review and editing, S.H. All authors have read and agreed to the published version of the manuscript.

Funding: This research was funded by the National Natural Science Foundation of China, grant number 51973057.

Institutional Review Board Statement: Not applicable.

Data Availability Statement: Not applicable.

Conflicts of Interest: The authors declare that the research was conducted in the absence of any commercial or financial relationships that could be construed as potential conflict of interest.

Nomenclature

SIR	Silicone rubber
FSIR	Fluorosilicone rubber
HSO	Hydroxy silicone oil
ATH	Aluminum hydroxide
VTMS	Vinyltrimethoxysilane
DBPH	2,5-dimethyl-2,5-di (tert-butylperoxy) hexane
SEM	Scanning electron microscope
DMA	Dynamic mechanical properties
SCA	Static contact angle

References

1. Liao, Y.; Weng, Y.; Wang, J.; Zhou, H.; Lin, J.; He, S. Silicone Rubber Composites with High Breakdown Strength and Low Dielectric Loss Based on Polydopamine Coated Mica. *Polymers* **2019**, *11*, 2030. [CrossRef] [PubMed]
2. He, S.; Hu, J.; Zhang, C.; Wang, J.; Chen, L.; Bian, X.; Lin, J.; Du, X. Performance improvement in nano-alumina filled silicone rubber composites by using vinyl tri-methoxysilane. *Polym. Test.* **2018**, *67*, 295–301. [CrossRef]
3. Li, S.; Yu, S.; Feng, Y. Progress in and prospects for electrical insulating materials. *High Volt.* **2016**, *1*, 122–129. [CrossRef]

4. Hall, J.F. History and bibliography of polymeric insulators for outdoor applications. *IEEE Trans. Power Deliv.* **1993**, *8*, 376–385. [CrossRef]
5. Hillborgl, H.; Gedde, U.W. Hydrophobicity changes in silicone rubbers. *IEEE Trans. Dielectr. Electr. Insul.* **1999**, *6*, 703–717. [CrossRef]
6. Polmanteer, K.E. Current perspectives on silicone rubber technology. *Rubber Chem. Technol.* **1981**, *54*, 1051–1080. [CrossRef]
7. Reynders, J.P.; Jandrell, I.R.; Reynders, S.M. Review of Aging and Recovery of Silicone Rubber Insulation for Outdoor Use. *IEEE Trans. Dielectr. Electr. Insul.* **1999**, *6*, 620–631. [CrossRef]
8. He, S.; Wang, J.; Hu, J.; Zhou, H.; Nguyen, H.; Luo, C.; Lin, J. Silicone rubber composites incorporating graphitic carbon nitride and modified by vinyl tri-methoxysilane. *Polym. Test.* **2019**, *79*, 106005. [CrossRef]
9. Chen, X.; Wang, J.; Zhang, C.; Yang, W.; Lin, J.; Bian, X.; He, S. Performance of silicone rubber composites using boron nitride to replace alumina tri-hydrate. *High Volt.* **2020**, *6*, 480–486. [CrossRef]
10. Shit, S.C.; Shah, P. A Review on Silicone Rubber. *Natl. Acad. Sci. Lett.* **2013**, *36*, 355–365. [CrossRef]
11. Farzaneh, M.; Li, Y.; Zhang, J.; Shu, L.; Jiang, X.; Sima, W.; Sun, C. Electrical performance of ice-covered insulators at high altitudes. *IEEE Trans. Dielectr. Electr. Insul.* **2004**, *11*, 870–880. [CrossRef]
12. Wang, J.; Liang, X.; Gao, Y. Failure analysis of decay-like fracture of composite insulator. *IEEE Trans. Dielectr. Electr. Insul.* **2014**, *21*, 2503–2511. [CrossRef]
13. Amin, M.; Akbar, M.; Amin, S. Hydrophobicity of silicone rubber used for outdoor insulation (an overview). *Rev. Adv. Mater. Sci.* **2007**, *16*, 10–26.
14. Chen, Y.; Wang, K.; Zhang, C.; Yang, W.; Qiao, B.; Yin, L. The Effect of Various Fillers on the Properties of Methyl Vinyl Silicone Rubber. *Polymers* **2023**, *15*, 1584. [CrossRef] [PubMed]
15. Li, S.; Zhang, R.; Wang, S.; Fu, Y. Plasma treatment to improve the hydrophobicity of contaminated silicone rubber—The role of LMW siloxanes. *IEEE Trans. Dielectr. Electr. Insul.* **2019**, *26*, 416–422. [CrossRef]
16. Yu, J.; Liu, Y.; Huang, B.; Xia, L.; Kong, F.; Zhang, C.; Shao, T. Rapid hydrophobicity recovery of contaminated silicone rubber using low-power microwave plasma in ambient air. *Chem. Eng. J.* **2023**, *465*, 142921. [CrossRef]
17. Zhu, M.; Song, H.; Li, J.; Xue, J.; Yu, Q.; Chen, J.; Zhang, G. Superhydrophobic and high-flashover-strength coating for HVDC insulating system. *Chem. Eng. J.* **2021**, *404*, 126476. [CrossRef]
18. Emelyanenko, A.M.; Boinovich, L.B.; Bezdomnikov, A.A.; Chulkova, E.V.; Emelyanenko, K.A. Reinforced Superhydrophobic Coating on Silicone Rubber for Longstanding Anti-Icing Performance in Severe Conditions. *ACS Appl. Mater. Interfaces* **2017**, *9*, 24210–24219. [CrossRef]
19. Mendoza, A.I.; Moriana, R.; Hillborg, H.; Strömberg, E. Super-hydrophobic zinc oxide/silicone rubber nanocomposite surfaces. *Surf. Interfaces* **2019**, *14*, 146–157. [CrossRef]
20. Nazir, M.T.; Khalid, A.; Wang, C.; Baena, J.-C.; Phung, B.T.; Akram, S.; Wong, K.L.; Yeoh, G.H. Enhanced fire retardancy with excellent electrical breakdown voltage, mechanical and hydrophobicity of silicone rubber/aluminium trihydroxide composites by milled glass fibres and graphene nanoplatelets. *Surf. Interfaces* **2022**, *35*, 102494. [CrossRef]
21. Zhu, Y. Influence of corona discharge on hydrophobicity of silicone rubber used for outdoor insulation. *Polym. Test.* **2019**, *74*, 14–20. [CrossRef]
22. Khan, H.; Mahmood, A.; Ullah, I.; Amin, M.; Nazir, M.T. Hydrophobic, dielectric and water immersion performance of 9000 h multi-stresses aged silicone rubber composites for high voltage outdoor insulation. *Eng. Fail. Anal.* **2021**, *122*, 105223. [CrossRef]
23. Sheng, K.; Dong, X.; Chen, Z.; Zhou, Z.; Gu, Y.; Huang, J. Increasing the surface hydrophobicity of silicone rubber by electron beam irradiation in the presence of a glycerol layer. *Appl. Surf. Sci.* **2022**, *591*, 153097. [CrossRef]
24. Du, B.; Li, Z. Hydrophobicity surface charge and DC flashover characteristics of direct-fluorinated RTV silicone rubber. *IEEE Trans. Dielectr. Electr. Insul.* **2015**, *22*, 934–940. [CrossRef]
25. Qi, M.; Jia, X.; Wang, G.; Xu, Z.; Zhang, Y.; He, Q. Research on high temperature friction properties of PTFE/Fluorosilicone rubber/silicone rubber. *Polym. Test.* **2020**, *91*, 106817. [CrossRef]
26. Zhao, R.; Yin, Z.; Zou, W.; Yang, H.; Yan, J.; Zheng, W.; Li, H. Preparation of High-Strength and Excellent Compatibility Fluorine/Silicone Rubber Composites under the Synergistic Effect of Fillers. *ACS Omega* **2023**, *8*, 3905–3916. [CrossRef] [PubMed]
27. Xu, X.; Liu, J.; Chen, P.; Wei, D.; Guan, Y.; Lu, X.; Xiao, H. The effect of ceria nanoparticles on improving heat resistant properties of fluorosilicone rubber. *J. Appl. Polym. Sci.* **2016**, *133*, 44117. [CrossRef]
28. Sun, C.; Shi, K.; Wang, H.; Chen, B.; Liu, J. Reliability Study of Fluorosilicone Rubber Composite Insulators under Strong Radiation and Low Temperature. In Proceedings of the 2023 5th Asia Energy and Electrical Engineering Symposium, Chengdu, China, 23–26 March 2023; pp. 367–371.
29. Wei, Y.; Yang, J.; Li, G.; Zhou, X.; Hao, C.; Lei, Q.; Li, S. Influence of Molecular Chain Side Group on the Electrical Properties of Silicone Rubber and Mechanism Analysis. *IEEE Trans. Dielectr. Electr. Insul.* **2022**, *29*, 1465–1473. [CrossRef]
30. Metivier, T.; Cassagnau, P. Compatibilization of silicone/fluorosilicone blends by dynamic crosslinking and fumed silica addition. *Polymer* **2018**, *147*, 20–29. [CrossRef]
31. Khanra, S.; Ganguly, D.; Ghorai, S.K.; Goswami, D.; Chattopadhyay, S. The synergistic effect of fluorosilicone and silica towards the compatibilization of silicone rubber and fluoroelastomer based high performance blend. *J. Polym. Res.* **2020**, *27*, 96. [CrossRef]
32. Valentın, J.L.; Carretero-Gonzalez, J.; Mora-Barrantes, I.; Chasse, W.; Saalwachter, K. Uncertainties in the Determination of Cross-Link Density by Equilibrium Swelling Experiments in Natural Rubber. *Macromolecules* **2008**, *41*, 4717–4729. [CrossRef]

33. Zhang, X.; Zhang, Q.; Zheng, J. Effect and mechanism of iron oxide modified carbon nanotubes on thermal oxidative stability of silicone rubber. *Compos. Sci. Technol.* **2014**, *99*, 1–7. [CrossRef]
34. Fabiani, D.; Simoni, L. Discussion on application of the Weibull distribution to electrical breakdown of insulating materials. *IEEE Trans. Dielectr. Electr. Insul.* **2005**, *12*, 11–16. [CrossRef]

Disclaimer/Publisher's Note: The statements, opinions and data contained in all publications are solely those of the individual author(s) and contributor(s) and not of MDPI and/or the editor(s). MDPI and/or the editor(s) disclaim responsibility for any injury to people or property resulting from any ideas, methods, instructions or products referred to in the content.

Article

Adducts of Carbon Black with a Biosourced Janus Molecule for Elastomeric Composites with Lower Dissipation of Energy

Federica Magaletti [1], Fatima Margani [1], Alessandro Monti [1], Roshanak Dezyani [1], Gea Prioglio [1], Ulrich Giese [2], Vincenzina Barbera [1,*] and Maurizio Stefano Galimberti [1,*]

[1] Department of Chemistry, Materials and Chemical Engineering "G. Natta", Politecnico di Milano, Via Mancinelli 7, 20131 Milano, Italy; federica.magaletti@polimi.it (F.M.); fatima.margani@polimi.it (F.M.); alessandro16.monti@mail.polimi.it (A.M.); gea.prioglio@polimi.it (G.P.)

[2] Deutsches Institut für Kautschuktechnologie e. V., Eupener Straße 33, 30519 Hannover, Germany; ulrich.giese@dikautschuk.de

* Correspondence: vincenzina.barbera@polimi.it (V.B.); maurizio.galimberti@polimi.it (M.S.G.)

Citation: Magaletti, F.; Margani, F.; Monti, A.; Dezyani, R.; Prioglio, G.; Giese, U.; Barbera, V.; Galimberti, M.S. Adducts of Carbon Black with a Biosourced Janus Molecule for Elastomeric Composites with Lower Dissipation of Energy. *Polymers* 2023, 15, 3120. https://doi.org/10.3390/polym15143120

Academic Editor: Md Najib Alam

Received: 5 June 2023
Revised: 12 July 2023
Accepted: 20 July 2023
Published: 22 July 2023

Copyright: © 2023 by the authors. Licensee MDPI, Basel, Switzerland. This article is an open access article distributed under the terms and conditions of the Creative Commons Attribution (CC BY) license (https://creativecommons.org/licenses/by/4.0/).

Abstract: Elastomer composites with low hysteresis are of great importance for sustainable development, as they find application in billions of tires. For these composites, a filler such as silica, able to establish a chemical bond with the elastomer chains, is used, in spite of its technical drawbacks. In this work, a furnace carbon black (CB) functionalized with polar groups was used in replacement of silica, obtaining lower hysteresis. CBN326 was functionalized with 2-(2,5-dimethyl-1H-pyrrol-1-yl)-1,3-propanediol (serinol pyrrole, SP), and samples of CB/SP adducts were prepared with different SP content, ranging from four to seven parts per hundred carbon (phc). The entire process, from the synthesis of SP to the preparation of the CB/SP adduct, was characterized by a yield close to 80%. The functionalization did not alter the bulk structure of CB. Composites were prepared, based on diene rubbers—poly(1,4-*cis*-isoprene) from *Hevea Brasiliensis* and poly(1,4-*cis*-butadiene) in a first study and synthetic poly(1,4-*cis*-isoprene) in a second study—and were crosslinked with a sulfur-based system. A CB/silica hybrid filler system (30/35 parts) was used and the partial replacement (66% by volume) of silica with CB/SP was performed. The composites with CB/SP exhibited more efficient crosslinking, a lower Payne effect and higher dynamic rigidity, for all the SP content, with the effect of the functionalized CB consistently increasing the amount of SP. Lower hysteresis was obtained for the composites with CB/SP. A CB/SP adduct with approximately 6 phc of SP, used in place of silica, resulted in a reduction in $\Delta G'/G'$ of more than 10% and an increase in E' at 70 °C and in σ_{300} in tensile measurements of about 35% and 30%, respectively. The results of this work increase the degrees of freedom for preparing elastomer composites with low hysteresis, allowing for the use of either silica or CB as filler, with a potentially great impact on an industrial scale.

Keywords: carbon black; elastomers; nanocomposites; rubber; functionalization

1. Introduction

Global greenhouse emissions, measured in carbon dioxide equivalents over a 100-year timescale, were 54.59 billion tons in 2021 [1]. The United Nations (UN) has recognized the importance of sustainable transport since the 1992 Earth Summit [2,3]. In fact, the energy sector contributes more than 70% to global greenhouse emissions, with road transport accounting for about 12% [4]. According to the UN's mobility report, global freight volumes and annual passenger traffic are projected to grow by 50% and 70%, respectively, compared to 2015, with 2.4 billion cars on the road. In this context, tires play a major role, with 2.7 billion units forecast to be on the market by 2025 [5,6], contributing about 20% to the global warming potential, 90% of which is derived from the use of cars. So-called rolling resistance (RR), i.e., "the energy consumed per unit distance of travel as a tire rolls under load" [7–9] is mostly responsible for the dissipation of energy and, hence, for the environmental impact of a tire during its use. RR is influenced by the hysteresis of an elastomeric composite, which in turn is mainly due to the so-called filler networking phenomenon, that is, the reversible interactions of the reinforcing filler particles [10–13].

Furnace carbon black (CB) [14–19] and precipitated silica [20,21] are the traditional reinforcing fillers in tire compounds: they possess a high surface area and therefore high interfacial area with the rubber matrix and are nanostructured, which means their aggregates have voids able to occlude polymer chains, thereby promoting mechanical reinforcement. Silica is the preferred filler for reducing the hysteresis of a rubber compound: the silanols on the silica surface react with so-called coupling agents, which enable compatibility with the silica and establish chemical bonds with the polymer chains [20,21], reducing if not preventing the filler networking phenomenon. In order to reduce the hysteresis of elastomeric composites for a dynamic-mechanical application such as the one in tire, the filler–polymer chemical bond is a key feature, alongside the appropriate selection of materials. The preferred coupling agents are silanes containing sulfur atoms [22,23], such as bis(triethoxysilylpropyl)tetrasulfide (TESPT) [22]. Silica-based technology is well established and the use of silica in tire compounds is steadily increasing. However, silica presents remarkable drawbacks. It promotes the increase of compound viscosity, and this leads to worse processability and to a shorter storage time of the composites, which makes it necessary to change the planning of the production of the compounds as well as the procedures for storing and moving them, with a clear impact on logistics. Particular mixing equipment has to be used, which is more energy consuming and more expensive. Silica is corrosive and abrasive, and the silane causes increased adhesiveness to the metal parts of the mixing machines. Hence, the metal surfaces have to be treated with special substances, which requires a revision of maintenance procedures. Moreover, compounds based on silica suffer from a lack of electrical conductivity. All these technical problems are particularly relevant on an industrial scale. It would be highly desirable to use, as a reinforcing filler for an elastomeric composite with low hysteresis, a CB with the properties of traditional CB grades but with chemical reactivity with rubber chains. This objective can be achieved through introducing functional groups on the CB surface.

Some of the authors developed a functionalization method [24–29] which uses biosourced molecules, is performed in the absence of solvents and catalyts and is characterized by high yield (also higher than 95%) and carbon efficiency [30,31]. A solvent (acetone) was used only to allow for the easy mixing of CB and the pyrrole compound in the lab. This method is based on pyrrole compounds, with their general chemical structure shown in Figure 1.

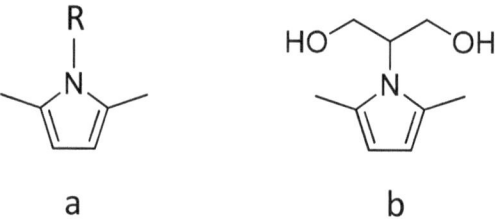

Figure 1. Pyrrole compounds. General chemical structure (**a**); 2-(2,5-dimethyl-1*H*-pyrrol-1-yl)-1,3-propanediol (serinol pyrrole, SP) (**b**).

The functionalization of CB [24,25], a high-surface-area graphite (HSAG) [26,27] and carbon nanotubes [28] was performed through simply mixing the carbon materials and a pyrrole compound (PyC), giving either thermal or mechanical energy. A domino reaction occurs, in which the steps are the carbocatalyzed oxidation of the pyrrole compound in the benzylic position and then the cycloaddition reaction with the carbon substrate [29]. It is worth highlighting the versatility of this functionalization reaction, which allows us to attach nearly any type of R group. A large part of the research was performed with 2-(2,5-dimethyl-1*H*-pyrrol-1-yl)-1,3-propanediol (serinol pyrrole, SP), whose chemical structure is in Figure 1b. SP was prepared from the Paal Knorr reaction of a biosourced molecule, a glycerol derivative such as 2-amino-1,3-propandiol (known as serinol), with 2,5-hexanedione, which could be prepared from dimethylfuran [32–35]. SP is a Janus molecule [27,36]: the pyrrole ring gives rise to a covalent bond with the carbon substrate and the serinol moiety changes the solubility parameter of the carbon allotrope [27,28] and promotes its chemical reactivity.

The CB/SP adduct appears to be an ideal filler for preparing rubber composites with low dissipation of energy, avoiding or substantially reducing the use of silica. The OH groups brought by SP on the CB surface can promote the interaction of CB with silica and with the coupling agent TESPT, and hence with the elastomer chains. In this work, CB/SP was used in replacement of a major amount of silica, in elastomeric composites based on diene rubbers, with a CB/silica hybrid filler system. The objective was to at least reproduce the properties of the silica-based compound, above all the low hysteresis. For the preparation of the CB/SP adducts, CB N326 was used. According to the ASTM-D1765 standard classification system for carbon blacks used in rubber products, this grade of CB has a surface area of about 80 m^2/g, remarkably lower than the one of the sp^2 carbon allotropes used by some of the authors for the functionalization with SP: 300 m^2/g for HSAG [26] and 250 m^2/g for CNT [28]. Hence, this work also had the objective to investigate the robustness of the methodology based on pyrrole compounds for the functionalization of CB. This technology appears of interest for its development to the industrial scale. A major player in the tire field has reported the use of a pyrrole compound, and in particular, of serinol pyrrole, for the industrial development [37].

The CB/SP adducts were characterized by means of thermogravimetric analysis (TGA) and X-ray diffraction.

In a first study, the CB/SP adduct was used for the partial replacement of silica in a composite based on poly(1,4-cis-isoprene) from *Hevea Brasiliensis* (natural rubber, NR) and poly(1,4-cis-butadiene) with high 1,4-cis content. This type of composite is typically used in an important tire compound such as the sidewall. In a second study, the CB/SP adduct was analogously used in partial replacement of silica, with poly(1,4-cis-isoprene) as the only rubber. Internal tire compounds, for carcass and belt, are typically based on this type of rubber. In most cases, NR is used. However, to ensure the reproducibility of the compounds' properties, the synthetic poly(1,4-cis-isoprene) (IR) from Ziegler-Natta catalysis is sometimes used in place of NR. In this study, the choice of IR in place of NR was aimed at avoiding the interaction of the polar CB/SP with the polar groups which are present at the NR chain ends, proteins and phospholipids. The compounds were crosslinked with a sulfur-based system and were characterized by means of dynamic-mechanical tests, both in the shear and in the axial mode, and tensile measurements.

2. Materials and Methods

2.1. Materials

2.1.1. Chemicals

All reagents and solvents were purchased and used without further purification: 2-amino-1,3-propanediol, 2,5-hexanedione, acetone from Sigma-Aldrich; TESPD 3,3'-bis (triethoxysilylpropyl)disulfide from Flexys, Ann Arbor, MI, USA.

The following chemicals have been used for the preparation of the elastomeric composites discussed in this paper: X50S (50% carbon black, 50% silane, Degussa, Milan, Italy), ZnO (Zincol Ossidi, Bellusco, MB, Italy), Stearic acid (Sogis, Milan, Italy), 1,3-dimethyl butyl)-N'-Phenyl-p-phenylenediamine (6PPD from Eastman, Kingsport, TN, USA), Sulfur (S from Solfotecnica, Cotignola, Italy), N-tert-butyl-2-benzothiazyl sulfenamide (TBBS from Lanxess Chemical, Shangai, China), N-(Cyclohexylthio)phthalimide (PVI, Brenntag, S.p.A., Milan, Italy).

2.1.2. Elastomers

Poly(1,4-cis-isoprene) from Hevea brasiliensis (NR) (EQR-E.Q. Rubber, BR-THAI, Eastern GR. Thailand—Chonburi) had the trade name SIR20 and 73 Mooney Units (MU) as

Mooney viscosity (ML(1 + 4)100 °C). Synthetic poly (1,4-butadiene) (BR) was Neocis BR 40 from Versalis, with a 43 Mooney Viscosity (ML(1 + 4)100 °C). Synthetic poly-1,4-*cis*-isoprene (IR) was from Nizhnekamskneftechim Export, with the trade name SKI3 and 70 Mooney Units (MU) as Mooney viscosity (ML (1 + 4) 100 °C).

2.1.3. Fillers

Carbon black N326 was kindly provided by Birla Carbon (Atlanta, GA, USA). Data from the technical data sheet are as follows. Oil absorption number (OAN): 72 mL/1000, nitrogen specific surface area (NSA): 78 m^2/g, statistical thickness surface area (STSA): 78 m^2/g.

Silica ZEOSIL 1165 MP were white micropearls from Solvay (Brussels, Belgium). Data in the technical data sheet are as follows. Specific surface area: 140–180 m^2/g, loss on drying (2 h @ 105 °C) ≤8.0%, soluble salts (as Na_2SO_4) ≤2.0%. In this work, the surface area was determined using the BET method. Samples were evacuated at 200 °C for 2 h and N_2 adsorption isotherms were recorded at 77 K in a liquid nitrogen bath using a MICROACTIVE TRISTAR ® II PLUS apparatus. The specific surface area (SSA) was found to be 160 m^2/g.

2.2. Preparation of SP and CB/SP Adducts

2.2.1. Synthesis of 2-2,5-Dimethyl-1*H*-pyrrol-1-yl-1,3-propanediol (SP)

In a round bottom flask, equipped with a condenser and magnetic stirrer, 16.23 g of serinol (0.1781 mol) was suspended in 21 mL of 2,5-hexanedione (0.1780 mol, 1 eq), the temperature was raised to 150 °C and stirring was performed for 2 h. The condenser was then removed, and the reaction mixture was stirred for a further 30 min. The water evaporated during the synthesis. The product was collected as a brown, viscous liquid without any further purification (27.73 g, 0.1639 mol). The yield was calculated using the following expression: 100 × (weighed mass of SP)/(theoretical mass of SP). It was calculated to be about 92%. NMR spectra are reported in the Supplementary Materials (Figure S1).

^{1}H NMR (CDCl$_3$, 400 MHz); δ (ppm) = 2.27 (s, 6H), 3.99 (m, 4H); 4.42 (quintet, 1H); 5.79 (s, 2H). ^{13}C NMR (DMSO-d_6, 100 MHz); δ (ppm) = 127.7, 105.9, 43.72, 71.6, 61.2, 13.9.

2.2.2. Preparation of Adducts of CB N326 with SP

Pristine CB N326 (1.0 g) and SP dissolved in acetone were poured in a 50 mL round flask, with a total amount of acetone suitable to completely cover the powder. The mixture was sonicated for 15 min. The solvent was then removed under reduced pressure and the resulting dry powder was heated through immersing the flask in a silicon bath.

The dry powder was then heated at different temperatures in order to obtain three different adducts, in particular: (i) at 150 °C functionalization temperature with SP at 10 phc (CB/SP-6); (ii) at 120 °C functionalization temperature with SP at 10 phc (CB/SP 5); (iii) at 120 °C functionalization temperature with SP at 5 phc (CB/SP-4).

After the reaction, the adduct CB-SP was cooled to room temperature and then washed with a Soxhlet extractor overnight, using acetone as the solvent. The wet powder was dried in the oven at 80 °C for 1 h. The dry powder and the SP possibly extracted were not weighed. The functionalization yield, reported in the following, was thus estimated by means of TGA measurements.

2.3. Preparation of Elastomer Composites

2.3.1. Elastomer Composite with CB/SP-6

The recipe is in Table 1. The reference composite, without CB/SP, is indicated as the "silica" composite. The CB/silica hybrid filler system, in the (30/35 parts) ratio, was adopted in previous works (results not reported) and appeared to be suitable, in the present study, to investigate the behavior of CB/SP. The processing procedure is shown in Figure 2.

Table 1. Recipes of NR/BR-based composites with CB/SP-6 [a].

Ingredient	Silica	CB/SP-6
NR (SIR-20)	70	70
BR	30	30
Silane/CB	5.6	5.6
N326	30	30.0
Silica	35	12
CB/SP-6	0	19.72

[a] Other ingredients (phr): stearic acid 2, ZnO 4, 6PPD 2, sulfur 2, TBBS 1.8, N-cyclohexylthiophthalimide 0.5.

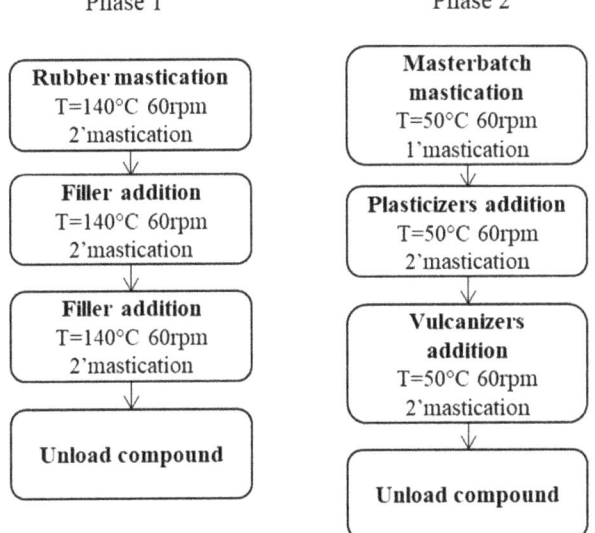

Figure 2. Block diagram of the process for the preparation of rubber composites.

2.3.2. Elastomer Composite with CB/SP-4 and CB/SP-5

The recipe is in Table 2. The reference composite, without CB/SP, is indicated as the "silica" composite. The processing procedure is shown in Figure 2.

Table 2. Recipe of IR-based composites with silica, CB N326 and CB/SP [a] as the fillers [b].

Ingredients	Silica	CB/SP [b]
IR	100.00	100.00
Silica	35.00	12.00
CB/TESPT	5.60	5.60
CB N326	30.00	30.00
CB—SP	0.00	19.72

[a] either CB/SP-5 or CB/SP-4. [b] Other ingredients (phr): stearic acid 2, ZnO 4, 6PPD 2, sulfur 2, TBBS 1.8, N-cyclohexylthiophthalimide 0.5.

2.4. Characterization Techniques

2.4.1. Thermogravimetry Analysis (TGA) of CB/SP Adducts

TGA analysis were performed with a TGA TA instrument Q500 (TA instrument, Newcastle, DE) according to the following method: heating under a nitrogen blanket from 30 °C to 300 °C at 10 °C/min, isothermal step at 300 °C for 15 min, heating to 550 °C at 20 °C/min, isothermal step for 15 min, heating to 900 °C at 10 °C/min, isothermal step for 3 min, shift from nitrogen to air, final isothermal step for 30 min.

In this work, the amount of SP in the CB/SP adduct is expressed in parts per hundred carbon (phc), which were estimated using Equation (1), where (weight loss)$_{150-900°C}$ = x.

$$SP\ (phc) = 100 \cdot \frac{x}{(100-x)} \quad (1)$$

The yield of the functionalization process, from the weight of reactants to the estimation of the mass loss of CB/SP adduct via TGA, was calculated by means of Equation (2).

$$Process\ yield\ (\%) = 100 \cdot \frac{SP\ mass\ \%\ in\ (CB/SP\ adducts)_{after\ washing}}{SP\ mass\ \%\ in\ CB-SP_{reactants\ mixture}} \quad (2)$$

2.4.2. Wide-Angle X-ray Diffraction

With a Bruker D8 Advance automatic diffractometer (Bruker Italia SRL, Milan, Italy) and nickel-filtered Cu-Kα radiation, wide-angle X-ray diffraction patterns were carried out in reflection. The range of 4.7° to 90° was used to record the patterns because these angles were the 2θ peak diffraction angles.

Spectra were developed using Origin Pro 2018.

From the Bragg law (Equation (3)), the distance between crystallographic planes was estimated.

$$d_{hkl} = \frac{n \cdot \lambda}{2 \cdot \sin \theta_{hkl}} \quad (3)$$

where n is an integer number, λ is the wavelength of the irradiating beam and θ_{hkl} is the diffraction angle.

The Scherrer equation (Equation (4)) was used to calculate the D_{hkl} crystallite dimensions using the chosen planes.

$$D_{hkl} = \frac{k \cdot \Lambda}{\beta_{hkl} \cdot \cos \theta_{hkl}} \quad (4)$$

where β_{hkl} is the width at half height, θ_{hkl} is the diffraction angle, K is the Scherrer constant and λ is the wavelength of the irradiating beam.

As shown in the following equation (Equation (5)), the number of layers was simply calculated through dividing the crystallite size of the (002) plane, which corresponds to the perpendicular dimension of the multi-layered material, by the interlayer distance.

$$number\ of\ layers = \frac{D_{hkl}}{d_{hkl}} \quad (5)$$

2.4.3. Crosslinking

It was performed in a rubber process analyzer (Monsanto R.P.A. 2000, Alpha Technologies Hudson, Ohio, USA). An amount of 5 g of rubber composite was weighed and put in the rheometer. Measurements were carried out at a frequency of 1.7 Hz and an oscillation angle of 6.98%. The sample, loaded at 50 °C, underwent a first strain sweep (0.2–25% strain) to cancel the thermo-mechanical history of the rubber composite; it was then maintained at 50 °C for 10 min and then underwent another strain sweep at 50 °C to measure the dynamic-mechanical properties at low deformations of the uncured sample. A crosslinking reaction was then carried out at 170 °C for 10 min. A torque–time curve was obtained. The minimum achievable torque (M_L), the maximum achievable torque (M_H), the time required to have a torque equal to $M_L + 1$ (t_{S1}) and the time required to reach 90% of the maximum torque (t_{90}) were measured.

2.4.4. Dynamic-Mechanical Analysis in the Shear Mode Strain Sweep Test

The shear dynamic-mechanical characteristics of the rubber compounds were evaluated through performing strain sweep tests in a rubber process analyzer (Monsanto R.P.A. 2000, Alpha Technologies Hudson, Ohio, USA). As reported in Section 2.4.3, the crude sample was subjected to a first strain sweep, was then held at 50 °C for ten minutes, followed by another strain sweep at 50 °C. Data from the second strain sweep were collected and are reported in the text below to discuss the behavior of uncured samples. The crosslinking was then carried out, as reported in Section 2.4.3, and the shear dynamic-mechanical properties of the cured samples were then assessed after 20 min at 50 °C using a 0.2–25% strain sweep at a frequency of 1 Hz. Shear storage and loss moduli (G′, G″), and subsequently Tan δ, were measured characteristics.

2.4.5. Dynamic-Mechanical Analysis in the Axial Mode

The elastomeric compound was rolled up to obtain a long cylinder. This cylinder was then cut into smaller cylinders and vulcanized (at 170 °C for 10 min) to produce cylindrical test pieces with dimensions of 25 mm in length and 12 mm in diameter. An Instron dynamic device (Instron, Buckinghamshire, UK) in traction–compression mode was employed to perform dynamic mechanical measurements and was maintained at the predetermined temperatures (10, 23 and 70 °C) throughout the entire experiment. The cylinder was preloaded to a 25% longitudinal deformation with respect to the original length. The compression was subjected to a dynamic sinusoidal strain in compression with an amplitude of around 3.5% regarding the length under pre-load, at a frequency of 100 Hz. This generated the values of dynamic storage modulus (E′) and loss modulus (E″), as well as loss factor (tan δ), calculated as the ratio between E″ and E′.

2.4.6. Tensile Test

Standard dumbbells made from 10 cm by 10 cm by 1 mm vulcanized compound plates were used to perform tensile tests at room temperature with a Zwick Roell Z010 and an optical extensometer. Measurements were performed at 1 mm/min. Stresses at different elongations (respectively σ_{50}, σ_{100}, σ_{300}), stress at break (σ_B), elongation at break (ε_B) and the energy required to break were measured according to Standard ISO [38].

2.4.7. Electrical Resistance

Keysight Technologies 34,450 A 5 $\frac{1}{2}$ Digital Multimeter (Keysight Technologies SRL, Milan, Italy) was used to measure the electrical resistance of rubber composites. The measurements were taken with a hand-applied four-point probe (FPP) with four distinct gold-coated points. (Each tip has an area of 18 mm^2). Only two of the FPP's points were used as crocodile clips for volumetric measurement. The samples are 3.5 cm diameter cured rubber compound disks with a thickness of 3 mm. For each composite, five measurements were taken.

2.4.8. Headspace Analyses

Headspace analyses were carried out using an Agilent HS-Autosampler 7697A, an Agilent GC 6890 N and an Agilent MS 5975C.

3. Results and Discussion

3.1. Preparation and Characterization of CB/SP Adducts

The synthetic pathway for the preparation of the CB/SP started from the synthesis of SP, which was carried out as described elsewhere [28]. Quantitative yield [39] was achieved (96%), with water as the only co-product and thus with very high carbon efficiency [30,31], i.e., with a high amount of the reagents' carbon atoms present in the final product.

For the preparation of the CB/SP adducts, CB N326 was used. As mentioned in the Introduction, this grade of carbon black has a surface area of about 80 m^2/g, remarkably lower than the one of the sp^2 carbon allotropes used by some of the authors for functionalization with SP.

The procedure for the preparation of the CB/SP adduct is in Figure 3.

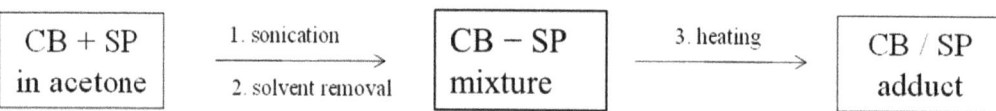

Figure 3. Procedure for the preparation of the CB/SP adduct.

Details are in the experimental section. In brief, CB and SP were mixed in acetone, and sonication was then performed to prepare a homogeneous dispersion. After the removal of the solvent, the temperature was increased, either to 120 °C or to 150 °C, and the reaction was carried out for 4 h. The adducts were then washed with acetone, to completely remove the unreacted SP. In this procedure, already used for preparing adducts with HSAG [27] and CNT [28], a solvent such as acetone was used, to allow for easy mixing at the lab scale. For the scale-up of the reaction, an organic solvent for the preparation of the physical mixture could be avoided, for example, through using an apparatus such as a spray dryer [40].

Adducts with different amounts of SP were prepared through adopting different reaction temperatures and different amounts of SP in the reaction mixture. When the reaction was performed at 150 °C, 10 parts of SP per hundred parts of CB (phc) was used. In the case of the reaction carried out at 120 °C, 5 and 10 phc of SP were used. The amount of SP in the adducts was estimated by means of TGA analysis. Data of mass losses are in Table 3. The thermographs are in Figure S2 in the Supplementary Material. Three main steps can be seen in the decomposition profile: at a temperature below 150 °C, between 150 °C and 800 °C and above 900 °C. Low-molar-mass substances, such as adsorbed water or the acetone used for washing, could be responsible for the mass loss at T < 150 °C. The decomposition of PyC and the alkenylic defects of CB can account for the mass loss between 150 °C and 900 °C. Residues at T > 900 °C were not observed or were present in a very low amount (about 0.5%).

Table 3. Mass losses of CB and CB/SP adducts from TGA analysis.

Sample	SP [a] (phc)	Reaction T (°C)	Mass Loss (%)			SP in CB/SP (phc)	F.Y. [b] (%)
			T < 150 °C	150 °C < T < 900 °C	T > 900 °C		
CB/SP-6	10	150	0.2	5.6	94.2	5.9	59
CB/SP-4	5	120	0.7	3.8	95.0 [c]	4.0	80
CB/SP-5	10	120	0.4	4.8	94.1 [d]	5.1	51

[a] amount of SP in the reaction mixture; [b] functionalization yield (see Equation (1)); [c] residue: 0.6%; [d] residue: 0.7%.

The content of SP in the adducts (in phc) was estimated on the basis of the mass loss between 150 °C and 900 °C, which, as mentioned above, could be also due to the defects of CB, which, however, are supposed to react with SP to form the adducts. The estimation of the SP content was made by means of Equation (1) (see Section 2.4.1) and is shown, with phc as the measure unit, in Table 3. In the following, the adducts are named with a round figure indicating the level of functionalization. The yield of the whole functionalization process was estimated by means of Equation (2) (see Section 2.4.1). Through performing the functionalization with a higher amount of SP (10 phc) at 150 °C, a yield of about 60% (at 150 °C) or about 50% (at 120 °C) was achieved, lower than the ones reported in the case of HSAG [26,27] and CNT [28], sp^2 carbon allotropes with remarkably higher surface area. Functionalization carried out with a lower amount of SP (5 phc) led to about 80% functionalization. The amount of SP in the reaction mixture seems to play a major role with respect to the functionalization temperature. The moderate surface area of CBN326 could account for these findings. In consideration of the yield for the SP synthesis, in the case of CB/SP-4, a yield close to 80% was obtained for the global process for preparing the CB/SP adduct.

X-ray analysis was performed on CB and on the adducts CB/SP-4 and CB/SP-5. CB is made up of poorly organized areas (disordered and amorphous carbons) and, mostly, of aggregates made by a few graphene layers, assembled in spherical particles. The patterns are in Figure 4, and the parameters indicating the size of crystallites in the direction orthogonal and parallel to the structural layers, the interlayer distance and the anisotropic index are in Table S1 in the Supplementary Material.

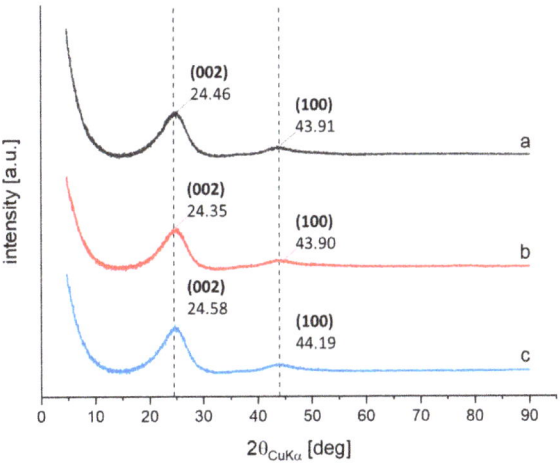

Figure 4. X-ray diffraction patterns of pristine CB N326 (**a**), CB/SP-4 (**b**), CB/SP-5 (**c**).

The diffraction peaks were observed at the following 2θ values: 24.46° (002) and 43.99° (100) for CB/SP-4 and at 24.41° (002) and 43.79° (100) for CB/SP-5. The values reported in Table S1 indicate that the functionalization did not alter the crystalline structure of CB, as already observed for HSAG [27] and CNT [28]. These results confirm that the pyrrole methodology for the functionalization of the sp^2 carbon allotropes modifies the surface properties without affecting the crystalline structure of CB. Traditional oxidation methods of CB, for example, with HNO_3, were shown to be able to destroy the CB structure, leading to graphene layers [41].

3.2. Preparation and Characterization of Rubber Composites

Two studies were performed, changing the rubber matrix and the extent of SP in the CB/SP adduct.

Composites based on BR/NR. In the first study, composites were based on BR and NR, and the hybrid filler system was made up of silica and CB (35 phr and 30 phr, respectively). As mentioned in the Introduction, this type of composite is suitable for a tire sidewall compound with a low hysteresis. CB/SP-6 was used in the partial replacement of silica: 66% by volume. Recipes are in Table 1. The same amount of the coupling agent, the silane TESPT, was used in both the composites. The compounding procedure is in Figure 2. In the first step, silica was mixed with the rubber in the presence of silane TESPT to allow the interaction of TESPT with silica and the occurrence of the silanization reaction, at least to some extent. CB/SP and CB were fed in the second step.

The crosslinking was performed with a sulfur-based system. The rheometric curves are in Figure 5 and data of minimum modulus M_L, maximum modulus M_H, induction crosslinking time t_{s1}, optimum crosslinking time t_{90} and curing rate are in Table S2 in the Supplementary Material.

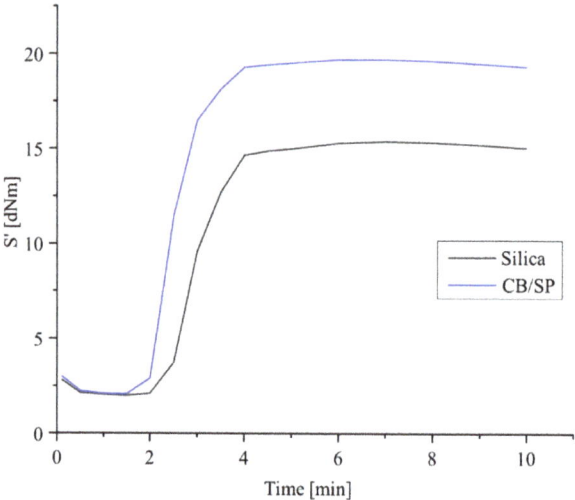

Figure 5. Rheometric curves for rubber composites of Table 1.

Similar values of M_L were obtained. M_L is an index of the viscosity of the composite and, for composites with the same rubber matrix and loaded with the same volume of reinforcing fillers, is essentially due to the extent of the filler network, which appears, thus, to be similar in the two composites. A remarkably higher M_H was obtained with CB/SP. The values of M_H are due to the extent of both the crosslinking network and the filler network. In fact, the maximum amplitude explored through the shear test is not enough to completely disrupt the filler network. Values of induction and optimum times of vulcanization are similar for the two composites, whereas a remarkably higher curing rate was obtained for the composite with CB/SP. In the presence of similar M_L, and taking also in consideration that the two composites contained the same amount of sulfur atoms, the higher value of M_H for the composite with CB/SP could be due to a more efficient vulcanization, with the formation of a higher crosslinking network density and/or shorter sulfur bridges. The higher M_H and curing rate could be due to the presence of sp^3 nitrogen atoms on the CB surface. This hypothesis is proposed assuming that TESPT plays its role for the functionalization of the filler(s) rather than as sulfur donor. The silanization of CB/SP with a silane is discussed below in the text. However, further studies are needed to clarify this point.

Dynamic-mechanical properties were determined in the shear mode through performing strain sweep experiments in the range from 0.2% to 25%, as described in the experimental section. Data are in Table 4, and curves of G' and tan delta are in Figures S3 and S4 in the Supplementary Material.

Table 4. Dynamic-mechanical properties from shear tests of composites of Table 1.

	Silica	CB/SP-6
$G'_{0.2\%}$ [MPa]	2.52	2.30
$G'_{25\%}$ [MPa]	0.94	0.94
$\Delta G'$ [MPa]	1.58	1.36
$\Delta G'/G'_{0.2\%}$	0.67	0.59
G''_{max} [MPa]	0.15	0.15
$Tan(\delta)_{max}$	0.11	0.11

With respect to the reference composite, CB/SP led to lower values of $\Delta G'$ and $\Delta G'/G'_{\gamma min}$, whereas the G''_{max} and the Tan δ_{max} are in line. The replacement of silica with CB/SP appears to reduce the Payne effect [42–45] and hence reduce the filler networking phenomenon.

Dynamic-mechanical properties were also determined in the axial mode through applying sinusoidal stress as described in the experimental section. Data of E′, E″ and Tan delta, measured at 10 °C, 23 °C and 70 °C, are in Table 5, and the dependence of E′ and Tan delta on temperature are in Figure 6, Figure 6a and Figure 6b, respectively.

Table 5. Axial dynamic-mechanical properties of composites of Table 1.

	T (°C)	Silica	CB/SP-6
E′ (Mpa)	10	5.33	7.20
	23	5.03	6.72
	70	4.27	5.72
E″ (MPa)	10	1.13	1.41
	23	0.86	1.04
	70	0.52	0.56
Tan δ	10	0.21	0.20
	23	0.17	0.16
	70	0.12	0.10

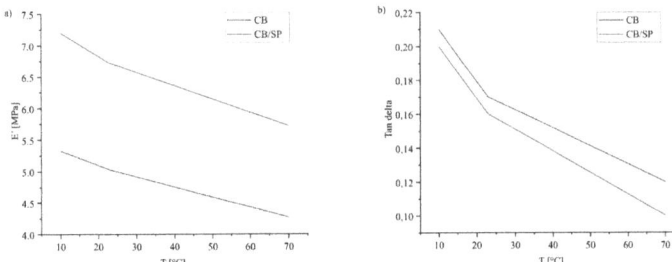

Figure 6. Dependence on temperature of E′ (**a**) and Tan delta (**b**) for the composites of Table 1.

With respect to the reference composite, CB/SP-6 led to remarkably higher values of E′, at every temperature, and lower values of Tan delta, particularly at high temperature. These results indicate a lower extent of the filler network and are in line with those from the shear tests.

Tensile properties were determined through quasi-static measurements. Data of stresses at different elongations and stress, strain and energy at break are in Table 6, and curves are in Figure S5 in the Supplementary Material.

Table 6. Tensile properties of composites of Table 1.

	Silica	CB/SP-6
σ_{100} (Mpa)	2.21 ± 0.03	3.39 ± 0.03
σ_{200} (MPa)	6.28 ± 0.12	9.05 ± 0.11
σ_{300} (Mpa)	13.61 ± 0.25	17.58 ± 0.22
$\sigma_{300}/\sigma_{100}$	6.15 ± 0.22	5.18 ± 0.13
σ_B (Mpa)	30.34 ± 1.15	25.91 ± 1.19
ε_B (%)	504.22 ± 10.14	392.94 ± 7.15
Energy (J/cm^3)	60.18 ± 2.54	41.16 ± 4.52

CB/SP led to remarkably higher values of stress at every strain and to lower stress, elongation and energy at break. These results appear to be in line with the higher values of dynamic rigidity. To account for the lower values of the ultimate properties, though for both the composites remarkable values were obtained, the ability of CB/SP to interact with silica and to give potentially rise to filler agglomerates should be investigated, in future studies. Moreover, the viscosity of the composite could play a role.

Composites based on IR. In the second study, composites were based on IR as the only rubber and, analogously to the first study, a hybrid silica/CB filler system (35 phr and 30 phr, respectively) was used. As reported in the Introduction, this type of composite, with NR as the rubber, is suitable for internal tire compounds with low hysteresis. The recipes of the composites are in Table 2. The replacement of silica with CB/SP was performed at the same level as in the first study (66% by volume), and the two samples of CB/SP adducts, CB/SP-4 and CB/SP-5, were used. The main objective of this study was to investigate the effect of CB/SP adducts with a lower level of SP. Composites were prepared via melt blending with the same procedure used for the first study, shown in Figure 2.

Sulfur-based crosslinking was investigated with a rheometric test. Rheometric curves are in Figure 7, and data of M_L and M_H, t_{90}, t_{s1} and curing rate are in Table S3 in the Supplementary Material.

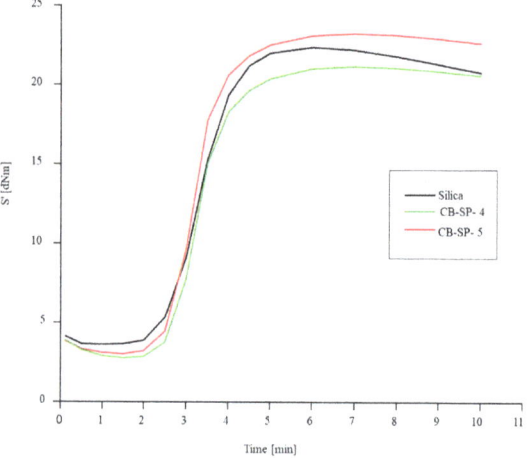

Figure 7. Vulcanization curves of IR-based compounds: torque versus time.

The crosslinking curve of the composite with the whole amount of silica reveals the reversion phenomenon, absent for the composites with CB/SP, which also give rise to lower M_L values, particularly with CB/SP-4. The highest value of M_H was obtained with CB/SP-5. These results indicate that even a subtle difference of the SP amount in the CB/SP adduct have an appreciable effect on the crosslinking behavior and suggest that a higher amount of SP leads to higher composite's viscosity and, particularly, to more efficient vulcanization. Moreover, they appear to be in line with those obtained in the first study.

Strain sweep experiments in the shear mode were carried out, as described in the experimental section, similar to the first study. Data collected from the experiments on both uncured and cured samples are considered. The curves of G' vs. strain amplitude for the uncured and cured samples of Table 2 are to be seen in Figure 8a and Figure 8b, respectively, and the curves of Tan delta vs. strain amplitude for the uncured and cured samples of Table 2 are shown in Figure 9a and Figure 9b, respectively. Data of $G'_{0.2\%}$, $G'_{25\%}$, $\Delta G'/G'_{0.2\%}$, G''_{max} and $Tan(\delta)_{max}$ for the uncured and cured samples are in Table S4 and in Table S5, respectively, in the Supplementary Material. The curves of G' for the uncured samples with CB/SP in Figure 8a overlap and lie below the curve of the reference silica sample, throughout the range of strain amplitudes. The lower values of $\Delta G'/G'_{0.2\%}$ for the CB/SP samples suggest that the replacement of silica with the functionalized CB leads to a lower Payne effect [41–44]. The dependence of G' on the strain amplitude for cured CB/SP-

4 and CB/SP-5 samples appears to be different, as shown in Figure 8b. Higher values of G′ were obtained with CB/SP-5, and a crossover of the curves due to the CB/SP-5 and the reference silica samples can be observed: higher values of G′ at high strain amplitude were obtained with the CB/SP-5 sample. The replacement of silica with CB/SP leads to the reduction of ΔG′ and ΔG′/G′, and, hence, of the Payne Effect, as well as to the reduction of Tan delta. The reduction of ΔG′/G′, in relation to the reference composite without CB/SP, can be correlated with the amount of SP in the adduct. It is worth observing that these results are in line with those obtained in the first study and that even a slight difference in the SP content in the CB/SP adduct leads to an appreciable different behavior in the strain sweep test. Although it is challenging to compare data from different studies, based on different rubber matrices, the numbers in Table 4 and Table S5 allow for the calculation of a percentage reduction in ΔG′/G′, of approximatively 5, 8 and 12% for composites with CB-4, CB-5 and CB-6, respectively, compared to the reference composites without CB/SP. This elaboration should be intended only as a qualitative indication of the correlation between the extent of CB functionalization with SP and the modification of the properties of the reference composite.

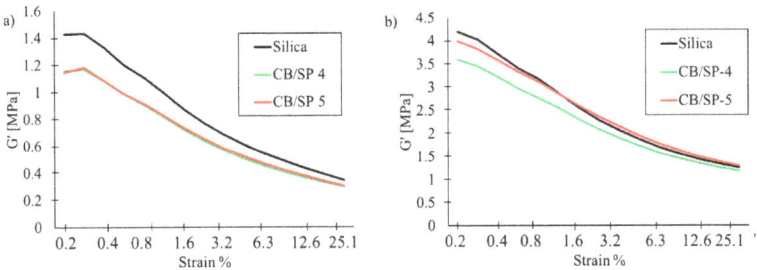

Figure 8. G′ vs. Strain curve for rubber composites of Table 2 (**a**) uncured (**b**) cured.

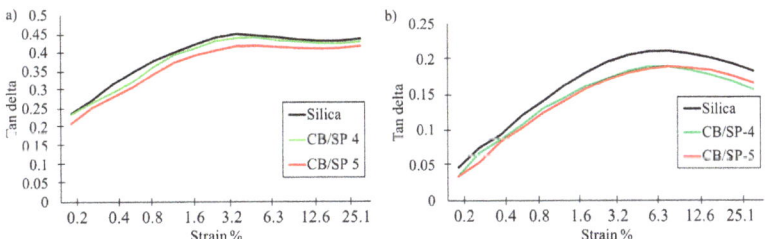

Figure 9. Tan delta vs. Strain curve for rubber composites of Table 2 (**a**) uncured (**b**) cured.

Axial dynamic-mechanical properties were obtained as outlined in the experimental section. Data of E′, E″ and tan δ, measured at 10 °C, 23 °C and 70 °C, are in Table 7, and the dependence on temperature of E′ and Tan delta is in Figure S6a,b in the Supplementary Material.

The replacement of silica with the CB/SP adducts led to higher dynamic rigidity: higher values were obtained with higher SP content. As observed for the dynamic-mechanical properties in the shear mode, the values of E′ from axial measurements can also be correlated with the amount of SP in the CB/SP adducts. As already commented, it is indeed hard to compare values obtained in different studies with different polymer matrices. However, it can be noticed that the increase of E′, with respect to the reference composite without CB/SP, was (%) 8, 13 and 35 at 10 °C and 3, 11 and 34 at 70 °C for CB/SP-4, CB/SP-5 and CB/SP-6, respectively. These findings indicate the prevailing effect of CB/SP on the dynamic-mechanical properties of the rubber composite, in spite of the different rubber matrices and the different grades (different providers) of CB. The reduction of hysteresis was achieved with the greater amount of SP.

Table 7. Axial dynamic-mechanical properties of composites of Table 2.

	T (°C)	Silica	CB/SP-4	CB/SP-5
E' [Mpa]	10	7.09	7.67	8.00
	23	6.51	6.91	7.22
	70	5.69	5.83	6.30
E'' [Mpa]	10	1.89	2.09	2.05
	23	1.49	1.64	1.60
	70	0.84	0.83	0.85
Tan (δ)	10	0.27	0.27	0.26
	23	0.23	0.24	0.22
	70	0.15	0.14	0.13

Quasi-static measurements were used to determine the composites' tensile characteristics. The tensile curves are in Figure 10 and data are reported in Table S6 in the Supplementary Material.

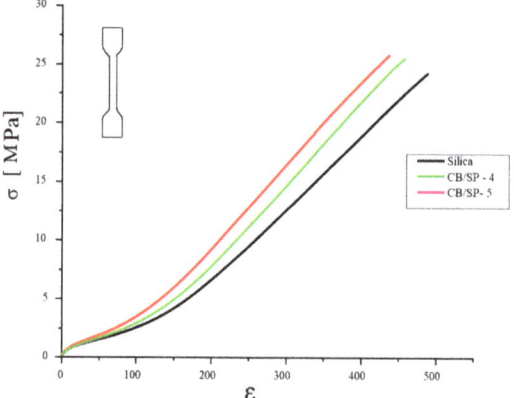

Figure 10. Tensile curves of composites of Table 2.

Higher values of stresses at every elongation and at break were obtained with CB/SP in the composites in place of silica. The highest values were for the sample with the highest amount of SP in the adduct. This larger mechanical reinforcement was accompanied by a lower elongation at break, though all the composites achieved high elongations. These findings are in line with what was observed in the first study. The ultimate properties appear to be closer to those of the reference composite in this second study. The higher viscosity of the rubber matrix could play a role. However, the ability of CB/SP to interact with silica and potentially to give rise to filler agglomerates could lead to lower elongation at break and it should be investigated in future studies. It is worth adding that data from optical microscopy revealed that the CB dispersion was not appreciably modified through the use of CB/SP in place of silica. Values are in Table S7 in the Supplementary Material.

As mentioned in the Introduction, the replacement of silica with CB/SP could have a positive effect on the electrical conductivity of the elastomer composite. In this work, the electrical resistance was measured, as reported in the experimental section, for the samples of composites in Table 2. In particular, the data reported in Table 8 refer to the electrical properties of the silica and the CB/SP 4 compounds.

Table 8. Electrical resistance of composites of Table 2: reference silica and CB/SP-4 composites.

Sample	Resistance [MΩ]
Silica	1.9 ± 0.3
CB—SP 4	$(5.4 \pm 1.0) \times 10^{-3}$

The measures show that a substitution of 66% in volume of silica with CB/SP 4 decrease the resistance by three orders of magnitude with respect to the reference.

3.3. Silanization of CB/SP

Preliminary studies have been conducted to investigate the reactivity of CB/SP with a sulfur-based silane such as bis(triethoxysilylpropyl)tetrasulfide (TESPD). Details are in the experimental section. In brief, headspace (HS) analyses were performed at 150 °C for 20 min, measuring the ethanol emission from the reaction of TESPD with CB, SP and CB/SP. In these reactions, the amount of TESPD and the molar TESPD/OH ratio were kept constant. The area of the ethanol GC peak was correlated with the extent of silanization [12]. In Figure 11, the chromatograms of the volatiles of the above-mentioned samples are reported and compared.

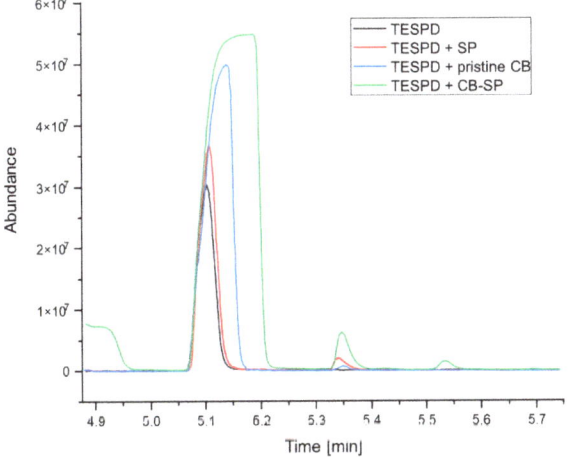

Figure 11. Chromatograms from HS-MS testing at 150 °C for 20 min of TESPD (black), TESPD + SP (red), TESPD + pristine CB (blue) and TESPD + CB/SP (green).

The area of the ethanol peak coming from the reaction of TESPD with SP is larger than the area of the pure TESPD. The area of the ethanol peak coming from the reaction of TESPD with CB/SP sample is the largest. These findings suggest that the OH of the serinol moiety of SP reacts with TESPD and that a silanization reaction of CB/SP can occur. The reactivity of the OH on the CB surface could explain the appreciable differences among the composites' properties, observed even in the presence of a subtle difference in SP content in the CB/SP adducts.

3.4. On the Reactivity of CB/SP

The results discussed above show that CB/SP can replace silica in elastomeric composites, resulting in higher dynamic rigidity, lower hysteresis and higher stresses at every elongation in tensile tests. Moreover, it is worth highlighting the appreciable differences among composites containing CB/SP adducts with a subtle difference of SP content. These findings appear to support the working hypothesis reported in the Introduction: CB functionalized with SP acts as a polar filler and is able to interact with both silica and the coupling agent TESPT. The results of the silanization study discussed in the previous paragraph confirm the ability of CB/SP to react with sulfur-based silane. In a recent paper [46], some of the authors demonstrated the ability of serinol pyrrole to react with silica. SP was used as coupling agent of silica in place of TESPT. The reactivity with silica was due to the serinol moiety, which is on CB's surface in the CB/SP adduct. Hence, it can be concluded that the OH on the CB surface can enhance the reactivity of CB with polymer chains, mediated either by silica or by the sulfur-based silane.

4. Conclusions

This work demonstrates that silica is not the sole nanostructured reinforcing filler for elastomer composites with low hysteresis. In fact, a furnace carbon black functionalized with SP was used in rubber composites based on diene elastomers and a CB/silica hybrid filler system, replacing 66% of silica by volume, thus obtaining more efficient vulcanization, a lower Payne Effect, higher dynamic rigidity, low hysteresis and increased stresses in tensile measurements. Thus, a furnace CB, even with relatively low surface area (80 m^2/g), when functionalized with a polar modifier such as serinol pyrrole, can replace a significant portion of silica, leading to composites with even lower hysteresis. The functionalization of CB with SP is achieved through a simple mixing and heating process, yielding a high carbon efficiency without the need for catalysts and, potentially, for solvents. Traditional formulation and melt blending can be applied. The chemical modification of CB appears to play a major role. The OH groups facilitate interaction with silica and with silane TESPT and, thus, with the polymer chains. The sp^3 nitrogen atom could be responsible for the efficient sulfur-based crosslinking and the high dynamic rigidity. The lower Payne effect and the lower hysteresis, compared to the silica-based composite, could be attributed to the synergy between chemical reactivity and the lower surface area of CB.

All the procedures here described can be easily scaled up to the industrial level. As mentioned in the introduction, a major player in the tire industry is currently undertaking this work.

This research highlights the potential of carbon fillers with polar groups in the development of elastomer composites with low environmental impact. Furnace CB is oil based and is not considered a sustainable material. CB from different sources are under development: bio char from renewable sources and circular char from end-of-life polymer products. These chars typically have low surface area and are not nanostructured. However, this work shows that these drawbacks could be overcome through chemical reactivity. Functionalization with pyrrole compounds appears to pave the way for the development of new, more sustainable generations of carbon materials.

Supplementary Materials: The following supporting information can be downloaded at: https://www.mdpi.com/article/10.3390/polym15143120/s1. Figure S1: ^{1}H (a, CDCl$_3$, 400 MHz) and ^{13}C (b, CDCl$_3$, 100 MHz) NMR spectra of 2-(2,5-dimethyl-1H-pyrrol-1-yl)-1,3-propanediol (serinol pyrrole, SP). Figure S2: Thermograph from TGA analysis of CB/SP-4, CB/SP-5 and CB/SP-6. Figure S3: G' vs strain for composites of Table 1. Figure S4: Tan delta vs strain for composites of Table 1. Figure S5: Tensile properties for composites of Table 1. Figure S6: Storage modulus (a) and tan δ (b) curves of IR composites. Table S1: Structural parameters from X-ray analysis for pristine CB N326 and CB/SP-6 adduct. Table S2: Data from the crosslinking of composites of Table 1[a,b]. Table S3: Data from the crosslinking reaction of composites of Table 2[a,b]. Table S4: Dynamic-mechanical properties from shear tests of the uncured composites of Table 2. Table S5: Dynamic-mechanical properties from shear tests of the cured composites of Table 2. Table S6: Tensile properties of composites of Table 2. Table S7: Optic microscopy CB dispersion results of IR-based rubber composites.

Author Contributions: Conceptualization, V.B. and M.S.G.; Methodology, U.G., V.B. and M.S.G.; Validation, V.B. and M.S.G.; Investigation, F.M. (Federica Magaletti), F.M. (Fatima Margani), A.M., R.D., G.P., U.G., V.B. and M.S.G.; Data curation, F.M. (Federica Magaletti), F.M. (Fatima Margani), G.P., V.B. and M.S.G.; Writing—original draft, F.M. (Federica Magaletti), F.M. (Fatima Margani), G.P., V.B. and M.S.G.; Supervision, U.G., V.B. and M.S.G. All authors have read and agreed to the published version of the manuscript.

Funding: This research received no external funding.

Institutional Review Board Statement: Not applicable.

Data Availability Statement: Not applicable.

Acknowledgments: Pirelli Tyre is gratefully acknowledged for financing the PhD activities of Federica Magaletti, Fatima Margani and Gea Prioglio.

Conflicts of Interest: The authors declare no conflict of interest.

References

1. Available online: https://ourworldindata.org/greenhouse-gas-emissions#annual-greenhouse-gas-emissions-how-much-do-we-emit-each-year (accessed on 15 May 2023).
2. Available online: https://www.un.org/en/conferences/environment/rio1992/ (accessed on 2 June 2023).
3. Available online: https://www.un.org/sustainabledevelopment/development-agenda/ (accessed on 2 June 2023).
4. Available online: https://ourworldindata.org/emissions-by-sector (accessed on 16 May 2023).
5. *Global Mobility Report 2017, Tracking Sector Performance*; Sustainable Mobility for All: Washington, DC, USA, 2017.
6. Available online: https://www.reportlinker.com/p05379599/?utm_source=GNW (accessed on 2 June 2023).
7. Hall, D.E.; Moreland, J.C. Fundamentals of Rolling Resistance. *Rubber Chem. Technol.* **2001**, *74*, 525–539. [CrossRef]
8. Warasitthinon, N.; Robertson, C.G. Interpretation of the tanδ peak height for particle-filled rubber and polymer nanocomposites with relevance to tire tread performance balance. *Rubber Chem. Technol.* **2018**, *91*, 577–594. [CrossRef]
9. Tao, Y.C.; Dong, B.; Zhang, L.Q.; Wu, Y.P. Reactions of silica–silane rubber and properties of silane–silica/solution-polymerized styrene–butadiene rubber composite. *Rubber Chem. Technol.* **2016**, *89*, 526–539. [CrossRef]
10. Tunnicliffe, L.B.; Busfield, J.J. Reinforcement of rubber and filler network dynamics at small strains. In *Designing of Elastomer Nanocomposites: From Theory to Applications*; Advances in Polymer Science; Stöckelhuber, K., Das, A., Klüppel, M., Eds.; Springer: Cham, Switzerland, 2017; Volume 275, pp. 71–102.
11. Leblanc, J.L. Rubber-filler interactions and rheological properties in filled compounds. *Progr. Polym. Sci.* **2002**, *27*, 627–687. [CrossRef]
12. Donnet, J.B.; Custodero, E. Reinforcement of Elastomers by Particulate Fillers. In *The Science and Technology of Rubber*, 3rd ed.; Mark, J.E., Erman, B., Eirich, F.R., Eds.; Academic Press: Cambridge, MA, USA, 2005; pp. 367–400.
13. Fröhlich, J.; Niedermeier, W.; Luginsland, H.D. The effect of filler-filler and filler-elastomer interaction on rubber reinforcement. *Compos. Part A* **2005**, *36*, 449–460. [CrossRef]
14. Kraus, G. Reinforcement of Elastomers by Carbon Black. In *Fortschritte der Hochpolym*; Springer: Berlin, Heidelberg, 1971; pp. 155–237.
15. Medalia, A.I. Effect of carbon black on dynamic properties of rubber vulcanizates. *Rubber Chem. Technol.* **1978**, *51*, 437–523. [CrossRef]
16. Voll, M.; Kleinschmit, P. Carbon, 6. Carbon Black. In *Ullmann's Encyclopedia of Industrial Chemistry*; Wiley: Hoboken, NJ, USA, 2010.
17. Donnet, J.B.; Bansal, R.C.; Wang, M.J. *Carbon Black: Science and Technology*, 2nd ed.; Dekker: New York, NY, USA, 1993.
18. Fan, Y.; Fowler, D.G.; Zhao, M. The past, present and future of carbon black as a rubber reinforcing filler—A review. *J. Clean. Prod.* **2020**, *247*, 119115. [CrossRef]
19. Robertson, C.G.; Hardman, N.J. Nature of carbon black reinforcement of rubber: Perspective on the original polymer nanocomposite. *Polymers* **2021**, *13*, 538. [CrossRef]
20. Chevalier, Y.; Morawski, J.C. Precipitated Silica with Morphological Properties, Process for Producing It and Its Application, Especially as a Filler. European Patent EP 0157703 B1, 31 May 1989.
21. Legrand, A.P. On the silica edge. In *The Surface Properties of Silicas*; Legrand, A.P., Ed.; Wiley and Sons: New York, NY, USA, 1998; pp. 1–20.
22. Ten Brinke, J.W.; Debnath, S.C.; Reuvekamp, L.A.; Noordermeer, J.W. Mechanistic aspects of the role of coupling agents in silica–rubber composites. *Compos. Sci. Technol.* **2003**, *63*, 1165–1174. [CrossRef]
23. Lee, S.Y.; Kim, J.S.; Lim, S.H.; Jang, S.H.; Kim, D.H.; Park, N.H.; Jung, J.W.; Choi, J. The investigation of the silica-reinforced rubber polymers with the methoxy type silane coupling agents. *Polymers* **2020**, *12*, 3058. [CrossRef]
24. Galimberti, M.; Barbera, V.; Sebastiano, R.; Citterio, A.; Leonardi, G.; Valerio, A.M. Adducts between Carbon Allotropes and Serinol Derivatives. U.S. Patent 10,160,652 B2, 25 December 2018.
25. Galimberti, M.; Barbera, V. Adducts of Pyrrole Derivatives to Carbon Allotropes. U.S. Patent 11,098,012 B2, 24 August 2018.
26. Galimberti, M.; Barbera, V.; Guerra, S.; Conzatti, L.; Castiglioni, C.; Brambilla, L.; Serafini, A. Biobased Janus Molecule for the Facile Preparation of Water Solutions of Few Layer Graphene Sheets. *RSC Adv.* **2015**, *5*, 81142–81152. [CrossRef]
27. Barbera, V.; Bernardi, A.; Palazzolo, A.; Rosengart, A.; Brambilla, L.; Galimberti, M. Facile and Sustainable Functionalization of Graphene Layers with Pyrrole Compounds. *Pure Appl. Chem.* **2018**, *90*, 253–270. [CrossRef]
28. Locatelli, D.; Barbera, V.; Brambilla, L.; Castiglioni, C.; Sironi, A.; Galimberti, M. Tuning the Solubility Parameters of Carbon Nanotubes by Means of Their Adducts with Janus Pyrrole Compounds. *Nanomaterials* **2020**, *10*, 1176. [CrossRef] [PubMed]

29. Barbera, V.; Brambilla, L.; Milani, A.; Palazzolo, A.; Castiglioni, C.; Vitale, A.; Bongiovanni, R.; Galimberti, M. Domino reaction for the sustainable functionalization of few-layer graphene. *Nanomaterials* **2018**, *9*, 44. [CrossRef]
30. Sheldon, R.A. The E factor 25 years on: The rise of green chemistry and sustainability. *Green Chem.* **2017**, *19*, 18–43. [CrossRef]
31. Sheldon, R.A. Metrics of green chemistry and sustainability: Past, present, and future. *ACS Sustain. Chem. Eng.* **2018**, *6*, 32–48. [CrossRef]
32. Zhang, Y.; Li, W.L.; Zong, S.; Du, H.X.; Shi, X.X. Clean synthesis process of 2,5-hexanedione. *Adv. Mat. Res.* **2012**, *518*, 3947–3950. [CrossRef]
33. Waidmann, C.R.; Pierpont, A.W.; Batista, E.R.; Gordon, J.C.; Martin, R.L.; West, R.M.; Wu, R. Functional group dependence of the acid catalyzed ring opening of biomass derived furan rings: An experimental and theoretical study. *Catal. Sci. Technol.* **2013**, *3*, 106–115. [CrossRef]
34. Gilkey, M.J.; Vlachos, D.G.; Xu, B. Poisoning of Ru/C by homogeneous Brønsted acids in hydrodeoxygenation of 2,5-dimethylfuran via catalytic transfer hydrogenation. *Appl. Catal. A Gen.* **2017**, *542*, 327–335. [CrossRef]
35. Galimberti, M.; Barbera, V.; Giannini, L.; Naddeo, S. Processo per la Preparazione di Dichetoni e Derivati Pirrolici. Italian Patent Application n. 102021000032138, 1 January 2021.
36. de Gennes, P.G. Soft matter (nobel lecture). *Angew. Chem. Int. Ed. Engl.* **1992**, *31*, 842–845. [CrossRef]
37. Available online: https://corporate.pirelli.com/var/files2020/EN/PDF/PIRELLI_ANNUAL_REPORT_2020_ENG.pdf (accessed on 2 June 2023).
38. *Standard ISO37/UNI 6065*; Rubber—Tests on Vulcanized or Thermoplastic Rubber—Tensile Test. Ente Nazionale Italiano di Unificazione (UNI): Milano, Italy, 2001.
39. Vogel, A.I.; Tatchell, A.R.; Furnis, B.S.; Hannaford, A.J.; Smith, P.W.G. (Eds.) *Vogel's Textbook of Practical Organic Chemistry*; Prentice Hall: Hoboken, NJ, USA, 1996; ISBN 978-0-582-46236-6.
40. McCabe, W.L.; Julian, C.S.; Harriot, P. *Unit Operations of Chemical Engineering*; McGraw-Hill: New York, NY, USA, 1993; Volume 5.
41. Alfe, M.; Gargiulo, V.; Di Capua, R.; Chiarella, F.; Rouzaud, J.N.; Vergara, A.; Ciajolo, A. Wet chemical method for making graphene-like films from carbon black. *ACS Appl. Mater. Interfaces* **2012**, *4*, 4491–4498. [CrossRef] [PubMed]
42. Dillon, J.H.; Prettyman, I.B.; Hall, G.L. Hysteretic and elastic properties of rubberlike materials under dynamic shear stresses. *J. Appl. Phys.* **1944**, *15*, 309–323. [CrossRef]
43. Fletcher, W.P.; Gent, A.N. Nonlinearity in the dynamic properties of vulcanized rubber compounds. *Trans. Inst. Rubber Ind.* **1953**, *29*, 266–280. [CrossRef]
44. Payne, A.R. The dynamic properties of carbon black-loaded natural rubber vulcanizates. Part I. *J. Appl. Polym. Sci.* **1962**, *6*, 57–63. [CrossRef]
45. Warasitthinon, N.; Genix, A.C.; Sztucki, M.; Oberdisse, J.; Robertson, C.G. The Payne effect: Primarily polymer-related or filler-related phenomenon? *Rubber Chem. Technol.* **2019**, *92*, 599–611. [CrossRef]
46. Locatelli, D.; Bernardi, A.; Rubino, L.R.; Gallo, S.; Vitale, A.; Bongiovanni, R.; Barbera, V.; Galimberti, M. Biosourced Janus Molecules as Silica Coupling Agents in Elastomer Composites for Tires with Lower Environmental Impact. *ACS Sustain. Chem. Eng.* **2023**, *11*, 2713–2726. [CrossRef]

Disclaimer/Publisher's Note: The statements, opinions and data contained in all publications are solely those of the individual author(s) and contributor(s) and not of MDPI and/or the editor(s). MDPI and/or the editor(s) disclaim responsibility for any injury to people or property resulting from any ideas, methods, instructions or products referred to in the content.

Article

Optimized End Functionality of Silane-Terminated Liquid Butadiene Rubber for Silica-Filled Rubber Compounds

Sanghoon Song [1,†], Haeun Choi [2,†], Junhwan Jeong [3], Seongyoon Kim [2], Myeonghee Kwon [4], Minji Kim [4], Donghyuk Kim [5], Heungbae Jeon [4,*], Hyun-jong Paik [2,*], Sungwook Chung [1,*] and Wonho Kim [1,*]

1. School of Chemical Engineering, Pusan National University, Busan 46241, Republic of Korea; thdtkd1111@gmail.com
2. Department of Polymer Science and Engineering, Pusan National University, Busan 46241, Republic of Korea; fexox06112@gmail.com (H.C.); kimsy9780@gmail.com (S.K.)
3. School of Chemical and Biomolecular Engineering, Pusan National University, Busan 46241, Republic of Korea; wnsghks4192@gmail.com
4. Department of Chemistry, Kwangwoon University, Seoul 139-701, Republic of Korea; myeonghee814@gmail.com (M.K.); annie73329@gmail.com (M.K.)
5. Hankook Tire & Technology Co., Ltd., R&D Center, 50 Yuseong-daero 935beon-gil, Daejeon 34127, Republic of Korea; ehdgurzxc@gmail.com
* Correspondence: hbj@kw.ac.kr (H.J.); hpaik@pusan.ac.kr (H.-j.P.); sungwook.chung@pusan.ac.kr (S.C.); whkim@pusan.ac.kr (W.K.); Tel.: +82-51-510-3190 (W.K.)
† These authors contributed equally to this work.

Abstract: As the world is shifting from internal combustion engine vehicles to electric vehicles in response to environmental pollution, the tire industry has been conducting research on tire performance to meet the requirements of electric vehicles. In this experiment, functionalized liquid butadiene rubber (F-LqBR) with triethoxysilyl groups at both ends was introduced into a silica-filled rubber compound as a substitute for treated distillate aromatic extract (TDAE) oil, and comparative evaluation was conducted according to the number of triethoxysilyl groups. The results showed that F-LqBRs improved silica dispersion in the rubber matrix through the formation of chemical bonds between silanol groups and the base rubber, and reduced rolling resistance by limiting chain end mobility and improving filler–rubber interaction. However, when the number of triethoxysilyl groups in F-LqBR was increased from two to four, self-condensation increased, the reactivity of the silanol groups decreased, and the improvement of properties was reduced. As a result, the optimized end functionality of triethoxysilyl groups for F-LqBR in silica-filled rubber compound was two. The 2-Azo-LqBR with the optimized functionality showed an improvement of 10% in rolling resistance, 16% in snow traction, and 17% in abrasion resistance when 10 phr of TDAE oil was substituted

Keywords: liquid butadiene rubber; radical polymerization; silica-filled compound; rubber compounding; passenger car radial tire tread

1. Introduction

In response to greenhouse gas emissions and various environmental pollution problems around the world, interest in sustainability and eco-friendly transportation, for which the power source does not emit pollutants, is growing [1,2]. Accordingly, the automotive industry has focused on the transition from internal combustion engine vehicles to electric vehicles [3,4], and the tire industry is researching tire designs that meet the performance requirements of electric vehicles. Because of the nature of the electric power source, electric vehicles apply high torque to the tires from the beginning of acceleration, and the high battery load requires a better wear resistance for the tires [5,6]. In addition, because of the capacity limitations of the batteries, a dramatic reduction in rolling resistance is required to ensure a sufficiently long driving range [5,7]. Therefore, tire manufacturers are actively trying to apply new materials to improve the performance of tires for electric vehicles [8].

Tire tread is the most crucial element in determining tire performance [9] as it contributes the largest volume fraction [10], protects the carcass and belt layers, and is in direct contact with the road surface. In passenger car radial tires, a blend of solution styrene-butadiene rubber (SSBR) and butadiene rubber (BR) is used as the base rubber [11,12]. To achieve better traction and rolling resistance, conventional carbon black fillers are being replaced with silica/silane systems [13,14]. To enhance the compound's processability, manufacturers add processing aids like TDAE oil [15]. However, adding processing oil is unfavorable for the abrasion resistance and rolling resistance of the compound [16,17]. Additionally, when the tire operates for a long time, a migration problem occurs in which processing oil migrates to the tire surface. This, in turn, reduces the suppleness of the tire tread and its physical properties [18]. Therefore, the demand for liquid butadiene rubber (LqBR) as an alternative to processing oils has begun to grow.

LqBR has a higher molecular weight (1000 to 50,000 g/mol) than processing oils. Due to the presence of double bonds, LqBR acts as a co-vulcanizable plasticizer that can be crosslinked with base rubber [19]. Thus, LqBR shows less migration compared to processing oils, which provides stable long-term performance over extended driving times [20]. Furthermore, by controlling the microstructure of LqBR, the compound's glass transition temperature (T_g) can be regulated. The addition of LqBR with a low T_g reduces the T_g of the compound, enhancing its abrasion resistance [21].

Kuraray Co., Ltd. (Tokyo, Japan) confirmed that the co-vulcanization of non-functionalized LqBR (N-LqBR) with base rubber is possible through toluene extraction experiments [22]. It was also confirmed that the viscoelastic properties in the low-temperature region and abrasion resistance were concurrently enhanced. Continental AG in Germany applied N-LqBR to enhance the abrasion resistance while maintaining the wet traction of tire tread [23]. In contrast, Kitamura et al. reported a deterioration in fuel efficiency due to hysteresis loss, resulting from the free chain ends of LqBR when N-LqBR was applied into the compound [24]. This implies the need to introduce functional groups to limit the chain-end mobility of LqBR.

Responding to the need for F-LqBR, Cray Valley Co., Ltd. (Exton, PA, USA) and Evonik Industries AG (Essen, Germany) have conducted studies on silane-functionalized LqBR and successfully commercialized them [25,26], and the tire industry has reported studies on its application in rubber compounds [27–29]. Kim et al. synthesized F-LqBRs with different functional group positions and applied them to silica-filled rubber compounds to determine the effects of the functional group positions and free chain ends of LqBR on the properties of tire tread compounds [30]. As a result, the functional group of F-LqBRs were bonded on the silica surface, so there was no free chain end effect. Thus, it was confirmed that the addition of F-LqBR, which functionalizes at both ends of the chain, can concurrently enhance the fuel efficiency and abrasion resistance of a compound.

Prior studies have investigated the applicability of LqBR and the effects of functionalization and the macro- and microstructure of the polymer and verified the superiority of both chain-end functionalized LqBR. However, since the synthesis of telechelic polymers requires advanced technics related to initiator design and molecular weight control [31,32], there have been no previous studies on the effect of the number of end-functional groups of F-LqBR on the compound properties.

Therefore, we designed and synthesized a functional initiator to operate during the polymerization step [33,34], which is one of the methods for synthesizing polymers with a telechelic structure. Considering that the initiation mechanism, initiation efficiency, concentration, and solubility of the initiator can affect polymerization [35–37], we designed an initiator structure suitable for the polymerization of 1,3-butadiene. Furthermore, we synthesized F-LqBRs with end functionalities of two or four triethoxysilyl groups. These F-LqBRs were then used as processing aids for preparing silica-filled rubber compounds to evaluate the effect of the number of end-functional groups of LqBR on the compound's properties. This research provides a basis for the selection of optimized end-functional groups and F-LqBR structures, toward improving both the wear resistance and fuel efficiency of tires required for electric vehicles, while ensuring long-term performance.

2. Materials and Methods

2.1. Synthesis of the Functional Initiator

2.1.1. Synthesis of (E)-4,4′-(diazene-1,2-diyl)bis-(4-cyano-N-(3-(triethoxysilyl)propyl)pentanamide) (difunctional initiator)

4,4′-azobis(4-cyanovaleric acid) (ACVA, 98%, Sigma-Aldrich Corp.; Seoul, Republic of Korea), (3-aminopropyl) triethoxysilane (98%, Sigma-Aldrich Corp.; Seoul, Republic of Korea), and phosphorus pentachloride (98%, SigmaAldrich Corp, Seoul, Republic of Korea) were used as reagents, and dichloromethane (DCM, 99%, Duksan General Chemical Co.; Seoul, Republic of Korea) was used as solvent. For purification, n-hexane (95%, Duksan General Chemical Co.; Seoul, Republic of Korea) and diethyl ether (99%, Daejung Chemicals & Metals Co.; Siheung, Republic of Korea) were used.

2.1.2. Synthesis of (E)-4,4′-(diazene-1,2-diyl)bis(4-cyano-N,N-bis(3-(triethoxysilyl)propyl) pentanamide) (tetrafunctional initiator)

ACVA, bis[3(triethoxysilyl)propyl]amine (95%, Gelest Inc.; Morrisville, PA, USA), phosphorus pentachloride (98%, Sigma-Aldrich Corp, Seoul, Republic of Korea), and triethylamine (TEA, 98%, Duksan General Chemical Co.; Seoul, Republic of Korea) were used as reagents, and dichloromethane (DCM, 99%, Duksan General Chemical Co.; Seoul, Republic of Korea) was used as a solvent. N-hexane (95%, Duksan General Chemical Co.; Seoul, Republic of Korea) and diethyl ether (99%, Daejung Chemicals & Metals Co.; Republic of Korea) were used for purification.

2.1.3. Polymerization

For polymerization, the reaction mixture was dissolved in tetrahydrofuran (THF, 99.9%, Samchun Chemical Co.; Seoul, Republic of Korea). The functional initiators were prepared as described above. Radical polymerization was performed using 1,3-butadiene (Kumho Petrochemical Co., Ltd.; Daejeon, Republic of Korea), which was used as received without further purification.

2.1.4. Compounding

SSBR (SOL-5220M, Kumho Petrochemical Co. Daejeon, Republic of Korea, styrene content: 26.5 wt%, vinyl content: 26 wt%, non-oil extended) and high-cis butadiene rubber (CB24, Lanxess Chemical Industry Co., Ltd.; Cologne, Germany; cis content: 96 wt%) were used as the base rubber. Silica (ZEOSIL 195MP, Solvay Silica Korea Co., Ltd.; Gunsan, Republic of Korea) was used as a filler. The silane coupling agent was X50-S (Evonik Industries AG, Essen, Germany; bis-[3-(triethoxysilyl)propyl]tetrasulfide (TESPT) 50%, carbon black N330 50%). Zinc oxide (Sigma-Aldrich Corp.; Seoul, Republic of Korea) and stearic acid (Sigma-Aldrich Corp.; Seoul, Republic of Korea) were used as crosslinking activators. N-(1,3-dimethybutyl)-N′-phenyl-phenylenediamine (6PPD, Kumho Petrochemical Co., Ltd.; Daejeon, Republic of Korea) and 2,2,4-trimethyl-1,2-dihydroquinoline (TMQ, Sinopec Corp.; Beijing, China) were used as antioxidants. Sulfur (Daejung Chemicals & Metals Co.; Siheung, Republic of Korea) was used as a crosslinking agent, while cyclohexyl benzothiazole-2-sulfenamide (CBS; 98%, Tokyo Chemical Industry Co., Ltd.; Tokyo, Japan) and 1,3-diphenylguanidine (DPG, 98%, Tokyo Chemical Industry Co., Ltd.; Tokyo, Japan) were used as vulcanization accelerators. For processing aids, four different substances were prepared: TDAE oil (Kukdong Oil & Chemicals Co., Yangsan, Republic of Korea) and N-LqBR (LBR-307, Kuraray Co., Ltd.; Tokyo, Japan), which are commercially available, and 2-Azo-LqBR and 4-Azo-LqBR, which were synthesized using functional initiators (as explained in detail in Section 2.4).

2.2. Characterization

2.2.1. Gel Permeation Chromatography (GPC)

GPC (Shimadzu, Kyoto, Japan) was used to determine the number average molecular weight of the synthesized polymer. A solvent delivery system, a refractive index detector, and columns were used for the GPC experiment. The columns used were as follows: a HT 6E (10 µm, 7.8 mm × 300 mm), a HMW 7 column (15–20 µm, 7.8 mm × 300 mm), and a HMW 6E column (15–20 µm, 7.8 mm × 300 mm). Polybutadiene standards (Kit Poly(1,3-butadiene) number average molecular weight (M_n) standard, WAT035709, Waters Corp.; Eschborn, Germany) were used for molecular weight calibration.

2.2.2. Proton Nuclear Magnetic Resonance Spectroscopy (^{1}H NMR)

The molecular structure of LqBR was confirmed by ^{1}H NMR spectroscopy (Varian, Unity Plus 300 spectrometer, Garden State Scientific, Morristown, NJ, USA), using deuterochloroform ($CDCl_3$, Cambridge Isotope Laboratories, Inc.; Andover, MA, USA) as the solvent.

2.2.3. Differential Scanning Calorimetry (DSC)

T_g was confirmed using a DSC (DSC-Q10, TA Instruments, New Castle, DE, USA). DSC measurements were performed at temperatures ranging between -100 and $100\ ^\circ C$ at a heating rate of $10\ ^\circ C$/min. After the first heating-cooling cycle, T_g was obtained during the second heating.

2.2.4. Payne Effect

Strain sweep tests (0.28–40.04%) at $60\ ^\circ C$ were performed using a rubber processing analyzer (RPA 2000, Alpha Technologies, Hudson, Ohio, USA) to analyze the degree of filler network formation in the unvulcanized compound. As the strain increases, the filler network is destroyed and the storage modulus (G') decreases. Thus, $\Delta G'$ (G' at 0.28% minus G' at 40.04%) shows the degree of filler-filler interaction, as described by the Payne effect [38].

2.2.5. Mooney Viscosity

A Mooney rotatory viscometer (Vluchem IND Co.; Seoul, Republic of Korea) was used for measuring the Mooney viscosity, which is correlated to the processability of a compound. Measurements were performed using a large rotor (diameter 38.10 ± 0.03 mm, thickness 5.54 ± 0.03 mm) at $100\ ^\circ C$ and 2 rpm for 4 min after preheating for 1 min according to the ASTM D1646 conditions.

2.2.6. Curing Characteristics

A moving die rheometer (RLR-3-rotorless rheometer, Toyoseiki, Tokyo, Japan) was used for the analysis of the curing characteristics of the compounds at $160\ ^\circ C$ for 30 min under an angular displacement of $\pm 1^\circ$. In this experiment, the minimum torque (T_{min}), maximum torque (T_{max}), and optimal cure time (t_{90}) were obtained. Then, t_{90} was used for preparing the vulcanizates in a heating press at $160\ ^\circ C$.

2.2.7. Solvent Extraction and Crosslink Density

The vulcanizates, with pieces of 10 mm (length) × 10 mm (width) × 2 mm (thickness), were prepared and weighed. The specimen was immersed in THF and n-hexane at $25\ ^\circ C$ for 2 days each to eliminate the organic additives. The samples were then dried at $25\ ^\circ C$ for 1 day and weighed, and the mass fraction of the extracted organic additives was calculated. The samples were then swollen in toluene at $25\ ^\circ C$ for 1 day and weighed again. The crosslink density was calculated from the measured weight of the sample using the Flory–Rehner equation below [39–41]:

$$\nu = \frac{1}{2M_c} = -\frac{\ln(1-V_1) + V_1 + \chi V_1^2}{2\rho_r V_0 \left(V_1^{\frac{1}{3}} - \frac{V_1}{2}\right)} \quad (1)$$

where ν is the crosslink density of vulcanizates (mol/g), M_C is the average molecular weight between crosslink points (g/mol), V_1 is the volume fraction of rubber in the swollen gel at equilibrium, V_0 is the molar volume of solvent (cm^3/mol), ρ_r is the density of the rubber (g/cm^3), and χ is the polymer–solvent interaction parameter.

2.2.8. Mechanical Properties

The properties of dumbbell-shaped specimens of 100 mm (length) × 25 mm (width) made according to ASTM D 412 were estimated using a universal testing machine (UTM, KSU-05M-C, KSU Co., Ansan, Republic of Korea). The tensile test for evaluating the mechanical properties of the vulcanizates was conducted at 500 mm/min and repeated three times, and the median value was used according to the ASTM standard.

2.2.9. Abrasion Resistance

The abrasion resistance test was conducted using a cylindrical sample with diameters of 16 mm and thicknesses of 8 mm. First, the initial mass was measured, and the specimen was ground for 40 s at a speed of 40 rpm under a load of 5 N using a Deutsche Industrie Normen (DIN) abrasion tester. The mass loss was then calculated by measuring the mass of the specimen.

2.2.10. Dynamic Viscoelastic Properties

The dynamic viscoelastic properties were obtained using a dynamic mechanical analyzer (DMA; Q800, TA Instrument, New Castle, DE, USA) and a dynamic mechanical thermal spectrometer (DMTS; EPLEXOR 500N, GABO Instruments GmbH, Ahlden, Germany). First, the T_g and dynamic viscoelastic properties of the vulcanizates were measured over a temperature range of -80 °C to 80 °C using DMA. Then, the DMTS was used to measure tan δ through strain sweeps from 0.5% to 10% in tension mode at a temperature of 60 °C and a frequency of 10 Hz.

2.3. Synthesis of the Functionalized Initiator

2.3.1. Di-functional Initiator (2-Azo-initiator)

The synthesis of the di-functional initiator substituted on both sides by silane is shown in Scheme 1. One equivalent of ACVA (5 g, 17.8 mmol) was dissolved in 150 mL of dichloromethane under an Ar atmosphere at 25 °C, and then cooled to 0 °C. Then, 2.5 equivalents of phosphorus pentachloride (9.3 g, 44.6 mmol) were added for 30 min and then stirred at 25 °C for 1 h. A measure of 120 mL of dichloromethane was removed using vacuum distillation, 70 mL of hexane was added, and recrystallization at 0 °C produced €-′,4′-(diazene-1,2-diyl)bis(4-cyanopentanoyl chloride) as a white solid. Then, 1 equivalent (4.5 g, 14.2 mmol) of the previously obtained €(′)-4,4′(diazene-1,2-diyl)bis(4-cyanopentanoyl chloride) was dissolved in 35 mL of dichloromethane under Ar at 25 °C. Furthermore, 5 equivalents of (3-aminopropyl)triethoxysilane (15.7 g, 70.9 mmol) dissolved in 35 mL of dichloromethane were added dropwise over 30 min at 0 °C and then stirred at 25 °C for 16 h. The resulting solid was washed with diethyl ether to remove dichloromethane using vacuum distillation to produce a white, solid product (2-Azo-initiator).

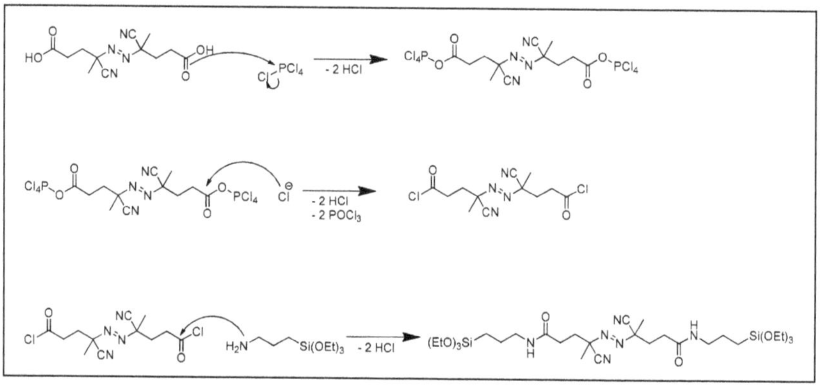

Scheme 1. Synthesis of difunctional initiator.

2.3.2. Tetra-Functional Initiator (4-Azo-initiator)

The synthesis of the tetra-functional initiator, substituted on both sides by two pairs of silanes, is shown in Scheme 2. One equivalent of ACVA (5 g, 17.8 mmol) was dissolved in 150 mL of dichloromethane at 25 °C under Ar and then cooled to 0 °C. Then, 2.5 equivalents of phosphorus pentachloride (9.3 g, 44.6 mmol) were added over 30 min, and the mixture was then stirred at 25 °C for 1 h. Using vacuum distillation, 120 mL of dichloromethane was removed using vacuum distillation, 70 mL of hexane was added, and recrystallization at 0 °C produced (E)-4,4′-(diazene-1,2-diyl)bis(4cyanopentanoyl chloride) as a white solid. Then, one equivalent (4.5 g, 14.2 mmol) of the previously obtained (E)-4,4′-(diazene-1,2-diyl)bis(4-cyanopentanoyl chloride) was dissolved in 30 mL of dichloromethane under Ar at 25 °C. Next, 2.1 equivalents (12.7 g, 29.8 mmol) of bis[3-(triethoxysilyl)propyl]amine in 35 mL of dichloromethane were slowly added dropwise at 0 °C, followed by the dropwise addition of four equivalents (5.77 g, 57 mmol) of triethylamine. Then, the mixture was stirred at 25 °C for 14 h. Using vacuum distillation, dichloromethane was evaporated, and the resulting salt was removed with diethyl ether. Finally, n-hexane was added to remove the resulting salt to produce the product (4-Azo-initiator) as an orange liquid.

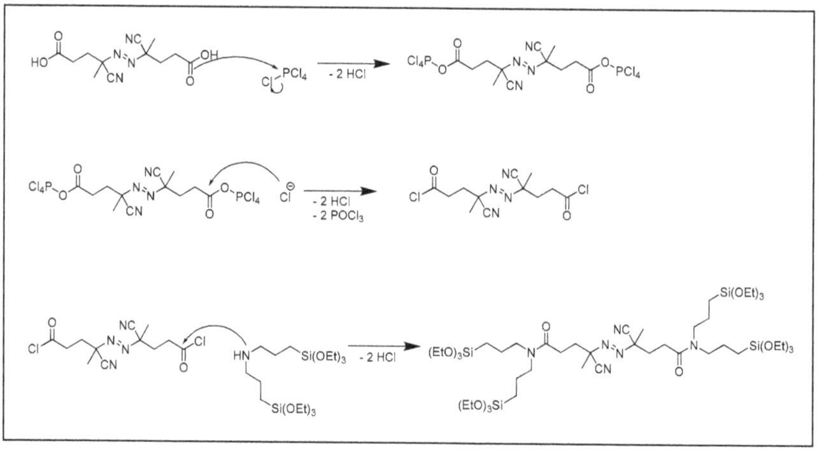

Scheme 2. Synthesis of tetrafunctional initiator.

2.4. Synthesis of Functionalized LqBR

Radical Polymerization

F-LqBRs were polymerized by radical polymerization in a high-pressure stainless-steel reactor (1L), and the formulation is shown in Table 1. First, the functional initiator (2.183 g of 2-Azo-LqBR or 1.217 g of 4-Azo-LqBR) and THF were filled in the reactor. Then, 60 g of 1,3-butadiene was charged into the reactor using a chamber under an atmosphere of N_2. The 2-Azo-LqBR and 4-Azo-LqBR were polymerized at 75 °C for 8 h and at 80 °C for 3 h, respectively. The target molecular weight of 4-Azo-LqBR was set considering the molecular weight of the functional groups and the consequent increase in hydrodynamic volume. To achieve this, 4-Azo-LqBR was polymerized at a slightly higher temperature than 2-Azo-LqBR. Subsequently, the mixture was cooled, and the unreacted 1,3-butadiene was discharged using the eject line of the reactor. The solvent was removed through evaporation and precipitated in ethanol to remove residual initiator. As a final step, centrifugation was used to collect the polymers.

Table 1. Formulation of F-LqBRs (unit: g).

	Organic Compounds	2-Azo-LqBR	4-Azo-LqBR
Di-functional initiator	2-Azo-initiator	2.183	-
Tetra-functional initiator	4-Azo-initiator	-	1.217
Monomer	1,3-butadiene	60	60
Solvent	THF	420	300

The growing chains can be terminated by coupling or disproportionation in the absence of a chain transfer agent. Since 1,3-butadiene is mainly terminated via coupling [33,34], triethoxysilyl groups were introduced at both ends of a radical initiator. Consequently, silane-terminated polybutadiene was synthesized by using the triethoxysilyl-functionalized initiator (Scheme 3) and the molecular structures of the LqBRs were confirmed using GPC and a ^{1}H NMR spectra.

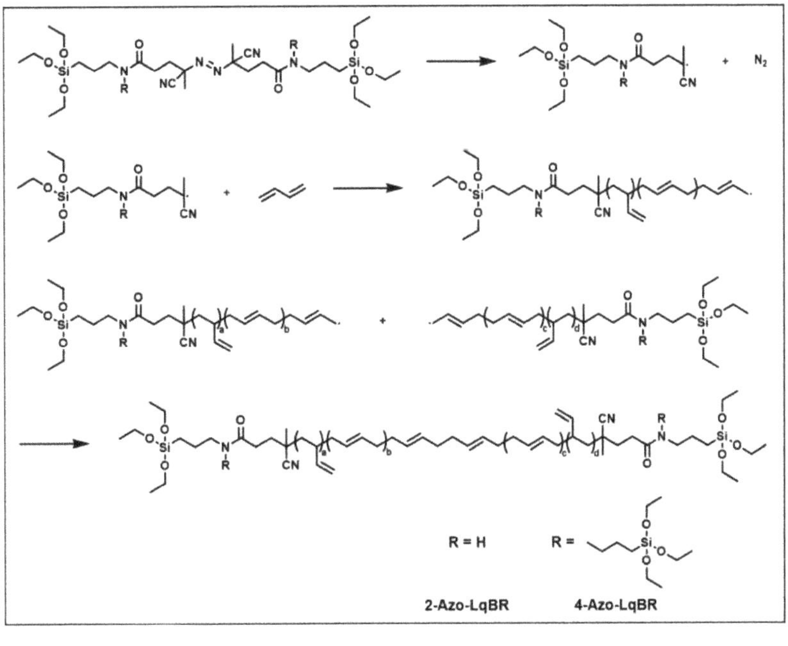

Scheme 3. Polymerization of F-LqBRs.

2.5. Manufacture of Rubber/Silica Compounds and Vulcanizates

To determine the effect of LqBRs substituted with TDAE oil on the compound's mechanical and dynamic properties, samples were prepared using the formulations shown in Table 2. In this experiment, the compounds were named according to the processing aid used (TDAE oil, N-LqBR, 2-Azo-LqBR, 4-Azo-LqBR). Compounds were manufactured using an internal kneader (300cc, MIRAESI Company, Gwangju, Republic of Korea) with a fill factor of 0.7, and the three kinds of LqBRs were introduced by replacing 10 phr of the 40 phr of TDAE oil. The kneading process began at 110 °C and continued for 7 min and 40 s. Mixing was then started at 50 °C and continued for 2 min. After each stage of mixing, the compound was sheeted using a two-roll mill. A moving die rheometer was used to measure the optimal cure times of the compounds at 160 °C, and the optimal cure time was used to produce vulcanizates in a press at 160 °C. The detailed process is shown in Table 3.

Table 2. Formulation of compounds (unit: phr).

Sample	TDAE Oil	N-LqBR	2-Azo-LqBR	4-Azo-LqBR
SSBR	80	80	80	80
BR	20	20	20	20
Silica	120	120	120	120
X50S	20	20	20	20
DPG	2	2	2	2
TDAE oil	40	30	30	30
N-LqBR	-	10	-	-
2-Azo-LqBR	-	-	10	-
4-Azo-LqBR	-	-	-	10
Wax	1	1	1	1
TMQ	1	1	1	1
ZnO	3	3	3	3
Stearic acid	1	1	1	1
6PPD	2	2	2	2
Sulfur	1.3	1.3	1.3	1.3
CBS	1.6	1.6	1.6	1.6
ZBEC	0.1	0.1	0.1	0.1

Notes: phr, parts per hundred rubber.

Table 3. Mixing procedures.

	Time, mins	RPM	Action
Silica masterbatch (SMB) mixing	00:00–00:40	20	Add rubber (initial temp.: 110 °C)
	00:40–01:40	40	Add silica $^1/_2$, X50S $^1/_2$, DPG $^1/_2$, processing aid $^1/_2$
	01:40–02:40	40	Add silica $^1/_2$, X50S $^1/_2$, DPG $^1/_2$, processing aid $^1/_2$
	02:40–05:00	60	Add additives
	05:00–05:30	60	Ram up
	05:30–07:40	50	Extra mixing and dump (dump temp.: 150–155 °C)
Final masterbatch (FMB) mixing	00:00–00:20	20	Add SMB (initial temp.: 50 °C)
	00:20–02:00	40	Add sulfur, CBS, ZBEC and dump (dump temp.: 80–90 °C)

3. Results and Discussion

3.1. Synthesis of Functional Initiators

The ^{1}H NMR peak assignments of the di-functional and tetra-functional initiators are shown in Figure 1a,b, respectively. The chemical shift (δ) was determined using tetramethylsilane as an internal standard, and the following peaks were assigned.

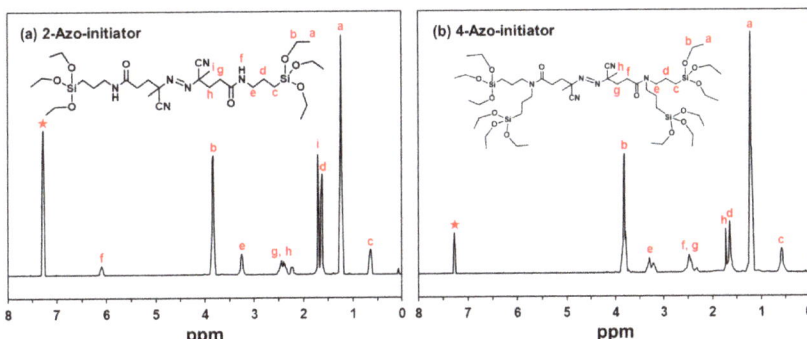

Figure 1. ^{1}H NMR spectra of the (**a**) 2-Azo-initiator, (**b**) 4-Azo-initiator (★ CDCl$_3$ solvent).

Di-functional initiator: ^{1}H NMR (CDCl$_3$): δ ppm 0.64 (t, 4H, Si-**CH$_2$**-), 1.23 (t, 18H, SiO-CH$_2$-**CH$_3$**), 1.64 (q, 4H, Si-CH$_2$-**CH$_2$**), 1.70 (s, 6H, C-**CH$_3$**), 2.20–2.52 (m, 8H, CO-CH$_2$-**CH$_2$**, CO-**CH$_2$**), 3.25 (m, 4H, NH-**CH$_2$**), 3.84 (q, 12H, SiO-**CH$_2$**-), 6.09 (t, 2H, CO-**NH**).

Tetra-functional initiator: ^{1}H NMR (CDCl$_3$): δ ppm 0.64 (t, 4H, Si-**CH$_2$**-), 1.23 (t, 18H, SiO-CH$_2$-**CH$_3$**), 1.62 (q, 4H, Si-CH$_2$-**CH$_2$**), 1.70 (s, 6H, C-**CH$_3$**), 2.16–2.52 (m, 8H, CO-CH$_2$-**CH$_2$**, CO-**CH$_2$**), 3.52 (m, 4H, N-**CH$_2$**), 3.84 (q, 12H, SiO-**CH$_2$**-).

3.2. Synthesis of F-LqBRs

The GPC curves and ^{1}H NMR spectra of the F-LqBRs are shown in Figures 2 and 3 and are summarized with the DSC results in Table 4. The GPC measurements showed that the 2-Azo-LqBR and 4-Azo-LqBR have M$_n$ values of 4,000 and 5,100 (g/mol) and polydispersity indexes of 1.20 and 1.32, respectively. We regulated the reaction time and temperature conditions according to the target molecular weight, and as a result, the M$_n$ and polydispersity index of 4-Azo-LqBR increased compared to 2-Azo-LqBR at the higher temperature. Resonance peaks corresponding to the 1,4-addition (cis, trans), 1,2-addition (vinyl) structure of 1,3-butadiene appear at 5.3–5.5 ppm, 5.5–5.6 ppm, and 4.8–5.0 ppm, respectively. The vinyl contents of the LqBRs were determined according to the proportion of the vinyl peaks to the total 1,3-butadiene peaks. In addition, the chemical shift of the ethoxy groups in alkoxysilane showed a triplet peak at 1.2–1.3 ppm (SiO-CH$_2$-**CH$_3$**) and a quartet peak at 3.7–3.8 ppm (SiO-**CH$_2$**-) [42]. Hence, the functionality of F-LqBRs, which represents the proportion of the chain containing the triethoxysilyl groups to the total number of chains, can be determined by calculating the proportion of the 1,2-vinyl-H integrals in polybutadiene and SiO-**CH$_2$**- peak integrals [43]:

$$\frac{S_{Vinyl\text{-}H}}{S_{Alkoxysilane\text{-}H}} = \frac{2 \times \left(R_{Vinyl}\right) \times \left(\frac{M_n}{M_B}\right)}{n_{Alkoxysilane} \times F} \quad (2)$$

where $S_{Vinyl\text{-}H}$ and $S_{Alkoxysilane\text{-}H}$ are the peak integrals of 1,2-vinyl-H and alkoxysilane-H, respectively; R_{Vinyl} is the vinyl content of LqBR; M_n is the number average molecular weight of LqBR; M_B is the molecular weight of 1,3-butadiene; $n_{alkoxysilane}$ is the hydrogen atom number in alkoxysilane, i.e., -Si-(OCH$_2$CH$_3$)$_3$; and F is the functionality (triethoxysilyl groups per chain), i.e., "2" means di-functionalized at the ends of macromolecular chains and "4" means tetra-functionalized at the ends of macromolecular chains.

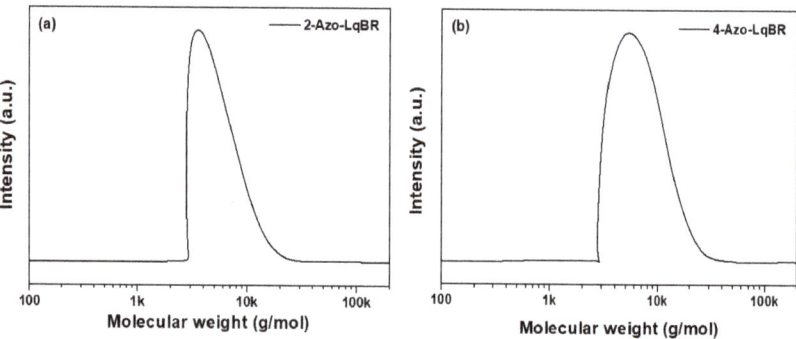

Figure 2. Gel permeation chromatograms of LqBRs: (**a**) 2-Azo-LqBR, (**b**) 4-Azo-LqBR.

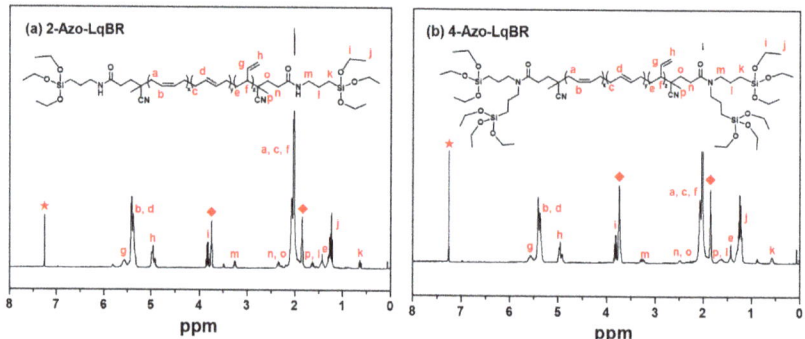

Figure 3. ^{1}H NMR spectra: (**a**) 2-Azo-LqBR, (**b**) 4-Azo-LqBR (★ CDCl$_3$, ♦ THF).

Table 4. Characteristics of LqBRs.

Property	Unit	N-LqBR	2-Azo-LqBR	4-Azo-LqBR
Sample M_n	g/mol	4400	4000	5100
Polydispersity index (Đ)	-	1.04	1.20	1.32
Vinyl content (% in BD)	-	15	19	19
T_g	°C	−95	−86	−78
End functionality (Si per chain)	-	0	2.3	4.3

3.3. Payne Effect

The Payne effect is used to measure the degree of formation of a filler network. When the filler network is destroyed due to increased strain, the smaller difference (ΔG′) in G′ indicates weaker filler-filler interaction. Using this test, filler dispersion in the rubber matrix can be determined. Furthermore, the hydrodynamic effect, filler-polymer interaction, polymer network, and rubber structure affect the G′ of the compound under high strain where the filler network breaks [44].

Figure 4 and Table 5 exhibit the Payne effect measurements of the compounds. LqBR exhibited better mixing efficiency, due to its higher viscosity, than that of TDAE oil [19,30]. The addition of LqBR to compounds led to a reduced Payne effect compared to those containing TDAE oil due to improved silica dispersion. In addition, 2-Azo-LqBR and 4-Azo-LqBR have functional groups which can react with silanol groups and hydrophobize the silica surface. Therefore, the compounds with F-LqBR showed significantly lower Payne effect values than the compounds with N-LqBR.

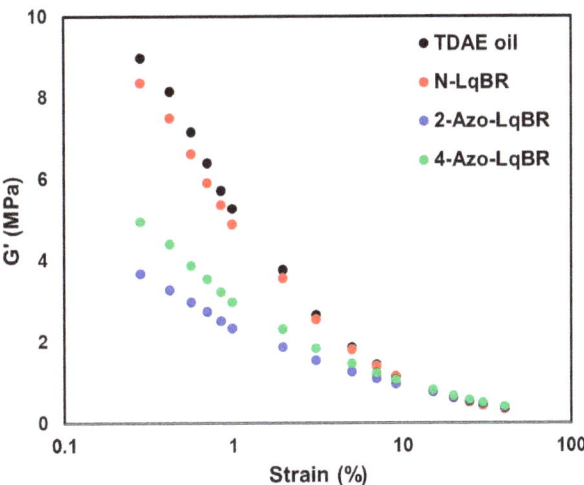

Figure 4. Payne effect in silica masterbatches.

Table 5. G' values (MPa) of compounds.

Property	TDAE Oil	N-LqBR	2-Azo-LqBR	4-Azo-LqBR
G' (at 0.28% strain)	8.98	8.37	3.69	4.96
G' (at 40.04% strain)	0.36	0.36	0.40	0.41
ΔG' (at 0.28–40.04% strain)	8.62	8.01	3.29	4.55

The self-condensation of F-LqBR indicates the formation of a Si-O-Si bond through condensation between alkoxy groups, and there are two types: intramolecular and intermolecular self-condensation. Increasing the number of triethoxysilyl groups in F-LqBR can increase both the reactivity to the filler and to intramolecular and intermolecular self-condensation [45]. Intramolecular condensation can occur within a molecule as the distance between alkoxy groups becomes smaller [46]. The 4-Azo-LqBR compound demonstrated a higher Payne effect than the 2-Azo-LqBR compound. This was due to the higher number of neighboring alkoxy groups within the 4-Azo-LqBR molecule. As a result, the reactivity of 4-Azo-LqBR with silanol groups of the filler is lower than that of 2-Azo-LqBR due to self condensation. The self-condensation effect in 4-Azo-LqBR was also confirmed by the higher G' value under high strain (40.04% strain) compared to that of the 2-Azo-LqBR compound.

3.4. Cure Characteristics and Mooney Viscosity of the Compounds

Table 6 presents the Mooney viscosity results, a measure of compound processability. The introduction of LqBR enhanced silica dispersion in the rubber matrix and reduced Mooney viscosity compared to the TDAE oil compound. Furthermore, F-LqBR addition resulted in even lower Mooney viscosity values due to the surface hydrophobation of silica that further improved its dispersion in the compound. 4-Azo-LqBR is less reactive with silanol groups than 2-Azo-LqBR due to its self-condensation, which results in a smaller improvement in silica dispersion, and a higher Mooney viscosity compared to that of 2-Azo-LqBR.

Figure 5 and Table 6 show the cure characteristics obtained using a moving die rheometer. The cure characteristics results exhibit a remarkably low error of less than 1%. The Δtorque ($T_{max} - T_{min}$) value is closely correlated with the crosslink density of a compound [47,48]. The N-LqBR compound exhibited a lower Δtorque value than the TDAE oil compound because N-LqBR consumed the sulfur required for the crosslinking of the base rubber. In contrast, F-LqBRs not only react with silanol groups on the silica surface, but also strengthen filler–rubber interactions by crosslinking with the base rubber. Accordingly, compounds containing the F-LqBRs exhibited higher Δtorque values than those containing N-LqBR. On the other hand, the 4-Azo-LqBR compound exhibited a lower Δtorque than the 2-Azo LqBR compound, which can be attributed to the lower reactivity of the end-functional groups of 4-Azo-LqBR with silanol groups compared to 2-Azo-LqBR.

Table 6. Cure characteristics and Mooney viscosities of the compounds.

Property	Unit	TDAE Oil	N-LqBR	2-Azo-LqBR	4-Azo-LqBR
Mooney viscosity (ML$_{1+4}$ at 100 °C, FMB)	MU	153	147	116	119
T$_{min}$	N-m	0.88	0.82	0.68	0.77
T$_{max}$	N-m	3.02	2.65	2.88	2.85
Δtorque	N-m	2.14	1.83	2.20	2.09
t$_{10}$	min:sec	1:11	1:05	1:09	1:15
t$_{90}$	min:sec	10:02	11:55	10:40	11:51

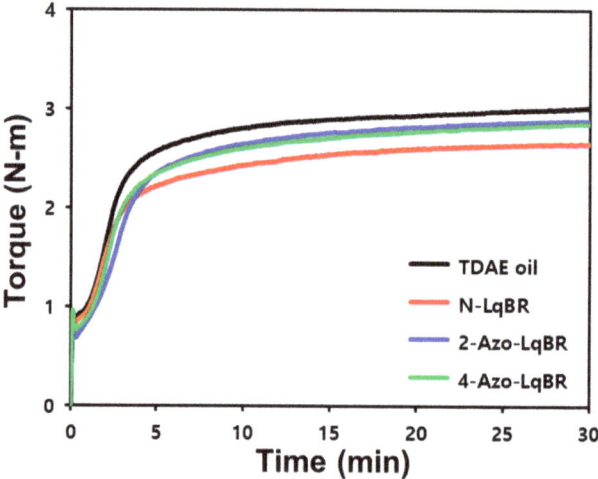

Figure 5. Cure curves of the compounds.

3.5. Solvent Extraction and Crosslink Density

Two organic solvents were used to extract organic compounds from vulcanizates, and the amount of extracted organic compounds in the vulcanizates was obtained. First, THF was used to extract the oil and low molecular weight chemicals, followed by n-hexane to extract un-crosslinked LqBR. Figure 6a and Table 7 show the amount of organic compound extracted by the two solvents. The results of solvent extraction exhibit a remarkably low relative standard deviation of 0.06–1.79%. The extraction rate was highest in the TDAE oil compound, which was 15.66 wt%, and it only acts as a plasticizer and does not form chemical bonds in the compound. However, compounds with LqBRs showed lower extraction rates than the TDAE oil compound because LqBRs form a crosslink with the base rubber and bond to the polymer network. Furthermore, the F-LqBR compounds exhibited lower extraction rates compared to those of the N-LqBR compound because they are more strongly bonded to the silica surface by forming chemical bond. However, the 4-Azo-LqBR compound showed a slightly higher extraction rate than the 2-Azo-LqBR compound due to its lower reactivity with silanol groups. Assuming that the 10 phr (3.41 wt%) of TDAE oil was completely extracted and that other additives were equally extracted (30 phr of TDAE oil 10.23 wt% + 2.02 wt%), 79.5% of N-LqBR, 2.2% of 2-Azo-LqBR, and 4.8% of 4-Azo-LqBR were extracted compared to TDAE oil. This indicates that the addition of F-LqBR could prevent migration and maintain the suppleness of the tire to prevent the degradation of its properties.

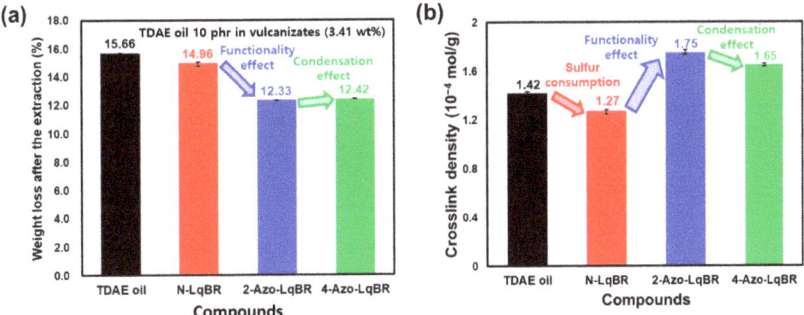

Figure 6. (**a**) Weight loss after solvent extraction and (**b**) crosslink density of the vulcanizates.

Table 7. Weight loss after solvent extraction and crosslink density of the vulcanizates.

Property	Unit	TDAE Oil	N-LqBR	2-Azo-LqBR	4-Azo-LqBR
Weight loss after extraction	%	15.66	14.96	12.33	12.42
Weight loss after extraction in 10 phr of oil and LqBR	%	100	79.5	2.2	4.8
Crosslink density	10^{-4} mol/g	1.42	1.27	1.75	1.65

Table 7 and Figure 6b show the crosslink density of the vulcanizates. The results of crosslink density exhibit a remarkably low relative standard deviation of 0.36–1.88%. The N-LqBR compound exhibited a lower crosslink density than the TDAE oil compound because N-LqBR consumes the sulfur needed for crosslinking with the base rubber instead. However, F-LqBR compounds exhibited a higher crosslink density compared to the TDAE oil compound because of the improved filler–rubber interaction resulting from covalent bonding with silanol groups. The lower crosslink density of the 4-Azo-LqBR compound compared to the 2-Azo-LqBR compound can be attributed to weaker filler–rubber interaction due to self-condensation of 4-Azo-LqBR.

3.6. Mechanical Properties and DIN Abrasion Loss

The tensile modulus determines rubber stiffness, which is higher for a compound when the crosslink density is high [49]. In Figure 7 and Table 8, the N-LqBR compound exhibited lower M_{100} (modulus at 100% elongation) and M_{300} (modulus at 300% elongation) values than the TDAE oil compound due to lower crosslink density resulting from sulfur consumption of N-LqBR. However, the F-LqBR compounds exhibited higher crosslink density than the TDAE oil compound because of improved filler–rubber interactions resulting from chemical bonding with silanol groups, resulting in higher M_{100} and M_{300} values. In contrast, the 4-Azo-LqBR compound exhibited a lower modulus due to lower filler–rubber interactions compared to 2-Azo-LqBR.

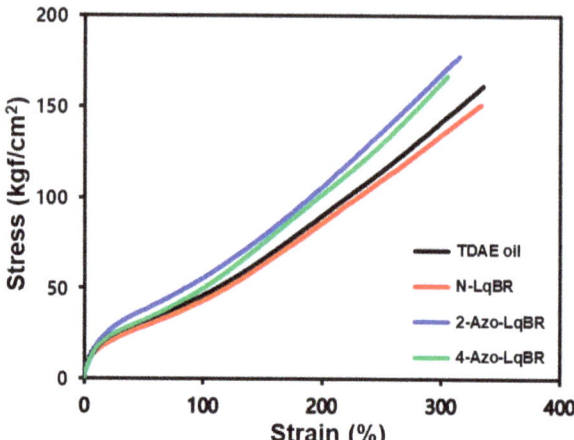

Figure 7. Stress–strain curves of vulcanizates.

Table 8. Mechanical properties and abrasion resistance of vulcanizates.

Property	Unit	TDAE Oil	N-LqBR	2-Azo-LqBR	4-Azo-LqBR
M_{100}	kgf/cm^2	46	43	55	50
M_{300}	kgf/cm^2	142	135	168	163
Elongation at break	%	335	334	315	305
Tensile strength	kgf/cm^2	161	151	178	167
DIN abrasion loss	mg	109	96	91	93

DIN abrasion tests demonstrated that stronger filler–rubber interactions in the compound and lower T_g values of the polymer result in the compound having higher abrasion resistance [50,51]. The compounds were manufactured using the same base rubber with the addition of processing aids with different T_g values. Thus, the compound with TDAE oil, which demonstrated no chemical reaction and had the highest T_g (−50 °C to −44 °C [52]), had the worst abrasion resistance. In contrast, the LqBRs, with lower T_g values than TDAE oil, resulted in the compound having superior abrasion resistance. F-LqBRs, which can interact with silanol groups, had higher T_g values than the N-LqBR, but the corresponding compounds showed excellent abrasion resistance due to improved filler–rubber interactions. In particular, the compound with 2-Azo-LqBR showed the best abrasion resistance.

3.7. Dynamic Viscoelastic Properties

A tire undergoes repeated deformation and recovery when it rotates under the load of a vehicle, resulting in energy loss called hysteresis loss. On icy roads, the tire tread should deform easily, and a large contact area is required for good snow traction [53]. The dynamic viscoelastic properties show that a lower E' in the low-temperature region results in the tire having better snow traction [54,55]. Tan δ at 60 °C indicates the hysteresis loss and is a measure of the destruction and reformation of the filler network under strain. A higher tan δ at 60 °C is disadvantageous for the rolling resistance of the compound [56]. When the filler is well dispersed and filler–rubber interaction is stronger, it results in a reduced tan δ at 60 °C [19,56].

Figure 8 and Table 9 show tan δ values as a function of temperature, measured in our samples using DMA at 0.2% strain. The results indicate that LqBR compounds exhibit lower E' at −30 °C than the TDAE oil compound. This is due to the effective filler volume fraction being reduced by the improved silica dispersion in the LqBR-containing compounds, as confirmed by the Payne effect results [57]. The N-LqBR compound exhibited a higher tan δ at 60 °C than the TDAE oil compound due to the hysteresis loss caused by the free chain ends of N-LqBR. On the other hand, F-LqBR compounds exhibited lower tan δ at 60 °C values than the TDAE oil compound due to forming chemical bonds between the F-LqBR chain ends and the silica surface. In particular, the 2-Azo-LqBR compound exhibited the lowest tan δ at 60 °C because it had the lowest degree of filler network formation and the strongest bonding to the silica surface due to superior silica hydrophobization compared to the 4-Azo-LqBR compound.

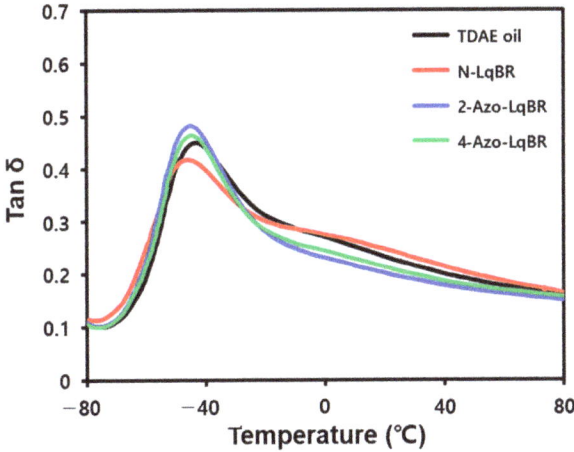

Figure 8. Temperature-dependent tan δ at 0.2% strain curves for the various vulcanizates.

Table 9. Viscoelastic properties of the vulcanizates.

Property	Unit	TDAE Oil	N-LqBR	2-Azo-LqBR	4-Azo-LqBR
Number end functional groups	-	N/A	0	2	4
T_g	°C	−43.1	−45.9	−44.6	−44.8
E' at −30 °C	MPa	164	163	137	152
Tan δ at 60 °C (0.2% strain, temperature sweep)	-	0.177	0.188	0.165	0.170
Tan δ at 60 °C (5% strain, strain sweep)	-	0.196	0.207	0.176	0.190

Figure 9 and Table 9 show tan δ values as a function of strain, measured using DMTS at 60 °C. In the high-strain region, the filler network is destroyed to a greater degree than in the low-strain region, resulting in a higher degree of hysteresis [58,59]. Therefore, better silica dispersion results in less filler network formation, thus decreasing the tan δ at 60 °C in the high-strain region. Moreover, in general, compounds with high crosslink density exhibit low tan δ values at 60 °C [48]. Compounds with LqBR formed fewer filler networks due to better silica dispersion compared to the TDAE oil compound. However, the N-LqBR compound exhibited a lower crosslink density due to sulfur consumption, resulting in a higher tan δ at 60 °C compared to the TDAE oil compound. In contrast, the F-LqBR compounds exhibited lower tan δ values at 60 °C in the high-strain region due to a lower hysteresis loss. This is attributed to their excellent silica dispersion, and the 2-Azo-LqBR compound exhibited the lowest tan δ of all samples.

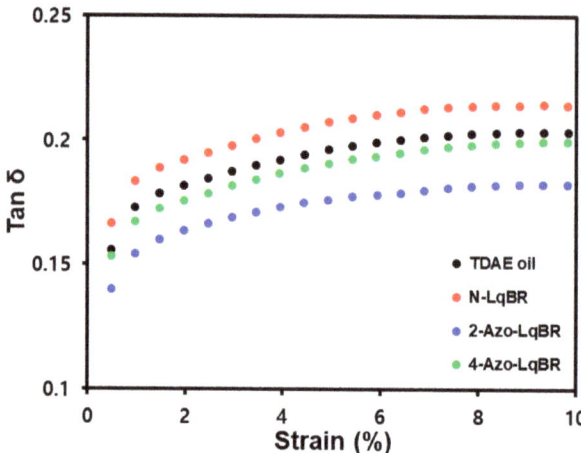

Figure 9. Strain-dependent tan δ at 60 °C curves of the various vulcanizates.

4. Conclusions

The objective of this experiment was to investigate the effect of the number of end-functional groups of LqBR used as a processing aid in silica-filled rubber compounds on the compound properties. To achieve this, N-LqBR and F-LqBRs were used as substitutes for TDAE oil to prepare silica-filled rubber compounds, and the properties were evaluated.

The experimental results showed that LqBR acts as a processing aid and forms crosslinks with the base rubber, improving migration resistance. In particular, 2-Azo-LqBR and 4-Azo-LqBR with functional groups formed chemical bonds with the silica surface, showing better silica dispersion for N-LqBR. However, the 4-Azo-LqBR compound, with an end-functionality of four, demonstrated poor silica dispersion compared to the 2-Azo-LqBR compound due to increased self-condensation.

As a result of the sulfur consumption of LqBR, the crosslink density of the N-LqBR compound was lower compared to that of the TDAE oil compound. Despite the sulfur consumption, 2-Azo-LqBR and 4-Azo-LqBR showed a strong interaction between the filler and base rubber due to the functional groups, resulting in higher crosslink density than that of the TDAE oil compound. Crosslink density had a strong correlation with the mechanical properties of the compounds. The T_g of the processing aids and the strength of the filler–rubber interaction determined the abrasion resistance. The dynamic viscoelastic properties showed that improvements in the snow traction and rolling resistance of the compound could be expected by applying the developed F-LqBRs. Rolling resistance is mainly dependent on hysteresis loss due to the free chain end of the polymer, the filler network destruction and reformation, and the filler–rubber interaction. Consequently, the compound with 2-Azo-LqBR, which had the strongest reactivity with the silica surface, exhibited the lowest tan δ at 60 °C because of the improved silica dispersion and filler–rubber interaction. Accordingly, 2-Azo-LqBR showed the best performance among the studied processing aids, and the optimized end functionality of triethoxysilyl group of F-LqBR is two.

This study was novel in that we controlled the end functionality of telechelic polymer, which requires a high level of technics, and confirmed the effect of reactivity according to the number of end-functional groups of LqBR on the physical properties of the tire tread compound. The results obtained in this study suggest the guideline of designing F-LqBR structures that concurrently improve fuel efficiency and wear resistance for electric vehicles that require improved performances and performance sustainment and prevent deterioration of physical properties over time.

Author Contributions: Data curation, S.S., H.C., J.J. and S.K.; formal analysis, S.S., H.C., M.K. (Myeonghee Kwon), M.K. (Minji Kim) and D.K.; investigation, S.S., H.C., J.J., S.K. and D.K.; methodology, S.S., H.C., M.K. (Myeonghee Kwon) and M.K. (Minji Kim); visualization, S.S., H.C., M.K. (Myeonghee Kwon), M.K. (Minji Kim) and D.K.; supervision, W.K.; validation, H.J., H.-j.P., S.C. and W.K.; writing—original draft, S.S., H.C., J.J. and S.K.; writing—review and editing, H.J., H.-j.P., S.C. and W.K. All authors have read and agreed to the published version of the manuscript.

Funding: This research received no external funding.

Institutional Review Board Statement: Not applicable.

Informed Consent Statement: Not applicable.

Data Availability Statement: Data presented in this study are available upon request from the corresponding author.

Acknowledgments: This work was supported by the Minister of Trade, Industry, and Energy Grant funded by Republic of Korean Government. (project number 20010851) H.J. also thanks to the research grant by the Excellent researcher support project of Kwangwoon University in 2021.

Conflicts of Interest: Author Donghyuk Kim was employed by the company Hankook Tire & Technology Co., Ltd. The remaining authors declare that the research was conducted in the absence of any commercial or financial relationships that could be construed as a potential conflict of interest.

References

1. Sadeghian, S.; Wintersberger, P.; Laschke, M.; Hassenzahl, M. Designing Sustainable Mobility: Understanding Users' Behavior. In Proceedings of the 14th International Conference on Automotive User Interfaces and Interactive Vehicular Applications, Ingolstadt, Germany, 17–20 September 2022; pp. 34–44.
2. Köhler, J.; Whitmarsh, L.; Nykvist, B.; Schilperoord, M.; Bergman, N.; Haxeltine, A. A transitions model for sustainable mobility. *Ecol. Econ.* **2009**, *68*, 2985–2995. [CrossRef]
3. Geronikolos, I.; Potoglou, D. An exploration of electric-car mobility in Greece: A stakeholders' perspective. *Case Studies Trans. Pol.* **2021**, *9*, 906–912. [CrossRef]
4. Kley, F.; Lerch, C.; Dallinger, D. New business models for electric cars—A holistic approach. *Energy Policy* **2011**, *39*, 3392–3403. [CrossRef]
5. Whitby, R.D. Electric cars and winter tires. *Tribol. Lubr. Technol.* **2022**, *78*, 80.
6. Mirzanamadi, R.; Gustafsson, M. *Users' Experiences of Tyre Wear on Electric Vehicles: A Survey and Interview Study*; Statens väg-och transportforskningsinstitut: Linköping, Sweden, 2022.
7. Chan, C.C. An overview of electric vehicle technology. *Proc. IEEE* **1993**, *81*, 1202–1213. [CrossRef]
8. Paik, H.J.; Kim, W.; Yeom, G.; Kim, D.; Choi, H.; Song, S. Rubber Composition for Manufacturing Tires Comprising Terminally Modified Liquid Polybutadienes. K.R. Patent 102022-0032425, 16 March 2022.
9. Han, S.C.; Choe, S.J.; Han, M.H. Compounding technology of silica-filled rubber. *Rubber Technol.* **2001**, *2*, 100–116.
10. Lee, J.W.; Chung, C.B.; Choi, I.C. Severity factors affecting tire wear. *Polymer* **2005**, *29*, 48–53.
11. Malas, A.; Pal, P.; Das, C.K. Effect of expanded graphite and modified graphite flakes on the physical and thermo-mechanical properties of styrene butadiene rubber/polybutadiene rubber (SBR/BR) blends. *Mat. Des.* **2014**, *55*, 664–673. [CrossRef]
12. Waddell, B.W.; Lee, S.; Lin, T.Y.; Yang, E. Factors influencing silica's effectiveness in PCR tires. *Rubber Plast. News* **2019**, 16–19.
13. Cichomski, E.M. Silica-Silane Reinforced Passenger Car Tire Treads. Ph.D. Thesis, University of Twente, Enschede, The Netherlands, 2015.
14. Ahmad, S.; Schaefer, R.J. The B. F. Goodrich Company. U.S. Patent 4,519,430, 28 May 1985.
15. Ezzoddin, S.; Abbasian, A.; Aman-Alikhani, M.; Ganjali, S.T. The influence of noncarcinogenic petroleum-based process oils on tire compounds' performance. *Iran. Polym. J.* **2013**, *22*, 697–707. [CrossRef]
16. Sökmen, S.; Oßwald, K.; Reincke, K.; Ilisch, S. Influence of treated distillate aromatic extract (TDAE) content and addition time on rubber-filler interactions in silica filled SBR/BR blends. *Polymers* **2021**, *13*, 698. [CrossRef] [PubMed]
17. Hwang, K.; Kim, W.; Ahn, B.; Mun, H.; Yu, E.; Kim, D.; Ryu, G.; Kim, W. Effect of surfactant on the physical properties and crosslink density of silica filled ESBR compounds and carbon black filled compounds. *Elast. Comp.* **2018**, *53*, 39–47.
18. Corman, B.G.; Deviney, M.L., Jr.; Whittington, L.E. The migration of extender oil in natural and synthetic rubber. IV. Effect of saturates geometry and carbon black type on diffusion rates. *Rubber Chem. Technol.* **1970**, *43*, 1349–1358. [CrossRef]
19. Kim, D.; Ahn, B.; Kim, K.; Lee, J.; Kim, I.J.; Kim, W. Effects of molecular weight of functionalized liquid butadiene rubber as a processing aid on the properties of SSBR/silica compounds. *Polymers* **2021**, *13*, 850. [CrossRef]
20. Ikeda, K. Bio liquid polymer for winter tires. In Proceedings of the Tire Technology EXPO 2018, Hannover, Germany, 14–16 February 2017; pp. 105–127.
21. Iz, M.; Kim, D.; Hwang, K.; Kim, W.; Ryu, G.; Song, S.; Kim, W. The effects of liquid butadiene rubber and resins as processing aids on the physical properties of SSBR/silica compounds. *Elast. Comp.* **2020**, *55*, 289–299.

22. Gruendken, M. Liquid rubber for safer and faster tires. In Proceedings of the Tire Technology EXPO 2018, Hannover, Germany, 14–16 February 2017.
23. Sierra, V.P.; Wagemann, J.; Van De Pol, C.; Kendziorra, N.; Herzog, K.; Recker, C.; Mueller, N. Rubber Blend with Improved Rolling Resistance Behavior. U.S. Patent 9,080,042, 14 July 2015.
24. Kitamura, T.; Lawson, D.F.; Morita, K.; Ozawa, Y. Anionic Polymerization Initiators and Reduced Hysteresis Products Therefrom. U.S. Patent 5,393,721, 28 February 1995.
25. Cray Valley, Technical Data Sheet, Recon. Available online: http://www.crayvalley.com/docs/tds/ricon-603-.pdf?sfvrsn=2 (accessed on 9 April 2018).
26. Evonik. Less Fuel and Lower CO_2 Emissions with POLYVEST ST Tires. Available online: https://coatings.evonik.com/en/less-fuel-andlower-CO2-emissions-with-polyvest-st-tires100233.html (accessed on 13 February 2017).
27. Herpich, R.; Fruh, T.; Heiliger, L.; Schilling, K. Silica Gel-Containing Rubber Compounds with Organosilicon Compounds as Compounding Agent. U.S. Patent 6,593,418, 15 July 2003.
28. Takuya, H.; Tochiro, M. Tire Tread Rubber Composition. J.P. Patent 2,005,146,115, 14 November 2003.
29. Satoyuki, S.; Chikashi, Y. Rubber Composition Containing Compound Having Organosilicon Function Group through Urethane Bond at Terminal. J.P. Patent 2,005,350,603, 22 December 2005.
30. Kim, D.; Yeom, G.; Joo, H.; Ahn, B.; Paik, H.J.; Jeon, H.; Kim, W. Effect of the functional group position in functionalized liquid butadiene rubbers used as processing aids on the properties of silica-filled rubber compounds. *Polymers* **2021**, *13*, 2698. [CrossRef]
31. Tezuka, Y. Telechelic polymers. *Prog. Polymer Sci.* **1992**, *17*, 471–514. [CrossRef]
32. Lai, J.T.; Filla, D.; Shea, R. Functional polymers from novel carboxyl-terminated trithiocarbonates as highly efficient RAFT agents. *Macromolecules* **2002**, *35*, 6754–6756. [CrossRef]
33. Hoffman, R.F.; Gobran, R.H. Liquid carboxyl-terminated poly(butadiene). *Rubber Chem. Technol.* **1973**, *46*, 139–147. [CrossRef]
34. Berenbaum, M.B.; Bulbenko, G.F.; Gobran, R.H.; Hoffman, R.F. Liquid Carboxy-Terminated Polymers and Preparation Thereof with Dicarboxylic Acid Peroxides. U.S. Patent 3,235,589, 15 February 1966.
35. Buback, M.; Huckestein, B.; Kuchta, F.D.; Russell, G.T.; Schmid, E. Initiator efficiencies in 2, 2′-azoisobutyronitrile-initiated free-radical polymerizations of styrene. *Macromol. Chem. Phys.* **1994**, *195*, 2117–2140. [CrossRef]
36. Choi, Y.T.; El-Aasser, M.S.; Sudol, E.D.; Vanderhoff, J.W. Polymerization of styrene miniemulsions. *J. Polymer Sci. Polymer Chem. Ed.* **1985**, *23*, 2973–2987. [CrossRef]
37. Pesetskii, S.S.; Jurkowski, B.; Krivoguz, Y.M.; Kelar, K. Free-radical grafting of itaconic acid onto LDPE by reactive extrusion: I. Effect of initiator solubility. *Polymer* **2001**, *42*, 469–475. [CrossRef]
38. Ramier, J.; Gauthier, C.; Chazeau, L.; Stelandre, L.; Guy, L. Payne effect in silica-filled styrene-butadiene rubber: Influence of surface treatment. *J. Polymer Sci. B* **2007**, *45*, 286–298. [CrossRef]
39. Lee, J.Y.; Park, N.; Lim, S.; Ahn, B.; Kim, W.; Moon, H.; Paik, H.J.; Kim, W. Influence of the silanes on the crosslink density and crosslink structure of silica-filled solution styrene butadiene rubber compounds. *Comp. Interfaces* **2017**, *24*, 711–727. [CrossRef]
40. Boonstra, B.B.; Taylor, G.L. Swelling of filled rubber vulcanizates. *Rubber Chem. Technol.* **1965**, *38*, 943–960. [CrossRef]
41. Verbruggen, M.A.L.; Van Der Does, L.; Noordermeer, J.W.; Van Duin, M.; Manuel, H.J. Mechanisms involved in the recycling of NR and EPDM. *Rubber Chem. Technol.* **1999**, *72*, 731–740. [CrossRef]
42. Derouet, D.; Forgeard, S.; Brosee, J. Synthesis of alkoxysilyl-terminated polyisoprenes by means of 'living'anionic polymerization, 2. Synthesis of trialkoxysilyl-terminated 1, 4polyisoprenes by reaction of polyisoprenyllithium with various functional trialkoxysilanes selected as end-capping reagents. *Macromol. Chem. Phys.* **1999**, *200*, 10–24.
43. Liu, X.; Zhao, S.; Zhang, X.; Li, X.; Bai, Y. Preparation, structure, and properties of solutionpolymerized styrene-butadiene rubber with functionalized end-groups and its silica-filled composites. *Polymer* **2014**, *55*, 1964–1976. [CrossRef]
44. Jayalakshmy, M.S.; Mishra, R.K. Applications of carbon-based nanofillerincorporated rubber composites in the fields of tire engineering, flexible electronics and EMI shielding. In *Carbon-Based Nanofillers and Their Rubber Nanocomposites*; Elsevier: Amsterdam, The Netherlands, 2019; pp. 441–472.
45. de Monredon-Senani, S.; Bonhomme, C.; Ribot, F.; Babonneau, F. Covalent grafting of organoalkoxysilanes on silica surfaces in water-rich medium as evidenced by ^{29}Si NMR. *J. Sol-Gel Sci. Technol.* **2009**, *50*, 152–157. [CrossRef]
46. Sugiyama, T.; Shiba, H.; Yoshikawa, M.; Wada, H.; Shimojima, A.; Kuroda, K. Synthesis of polycyclic and cage siloxanes by hydrolysis and intramolecular condensation of alkoxysilylated cyclosiloxanes. *Chem. Eur. J.* **2019**, *25*, 2764–2772. [CrossRef]
47. Choi, S.S.; Kim, I.S.; Woo, C.S. Influence of TESPT content on crosslink types and rheological behaviors of natural rubber compounds reinforced with silica. *J. Appl. Polymer Sci.* **2007**, *106*, 2753–2758. [CrossRef]
48. Hwang, K.; Song, S.; Kang, Y.Y.; Suh, J.; Jeon, H.B.; Kwag, G.; Paik, H.J.; Kim, W. Effect of emulsion SBR prepared by asymmetric reversible addition fragmentation transfer agent on properties of silica-filled compounds. *Rubber Chem. Technol.* **2021**, *94*, 735–758. [CrossRef]
49. Zhao, F.; Bi, W.; Zhao, S. Influence of crosslink density on mechanical properties of natural rubber vulcanizates. *J. Macromol. Sci. B* **2011**, *50*, 1460–1469. [CrossRef]
50. Ryu, G.; Kim, D.; Song, S.; Hwang, K.; Kim, W. Effect of the epoxide contents of liquid isoprene rubber as a processing aid on the properties of silica-filled natural rubber compounds. *Polymers* **2021**, *13*, 3026. [CrossRef] [PubMed]
51. Halasa, A.F.; Prentis, J.; Hsu, B.; Jasiunas, C. High vinyl high styrene solution SBR. *Polymer* **2005**, *46*, 4166–4174. [CrossRef]
52. Isitman, N.A.H.; Thielen, G.M.V. Rubber Composition and Pneumatic Tire. E.U. Patent 3450490A1, 6 August 2018.

53. Hirata, K.; Moriguchi, M. Bio-based liquid rubber for tire application. *Rubber World* **2017**, *256*, 50–55.
54. Derham, C.F.; Newell, R.; Swift, P.M. The use of silica for improving tread grip in winter tires. *NR Technol.* **1988**, *19*, 1–9.
55. Dörrie, H.; Schröder, C.; Wies, B. Winter tires: Operating conditions, tire characteristics and vehicle driving behavior. *Tire Sci. Technol.* **2010**, *38*, 119–136. [CrossRef]
56. Wang, M.J. Effect of polymer-filler and filler-filler interactions on dynamic properties of filled vulcanizates. *Rubber Chem. Technol.* **1998**, *71*, 520–589. [CrossRef]
57. Warasitthinon, N.; Robertson, C.G. Interpretation of the tan δ peak height for particle-filled rubber and polymer nanocomposites with relevance to tire tread performance balance. *Rubber Chem. Technol.* **2018**, *91*, 577–594. [CrossRef]
58. Wang, M.J.; Kutsovsky, Y.; Zhang, P.; Murphy, L.J.; Laube, S.; Mahmud, K. New generation carbon-silica dual phase filler part I. Characterization and application to passenger tire. *Rubber Chem. Technol.* **2002**, *75*, 247–263. [CrossRef]
59. Wang, M.J.; Zhang, P.; Mahmud, K. Carbon-silica dual phase filler, a new generation reinforcing agent for rubber: Part IX. Application to truck tire tread compound. *Rubber Chem. Technol.* **2001**, *74*, 124–137. [CrossRef]

Disclaimer/Publisher's Note: The statements, opinions and data contained in all publications are solely those of the individual author(s) and contributor(s) and not of MDPI and/or the editor(s). MDPI and/or the editor(s) disclaim responsibility for any injury to people or property resulting from any ideas, methods, instructions or products referred to in the content.

Article

Simple and Efficient Synthesis of Oligoetherdiamines: Hardeners of Epoxyurethane Oligomers for Obtaining Coatings with Shape Memory Effect

Daria Slobodinyuk [1], Alexey Slobodinyuk [1,2,*], Vladimir Strelnikov [1] and Dmitriy Kiselkov [1,2]

[1] Institute of Technical Chemistry Ural Branch of the Russian Academy of Sciences, Academic Korolev 3, 614330 Perm, Russia; selivanovadg@gmail.com (D.S.); svn@itcras.ru (V.S.); dkiselkov@yandex.ru (D.K.)
[2] Department of Chemical Engineering, Perm National Research Polytechnic University, Komsomolsky Prospekt, 29, 614990 Perm, Russia
* Correspondence: slobodinyuk.aleksey.ktn@mail.ru; Tel.: +7-(342)-2378256

Abstract: In this work, new polymers with a shape memory effect for self-healing coatings based on oligomers with terminal epoxy groups, synthesized from oligotetramethylene oxide dioles of various molecular weights, were developed. For this purpose, a simple and efficient method for the synthesis of oligoetherdiamines with a high yield of the product, close to 94%, was developed. Oligodiol was treated with acrylic acid in the presence of a catalyst, followed by the reaction of the reaction product with aminoethylpiperazine. This synthetic route can easily be upscaled. The resulting products can be used as hardeners for oligomers with terminal epoxy groups synthesized from cyclic and cycloaliphatic diisocyanates. The effect of the molecular weight of newly synthesized diamines on the thermal and mechanical properties of urethane-containing polymers has been studied. Elastomers synthesized from isophorone diisocyanate showed excellent shape fixity and shape recovery ratios of >95% and >94%, respectively.

Keywords: self-healing coating; shape memory polymer; urethane; epoxyurethane oligomer; amino termination oligomer

1. Introduction

Corrosion of metals has long been a serious problem for industries around the world. The global cost of corrosion is USD 2.5 trillion—3.4% of the global Gross Domestic Product [1]. In China, the annual damage caused by corrosion is over USD 390 billion, which is equivalent to 4.2% of the Gross Domestic Product [1]. Corrosion costs can be reduced by up to 15% to 35% by corrosion prevention measures [2].

As a rule, one of the main corrosion prevention methods is applying a protective organic polymer coating on a metal surface. However, most synthetic polymers are susceptible to degradation and cross-linking when subjected to light and in an oxygen medium, i.e., in environmental conditions. In addition, a polymer coating can be destroyed by physical impact. In all these cases, the coating should be repaired properly; otherwise, the metal would be subjected to corrosion.

The repair of coatings by conventional methods, such as removal of the defective coating and recoating, is costly and labor-intensive. Therefore, there is a constant market demand for polymer materials for self-healing coating. Self-healing polymers are a class of materials that have the ability to repair micro-scale damage in the coating and restore the passive state of the metal substrate. In this case, the coating healing process needs very little human "assistance", if any.

The promising technology of self-healing polymers provides new opportunities for coatings to restore their protective characteristics with minimal or no intervention [3,4]. Thus, shape memory polymers (SMPs) represent a new class of smart materials that can

Citation: Slobodinyuk, D.; Slobodinyuk, A.; Strelnikov, V.; Kiselkov, D. Simple and Efficient Synthesis of Oligoetherdiamines: Hardeners of Epoxyurethane Oligomers for Obtaining Coatings with Shape Memory Effect. *Polymers* 2023, 15, 2450. https://doi.org/10.3390/polym15112450

Academic Editor: Md Najib Alam

Received: 25 April 2023
Revised: 23 May 2023
Accepted: 24 May 2023
Published: 25 May 2023

Copyright: © 2023 by the authors. Licensee MDPI, Basel, Switzerland. This article is an open access article distributed under the terms and conditions of the Creative Commons Attribution (CC BY) license (https://creativecommons.org/licenses/by/4.0/).

recover from temporary deformation to their original shape due to entropy processes facilitated by external stimuli such as heat [5,6]. Shape memory polymers return to their original shape after being altered by external stimulation, such as electric and magnetic fields, temperature, pH, and mechanical stress [5,6].

Shape memory polymers return to their original shape after being altered by external stimulation, such as electric and magnetic fields, temperature, pH, and mechanical stress [7–17].

Shape memory polymers have many applications in the aerospace industry [18], medicine [19–21], self-processing smart textiles [22,23], electronic devices [24], and structural assembly [25–27].

Recently, shape memory polymers (SMPs) have been used to produce self-healing coatings with corrosion protection. For example, there are several reports on the use of urethane-containing SMPs based on poly(ε-caprolactone) (PCL) as self-healing coatings on metal substrates [28–30]. These studies show that the shape memory effect can significantly narrow the crack, thereby reducing the area of metal in contact with aggressive media.

Polyurethanes are block copolymers consisting of alternating soft and hard segments, SS and HS, respectively [31,32]. These polymers are usually prepared in a two-step process. At the first stage, a prepolymer with isocyanate-terminal groups is synthesized by the interaction of oligodiol and diisocyanate, taken in a double excess. In the second stage, the prepolymer is cured with low molecular weight chain extenders to form polyurethanes (hardener–diol) or polyurethane ureas (hardener diamine). The structure of the soft segments is determined by oligodiols used for the synthesis of polyurethane foam [33]. The structure of the hard segments is determined by diisocyanates used in the synthesis of oligodiisocyanates and by low molecular weight chain extenders, diamines, or diols [34].

Recently, some approaches for the preparation of self-healing PUs have been developed. This research trend can be illustrated by the following examples: encapsulation [35–39], healing in the polymer bulk via dynamic interactions, including non-covalent π–π bonding [40–42], ionic interactions [43], and thermoreversible covalent bonds [44–53]. Moreover, the shape memory effect caused by the thermal phase transition of the polymer is successfully used in the design of polyurethane materials.

In [54], the preparation of shape memory polyurethane was reported. In this synthetic route, a crystallizing polyester, polycaprolactone diol, was used as the main initial component. The resulting polymer was considered to be suitable for self-healing polymer coating.

Thus, Gonzalez-Garcia et al. developed a self-healing polymer coating film based on polycaprolactone diol (PCL) [55]. In case the coating is damaged, the relaxation of the soft polymer phase takes place. This process triggers temperature-induced self-healing, followed by the restoration of the barrier properties. This way, the damaged area is self-recoated.

A similar approach was used by Jorcin et al. [56]. In this study, a self-healing polyurethane coating with polycaprolactone soft segments was developed. However, a self-healing phenomenon was observed only at a hard segment content of 12%.

In [57–59], self-healing coatings with a shape memory effect were suggested.

Thus, the coatings described in [55–59] are characterized by the presence of a crystalline phase. It is a necessary condition for the preparation of this type of self-healing material. In addition, the presence of the crystalline phase contributes to an increase in the barrier properties of the coatings, i.e., their stability with respect to aggressive media [60].

It is noteworthy that for high-quality coatings, the adhesive properties of the polymer should be rather high. However, the strength characteristics of polyurethanes and polyurethane ureas depend on humidity, as the isocyanate group of oligodiisocyanates can react with moisture.

To solve this problem and to reduce the toxicity of isocyanate-terminated compounds, shielding the isocyanate groups of oligodiisocyanates with epoxy alcohol is the most efficient approach. For example, 2,3-epoxy-1-propanol can be used for this purpose [61]. In this case, the isocyanate groups of the oligodiisocyanate and the hydroxyl group of 2,3-epoxy-1-propanol react to give epoxyurethane oligomer (EUO). The deformation and strength properties of the elastomers, based on these oligomers, are only slightly dependent on the presence of moisture.

The elastomers based on epoxyurethane oligomers have good mechanical and dielectric properties. In addition, the adhesive properties of these elastomers are much higher than those of polyurethanes and polyurethane ureas. Thus, these polymers are widely used as adhesives, polymer matrices for casting low-modulus compounds for various purposes, and biomedical materials.

The structure of elastomers, based on epoxy urethane oligomers, consists of alternating soft and hard urethane hydroxyl segments. The microphase separation and the formation of separate phases, or domains, is due to the difference in the polarity of the structural units, soft and hard segments. Domains play the role of a reinforcing nanofiller or can be considered as the nodes of a physical network. This phenomenon is the reason for the high-strength characteristics of the developed materials. In the domains, hydrogen bonds play a decisive role in stabilizing the structure of the hard phase. In this case, the structure of hard segments can affect the morphology of the polymer in general.

To date, the crystalline elastomers, based on epoxyurethane oligomers, were prepared. Most of these elastomers were synthesized from polyesters [62,63]. However, these coatings are known to be unstable in water and, hence, they cannot be used for anti-corrosion treatment of metals. For these purposes, coatings based on polyethers, for example, oligotetramethylene oxide diols, can be useful.

In our previous studies, the attempts to obtain polyether-based elastomers were described [64–66]. A curing agent with terminal amino groups was synthesized via nucleophilic substitution of the hydroxy groups of oligotetramethylene oxide diols with amino groups. The authors assume that a material with improved stress-deformation and strength characteristics can be obtained. This can be realized by increasing the polarity of the hard segments, which is proportional to the degree of microphase separation of soft and hard segments in the polymer.

The present study aims to obtain shape memory urethane-containing polymers based on epoxyurethane oligomers synthesized from polyethers.

2. Materials and Methods

2.1. Materials and Synthesis

2.1.1. Materials

Acrylic acid (Merck; Darmstadt, Germany), N-(2-aminoethyl)piperazine (AEP) (Merck; Darmstadt, Germany), 2,4-toluene diisocyanate (TDI) (BASF; Ludwigshafen, Germany), isophorone diisocyanate (IPDI) (Evonik Chemistry Ltd.; Essen, Germany), oligotetramethylene oxide diol with M_n~1008 g·mol^{-1}, M_n~1400 g·mol^{-1}, and M_n~2000 g·mol^{-1} (OTMO-1000; OTMO-1400; OTMO-2000)(BASF; Ludwigshafen, Germany), glycidol (grade pure, 99.0%)(Research Institute of Polymer Materials; Perm, Russia), and dibutyltin dilaurate (grade pure, 99.8%) (BASF; Ludwigshafen, Germany) were used without purification.

2.1.2. Synthesis of OTMO-diAEP

OTMO-diAEPs were prepared in two steps, shown in Figure 1.

In the first step, oligotetramethylene oxide diols (OTMO) were reacted with acrylic acid. Hydroquinone was added to prevent the copolymerization of acrylic acid with OTMO [67].

OTMO (0.02 mol), acrylic acid (0.052 mol), p-toluenesulfonic acid (0.002 mol), hydroquinone (0.0006 mol), and cyclohexane (320 mL) were placed in a round bottom flask with a Dean–Stark trap.

The reaction time was determined by the amount of water condensed in the Dean–Stark trap. The catalyst and acrylic acid residue was removed by adding potassium carbonate to the reaction mixture. The resulting mass was stirred at room temperature for 3 h. Then, the solution was filtered, and the solvent was distilled off on a rotary evaporator. The final products, OTMO diacrylates, were obtained in high yields (94%).

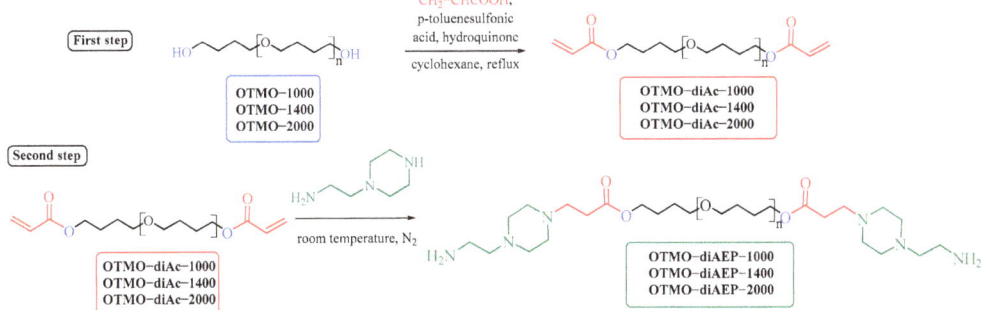

Figure 1. Synthesis of AEP functionalized OTMOs.

The second step of OTMO-diAEP preparation technique is the conjugated addition reaction of a cycloaliphatic diamine, aminoethylpiperazine, to a Michael acceptor, acrylate-terminated oligotetramethylen oxides. The addition reaction via the secondary amino group of aminoethylpiperazine is explained by the higher nucleophilicity of the secondary amino group compared to the primary [68].

OTMO-diAc (0.014 mol) was placed into a round bottom flask equipped with a magnetic stirrer. Further, aminoethylpiperazine (0.028 mol) was added to it. The reaction was carried out in a round-bottom flask equipped with a magnetic stirrer. The reaction mass was stirred in nitrogen medium for 2 h at room temperature. The product was isolated without purification. The product yield was 100%.

2.1.3. Preparation of Epoxyurethane Oligomers

The synthesis of epoxyurethane oligomers P−1 and P−2 was carried out in two steps (Figure 2).

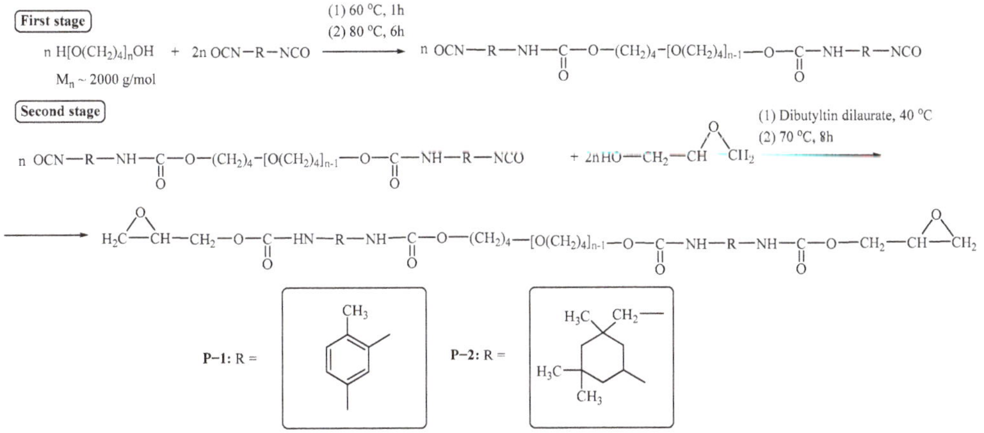

Figure 2. Preparation of P−1 and P−2.

OTMO was dried from water with constant stirring for 8 h at a temperature of 88–92 °C and absolute pressure in the reactor of 0.2–0.4 kPa. In the first stage, an oligomer with terminal epoxy groups was obtained by reacting the dried diol and diisocyanate, taken in a double excess at a temperature of 80 °C. The weight fraction of free isocyanate groups was determined according to ASTM D 2572-97. The second stage consisted of the treatment of the resulting oligomer with epoxy alcohol–glycidol in the presence of 0.03% urethane formation catalyst-di-n-butyl tin dilaurate. The reaction was carried out at a temperature of 70 ± 1 °C. Completion of the synthesis was determined by FTIR spectroscopy by the absence of an absorption band at 2270 cm^{-1} characteristic of the isocyanate group [69]. The mass fraction of free epoxy groups in the resulting oligomers was determined by the method described elsewhere [70].

The characteristics of the synthesized oligomers with various terminal groups are presented in Table 1.

Table 1. Characteristics of the epoxyurethane oligomers P−1 and P−2.

EUO Code	Molecular Weight of Initial OTMO	Diisocyanate	Content of Free NCO Groups, wt %		Content of Free Epoxy Groups, wt %	
			Experimental	Theoretical	Theoretical	Experimental
P−1	2000	2,4-toluene diisocyanate	3.57	3.51 ± 0.03	3.51	3.47 ± 0.03
P−2	2000	isophorone diisocyanate	3.43	3.42 ± 0.03	3.54	3.49 ± 0.03

2.1.4. Preparation of Polymers

Oligomers with terminal epoxy groups synthesized by a two-step method were cured with the developed diamines. To do this, the oligomer was stirred with diamine at a temperature of 50 °C for 5 min. The resulting mixture was thermostated at a temperature of 90 °C for 12 h. The disappearance of the absorption band at 910 cm^{-1} indicated the completeness of the transformation of the epoxy group [71]. The synthetic route is shown in Figure 3.

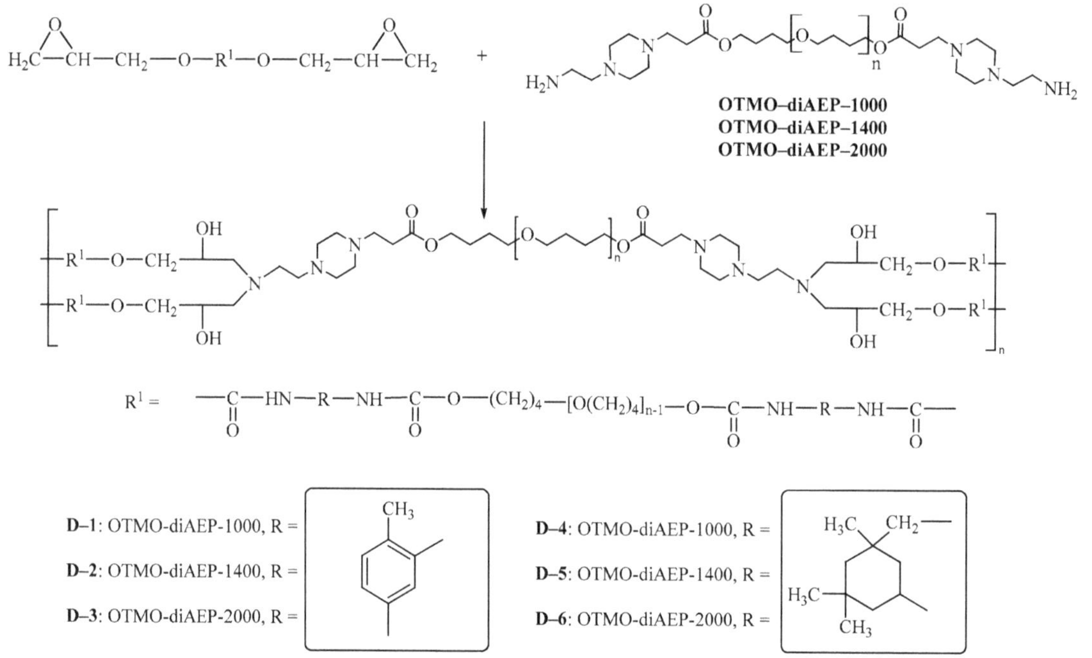

Figure 3. Synthetic route of polymers D−1–D−6.

The polymer compositions are given in Table 2.

Table 2. Compositions of polymers D−1, D−2, D−3, D−4, D−5, and D−6.

Polymer Code	EUO Code	Hardener
D−1	P−1	OTMO-diAEP−1000
D−2	P−2	OTMO-diAEP−1000
D−3	P−1	OTMO-diAEP−1400
D−4	P−2	OTMO-diAEP−1400
D−5	P−1	OTMO-diAEP−2000
D−6	P−2	OTMO-diAEP−2000

2.2. Methods

2.2.1. ^{1}H- and ^{13}C-NMR Spectra

^{1}H-NMR and ^{13}C-NMR spectra were registered on a Bruker Avance Neo III spectrometer (^{1}H: 400 MHz, ^{13}C: 75 MHz) using tetramethyl silane as an internal standard. The chemical shift was calibrated with respect to the deuterium signal of CDCl$_3$ at 7.26 ppm for ^{1}H-NMR and 77.16 ppm for ^{13}C-NMR.

2.2.2. Elemental Analysis

Elemental analysis of CHN was performed on a Vario EL cube analyzer.

2.2.3. Gel Permeation Chromatography (GPC)

The number average molecular weight of the oligomers obtained was measured using ULTIMATE 3000 HPLC chromatograph (Dionix Thermo Scientific, Moscow, Russia) equipped with a RefractoMax 521 refractometric detector, according to [72].

2.2.4. FTIR Spectra

FTIR spectra were recorded using Vertex 80v spectrometer (Bruker, Moscow, Russia) equipped with A225/Q Platinum ATR unit at a 6 mm aperture, with a spectral resolution of 1 cm^{-1}. The spectra were normalized using the band at 2860 cm^{-1}, corresponding to symmetric vibrations of aliphatic–CH$_2$ groups [73].

2.2.5. Differential Scanning Calorimetry (DSC)

Heat effects in the samples within the temperature range from −100 °C to +100 °C were measured using a Mettler Toledo DSC 3+ calorimeter. Heating and cooling rates were 5 K min^{-1}.

2.2.6. Mechanical Tests

Mechanical tests of samples of the obtained materials were carried out on an Instron 3365 testing machine at the extension velocity v = 0.417 s^{-1} and a temperature of 25 ± 1 °C according to the standard procedure. Based on the results of mechanical testing, the following characteristics were determined: σ_k (MPa)—the nominal strength (the maximal stress per initial specimen cross section); ε_k (%)—the relative critical strain; E_{100} (stress at the relative strain ε = 100%)—the nominal elastic modulus; and f_r—the true tensile strength ($f_r = \sigma_k \cdot \lambda_k$, where $\lambda_k = (\varepsilon_k + 100)/100$). The polymer was subjected to 5 tests.

2.2.7. Shape Memory Properties

The shape memory properties were measured by the bending test. The test was conducted in four steps, as described below. First, the sample was deformed to the U shape in a 50 °C lab oven. Then, the U shape sample was fixed in a freezing chamber at −10 °C for 5 min. After that, the sample was placed in a refrigerator at 0 °C for 5 min, and the angle after fixing (θ_A) was measured. Finally, the sample was heated to 50 °C in the lab oven, and the angle after recovery (θ_B) was measured. Before measuring θ_A and θ_B, the sample was held for 1 min at room temperature (25 °C). The shape fixing ratio and the shape recovery ratio are defined from θ_A and θ_B, respectively, as shown below:

$$\text{Shape fixity ratio (\%)} = \frac{\theta_A}{180} \cdot 100$$

$$\text{Shape recovery ratio (\%)} = \frac{180 - \theta_B}{180} \cdot 100$$

3. Results and Discussion

3.1. NMR Measurement of OTMO, OTMO-diAc, and OTMO-diAEP

The ^{1}H and ^{13}C NMR spectra of OTMOs, OTMO-diAcs, and OTMO-diAEPs are presented in Figure 4 and in Table S1 (Supplementary Materials).

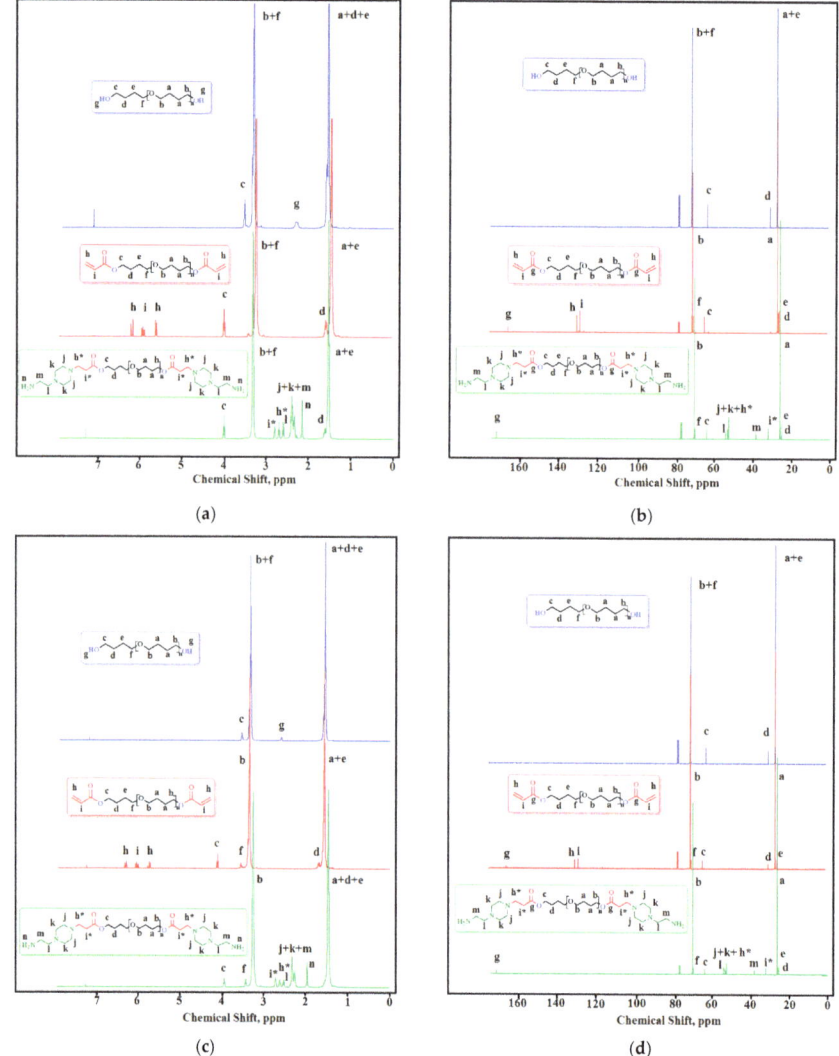

Figure 4. Cont.

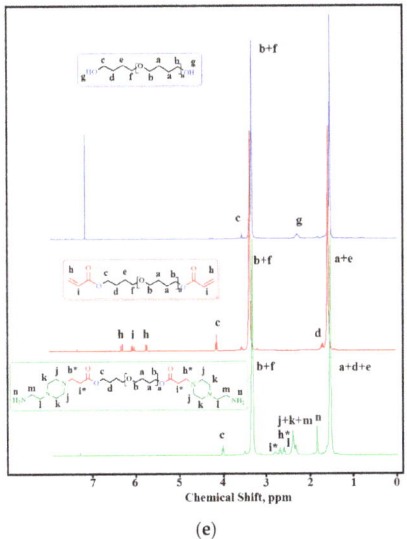
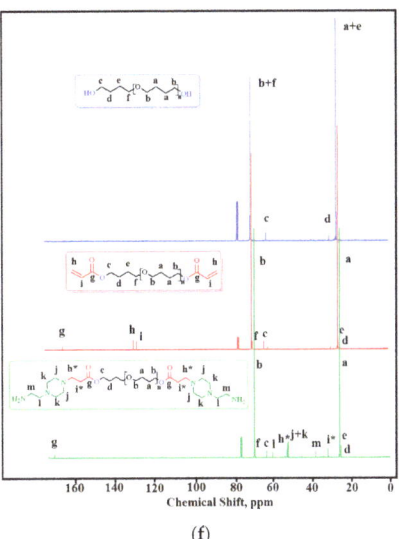

Figure 4. Nuclear magnetic resonance spectra (^{1}H-NMR and ^{13}C-NMR): (**a**,**b**) OTMO-1000, OTMO-diAc-1000, and OTMO-diAEP−1000; (**c**,**d**) OTMO-1400, OTMO-diAc-1400, and OTMO-diAEP−1400; (**e**,**f**) OTMO-2000, OTMO-diAc-2000, and OTMO-diAEP−2000.

In the ^{1}H-NMR spectra, the signals of the methylene protons of oligotetramethylene oxide diols are at 1.51–1.62 (a + d + e) and 3.29–3.38 (b + f)) ppm. The protons of the hydroxyl groups are observed at 2.31 ppm (OTMO-1000), 2.60 ppm (OTMO-1400), and 2.29 ppm (OTMO-2000). The hydroxyl group signals disappeared upon the acrylation of the oligotetramethyleneoxide diols, and vinyl protons of terminal acrylate groups (h + i) appeared in the region of 5.75–6.35 ppm. The shift is caused by the change in the chemical environment of these protons, as the terminal hydroxyls of the oligomer chain were converted into acrylate groups. Upon the further reaction of OTMO-diAcs with aminoethylpiperazine, no vinyl protons of terminal acrylate groups were detectable. As a consequence, the triplets at 2.60 and 2.70 ppm (h* and i*) appeared. In addition, the signals of amino group protons could be observed at 2.16, 1.94, and 1.92 ppm (n). Both the multiplet at 2.23–2.43 ppm and the signal at 2.50 ppm were assigned to the protons of the aminoethylpiperazine fragment (j + k + m + l).

In the ^{13}C-NMR spectra, there is a shift in the signals of terminal methylene carbons (c + d) during the sequential transformation of oligotetramethylene oxide diols into OTMO-diAEP. Additionally, for OTMO-diAc, the signals of the vinyl carbons of the acrylate groups appear in the downfield region (δ = 128 ppm (i), 130 ppm (h)). These signals shift upfield after the aza addition of the aminoethylpiperazine is completed (δ = 32 ppm (i*) and 53 (h*) ppm). In the ^{13}C NMR spectra of OTMO-diAEP, the signals of the methylene carbons of aminoethylpiperazine fragment are observed at 38, 52, 53, and 61 ppm.

3.2. Elemental Analysis

The results of the elemental analysis are presented in Table 3. The experimental values turned out to be close to the theoretical values. The obtained data confirm the structure of all new compounds.

Table 3. Elemental analysis data of OTMO-diAc and OTMO-diAEP.

	C, %		H, %		N, %	
	Experimental	Theoretical	Experimental	Theoretical	Experimental	Theoretical
OTMO-diAc (M_n = 1116 g/mol)	65.82	65.60	10.45	10.40	-	-
OTMO-diAc (M_n = 1508 g/mol)	66.03	65.88	10.88	10.58	-	-
OTMO-diAc (M_n = 2108 g/mol)	66.24	66.10	10.86	10.74	-	-
OTMO-diAEP (M_n = 1374 g/mol)	63.92	63.76	10.65	10.63	6.32	6.12
OTMO-diAEP (M_n = 1766 g/mol)	64.52	64.39	10.86	10.76	4.92	4.76
OTMO-diAEP (M_n = 2366 g/mol)	65.05	64.97	10.95	10.87	3.81	3.55

3.3. Gel Permeation Chromatography of OTMO-diAc and OTMO-diAEP

According to the results of gel permeation chromatography, the retention time of the synthesized compounds was determined, the values of which were from 5.04 to 6.58 min (Table 4). The obtained data indicate a narrow molecular weight distribution of compounds. Table 4 shows the average molecular weight of compounds determined by ^{1}H-NMR spectroscopy and gel permeation chromatography.

Table 4. The number average molecular weight of compounds determined via ^{1}H-NMR spectroscopy and gel permeation chromatography.

	M_n^1		M_n^2		M_n^3	
	GPC	^{1}H-NMR **	GPC	^{1}H-NMR **	GPC	^{1}H-NMR **
OTMO-diAc	1104 (5.041 *)	1116	1479 (5.491 *)	1508	2079 (6.211 *)	2108
OTMO-diAEP	1391 (5.385 *)	1374	1747 (5.812 *)	1766	2387 (6.580 *)	2366

M_n^1—the number average molecular weight of OTMO-diAc-1000 and OTMO-diAEP−1000; M_n^2—the number average molecular weight of OTMO-diAc-1400 and OTMO-diAEP−1400; M_n^3—the number average molecular weight of OTMO-diAc-2000 and OTMO-diAEP−2000; *—retention time, min; and **—by ^{1}H NMR, using the intensity of the signals of the terminal groups and repeating units.

3.4. FTIR Spectroscopy

3.4.1. FTIR Spectroscopy of OTMO, OTMO-diAc, and OTMO-diAEP

The FTIR spectra of OTMOs, OTMO-diAc, andOTMO-diAEPs are presented in Figure 5.

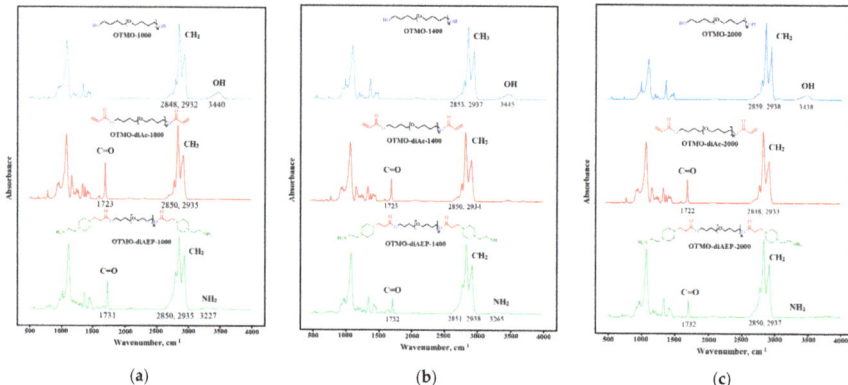

Figure 5. The Fourier-transform infrared spectroscopy spectra: (**a**) OTMO-1000, OTMO-diAc-1000, and OTMO-diAEP−1000; (**b**) OTMO-1400, OTMO-diAc-1400, and OTMO-diAEP−1400; and (**c**) OTMO-2000, OTMO-diAc-2000, and OTMO-diAEP−2000.

Acrylation of oligotetramethylene oxide diols leads to the appearance of a peak of carbonyl groups in the region of 1723 cm^{-1}, while the absorption band of hydroxyl groups disappears. Further interaction of OTMO-diAc with aminoethylpiperazine leads to the appearance of an absorption band of amino groups, as well as to a slight shift in the peak of carbonyl groups. The remaining bands of intermediates and final products of OTMO-diAEP are identical to the bands of the starting oligotetramethylene oxide diols. According to the results of FTIR spectroscopy, it was proved that during the two-stage synthesis of OTMO-diAEP, only the end groups of oligomers changed.

3.4.2. FTIR Spectroscopy of the Synthesized Elastomers

The overall spectrum of synthesized elastomers is shown in Figure 6. Elastomers synthesized from different diisocyanates are characterized by the same absorption bands: 3350 cm^{-1}—NH band of urethane; 1542, 1454, 1412 cm^{-1}—amide-NH stretching; 2860 cm^{-1}—CH$_2$ group; and also, 2950 cm^{-1}—CH asymmetric stretching. However, there are also differences: for elastomers based on cyclic diisocyanate (2,4-toluylene diisocyanate), absorption bands at 1600 cm^{-1} and 1612 cm^{-1} appear in the spectrum. More detailed differences in the supramolecular structure of elastomers will be analyzed below when describing the FTIR spectra in the absorption region of carbonyl at 1600–1760 cm^{-1}.

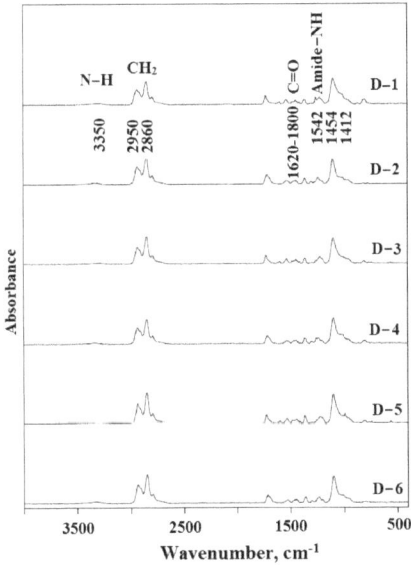

Figure 6. The Fourier-transform infrared spectroscopy spectra D−1, D−2, D−3, D−4, D−5, and D−6.

Revealing the features of the supramolecular structure of elastomers is possible by analyzing the FTIR spectra, namely, the region of carbonyl stretching vibrations (1600–1800 cm^{-1}). It is known that the absorption band at 1695 cm^{-1} characterizes self-associates of rigid urethane hydroxyl blocks based on isophorone diisocyanate, and at 1705 cm^{-1} based on cyclic diisocyanate-2,4-toluene diisocyanate [64,74]. The intensity of these absorption bands determines the degree of microphase separation in elastomers.

Figure 7 shows the analysis of IR spectra in the range of carbonyl stretching vibrations (1620–1800 cm^{-1}). It is shown that elastomers based on isophorone diisocyanate have two pronounced absorption bands—at 1695 cm^{-1} and 1730 cm^{-1}.

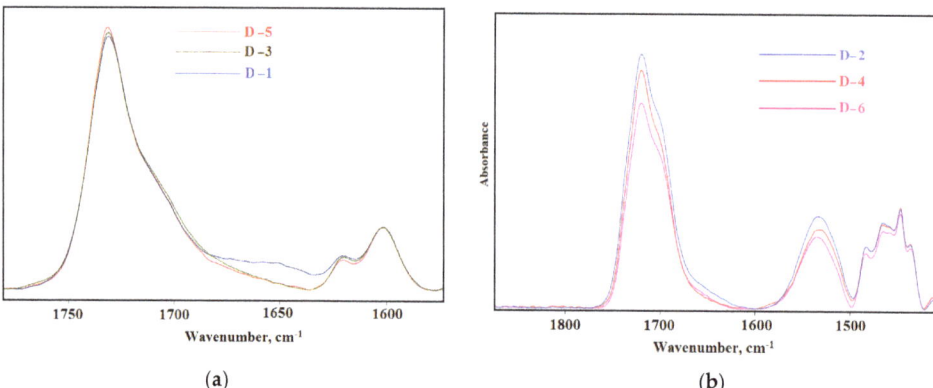

Figure 7. The Fourier-transform infrared spectroscopy spectra in the range of carbonyl stretching vibrations: (**a**) D−1, D−3, and D−5; (**b**) D−2, D−4, and D−6.

Elastomers based on 2,4-toluylene diisocyanate have one strong absorption band at 1730 cm^{-1} (Figure 7a). The absorption band at 1705 cm^{-1} appears weakly in the form of a shoulder. The lower intensity of this band indicates a lower degree of microphase separation of soft and hard segments in elastomers based on 2,4-toluylene diisocyanate.

3.5. Thermal Properties of the Synthesized Elastomers

The DSC curves for elastomers D−1, D−2, D−3, D−4, D−5, and D−6 are shown in Figure 8. The thermal effects of elastomers were recorded according to the regime indicated in the work [64]: first, the samples were heated to 150 °C, then cooled to 100 °C below zero, kept for 30 min, and heated at a heating rate of 5 °C/min.

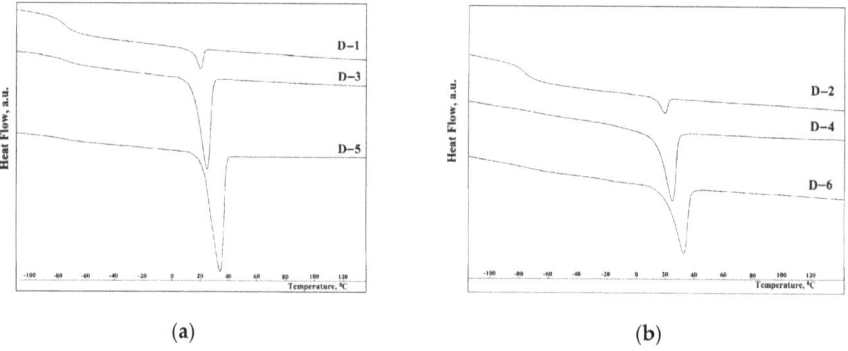

Figure 8. Differential scanning calorimetry data: (**a**) D−1, D−3, and D−5; (**b**) D−2, D−4, and D−6.

The DSC results show that with an increase in the molecular weight of the hardener from 1374 to 1766 g/mol, the melting enthalpy of the elastomer increases by more than 10 times. A further increase in molecular weight does not lead to such a significant effect. It should be noted that higher melting enthalpies are realized on samples synthesized from 2,4-toluylene diisocyanate. This is due to the steric hindrance of the bulkier hard segment structure based on isophorone diisocyanate.

A higher degree of microphase separation of soft and hard segments of elastomers based on isophorone diisocyanate (Figure 8) makes it possible to obtain polymers with a lower glass transition temperature. At the same time, the value of the glass transition temperature less than minus 70 °C makes it possible to use these polymers under the extreme conditions of the Arctic and the Far North.

The derivative thermogravimetric curves of the elastomers D−1, D−2, D−3, D−4, D−5, and D−6 are shown in Figures S1–S6 (Supplementary Materials) and in Table 5.

Table 5. Thermal properties of the elastomers D−1, D−2, D−3, D−4, D−5, and D−6.

Composition Code	Glass Transition Temperature of the Soft Phase, °C	Melting Temperature of the Soft Phase, °C	Enthalpy of Melting ΔH_m, J/g
D−1	−76	20	4.59
D−2	−76	18	3.41
D−3	−72	24	45.49
D−4	−74	24	40.88
D−5	−72	35	58.66
D−6	−74	35	48.63

The process of decomposition of samples, regardless of the molecular weight of the hardener, and the diisocyanate used in the synthesis of urethane-containing oligomers with terminal epoxy groups, occurs in two stages. The first stage occurs in the temperature range of 220–300 °C, and the second stage at temperatures of 320–360 °C, at a maximum weight loss temperature of about 310–320 °C.

3.6. Physical-Mechanical Characteristics of the Elastomers

According to the results of mechanical tests, it was proved that the molecular weight of the hardeners synthesized for the first time in this work affects the deformation properties of elastomers (Table 6). It has been established that elastomers synthesized on the basis of isophorone diisocyanate are characterized by the highest values of the relative critical strain.

Table 6. Deformation and strength characteristics of the elastomers D−1, D−2, D−3, D−4, D−5, and D−6.

Elastomer Code	σ_k, MPa	ε_k, %	E_{100}, MPa	f_r, MPa
D−1	10.20 ± 0.10	1150 ± 20	1.22 ± 0.05	127.5
D−2	11.50 ± 0.10	1023 ± 20	0.60 ± 0.05	129.2
D−3	9.10 ± 0.10	928 ± 20	2.28 ± 0.05	93.6
D−4	9.04 ± 0.10	911 ± 20	1.15 ± 0.05	91.7
D−5	8.96 ± 0.10	890 ± 20	3.28 ± 0.05	88.7
D−6	9.42 ± 0.10	830 ± 20	2.18 ± 0.05	87.6

It should be noted that with an increase in the molecular weight of the hardener from 1374 to 1766 g/mol, an increase in the nominal elastic modulus of elastomers is observed.

3.7. Shape Memory Properties of the Elastomers

The ratios of the shape memory fixity (performed at −10 °C) and the shape memory recovery (performed at 50 °C) measured at 25 °C (Figure S7 in Supplementary Materials) are listed in Table 7.

Table 7. Shape fixity ratios and shape recovery ratios for elastomers.

Elastomer Code	Shape Fixity Ratio, %	Shape Recovery Ratio, %
D−3	95 ± 3	87 ± 3
D−4	96 ± 3	94 ± 3
D−5	95 ± 3	91 ± 3
D−6	97 ± 2	96 ± 3

The results show that polymers cured with 1400 and 2000 molecular weight oligoamines have a high shape fixity ratio. At the same time, the Shape recovery ratio is higher for the samples synthesized from isophorone diisocyanate. The recovery process of all scaffolds at 50 °C was finished in 60 s.

3.8. Self-Healing Capability of the Prepared Coatings

The self-healing capability of the prepared coatings was visualized using an optical microscope Olympus BX-51 (Figure 9).

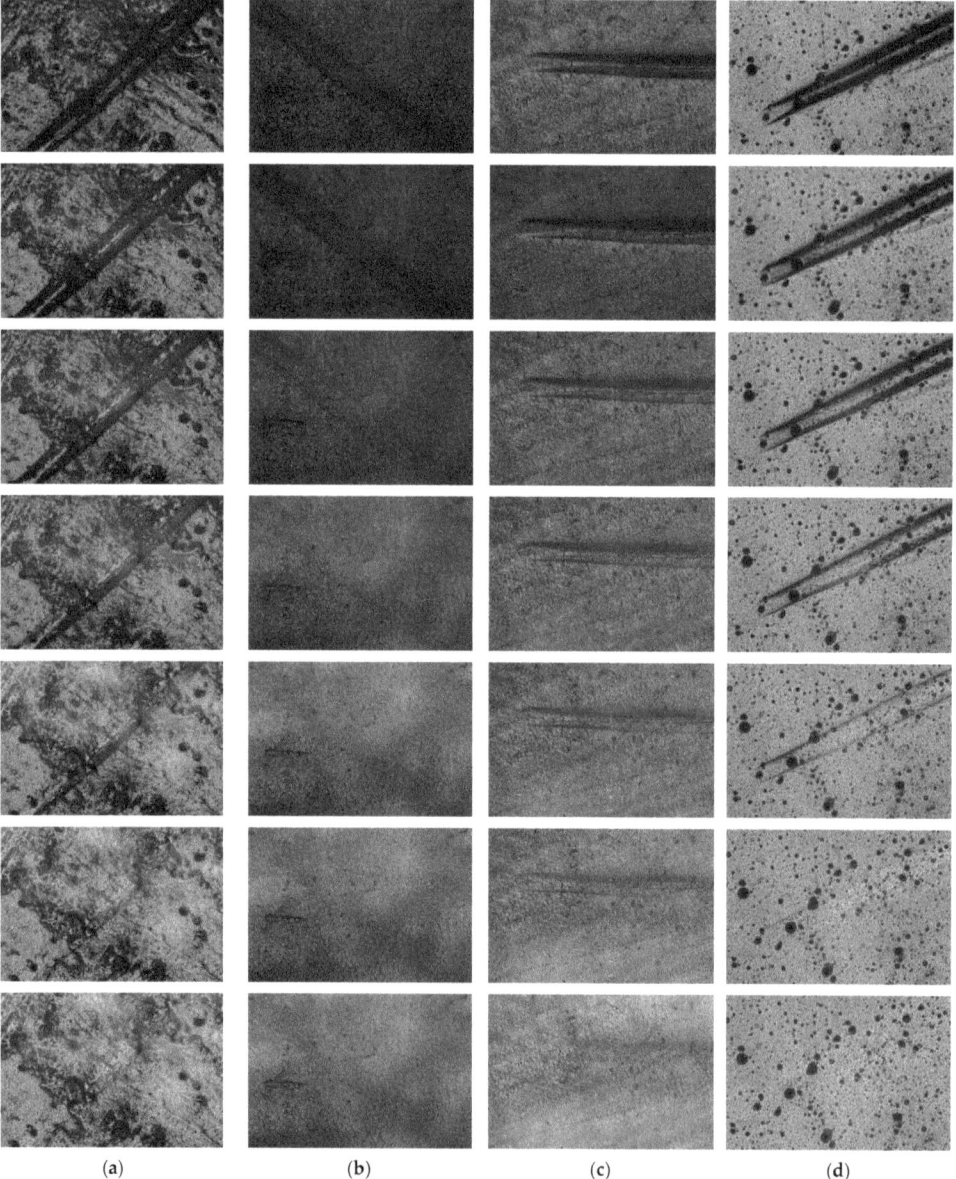

Figure 9. Optical micrographs for the samples: (**a**) D−3; (**b**) D−4; (**c**) D−5; and (**d**) D−6.

The prepared coatings were scratched with a spatula (d = 1 mm). As a result, plastic deformation of the coatings occurred without the formation of cracks [75]. Next, the scratched coating was heated at a rate of 2 °C/min to 60 °C. As can be seen from Figure 9, with such heating, the plastic deformation is completely restored, and the scratch is healed.

Since no cracks formed on the coatings when using a spatula, the result obtained indicates the creation of materials with a reversible shape memory effect.

4. Conclusions

A simple and efficient method for the synthesis of amino-terminated oligotetramethylene oxides was developed, including the initial preparation of oligodiacrylates by reacting oligotetramethylene oxide diols with acrylic acid in the presence of a catalytic amount of p-toluenesulfonic acid and an inhibitor of acid polymerization, hydroquinone, and further reaction of the oligodiacrylate with a cycloaliphatic diamine by the conjugated addition method. The yield of target oligoetherdiamines was 94%.

Two epoxyurethane oligomers were prepared using the oligotetramethylene oxide diol with Mn~2000 g·mol^{-1}, isophorone diisocyanate, 2,4-toluenediisocyanate, and epoxy alcohol–glycidol.

Six elastomers from urethane-containing elastomers with terminal epoxy groups were prepared using synthesized amines.

The degree of microphase separation is higher for samples synthesized from isophorone diisocyanate.

It has been proven that the lower cleavage temperature of elastomers based on isophorone diisocyanate is realized due to a higher degree of microphase separation compared to elastomers based on 2,4-toluene diisocyanate.

At the same time, the glass transition temperature of elastomers less than 72 °C allows them to be used in the extreme conditions of the Far North.

The ability of the coating based on the developed polymers to self-healing has been demonstrated.

Supplementary Materials: The following supporting information can be downloaded at https://www.mdpi.com/article/10.3390/polym15112450/s1. Figure S1: DTG curves D−1; Figure S2: DTG curves D−2; Figure S3: DTG curves D−3; Figure S4: DTG curves D−4; Figure S5: DTG curves D−5; Figure S6: DTG curves D−6; Figure S7: Shape Memory Properties of the elastomers D−3, D−4, D−5, and D−6; and Table S1: ^{1}H-NMR and ^{13}C-NMR data in CDCl$_3$ for OTMO-diAc and OTMO-diAEP.

Author Contributions: Conceptualization, A.S.; methodology, D.K.; investigation, A.S. and D.S.; writing—original draft preparation, A.S.; writing—review and editing, A.S. and D.S.; funding acquisition, V.S. All authors have read and agreed to the published version of the manuscript.

Funding: The reported study was funded by RFBR and Perm Territory, project number No. 20-43-596010. The work was carried out within the framework of the State Assignment (theme state registration number 122011900165-2) using the equipment of the Center for Collective Use "Investigations of materials and substances" of the Perm Federal Research Center of the Ural Branch of the Russian Academy of Sciences. This work was financially supported by the Perm Scientific and Educational Center "Rational Subsoil Use".

Institutional Review Board Statement: Not applicable.

Informed Consent Statement: Not applicable.

Data Availability Statement: The most significant data generated or analyzed during this study are included in this published article. Further results obtained during the current study are available from the corresponding author upon reasonable request.

Acknowledgments: The study was performed using the equipment of the Center for Shared Use Studies of Materials and Substances at the Perm Federal Research Center Ural Branch of the Russian Academy of Sciences.

Conflicts of Interest: The authors declare no conflict of interest.

References

1. Koch, G.; Varney, J.; Thompson, N.; Moghissi, O.; Gould, M.; Payer, J. International measures of prevention, application, and economics of corrosion technologies study. *NACE Int.* **2016**, *216*, 2–3.
2. Rajendran, S.; Nguyen, T.A.; Kakooei, S.; Yeganeh, M.; Li, Y. *Corrosion Protection at the Nanoscale*; Elsevier: Amsterdam, The Netherlands, 2020.
3. Garcia, S.J.; Fischer, H.R.; Zwaag, S. A critical appraisal of the potential of self healing polymeric coatings. *Prog. Org. Coat.* **2011**, *72*, 211–221. [CrossRef]
4. Montemor, M.F. Functional and smart coatings for corrosion protection: A review of recent advances. *Surf. Coat. Technol.* **2014**, *258*, 17–37. [CrossRef]
5. Behl, M.; Razzaq, M.Y.; Lendlein, A. Multifunctional shape-Memory polymers. *Adv. Mater.* **2010**, *22*, 3388–3410. [CrossRef]
6. Xie, T. Recent advances in polymer shape memory. *Polymer* **2011**, *52*, 4985–5000. [CrossRef]
7. Liu, F.; Urban, M.W. Recent advances and challenges in designing stimuli-responsive polymers. *Prog. Polym. Sci.* **2010**, *35*, 3–23. [CrossRef]
8. Deng, G.; Tang, C.; Li, F.; Jiang, H.; Chen, Y. Covalent cross-linked polymer gels with reversible sol- gel transition and self-healing properties. *Macromolecules* **2010**, *43*, 1191–1194. [CrossRef]
9. Yoon, J.A.; Kamada, J.; Koynov, K.; Mohin, J.; Nicolay, R.; Zhang, Y.; Balazs, A.C.; Kowalewski, T.; Matyjaszewski, K. Self-healing polymer films based on thiol–disulfide exchange reactions and self-healing kinetics measured using atomic force microscopy. *Macromolecules* **2011**, *45*, 142–149. [CrossRef]
10. Amamoto, Y.; Kamada, J.; Otsuka, H.; Takahara, A.; Matyjaszewski, K. Repeatable photoinduced self-healing of covalently cross-linked polymers through reshuffling of trithiocarbonate units. *Angew. Chem. Int. Ed.* **2011**, *50*, 1660–1663. [CrossRef]
11. Canadell, J.; Goossens, H.; Klumperman, B. Self-healing materials based on disulfide links. *Macromolecules* **2011**, *44*, 2536–2541. [CrossRef]
12. Burnworth, M.; Tang, L.; Kumpfer, J.R.; Duncan, A.J.; Beyer, F.L.; Fiore, G.L.; Rowan, S.J.; Weder, C. Optically healable supramolecular polymers. *Nature* **2011**, *472*, 334–337. [CrossRef]
13. Holten-Andersen, N.; Harrington, M.J.; Birkedal, H.; Lee, B.P.; Messersmith, P.B.; Lee, K.Y.C.; Waite, J.H. pH-induced metal-ligand cross-links inspired by mussel yield self-healing polymer networks with near-covalent elastic moduli. *Proc. Natl. Acad. Sci. USA* **2011**, *108*, 2651–2655. [CrossRef]
14. Yang, Y.; Urban, M.W. Self-repairable polyurethane networks by atmospheric carbon dioxide and water. *Angew. Chem. Int. Ed.* **2014**, *53*, 12142–12147. [CrossRef]
15. Lu, L.; Fan, J.; Li, G. Intrinsic healable and recyclable thermoset epoxy based on shape memory effect and transesterification reaction. *Polymer* **2016**, *105*, 10–18. [CrossRef]
16. Lai, J.C.; Mei, J.F.; Jia, X.Y.; Li, C.H.; You, X.Z.; Bao, Z. A stiff and healable polymer based on dynamic-covalent boroxine bonds. *Adv. Mater.* **2016**, *28*, 8277–8282. [CrossRef]
17. Zechel, S.; Geitner, R.; Abend, M.; Siegmann, M.; Enke, M.; Kuhl, N.; Klein, M.; Vitz, J.; Gräfe, S.; Dietzek, B.; et al. Intrinsic self-healing polymers with a high E-modulus based on dynamic reversible urea bonds. *NPG Asia Mater.* **2017**, *9*, 1–8. [CrossRef]
18. Wang, K.; Strandman, S.; Zhu, X.X. A mini review: Shape memory polymers for biomedical applications. *Front. Chem. Sci. Eng.* **2017**, *11*, 143–153. [CrossRef]
19. Menon, A.V.; Madras, G.; Bose, S. The Journey of Self-Healing and Shape Memory Polyurethanes from Bench to Translational Research. *Polym. Chem.* **2019**, *10*, 4370–4388. [CrossRef]
20. Maksimkin, A.V.; Dayyoub, T.; Telyshev, D.V.; Gerasimenko, A.Y. Electroactive Polymer-Based Composites for Artificial Muscle-like Actuators: A Review. *Nanomaterials* **2022**, *12*, 2272. [CrossRef]
21. Zhao, W.; Liu, L.; Zhang, F.; Leng, J.; Liua, Y. Shape memory polymers and their composites in biomedical applications. *Mater. Sci. Eng. C* **2019**, *97*, 864–883. [CrossRef]
22. Hu, J.; Meng, H.; Li, G.; Ibekwe, S. A review of stimuli-responsive polymers for smart textile applications. *Smart Mater. Struct.* **2012**, *21*, 053001. [CrossRef]
23. Chan, V.; Yvonne, Y. Investigating smart textiles based on shape memory materials. *Textil. Res. J.* **2007**, *77*, 290–300. [CrossRef]
24. Lee, J.; Hinchet, R.; Kim, S.; Kim, S.; Kim, S. Shape memory polymer-based self-healing triboelectric nanogenerator. *Energy Environ. Sci.* **2015**, *8*, 3605–3613. [CrossRef]
25. Ge, Q.; Qi, H.; Dunn, M. Active materials by four-dimension printing. *Appl. Phys. Lett.* **2013**, *103*, 131901. [CrossRef]
26. Felton, S.; Tolley, M.; Demaine, E.; Rus, D.; Wood, R. A method for building self-folding machines. *Science* **2014**, *345*, 644–646. [CrossRef] [PubMed]
27. Zarek, M.; Layani, M.; Cooperstein, I.; Sachyani, E.; Cohn, D.; Magdassi, S. 3D printing of shape memory polymers for flexible electronic devices. *Adv. Mater.* **2015**, *28*, 4449–4454. [CrossRef]
28. Fan, W.; Zhang, Y.; Li, W.; Wang, W.; Zhao, X.; Song, L. Multi-level self-healing ability of shape memory polyurethane coating with microcapsules by induction heating. *Chem. Eng. J.* **2019**, *368*, 1033–1044. [CrossRef]
29. Gonzalez-Garcia, Y.; Mol, J.M.C.; Muselle, T.; De Graeve, I.; Van Assche, G.; Scheltjens, G.; Van Mele, B.; Terryn, H. SECM study of defect repair in self-healing polymer coatings on metals. *Electrochem. Commun.* **2011**, *13*, 169–173. [CrossRef]

30. Ren, D.; Chen, Y.; Li, H.; Rehman, H.U.; Cai, Y.; Liu, H. High-efficiency dual-responsive shape memory assisted self-healing of carbon nanotubes enhanced polycaprolactone/thermoplastic polyurethane composites. *Colloids Surf. A Physicochem. Eng. Asp.* **2019**, *580*, 123731. [CrossRef]
31. Urban, M.W. *Stimuli-Responsive Materials: From Molecules to Nature Mimicking Materials Design*; Royal Society of Chemistry: Cambridge, UK, 2016.
32. Liu, Y.; Du, H.; Liu, L.; Leng, J. Shape memory polymers and their composites in aerospace applications: A review. *Smart Mater. Struct.* **2014**, *23*, 023001. [CrossRef]
33. Lendlein, A.; Behl, M.; Hiebl, B.; Wischke, C. Shape-memory polymers as a technology platform for biomedical applications. *Expert Rev. Med. Dev.* **2010**, *7*, 357–379. [CrossRef]
34. Wache, H.; Tartakowska, D.; Hentrich, A.; Wagner, M. Development of a polymer stent with shape memory effect as a drug delivery system. *J. Mater. Sci. Mater. Med.* **2003**, *14*, 109–112. [CrossRef]
35. Li, J.; Feng, Q.; Cui, J.; Yuan, Q.; Qiu, H.; Gao, S.; Yang, J. Selfassembled graphene oxide microcapsules in Pickering emulsions for self-healing waterborne polyurethane coatings. *Compos. Sci. Technol.* **2017**, *151*, 282–290. [CrossRef]
36. Chung, U.S.; Min, J.H.; Lee, P.-C.; Koh, W.-G. Polyurethane matrix incorporating PDMS-based self-healing microcapsules with enhanced mechanical and thermal stability. *Colloids Surf. A* **2017**, *518*, 173–180. [CrossRef]
37. Tatiya, P.D.; Hedaoo, R.K.; Mahulikar, P.P.; Gite, V.V. Novel Polyurea Microcapsules Using Dendritic Functional Monomer: Synthesis, Characterization, and Its Use in Self-healing and Anticorrosive Polyurethane Coatings. *Ind. Eng. Chem. Res.* **2013**, *52*, 1562–1570. [CrossRef]
38. Huang, L.; Yi, N.; Wu, Y.; Zhang, Y.; Zhang, Q.; Huang, Y.; Ma, Y.; Chen, Y. Multichannel and Repeatable Self-Healing of Mechanical Enhanced Graphene-Thermoplastic Polyurethane Composites. *Adv. Mater.* **2013**, *25*, 2224–2228. [CrossRef]
39. Chaudhari, A.B.; Tatiya, P.D.; Hedaoo, R.K.; Kulkarni, R.D.; Gite, V.V. Polyurethane Prepared from Neem Oil Polyesteramides for Self-Healing Anticorrosive Coatings. *Ind. Eng. Chem. Res.* **2013**, *52*, 10189–10197. [CrossRef]
40. Feula, A.; Pethybridge, A.; Giannakopoulos, I.; Tang, X.; Chippindale, A.; Siviour, C.R.; Buckley, C.P.; Hamley, I.W.; Hayes, W.A. Thermoreversible Supramolecular Polyurethane with Excellent Healing Ability at 45 °C. *Macromolecules* **2015**, *48*, 6132–6141. [CrossRef]
41. Lin, Y.; Li, G. An intermolecular quadruple hydrogen-bonding strategy to fabricate self-healing and highly deformable polyurethane hydrogels. *J. Mater. Chem. B* **2014**, *2*, 6878–6885. [CrossRef]
42. Liu, J.; Liu, J.; Wang, S.; Huang, J.; Wu, S.; Tang, Z.; Guo, B.; Zhang, L. An advanced elastomer with an unprecedented combination of excellent mechanical properties and high self-healing capability. *J. Mater. Chem. A* **2017**, *5*, 25660–25671. [CrossRef]
43. Chen, S.; Mo, F.; Yang, Y.; Stadler, F.J.; Chen, S.; Yang, H.; Ge, Z. Development of zwitterionic polyurethanes with multi-shape memory effects and self-healing properties. *J. Mater. Chem. A* **2015**, *3*, 2924–2933. [CrossRef]
44. Zheng, K.; Tian, Y.; Fan, M.; Zhang, J.; Cheng, J. Recyclable, shape-memory, and self-healing soy oil-based polyurethane crosslinked by a thermoreversible Diels-Alder reaction. *J. Appl. Polym. Sci.* **2018**, *135*, 46049. [CrossRef]
45. Kazemi-Lari, M.A.; Malakooti, M.H.; Sodano, H.A. Active photo-thermal self-healing of shape memory polyurethanes. *Smart Mater. Struct.* **2017**, *26*, 055003. [CrossRef]
46. Van Herck, N.; Du Prez, F.E. Fast Healing of Polyurethane Thermosets Using Reversible Triazolinedione Chemistry and ShapeMemory. *Macromolecules* **2018**, *51*, 3405–3414. [CrossRef]
47. Zhang, L.; Chen, L.; Rowan, S.J. Trapping Dynamic Disulfide Bonds in the Hard Segments of Thermoplastic Polyurethane Elastomers. *Macromol. Chem. Phys.* **2017**, *218*, 1600320. [CrossRef]
48. Jian, X.; Hu, Y.; Zhou, W.; Xiao, L. Self-healing polyurethane based on disulfide bond and hydrogen bond. *Polym. Adv. Technol.* **2018**, *29*, 463–469. [CrossRef]
49. Yang, Y.; Lu, X.; Wang, W. A tough polyurethane elastomer with self-healing ability. *Mater. Design* **2017**, *127*, 30–36. [CrossRef]
50. Yarmohammadi, M.; Shahidzadeh, M.; Ramezanzadeh, B. Designing an elastomeric polyurethane coating with enhanced mechanical and self-healing properties: The influence of disulfide chain extender. *Prog. Org. Coat.* **2018**, *121*, 45–52. [CrossRef]
51. Ling, L.; Li, J.; Zhang, G.; Sun, R.; Wong, C.-P. Self-Healing and Shape Memory Linear Polyurethane Based on Disulfide Linkages with Excellent Mechanical Property. *Macromol. Res.* **2018**, *26*, 365–373. [CrossRef]
52. Du, W.; Yong, J.; Jiezhou, P.; Wuhou, F.; Shuangquan, L.; Xiaopeng, S. Thermal induced shape-memory and self-healing of segmented polyurethane containing diselenide bonds. *J. Appl. Polym. Sci.* **2018**, *135*, 46326. [CrossRef]
53. Yuan, C.; Rong, M.Z.; Zhang, M.Q. Self-healing polyurethane elastomer with thermally reversible alkoxyamines as crosslinkages. *Polymer* **2014**, *55*, 1782–1791. [CrossRef]
54. Lutz, A.; van den Berg, O.; Van Damme, J.; Verheyen, K.; Bauters, E.; De Graeve, I.; Du Prez, F.E.; Terryn, H. A ShapeRecovery Polymer Coating for the Corrosion Protection of Metallic Surfaces. *ACS Appl. Mater. Interfaces* **2015**, *7*, 175–183. [CrossRef]
55. Gonzalez-García, Y.; Mol, J.M.C.; Muselle, T.; De Graeve, I.; Van Assche, G.; Scheltjens, G.; Van Mele, B.; Terryn, H. A combined mechanical, microscopic and local electrochemical evaluation of selfhealing properties of shape-memory polyurethane coatings. *Electrochim. Acta* **2011**, *56*, 9619–9626. [CrossRef]
56. Jorcin, J.-B.; Scheltjens, G.; Van Ingelgem, Y.; Tourwe, E.; Van Assche, G.; De Graeve, I.; Van Mele, B.; Terryn, H.; Hubin, A. Investigation of the self-healing properties of shape memory polyurethane coatings with the 'odd random phase multisine' electrochemical impedance spectroscopy. *Electrochim. Acta* **2010**, *55*, 6195–6203. [CrossRef]

57. Chen, S.; Yuan, H.; Zhuo, H.; Chen, S.; Yang, H.; Ge, Z.; Liu, J. Development of liquid-crystalline shape-memory polyurethane composites based on polyurethane with semi-crystalline reversible phase and hexadecyloxybenzoic acid for self-healing applications. *J. Mater. Chem. C* **2014**, *2*, 4203–4212. [CrossRef]
58. Fang, W.; Liu, L.; Li, T.; Dang, Z.; Qiao, C.; Xu, J.; Wang, Y. Electrospun N-substituted polyurethane membranes with self-healing ability for self-cleaning and oil/water separation. *Chem.-Eur. J.* **2016**, *22*, 878–883. [CrossRef]
59. Cao, S.; Li, S.; Li, M.; Xu, L.; Ding, H.; Xia, J.; Zhang, M.; Huang, K. The thermal self-healing properties of phenolic polyurethane derived from polyphenols with different substituent groups. *J. Appl. Polym. Sci.* **2019**, *136*, 47039. [CrossRef]
60. Nazari, M.H.; Zhang, Y.; Mahmoodi, A.; Xu, G.; Yu, J.; Wu, J.; Shi, X. Nanocomposite organic coatings for corrosion protection of metals: A review of recent advances. *Prog. Org. Coat.* **2022**, *162*, 106573. [CrossRef]
61. Mathew, A.; Kurmvanshi, S.; Mohanty, S.; Nayak, S.K. Influence of diisocyanate, glycidol and polyol molar ratios on the mechanical and thermal properties of glycidyl-terminated biobased polyurethanes. *Polym. Int.* **2017**, *66*, 1546–1554. [CrossRef]
62. Yeganeh, H.; Lakouraj, M.M.; Jamshidi, S. Synthesis and characterization of novel biodegradable epoxy-modified polyurethane elastomers. *J. Polym. Sci. Part A Polym. Chem.* **2005**, *43*, 2985–2996. [CrossRef]
63. Slobodinyuk, A.; Strelnikov, V.; Senichev, V.Y.; Slobodinyuk, D. Preparation, Structure and Properties of Urethane-Containing Elastomers Based on Epoxy Terminal Oligomers. *Polymers* **2022**, *14*, 524. [CrossRef] [PubMed]
64. Slobodinyuk, A.; Strelnikov, V.; Elchisheva, N.; Kiselkov, D.; Slobodinyuk, D. Synthesis and Study of Physical and Mechanical Properties of Urethane-Containing Elastomers Based on Epoxyurethane Oligomers with Controlled Crystallinity. *Polymers* **2022**, *14*, 2136. [CrossRef] [PubMed]
65. Slobodinyuk, A.; Elchisheva, N.; Strelnikov, V.; Chernova, G.; Slobodinyuk, D. Modified oligoether-diamine synthesis for the preparation of crystallizable polymers based on epoxyurethane oligomers. *Z. Nat. B* **2023**, *78*, 17–23. [CrossRef]
66. Slobodinyuk, A.; Strelnikov, V.; Kiselkov, D.; Slobodinyuk, D. Synthesis of oligotetramethylene oxides with terminal amino groups as curing agents for an epoxyurethane oligomer. *Z. Nat. B* **2021**, *76*, 511–515. [CrossRef]
67. Zhang, K.; Li, L.; Chen, X.; Lu, C.; Ran, J. Controlled preparation and properties of acrylic acid epoxy-acrylate composite emulsion for self-crosslinking coatings. *J. Appl. Polym. Sci.* **2022**, *139*, 51441. [CrossRef]
68. Desmet, G.; D'hooge, D.; Omurtag, P.; Espeel, P.; Marin, G.; Du Prez, F.; Reyniers, M.-F. Quantitative first-principles kinetic modeling of the Aza-Michael addition to acrylates in polar aprotic solvents. *J. Org. Chem.* **2016**, *81*, 12291–12302. [CrossRef]
69. Matsushima, H.; Shin, J.; Bowman, C.N.; Hoyle, C.E. Thiol-isocyanate-acrylate ternary networks by selective thiol-click chemistry. *J. Polym. Sci. Part A Polym. Chem.* **2010**, *48*, 3255–3264. [CrossRef]
70. Lee, H.; Neville, K. *Handbook of Epoxy Resins*; McGraw-Hill: New York, NY, USA, 1967.
71. Guadagno, L.; Vertuccio, L.; Sorrentino, A.; Raimondo, M.; Naddeo, C.; Vittoria, V.; Iannuzzo, G.; Calvi, E.; Russo, S. Mechanical and barrier properties of epoxy resin filled with multi-walled carbon nanotubes. *Carbon* **2009**, *47*, 2419–2430. [CrossRef]
72. Stefani, P.M.; Moschiar, S.M.; Aranguren, M.I. Epoxy-urethane copolymers: Relation between morphology and properties. *J. Appl. Polym. Sci.* **2001**, *82*, 2544–2552. [CrossRef]
73. Socrates, G. *Infrared and Raman Characteristic Group. Frequencies: Tables and Charts*, 3rd ed.; Wiley: New York, NY, USA, 2004.
74. Strel'nikov, V.N.; Senichev, V.Y.; Slobodinyuk, A.I.; Savchuk, A.V.; Volkova, E.R.; Makarova, M.A.; Belov, Y.L.; Derzhavinskaya, L.F.; Selivanova, D.G. Preparation and Properties of Frost-Resistant Materials Based on Compounds of Oligoether Urethane Epoxides and Diglycidyl Urethane. *Russ. J. Appl. Chem.* **2018**, *91*, 1937–1944. [CrossRef]
75. Luo, X.; Mather, P.T. Shape memory assisted self-healing coating. *ACS Macro Lett.* **2013**, *2*, 152–156. [CrossRef]

Disclaimer/Publisher's Note: The statements, opinions and data contained in all publications are solely those of the individual author(s) and contributor(s) and not of MDPI and/or the editor(s). MDPI and/or the editor(s) disclaim responsibility for any injury to people or property resulting from any ideas, methods, instructions or products referred to in the content.

Article

Optimization of Polyolefin-Bonded Hydroxyapatite Graphite for Sustainable Industrial Applications

Ahmed A. Bakhsh

Department of Industrial Engineering, Faculty of Engineering, King Abdulaziz University, Jeddah 21589, Saudi Arabia; aabakhsh@kau.edu.sa

Abstract: As a means of introducing environmental responsibility to industrial applications, the usage of biobased composite materials has been encouraged in recent years. Polymer nanocomposites utilize polyolefins increasingly as a matrix, owing to the diversity in their features and prospective applications, even though typical polyester blend materials, such as glass and composite materials, have garnered greater attention from researchers. The mineral hydroxy-apatite, or $Ca_{10}(PO_4)_6(OH)_2$, is the primary structural component of bone and tooth enamel. Increased bone density and strength result from this procedure. As a result, nanohms are fabricated from eggshells into rods with very tiny particle sizes. Although there have been many papers written on the benefits of HA-loaded polyolefins, the reinforcing effect of HA at low loadings has not yet been taken into account. The purpose of this work was to examine the mechanical and thermal characteristics of polyolefin-HA nanocomposites. These nanocomposites were built out of HDPE and LDPE (LDPE). As an extension of this work, we investigated what would happen when HA is added to LDPE composites at concentrations as high as 40% by weight. Carbonaceous fillers, including graphene, carbon nanotubes, carbon fibers, and exfoliated graphite, all play significant roles in nanotechnology owing to the extraordinary enhancements in their thermal, electrical, mechanical, and chemical properties. The purpose of this study was to examine the effects of adding a layered filler, such as exfoliated graphite (EG), to microwave zones that might have real-world applications for their mechanical, thermal, and electrical characteristics. Mechanical and thermal properties were significantly enhanced by the incorporation of HA, notwithstanding a minor decrease in these attributes at a loading of 40% HA by weight. A higher load-bearing capability of LLDPE matrices suggests their potential usage in biological contexts.

Keywords: polyolefins; mechanical; thermal; nano fillers; industrial rubber; hydroxy-apatite; exfoliated graphite; electrical characteristics

Citation: Bakhsh, A.A. Optimization of Polyolefin Bonded Hydroxyapatite Graphite for Sustainable Industrial Applications. *Polymers* **2023**, *15*, 1505. https://doi.org/10.3390/polym15061505

Academic Editor: Md Najib Alam

Received: 28 February 2023
Revised: 11 March 2023
Accepted: 15 March 2023
Published: 17 March 2023

Copyright: © 2023 by the author. Licensee MDPI, Basel, Switzerland. This article is an open access article distributed under the terms and conditions of the Creative Commons Attribution (CC BY) license (https://creativecommons.org/licenses/by/4.0/).

1. Introduction

Applications for standard polymers have become more diverse as a result of polymer blends and grafts. Polymer blends include a wide range of products that combine two or more polymer components into a single blend or network [1]. Several types of materials may be described by just one term. It is not only conceivable but also extremely usual to combine two polymers into a single material, to improve the qualities that each polymer gives on its own. Many polymer blends exhibit phase separation, with the extent of phase separation depending on the specific mix. Depending on the composition, glass transition temperature, and phase continuity, these multiphase component polymer systems [2] may provide a broad spectrum of qualities, from toughened elastomers to high-impact plastics.

Polymer Blends

Polymer blends, which have similar qualities to metal alloys, are created when at least two different polymers are mixed. There are now many different kinds of polymeric materials that can be produced by combining different kinds of polymers. Although no new monomers or chemical processes [3] have been discovered, this has not stopped a revolution in the study of materials science. Combining existing polymers with a known characteristic set of commercialized qualities not satisfied by any of them alone offers the benefit of a new scale of research and development with minimal expenditure, as opposed to producing new monomers and polymers to generate a comparable property profile. This is because combining existing polymers with a confirmed set of desirable qualities [4] that cannot be achieved by any one of them alone opens up new research and development scales. Lower costs associated with expanding production and bringing a product to market are potential advantages. When compared to single polymers, blends of various polymers may occasionally offer a more appealing combination of property profiles. Unique monomer/polymer blends, on the other hand, are completely novel.

The macroscopic and microscopic morphology is formed, in part, by the thermodynamic and rheological characteristics [5] of the constituents and the techniques of compatibilization. The "macro morphology" of a polymer mix describes the size and structure of the macromolecular phases that occur. These two variables may be used to define the phenomena that occur during compounding or mixing. Due to their low entropy of mixing, most polymer mixes are incompatible with one another.

Whether or not two polymers may be combined without precipitation is determined by the free energy of mixing, which includes both entropic and enthalpic components [6].

$$\Delta G_{mx} = \Delta H_{mu} - T\Delta S_{mx} = \Delta E_{ma} + P\Delta V_{mx} - T\Delta S_{wx} \tag{1}$$

When at least two different polymers are combined to create a new material with varying properties, polymer blends are formed, which are similar to metal alloys. The blending of polymers has allowed for the synthesis of a plethora of one-of-a-kind polymeric materials, therefore, revolutionizing the area of materials science, without necessitating the discovery of a new monomer or the development of a new chemical procedure. The ability to combine existing polymers with proven characteristics that set the commercialized properties [7] not met by any of them singularly offers the advantage of a new scale of research and development at marginal expense, as compared to the cost of developing new monomers and polymers to yield a similar property profile. The fact that these polymers go together so well makes this a real possibility. During both the expansion phase and the transformation phase into a commercial enterprise, the low cost of financing is beneficial. In many cases, it is not possible to achieve the desired balance of properties in a composition by using a single polymer or monomer, but a blend of polymers may be able to do so. This is because numerous different monomers combine to form a polymer mix.

The macroscopic and microscopic morphologies are affected by the thermodynamic and rheological characteristics [8] of the components as well as the compatibilization procedures. In polymer blends, the "macro morphology" describes the size and form of the macromolecular phases. The formation of macroscopic phases during compounding or blending is referred to as the "macro morphology" of polymer blends. As the entropy of mixing most polymer combinations is so low, it is not practical to use them together.

Whether or not two polymers can be combined depends on their free energy of mixing [9]. Therefore, there might be both entropic and enthalpic energy:

$$\frac{\Delta G_{mat}}{RT} = \left(\frac{\phi_A}{N_A}\right)ln\phi_A + \left(\frac{\phi_G}{N_B}\right)ln\phi_B + \chi_{FN}\phi_A\phi_B \tag{2}$$

When two or more types of polymers are combined to form a new material with unique characteristics, a blend of polymers is an inevitable consequence. The resultant material has properties similar to metal alloys. The ability to blend polymers has allowed for the production of several novel polymeric materials, sparking a revolution in materials research. There was no need to find a new monomer or develop a new chemical process to achieve this. Combining existing polymers with a proven characteristic set to commercialize properties not met by any of them singularly offers the advantage of new scales of research and development at marginal expense, in comparison to the cost of developing new monomers and polymers to yield a similar property profile. This is a real possibility when existing polymers are combined with a well-established set of desirable characteristics that none can provide on their own. When starting business and expanding existing operations, low expenditure is beneficial. If the objective is to generate a more acceptable balance of qualities, a composition consisting of a single polymer or monomer might be better than a composition consisting of a mix of polymers.

The thermodynamic and rheological properties of the components and the compatibilization procedures influence both the macroscopic and microscopic morphologies [10]. A polymer blend's "macro morphology" characterizes the shape and size of its macromolecular phases. The creation of macroscopic phases during the compounding or blending process is referred to as "macro morphology" [11]. Unfortunately, most polymer mixtures cannot be employed together because of their poor mixing entropy.

To ascertain whether or not two polymers may be mixed, the free mixing energy [12] is used. Entropic and enthalpic energy is detectable in this system. The necessary condition for phase separations to occur is

$$\left(\frac{\partial^2 \Delta G_a}{\partial \phi_2^2}\right)_{T,\beta} = 0 \qquad (3)$$

where Gm is Gibb's mixing free energy. If it is established that the miscibility parameter Gm is zero, we obtain the mixing configuration entropy, Sn0. Specifically, an interaction of Mer 1 and Mer 2 is the only way to produce a uniform mixture [13]. These interactions may result in ionic or dipolar interactions. In general, polymer miscibility decreases with increasing temperature but improves with increasing pressure.

For mixtures to be miscible by boiling, the mixture quality must be stated.

$$P = P_l \Phi_1 + P_2 \Phi_2 + l \phi_1 \Phi_2 \qquad (4)$$

When two or more polymers are combined, the resulting material may be tailored to specific applications [14]. Producing metal alloys is similar to this procedure. Many novel polymeric materials may be created by the mixing of polymers, which has transformed the area of materials science without the requirement to discover a new monomer or new chemistry. Combining existing polymers with a proven characteristic set of commercialized properties not met by any of them singularly offers the benefit of a new scale of research and development at marginal expense, as opposed to the process of developing new monomers and polymers to produce a similar property profile. This differs from the standard practice of creating new types of monomers and polymers. The ability to grow operations with little initial investment is a major advantage. A composition consisting of a mixture of polymers, as opposed to a composition consisting of a single polymer or monomer, may be preferable to attain a more acceptable balance of properties.

The thermodynamic and rheological characteristics of the constituents, in addition to the compatibilization processes, affect the macroscopic and microscopic morphologies. Macro morphology describes the size and shape of the macromolecular phases formed in polymer blends during the compounding or blending processes. Due to their low entropy in mixing, the vast majority of polymer mixtures are useless when combined.

The free mixing energy [15] is used as a criterion to establish whether two polymers are compatible with one another. Both entropy and enthalpy may be thought of as kinds of energy that are present in the system.

2. Background Study

Yoshihiko Ohama conducted research into syntactic foam material model assessment [16]. They carried out their analysis by employing the RVE method that is provided by commercial FE software (ANSYS) and placing balloons at random. They created an inclusion microstructure that does not overlap and has six distinct volume fractions—0.1, 0.2, 0.3, 0.4, 0.5, and 0.6, respectively. They assessed both the viable Youthful's modulus and the powerful Poisson's proportion by exposing the RVE to occasional uniaxial limit conditions. This was so they could look at the RVE. They contrasted the analytical model with the FE models in order to precisely predict the effective elastic module of syntactic foams.

Utilizing Laguerre tessellation models, Michele T. Byrne and Yurii K. Gun'ko investigated the effect that cell wall thickness and size have on the strength of closed-cell foam [17]. In 2017, their study was published. In order to allow it to withstand both compressive and shear loads, the RVE model received two hard shells—one at the top and one at the bottom. The top shell was given a constant velocity in the downward and sideways directions for the compressive and shear loadings, respectively, while the bottom shell was held stationary. Using a commercial FE software package known as ABAQUS/Explicit, they were able to successfully resolve the issue in both modes. They discovered that as the diversity of cell size and cell wall thicknesses increased, so did the compressive and shear strengths. Additionally, they discovered that compressive strength was more susceptible to variation than shear strength.

Utilizing the RVE method with a void percentage of forty percent, Eric J. H. Chen and Benjamin S. Hsiao investigated the effect that particle clustering has on the tensile characteristics and failure processes of syntactic foams [18]. The numbers 0.2, 0.4, 0.6, and 0.8 stand for the four distinct levels of clustering they used. The extreme values of =0 and =1 are regarded as 39, respectively, indicating that the particles are fully grouped and evenly distributed. They published a stress–strain curve that was a homogenized response of RVE with five different clustering levels. At first, the response was linear elastic in a particular strain range, and then it became nonlinear. They reported this to us. Additionally, they demonstrated that as the degree of particle clustering rises, syntactic foams' tensile strength and fracture strain gradually decrease. They discovered that this was the situation. They stated that the simulation curves' nonlinear parts were not as smooth as the experimental curve, and they also reported the experimental stress–strain curve for comparison with the simulated results. Utilizing an RVE model with few particles is the direct cause of these significant reductions in stiffness.

A dynamic finite element method for modeling blow molding and thermoforming was developed using the Mooney–Rivlin hyperplastic material model [19]. The degree of freedom of the parison remained fixed on the solid boundary until the simulation's conclusion because they assumed that the contact between the parison and the mound was sticky. They used a unique lumping method to put the explicit central difference time integration scheme into action. Discretization was carried out for triangular components. Instances of blow shaping and thermoforming of convoluted objects, such as a container with a handle and modern box math, were utilized to show the newfound interaction.

Ref. [20] used an extrusion blow molding method to study the particle formation. To understand the thickness distribution of the blown part, they used the FE software program ANSYS Polyfold to simulate blow-molded components and then compared the results of the simulation to those of the experiments. Their examination depended on a similar investigation. They looked at the comparison between the amount of blowing agent and the speed of extrusion and found that increasing the amount of blowing agent increases the comparison's length and perimeter, while increasing the speed of extrusion makes the comparison's length slightly shorter. The blow molding simulation was carried out at blowing pressures of 0.3 and 0.7 MPa because the mound is not symmetrical. It was found that the container's thickness does not stay the same around its perimeter, regardless of the pressure used. The portion that was blown at 0.7 MPa had a surface quality that was extremely regular, homogeneous, smooth, and brilliant, despite having a thickness that was slightly lower than that of the portion that was blown at 0.3 MPa. Shahzad et al.'s study, which was published in 2015, utilized experimental data as an input into the FE software tool to carry out an analysis that resulted in the establishment of the hyperplastic material model (ABAQUS). In order to collect the necessary input data for the identification of the material model, they carried out tests with a total of four distinct loadings—uniaxial tensile, volumetric, planar, and biaxial—in order to collect the data. The coefficients of the models were retrieved so that they could be simulated using the curve fitting tool that is included in ABAQUS. After running simulations and comparing the results to the test data, they came to the conclusion that the Yeoh model was the best option.

Research Motivation

For polymers such as polypropylene (PP) that have crystalline structures and low melt strengths, the production of olefin-based foams using a variety of processing processes has proven to be a problematic subject for both academic researchers and industrial manufacturers. Even though researchers have developed and examined a wide variety of foams created by a variety of processes, and although these foams are routinely used in industries, there are still unknown areas of study that may be further expounded upon. In addition to these significant advancements, further developments in terms of both experimental and modeling studies are required for better understanding and control of process parameters, particularly on processing conditions, such as pressure drop rate, temperature, the content of blowing agent in a matrix, etc., which govern the nucleation, cell growth, cell size, structures, and cell coalescence. These conditions include pressure drop rate, temperature, the content of the blowing agent in a matrix, etc. The structure of the foam and the cell shape are both determined by the process parameters.

Yet, it has been established that the addition of a tiny number of nanoparticles may affect the structure of foams and increase their capabilities, hence, making it possible for foams to be used in applications that were not possible before. The manufacture of polymeric foams, which regulates the governing characteristics of the foams, still faces considerable problems, however. In addition, there is a pressing need for a deeper comprehension of the structure–property link in conjunction with the operating conditions. The creation of cellular structures on both the micro- and the nanoscales has been the focus of a significant amount of study and technical progress over the last several decades. Despite the advances in research and technology, there is still a deficiency in the simulation analysis of foam structure and the optimal processing condition of foam products, particularly in foam blow molding. This is especially true of the case in foam blow molding. Despite this, the technology of foam blow molding is still relatively new. While it has begun to make its way into the market, it is still in the early stages of development. The simulation analysis of these foams is an excellent tool for optimizing both the products and the process. It is feasible, via the use of mathematical models, to obtain a good agreement between the outcome of a simulation and the solution of a real-world situation. Yet, due to the intricate nature of the foam's structure and the nature of its two-phase system, it is difficult to imitate the foam. In addition, there is a scarcity of data and mathematical models, which further complicates the task of research in this area.

3. Materials and Methods

Table 1 lists the many different possible combinations of saturation temperature, pressure, and other process parameters that may be used to perform PIF foaming. According to Table 1, when the PIF temperature rises, the relative density of the foams decreases and their porosity rises. Relative density decreases in a way that is directly proportional to the increase in porosity and expansion ratio.

To have a better understanding of how temperature-induced foaming works, we compiled a summary of the relevant processing parameters in Table 1. The saturation time in the TIF technique was much longer than in the PIF method when the materials were solid and the temperature was kept at room temperature. Further, the water in the tub was hotter than the PIF. The "immersion time" refers to the length of time the solid samples spent in the hot glycerol bath before turning into foam. The relative density and porosity of these foams are also shown in the table, showing that the relative density of the final foams did not differ much over the range of TIF conditions. The data support this claim since it is included in the table. The next several sections will explain why there is so little fluctuation in the relative density of TIF foams and what effect this has on the final product. Table 2 indicates the temperature, saturation pressure, saturation time and their destiny involved in industrial rubber.

Table 1. Pressure induced in industrial rubber.

Sample	Pressure Induced Foaming				
	Saturation Pressure (MPa)	Saturation Temperature (°C)	Saturation Time (h)	Relative Density	Porosity
PP-00	–	–	–	1.0	–
PPF-01	7	150	2	0.431 ± 0.048	0.569
PPF-02	7	160	2	0.347 ± 0.050	0.653
PPF-03	7	170	2	0.205 ± 0.009	0.795
PPF-04	7	180	2	0.207 ± 0.039	0.793

Table 2. Temperature induced in industrial rubber.

Sample	Temperature Induced Foaming				
	Saturation Pressure (MPa)	Saturation Time (h)	Glycerol Temperature (°C)	Relative Density	Porosity
PPF-05	7	24	180	0.473 ± 0.017	0.527
PPF-06	7	24	190	0.512 ± 0.038	0.488
PPF-07	7	24	200	0.509 ± 0.026	0.491
PPF-08	7	24	210	0.518 ± 0.017	0.482

Polypropylene (PP), which is a versatile polymer, has seen a rapid increase in its use as a result of its excellent performance as well as improvements in its production costs. The commercial market has shown a significant amount of interest in PP and PE blends. The addition of PE to PP has the potential to improve the material's resistance to impact at low temperatures. Table 3 shows the properties and values of the material's normalized mechanical.

Table 3. Properties of HDPE.

Properties	Values	Test Methods
At 192C and 1.98 kg	0.06	ISO1132
Input density	921	ISO1181
Input tensile		ASTM 881
Elasticity behavior	1150	
Thermal range	110	
Processing range	210–240	ASTM 227

Weld lines are an inevitable part of the manufacturing process for most big injection-molded items, such as appliances, home wares, furniture, sports goods, toys, packaging, chemical processing equipment, and industrial components. Automobiles are a good illustration of this category. In the presence of weld lines, optimizing processing parameters and determining the optimal number of modifiers are crucial for achieving the best possible features. There is no way to ensure that PE and PP work together. In terms of the materials, PP/HOPE blends do not have the best standing with regard to mechanical characteristics.

Their final mechanical compatibilizations are lacking in comparison to the compatibilization of the individual parts.

If included, this component would act as a computerizing agent in the amorphous regions of both polymer species. PPIPE mixes' compatibility may be enhanced in a variety of ways, one of which is by adding the copolymer EPR.

The reaction's aftereffects include the breakdown of PP and the branching or cross-linking of poly(ethylene). The process has to be controlled with a high degree of precision to be optimized. Increased radiation or reactive compatibilization may decrease the mixture's crystallinity. To satisfy the need for thermoplastic materials with the characteristics of vulcanized rubber, rubber-toughened PPIHDPE mixes are essential. According to [21], when HOPE serves as the matrix, PP lamella migrates into the dispersed phase of EDPM, whereas in the case of HDPE, PE lamellae migrate into the PP matrix. PPIHDPE/EPDM refers to tertiary mixes of polypropylene, ethylene, and ethylene propylene. The PP crystallization was unaffected by the addition of PE. As a pair, PP and PE produce results that are in direct opposition to one another. A two-phase structure is produced by the combination of two different polymers. When the mix is put under extreme stress, its structure might compromise its performance. Dispersion and thermal and mechanical degradation rates may be enhanced by increasing the mixing duration or intensity. Excessive stretching causes incompatibility between the PP and PE, which degrades their mechanical qualities. Evidence suggests that the ethylene-propylene copolymer provides stronger adhesion than the separate monomers provide to one another.

The adhesion between the polymers is the most influential element in determining the ductility-related properties of immiscible blends. The adhesion between ductile and brittle parts might be improved to increase ductility. The Flore-Huggins interaction parameter, a, defines the extent to which individual segments diffuse over the interface between components.

$$a = c / (\chi_{A\varrho})^m \tag{5}$$

$$\rho_f = \frac{a}{a-b} \rho_{water} \tag{6}$$

where it is the density of the foam sample, a is the apparent mass of the sample in air, and b is the apparent mass of the sample completely immersed in water.

$$R_\rho = \frac{\rho_f}{\rho_s} \tag{7}$$

The relative density R is defined as the ratio of density of foam to the density of unfamed solid (ρ_s):

$$R_v = \frac{1}{R_\rho} \tag{8}$$

They produce brittle incompatible combinations when melted together and injected into molds, which is a process known as melt mixing. Because of the poor ductility of the material, injection molding an immiscible mix produces a weld line that is perpendicular to the direction in which the load is applied.

Numerous investigations have been conducted, and the results of those studies have uncovered a variety of elements that lead to the weaker weld lines of homopolymers [20–22]. Quantifying the contents of each bin allowed for the calculation of this size distribution. Using this size distribution as a guide, we were able to calculate the micrograph-based average cell diameter, denoted by the symbol D_v:

$$D_v = \left[\frac{\sum_{i=1}^n d_i^3}{n} \right]^{1/3} \tag{9}$$

The cell density, also known as N_f, is the number of cells that can be found in a given volume of foam. It can be calculated as follows: where n represents the total number of cells and d_i represents the perimeter equivalent diameter of each counted cell:

$$N_f = \left[\frac{nM^2}{A}\right]^{3/2} \times R_v \tag{10}$$

The number of cells that may be found in a certain volume of foam is referred to as the cell density, and it is also denoted by the symbol N_f. It is possible to compute it as follows: where n is the total number of cells and di is the perimeter equivalent diameter of each counted cell. This may be carried out by using the formula.

$$K_{1C} = \delta\left(\frac{E}{H}\right)^{0.5} \frac{P}{C^{1.5}} \tag{11}$$

The process of injection molding is distinguished by the introduction of a complicated and nonisothermal flow into a chamber that has been sealed and cooled. An anisotropic skin core structure is produced as a consequence of this procedure in most cases. Fountain flow is used to accomplish the task of filling the mold cavity. During the molding process, the polymer that is in touch with the cooled mold wall instantly freezes, forming the skin at the location where the shear will be at its greatest.

When obtaining a material's hardness and Young's modulus using instrumented nanoindentation, the Oliver and Pharr (O-P) model, which is the approach that is used the majority of the time, states that the hardness (H) is stated as the Oliver and Pharr (O-P) model:

$$H = P_{mad}/A_{ct} \tag{12}$$

where P_{mad} represents the maximum load that is applied and A_{ct} represents the actual contact area that exists between the indenter and the material. In Oliver and Pharr's work, it is stated that the polynomial form of A_{ct} may be represented as:

$$A_{ct} = 24.56\, h_e^{\,2} + C_1 h_c + C_2 h_e^{\,1/2} + C_3 h_e^{\,1/4} + \cdots\cdots + C_8 h_e^{\,1/128} \tag{13}$$

where C_1 through C_8 are constants that may be determined with the help of a standard calibration procedure, and he is the penetration depth, which can be determined with the help of the formula that is provided below:

$$h_c = h_{max} - k(P_{mad}/S) \tag{14}$$

where k is less than 0.75 in the case of a Berkovich indenter. Another way to express the contact stiffness (S), also known as the slope of the load versus depth of penetration plot shown in Equation (15), is as follows:

$$S = (dP/dh)_{h=max} = aC_A E^* \sqrt{A_{cr}} \tag{15}$$

where is 1.034, C_A equals 2/, and E^* is the effective Young's modulus for a Berkovich indenter. Following the O-P paradigm, E^* may be expressed as:

$$1/E^* = \left(1 - v_i^2\right)/E_i + \left(1 - v_s^2\right)/E_s \tag{16}$$

where the subscripts i and (s) denote the indenter i and sample s, respectively, and where (E) and (v) represent the Young's modulus and the Poisson's ratio, respectively. For a Berkovich indenter, the E_i and v_i pressures are commonly calibrated to be 1140 GPa each.

Particles that are flexible and scattered will have their shape stretched in the direction of the flow as the flow moves through them. The agitated core will experience less stress as a result, and it will have more time to cool down. The result of this labor-intensive process is a skin-core structure with two layers. There have been reports that PP/EPDM and HDPE/PA-6 blends have poor weld line strength due to the form of the skin cores of the blends' respective polymers. The influence of processing variables and compatibilizers on the behavior of HOPE and PP blends along the weld line. To create test specimens with and without weld lines, a semi-automatic plunger-type injection molding machine was used. The weld line coefficient (WLC) is the value that is derived by comparing the characteristics of the sample that contains the weld line to those of the sample that does not include the weld line [22]. This occurs when two samples are prepared under identical conditions. The temperature at which the material is processed has an effect on the yield stress as well as the elongation at break, as shown in Figures 1 and 2. Increasing the processing temperature of HDPE from 190 °C to 250 °C results in an improvement in the material's WLC.

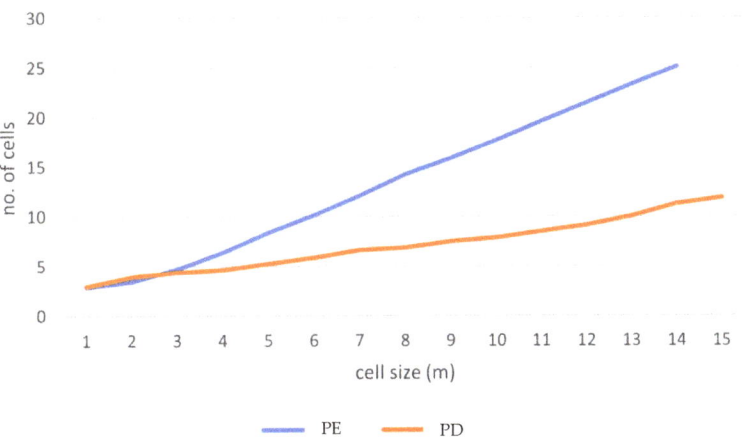

Figure 1. Cell size vs. number of cells.

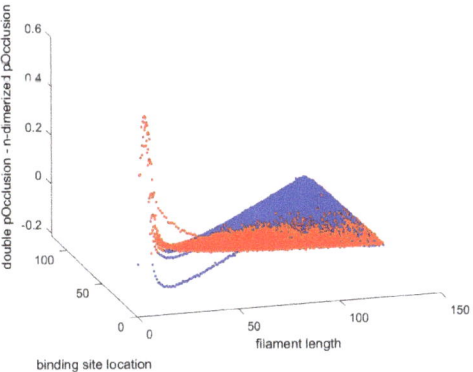

Figure 2. Variation in torque with time of HDPE-HA composites during mixing.

The rate of cooling is determined by several variables, one of which is the thermal diffusivity of the melt.

4. Results

All of the composites were made using a twin-roll mixer with rolls of 45 cm by 15 cm in diameter (a calendar) for ten minutes at a maximum temperature that remained constant at

140 °C. Following the melting of the polymer matrix for one minute, the fillers and additives were then combined and added to the slurry. After the mixing process, composites were recovered with a thickness of 1.5 mm.

The specific gravity of a substance is determined by comparing the mass of that substance to the mass of the same volume of deionized water at a temperature of 23 °C. The specimen is first measured and recorded while it is suspended in air, and then again when it is held submerged in distilled water at a temperature of 23 °C using a sinker and a wire. For each formulation, the density was determined using the standard ASTM D792 and then compared to the value that had been computed. This was so that we could be certain that the formulation had been correctly put together.

At room temperature, the tensile characteristics were evaluated using a Tinius Olsen H10KT dynamometer with an elongation speed of 250 mm/min. The length after stretching was 20 mm 0.5 mm, while the breadth and thickness of the specimens were, respectively, 3.0 mm and 2.0 mm 0.2 mm (according to the standard ISO 37 type 2).

Following the requirements of the international standard ISO 1133:1, the flow parameters were evaluated using an instrument called a Melt Flow Index (MFI).

The FEI Quanta 450 ESEM FEG (FELMI-ZFE, Graz, Austria) was used for the scanning electron microscopy (SEM) analysis that was performed so that the particle morphology of the fillers could be determined.

A laser diffraction technique was used to determine the particle size distribution of the fillers at the D50 level. For these studies, a Rasterizer 2000 manufactured by Malvern Analytical (Malvern, UK) was used.

An Olympus BTX 470 II diffractometer (Olympus, Shinjuku, Tokyo) was used to carry out an X-ray powder diffraction (XRD) analysis.

TA Instruments' TGA Q500 (New Castle, DE, USA) was used during all of the thermogravimetric tests that took place. Samples ranging from 10–15 mg were put in crucibles made of Al_2O_3, and the runs were performed in high-purity nitrogen. Over a range of 50–1000 °C, the rate of heating was 20 °C per minute.

All of the flame tests and LOI measurements were carried out using ASTM D2863 by using an SA ASSOCIATES Oxygen Index instrument on specimens with dimensions of 10 mm by 6 mm by 3 mm. A burner flame was applied to the top of a bar that was positioned vertically in a test column that runs with a combination of oxygen and nitrogen. The LOI value is the minimum percentage of oxygen (%) that must be present in the gas mixture to sustain the combustion of the item being measured. The starting concentration of oxygen is decided in a completely arbitrary manner.

HOPE grade 50 MA 180 from Indian Petrochemicals Limited (IPCL, Mumbai, India) and pp Repoll H 200 MA from Reliance Petrochemicals (Ahmedabad, India) were used in the mixing and molding process. A Thermos Hake Rheochord 600 mixer (Thermo Fisher Scientific Inc., Waltham, MA, USA) with a 69 ern' roller-type rotor chamber was also utilized in this process. Heredia Unmiters' EPDM grade 301 T, TMQ antioxidant, and dicotyl peroxide were used. After being combined in the mixing chamber, the substance was immediately transferred to a two-roll mill to have its particle size decrease.

4.1. Determine the Effect of Melt Temperature on the Strength of PP and HDPE Weld Lines

The weld line strength of semicrystalline materials, such as PP and HDPE, is impacted not only by the rate of crystallization but also by the amorphous sections that remain in the solidified samples. WLC, on the other hand, decreases as the temperature is elevated, most likely because processes that are detrimental to WLC begin to take place.

In terms of PP, the WLC for yield stress and elongation at break starts to dramatically fall at around 240 °C. This is the temperature at which this change occurs. Given that tertiary hydrogen enhances PP's vulnerability to degradation, this was, to some extent, to be expected. The values of the material's normalized mechanical properties are shown in Table 3. HDPE has a very high elongation at break, whereas PP has a much lower value.

Fractures in PP do not exhibit necking, and PP samples collected from weld lines exhibit a high degree of brittleness.

The cell size distributions that were generated from the micrographs and depicted in Figure 1 gave further validation of this behavior when they were examined more closely. Although there is substantial variety in cell sizes, the size distributions shown in Figure 1 demonstrate a clear and gradual shift to the right (bigger cell sizes) as the temperature rises. This occurs even though there is broad variation in cell sizes. This pattern is in agreement with the prediction that PIF cells would continue to an expansion in size as long as the temperature is increased, at least up to 170 °C.

According to these distributions, the average bubble size for PIF at each of the three temperatures is as follows: 3 m when the temperature is 150 °C (PPF-01), 4 m when the temperature is 160 °C (PPF-02), and 312 m when the temperature is 170 °C (PPF-03). The dramatic expansion of the bubbles that occurred when the temperature was raised from 160 to 170 °C may have been caused by a reduction in the viscosity of the polymer that occurred with an increase in temperature. This reduction in the viscosity of the polymer resulted in a lower barrier to the development of cells. Cells may also burst as a result of a drop in viscosity brought on by an increase in temperature, which is another potential cause of this phenomenon.

At a temperature of 180 °C, the average cell diameter was measured to be 115 m. (PPF-04). Further examination of the morphology of the PIF-made foam at 180 °C, as shown in Figure 3, reveals a significant gradient in the cell sizes, which range from the edge to the center of the foam. The cells on the edge are smaller in size in comparison to the cells in the middle. Because the foam morphology exhibits a significant departure at 180 °C, which is in contrast to the trend seen between 150 and 170 °C, it is possible to argue that PIF has a limiting temperature condition at 180 °C for the pressures that are used. This is because the trend seen between 150 and 170 °C can be attributed to the fact that the foam expands as the temperature rises. In addition, at temperatures higher than 180 °C, the solubility of carbon dioxide in the polypropylene matrix begins to noticeably decrease, while the diffusivity out of the polypropylene matrix begins to increase. This results in poor sorption and retention of carbon dioxide at the pressure that was used in this investigation. It is possible that the amplification of these effects at such temperatures is to blame for the lack of foam generation in specimens that were subjected to PIF temperatures higher than 180 °C.

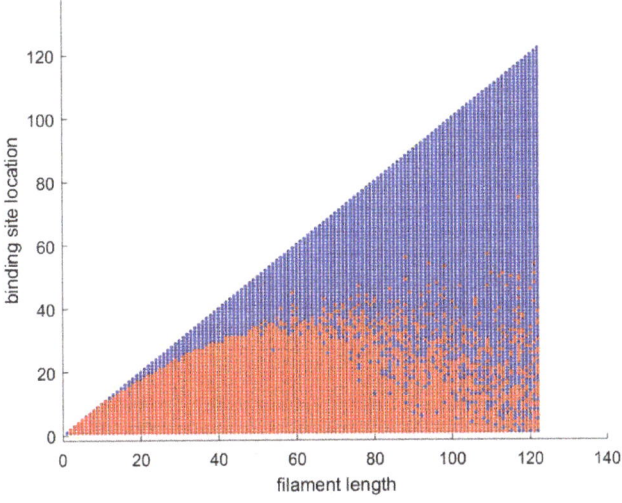

Figure 3. Variation in MFI of HDPE-HA composites with HA loading.

4.2. Torque Studies

As illustrated in Figure 2 the time-dependent torque variations that take place during the melting process of pure HDPE and HDPE-HA composites, respectively. In the illustration, the notation Hex refers to HDPE that has had an unspecified quantity of HA added to it. The torque rises as more polymer is injected, and then decreases while melting takes place, then stabilizing after around two minutes have passed. At this point, HA is added, and the mixture is agitated for a full ten minutes before being poured into the container. The torque is maintained at the same level during the whole mixing procedure, indicating a sufficient amount of filler dispersion across the matrix following the parameters described.

4.3. Melt Flow Index

Figure 3 illustrates how the amount of HA loading affects the pace at which the melt flows. MFI exhibits a minor decline when HA loading was stopped, but only slightly. This demonstrates that the inclusion of HA results in a decrease, although a minor one, in the flow characteristics. The melt flow index is a measurement of the entanglements in the polymer matrix, which may be caused by either chemical or physical cross-links. Since HA causes a small increase in the number of entanglements in the polymer chain, this leads to a decrease in the MFI.

4.4. Tensile Properties of HDPE-HA Composites

Figure 4 illustrates the stress–strain curves of both plain HDPE and composite materials. The elastic area of pure HDPE may be identified by its distinct yield point, which is followed by neck propagation and culminates in strain hardening. After it has been yielded, it can sustain extensive extension, which results in a long plastic area. This occurs as a direct consequence of the entanglements, which cause the molecular chains to lengthen. When HA was added to the mix, the structural properties of the stress–strain curve remained unchanged up to a loading of 1.5 weight %. After this, there is a discernible decrease in the plastic area that takes place. When the weight % of HA reaches three, the necking process fails without any strain hardening taking place. The loading of one weight % of HA is necessary to achieve the maximal amount of elongation. The capacity of the evenly scattered filler particles to perform the function of fracture stoppers is responsible for the improved ductility of the material. As a result, sufficient time is given the opportunity to pass. The tensile strength shows a small rise with HA loading, reaching its maximum value at a loading percentage of one weight percent. Further, also increasing with HA loading is the tensile modulus, which reaches the highest value of 32.7% in the HD1H composite.

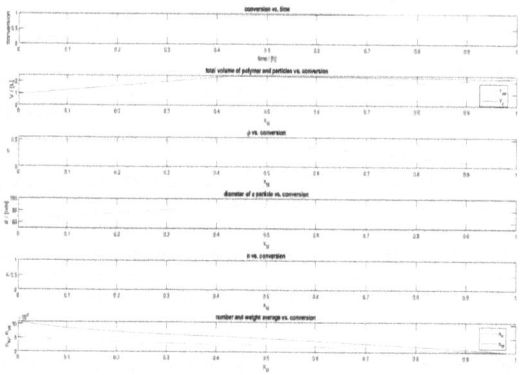

Figure 4. Stress–strain curves of HDPE-HA composites.

4.5. Compressive Properties

According to the findings shown in Figure 5, the compressive modulus increases with increasing HA loading up to a maximum of a 54.8% increase for HD1.5H. After reaching this peak, the rise almost levels out. The ability of the filler particles to close cracks and flaws that are perpendicular to the applied load is responsible for the increased compressive strength that is brought about by the addition of fillers. The increased compressive modulus is an indication of the increased load-bearing capacity of the material.

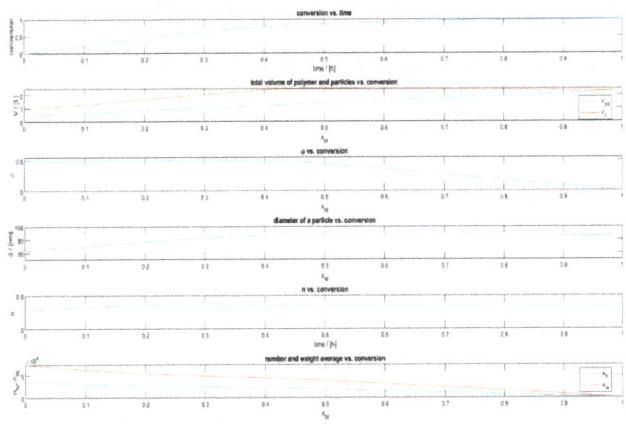

Figure 5. Variation in compressive modulus of HDPE-HA composites with HA loading.

4.6. Flexural Properties

The increased flexural property of the HDPE-HA composites is evident in Figure 6. Both flexural strength and modulus increased with increased HA content, substantiating the enhanced resistance toward bending forces. This is contributed by the stiffening effect of the HA particles.

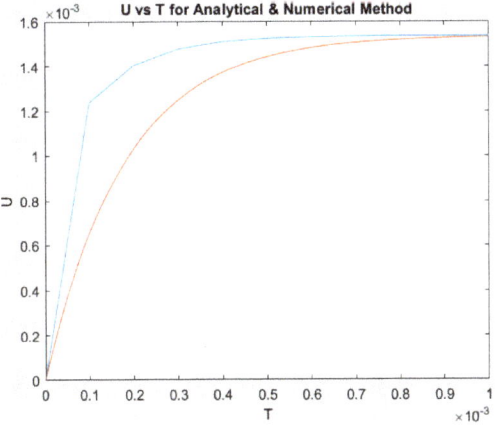

Figure 6. Variation in flexural strength.

4.7. Impact Strength

The effect of HA on the impact absorbance energy is obvious from Figure 7. Impact properties increase with increased HA content. The well-dispersed HA particles help to transfer stress effectively during the impact test.

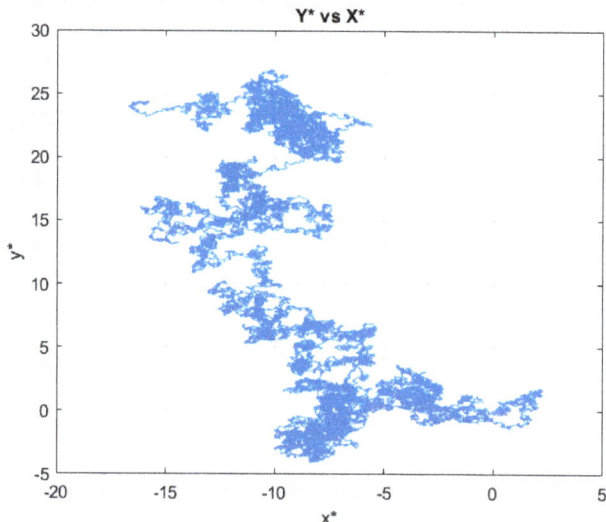

Figure 7. Variation in impact absorbance energy of HDPE-HA composites.

4.8. Dynamic Mechanical Analysis

A comparison of the dynamic mechanical characteristics of pure HDPE and HDPE-HA composites is shown in Figure 8. When the temperature is increased, there is a detectable decrease in the storage modulus across all of the samples. This finding lends credence to the hypothesis that the increased temperature is responsible for the increased mobility of polymer chains. Both storage modulus and loss modulus grow with increasing HA concentration; however, the values for HD1H's storage modulus and loss modulus are the greatest. This is evidence that the micro-HA rods and the polymer matrix are in close contact with one another and are transmitting stress to one another. It may be deduced from the fact that there is practically no movement in the tan delta peak that HA has very little to no effect on damping characteristics. When the amount of EG loading was increased, it was found that all of the nanocomposites' dielectric permittivity, dielectric loss, alternating current (AC) conductivity, and absorption coefficient all increased. The incorporation of conductive EG causes electromagnetic waves to be attenuated before they have the opportunity to enter the material, which results in a reduced depth of the material's skin. The dielectric heating co-efficient in polyolefin-EG composites experiences a significant decrease proportional to the increasing quantity of EG present in the material. When the frequency of the heating is increased, the depth of the skin and the heating coefficient both decrease as shown in Table 4.

Table 5 presents analysis of proposed work. It is indicated that composites made of high-density polyethylene (HDPE) and ethylene glycol (EG) demonstrated the largest improvement in dielectric properties when compared to composites made of low-density polyethylene (LLDPE) and polypropylene (PP)-EG, respectively. We were able to introduce HA into LLDPE at loadings of 10, 20, 30, and 40 weight percent with the use of melt mixing, and we then examined the impact of the HA at each of these different loadings. The inclusion of HA led to a notable improvement in the mechanical and thermal characteristics of the material as a whole, despite the fact that there was minor degradation in the properties when the loading was increased to 40% HA by weight. The higher load-bearing capability of the LLDPE matrix implies that it may be utilized in biomedical applications.

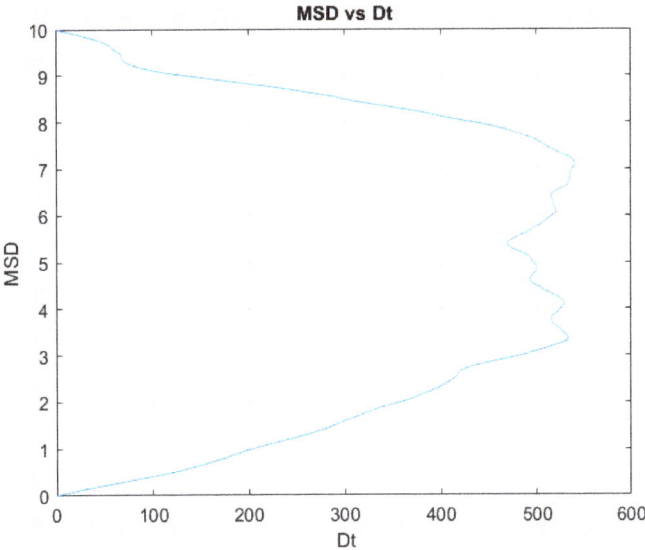

Figure 8. A comparison of the dynamic mechanical characteristics of pure HDPE and HDPE-HA composites.

Table 4. Concentration of proposed work.

Sample	Concentration μ bol.cm^{-2}	
	by UV	by Temperature
HDPE	0.92	0.9
LLDPE	0.42	0.4
HA	1.08	1.0

Table 5. TGA analysis of proposed work.

Sample	Temperature for Weight Loss		Peak Max Ma	Residue at 600 °C (%)	
	10%	25%			
HDPE	426.1	443.8	461.4	474.1	0.08
LLDPE	453.1	477.9	495.2	492.7	38.9
HA	473.2	490.8	505.3	500.6	40.6

Extensive biological testing led to this result. Melt mixing was used to create LLDPE-HA-EG hybrid composites, taking advantage of the synergy between the various fillers in the polymer matrix. Because of the enhanced properties provided by hybrid composites, LLDPE, which was previously solely usable for packaging, containers, and toys, may now be put to a far larger variety of uses. We evaluated the "in vitro" and "in vivo" possibilities of high HA-loaded polyolefin-HA composites and polyol-fin-HA-EG hybrid composites for orthopedic applications and ratio of their effects is shown in Table 6.

Table 6. Effect of ratio.

N	Effect of Ratio	α
1	715.813	3.7603
2	−369.475	3.9685
3	33,311.620	−3.4034

The materials' structures were characterized, and their crystallinity levels were evaluated, using wide-angle, two-angle x-ray microscopy (WAXM) from a range of 4 to 800. In addition to the HDPE peaks at 2 = 21.70, 24.10, 30.20, and 36.40, which correspond to reflections in the (100), (200), (210), and (020) planes, respectively, in all three samples and LLDPE, one can also observe the HA peaks at 2 = 25.90, 31.90 (highest-intensity peak), 39.90, 39.7, 46.80, and 49.50. Peak 260, which is distinctive of HA, is blended with the peak of HA, making the two peaks difficult to identify from one another.

Figure 9 defines the composites' dielectric constant drops with frequency, and the effect is particularly prominent at low frequencies. As the electric field is changed, the dipoles' orientations also change. Lower-frequency dipoles are capable of maintaining phases with the field's oscillation. Dipole orientation and polarization will lag behind the frequency rise, and this is represented by a decrease in the dielectric constant.

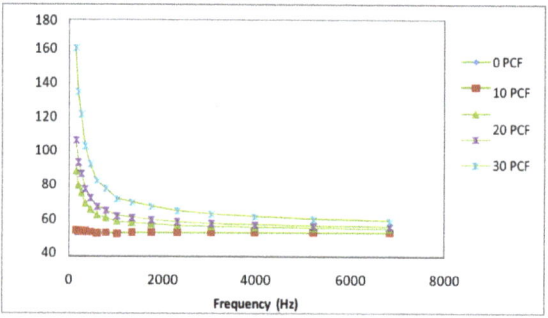

Figure 9. Variation in dielectric constant of PTH-coated fiber/PE composites with frequency.

4.9. Protection against Electromagnetic Interference

Figure 10 explain the variation with regard to the impact of HDPE-coated fiber content and frequency, dielectric loss also mimics the dielectric constant. With each cycle of the alternating electric field, the composites' dipoles swap positions. The friction between the dipole and the surrounding molecules causes the substance to heat up. The dielectric loss of composites may be traced back to this unavoidable thermal energy loss. Increases in fiber content led to an increase in dielectric loss and heat loss due to an increase in the number of dipoles. Lowered frictional heat dissipation manifests as decreased dielectric loss as the frequency at which the dipoles lag behind the alternating field rises.

Figure 11 displays the total EMSE of PCF/PE composites, while Figure 11 displays the EMSE caused by absorption. Inferences may be made about the primary mechanism of shielding, and they point to absorption, allowing reflection loss to play a minor part. Nevertheless, unlike HDPE/PE composites, PCF/PE composites have a very poor shielding effectiveness. This might be because the HDPE-coated fibers contain moisture, reducing conductivity, and/or because the coated fibers disturb the crystallites in the matrix, further reducing conductivity. For optimal shielding, sample thicknesses greater than 1 mm are being investigated. It has been shown that when the percentage of HDPE in a material rises, the frequency of the peak moves down.

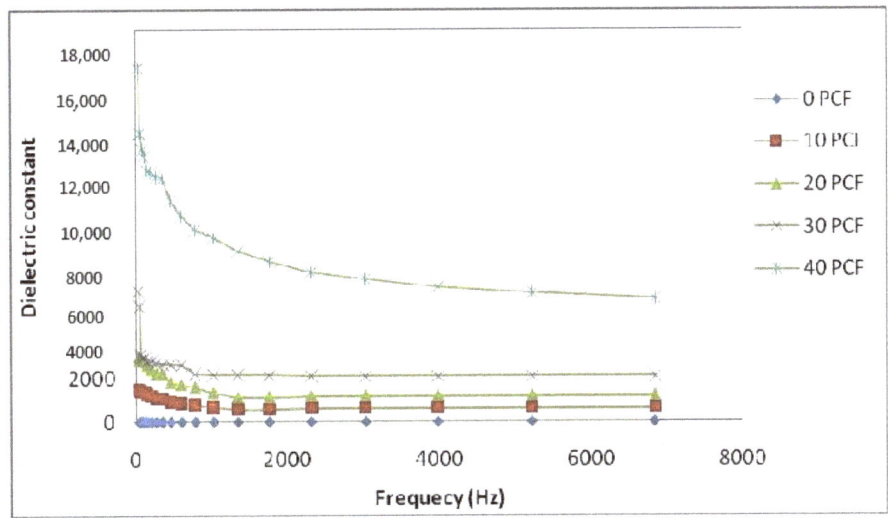

Figure 10. Variation in dielectric loss of HDPE-coated fiber/PE composites with frequency.

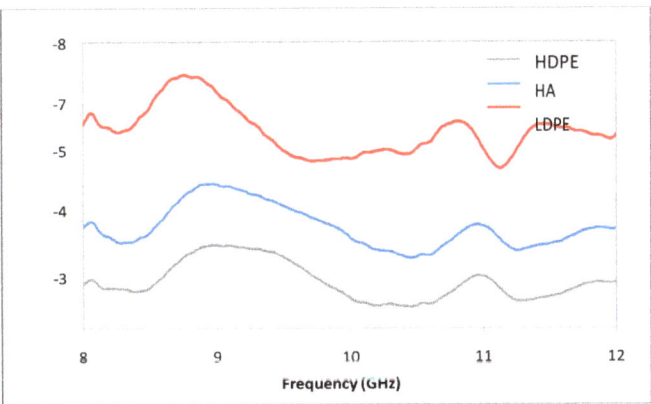

Figure 11. Total EMSE of PCF/PE composites in X band.

Protection against Electromagnetic Interference

Figure 12 illustrates that Polythiophene-coated cellulosic fibers might be incorporated into a common general-purpose thermoplastic polymer, high-density polyethylene, to create conducting composites. Scanning electron microscopy reveals weak matrix-coated fiber interaction due to polar–nonpolar incompatibility. As compared to HDPE/PE composites, the mechanical qualities of PCF/PE composites are inferior. Yet, a lower modulus is beneficial in applications that need adaptability. Moisture is trapped in the coated fibers, according to thermal experiments. Matrix thermal deterioration may be postponed with the use of coated fibers. Because of the presence of moisture in the cellulosic fibers, the conductivity of the HDPE coating and, by extension, the composites is considerably lower than that of HDPE/PE composites. With the use of dynamic mechanical analysis, we can see how a network of coated cellulose fibers forms. The composites are well-suited for use as capacitors due to their high dielectric constant and minimal dielectric loss. Due to their poorer conductivity, PCF/PE composites provide less protection from electromagnetic interference than HDPE/PE composites.

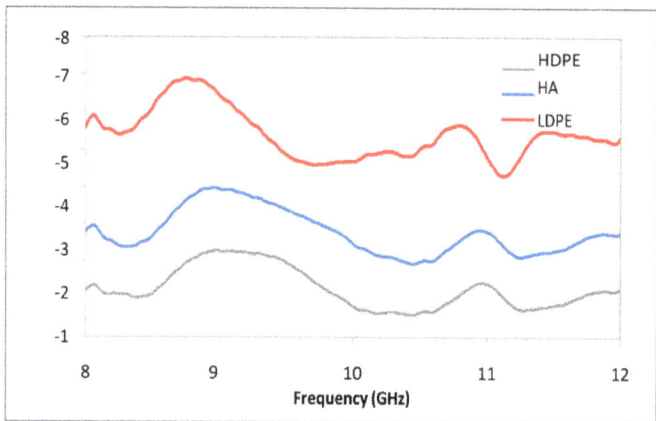

Figure 12. EMSE due to absorption of HDPE /PE composites in X band.

5. Conclusions

The research concluded that the mechanical properties of PP were enhanced by the addition of HA. Composites loaded with 1% HA showed the greatest increase in mechanical parameters, including tensile strength, flexural strength, and compressive modulus.

It would appear that public concern regarding the mechanical response of biocomposite materials is preventing their widespread use and production in industries, particularly in the primary load-carrying section. This is mainly due to the lack of technical data on characteristics, such as tension, compression, fatigue, impact, flammability, etc., on which a dependable engineering design relies significantly for the purpose of failure start and progression prediction. Research into the design and performance of industrial applications, such as product standards, conceptual breakthroughs, lab-scale concepts, durability studies, and degradation models, is crucial for their widespread use in industry. Growing the market for industrial applications might necessitate adjusting aspects, such as supplier–user relations and regulations controlling the use of nano-fillers, in industrial applications utilized in the packaging of food.

Increasing commercialization of these materials is anticipated as the public's awareness of environmental issues grows, as more cost-effective manufacturing methods become widely accessible and as new applications for these materials are identified. Commercialization has been delayed where these resources are most plentiful—in developing nations—because of a lack of adequate research activities. Researchers are challenged with addressing challenges connected to norms and product standards for these components, notwithstanding the renewability and recyclability of industrial applications. Industrial applications developed for structural usage should conform to regulations governing the disposal of massive garbage loads.

Both the scientific and business communities have a keen interest in industrial applications materials, but it has been challenging to find suitable substitutes for synthetic composites in this area. They would be a difficult replacement for conventional synthetic composites in industrial applications due to their low mechanical and thermal properties. Accomplishing the goal of high-performance industrial applications will necessitate careful consideration of the following factors: identification of raw material, extraction procedure, sustainable crop development, application of industrial applications interfacial qualities, material processing and product production, safe service life, and product design. Notwithstanding the potential benefits of multi-fiber reinforcements and mixes of diverse polymers, it seems that they have been the subject of very few studies. More study should be devoted to this area because of the interesting possibilities for new uses. The failure mechanisms caused by the thermo-mechanical–chemical processes of industrial applications, as well as the effects of environmental ageing, need further study. In order to be widely used, materials for industrial applications need to perform as expected, last as long as expected, be reliable, and be easily maintained.

Both the storage and loss moduli improved when HA was put into place. The thermal stability of the composites was improved by the addition of HA. Among these fillers, graphite stands out due to its usefulness as a nano-filler in the form of graphene layers or nano-scale layered stacks. These nanoscale stacked layers have the chemical properties of CNTs and the structural properties of layered silicates. This has the potential to greatly enhance the composites' conducting capabilities in addition to their mechanical and thermal properties.

Funding: This research received no external funding.

Institutional Review Board Statement: Not applicable.

Informed Consent Statement: Not applicable.

Data Availability Statement: Not applicable.

Conflicts of Interest: The authors declare no conflict of interest.

References

1. Pendhari, S.S.; Kant, T.; Desai, Y.M. Application of polymer composites in civil construction: A general review. *Compos. Struct.* **2008**, *84*, 114–124.
2. Thomas, S.; Joseph, K.; Malhotra, S.K.; Goda, K.; Sreekala, M.S. (Eds.) *Polymer Composites, Macro- and Microcomposites*; John Wiley & Sons: Hoboken, NJ, USA, 2012; Volume 1.
3. Andrews, R.; Weisenberger, M.C. Carbon nanotube polymer composites. *Curr. Opin. Solid State Mater. Sci.* **2004**, *8*, 31–37. [CrossRef]
4. Rizwan, A.; Abualsauod, E.H.; Othman, A.M.; Serbaya, S.H.; Shahzad, M.A.; Hameed, A.Z. A Multi-Attribute Decision-Making Model for the Selection of Polymer-Based Biomaterial for Orthopedic Industrial Applications. *Polymers* **2022**, *14*, 1020. [CrossRef] [PubMed]
5. Hsissou, R.; Seghiri, R.; Benzekri, Z.; Hilali, M.; Rafik, M.; Elharfi, A. Polymer composite materials: A comprehensive review. *Compos. Struct.* **2021**, *262*, 113640. [CrossRef]
6. Åström, B.T. *Manufacturing of Polymer Composites*; Routledge: Oxford, UK, 2018.
7. Balazs, A.C.; Emrick, T.; Russell, T.P. Nanoparticle polymer composites: Where two small worlds meet. *Science* **2006**, *314*, 1107–1110. [CrossRef] [PubMed]
8. Rahman, A.; Ali, I.; Al Zahrani, S.M.; Eleithy, R.H. A review of the applications of nanocarbon polymer composites. *Nano* **2011**, *6*, 185–203. [CrossRef]
9. Sathishkumar, T.P.; Satheeshkumar, S.; Naveen, J. Glass fiber-reinforced polymer composites–A review. *J. Reinf. Plast. Compos.* **2014**, *33*, 1258–1275. [CrossRef]
10. Guo, Y.; Ruan, K.; Shi, X.; Yang, X.; Gu, J. Factors affecting thermal conductivities of the polymers and polymer composites: A review. *Compos. Sci. Technol.* **2020**, *193*, 108134. [CrossRef]
11. Hollaway, L. *Polymer Composites for Civil and Structural Engineering*; Springer Science & Business Media: Berlin/Heidelberg, Germany, 2012.
12. Odegard, G.M.; Clancy, T.C.; Gates, T.S. Modeling of the mechanical properties of nanoparticle/polymer composites. In *Characterization of Nanocomposites*; Jenny Stanford Publishing: Dubai, United Arab Emirates, 2017; pp. 319–342.
13. Tong, L.; Mouritz, A.P.; Bannister, M.K. *3D Fibre Reinforced Polymer Composites*; Elsevier: Amsterdam, The Netherlands, 2002.
14. Nambiar, S.; Yeow, J.T. Polymer-composite materials for radiation protection. *ACS Appl. Mater. Interfaces* **2012**, *4*, 5717–5726. [CrossRef] [PubMed]
15. Saleem, M.; Rizwan, A. Development of Application Specific Intelligent Framework for the Optimized Selection of Industrial Grade Magnetic Material. *Polymers* **2021**, *13*, 4328. [CrossRef] [PubMed]
16. Ohama, Y. Recent progress in concrete-polymer composites. *Adv. Cem. Based Mater.* **1997**, *5*, 31–40. [CrossRef]
17. Byrne, M.T.; Gun'ko, Y.K. Recent advances in research on carbon nanotube–polymer composites. *Adv. Mater.* **2010**, *22*, 1672–1688. [CrossRef] [PubMed]
18. Chen, E.J.; Hsiao, B.S. The effects of transcrystalline interphase in advanced polymer composites. *Polym. Eng. Sci.* **1992**, *32*, 280–286. [CrossRef]
19. Bakis, C.E.; Bank, L.C.; Brown, V.; Cosenza, E.; Davalos, J.F.; Lesko, J.J.; Machida, A.; Rizkalla, S.H.; Triantafillou, T.C. Fiber-reinforced polymer composites for construction-state-of-the-art review. *J. Compos. Constr.* **2002**, *6*, 73–87. [CrossRef]
20. Zhang, Z.; Friedrich, K. Artificial neural networks applied to polymer composites: A review. *Compos. Sci. Technol.* **2003**, *63*, 2029–2044. [CrossRef]

21. Rizwan, A.; Saleem, M.; Serbaya, S.H.; Alsulami, H.; Ghazal, A.; Mehmood, M.S. Simulation of Light Distribution in Gamma Irradiated UHMWPE Using Monte Carlo Model for Light (MCML) Transport in Turbid Media: Analysis for Industrial Scale Biomaterial Modifications. *Polymers* **2021**, *13*, 3039. [CrossRef] [PubMed]
22. Serbaya, S.H.; Abualsauod, E.H.; Basingab, M.S.; Bukhari, H.; Rizwan, A.; Mehmood, M.S. Structure and Performance Attributes Optimization and Ranking of Gamma Irradiated Polymer Hybrids for Industrial Application. *Polymers* **2021**, *14*, 47. [CrossRef] [PubMed]

Disclaimer/Publisher's Note: The statements, opinions and data contained in all publications are solely those of the individual author(s) and contributor(s) and not of MDPI and/or the editor(s). MDPI and/or the editor(s) disclaim responsibility for any injury to people or property resulting from any ideas, methods, instructions or products referred to in the content.

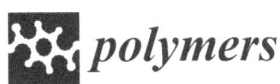

Article

4D Printing of Electroactive Triple-Shape Composites

Muhammad Yasar Razzaq [1,*], Joamin Gonzalez-Gutierrez [1], Muhammad Farhan [2], Rohan Das [1,3], David Ruch [1], Stephan Westermann [1] and Daniel F. Schmidt [1,*]

[1] Department of Materials Research and Technology, Luxembourg Institute of Science and Technology, ZAE Robert Steichen, L-4940 Hautcharage, Luxembourg
[2] Institute of Active Polymers, Helmholtz-Zentrum Hereon, D-14513 Teltow, Germany
[3] Department of Physics and Materials Science, University of Luxembourg, L-4365 Esch-sur-Alzette, Luxembourg
* Correspondence: yasar.razzaq@list.lu (M.Y.R.); daniel.schmidt@list.lu (D.F.S.)

Abstract: Triple-shape polymers can memorize two independent shapes during a controlled recovery process. This work reports the 4D printing of electro-active triple-shape composites based on thermoplastic blends. Composite blends comprising polyester urethane (PEU), polylactic acid (PLA), and multiwall carbon nanotubes (MWCNTs) as conductive fillers were prepared by conventional melt processing methods. Morphological analysis of the composites revealed a phase separated morphology with aggregates of MWCNTs uniformly dispersed in the blend. Thermal analysis showed two different transition temperatures based on the melting point of the crystallizable switching domain of the PEU ($T_m \sim 50 \pm 1$ °C) and the glass transition temperature of amorphous PLA ($T_g \sim 61 \pm 1$ °C). The composites were suitable for 3D printing by fused filament fabrication (FFF). 3D models based on single or multiple materials were printed to demonstrate and quantify the triple-shape effect. The resulting parts were subjected to resistive heating by passing electric current at different voltages. The printed demonstrators were programmed by a thermo-mechanical programming procedure and the triple-shape effect was realized by increasing the voltage in a stepwise fashion. The 3D printing of such electroactive composites paves the way for more complex shapes with defined geometries and novel methods for triggering shape memory, with potential applications in space, robotics, and actuation technologies.

Keywords: 4D printing; additive manufacturing; triple-shape effect; electro-active composites; shape-memory polymers

1. Introduction

Additive manufacturing, also known as three-dimensional printing (3DP), has become a topic of considerable interest in recent years due to its ability to realize complex structures at high resolution, which allows design flexibility and prototyping freedom [1–3]. The 3DP of smart materials with dynamically tunable shapes coupled with time as the fourth dimension is now referred to as "4D printing" (4DP). Within the family of smart materials, shape-memory polymers (SMPs), with their capability to change shapes upon exposure to various external stimuli such as heat, light, ultrasound, magnetic fields, or chemical substances, are the most investigated 4D-printed materials [4–9].

Thermo-sensitive SMPs (trSMPs), with their tailorable elastic properties and transition temperatures, have immense potential in the aerospace, biomedical, electronic and textile industries [10–13]. Most trSMPs reported to date display a dual-shape effect, changing from one shape to a second shape. At the molecular level, these polymers have crystallites or oriented polymeric chains that act as shape-switching domains associated with a transition temperature. In addition, these polymers contain physical or chemical crosslinks that are responsible for the stability of the permanent shape [14,15]. By introducing multiple types of switching domains with different transition temperatures (T_{trans}) into one polymer

Citation: Razzaq, M.Y.; Gonzalez-Gutierrez, J.; Farhan, M.; Das, R.; Ruch, D.; Westermann, S.; Schmidt, D.F. 4D Printing of Electroactive Triple-Shape Composites. *Polymers* **2023**, *15*, 832. https://doi.org/10.3390/polym15040832

Academic Editor: Md Najib Alam

Received: 17 January 2023
Revised: 2 February 2023
Accepted: 2 February 2023
Published: 7 February 2023

Copyright: © 2023 by the authors. Licensee MDPI, Basel, Switzerland. This article is an open access article distributed under the terms and conditions of the Creative Commons Attribution (CC BY) license (https://creativecommons.org/licenses/by/4.0/).

system, it is possible to observe a triple- or a multi-shape effect [16–22]. The first triple-shape effect was reported in multiphase polymer networks with two types of switching domains, either both crystalline or one crystalline and one amorphous [23]. Furthermore, by using polymer systems with a broad transition temperature or multiple transition temperatures, it was possible to enable a quadruple or quintuple shape-memory effect (SME) [24,25]. Additionally, incorporating functional fillers such as magnetic nanoparticles (MNPs) into triple-shape polymer (TSP) networks, a magnetically controlled triple-shape effect (TSE) was reported [26,27]. These magnetic nanocomposites could be inductively heated by exposure to an alternating magnetic field (AMF), thus enabling a remote-control triple-shape effect. Nevertheless, the processing of network architectures with covalent crosslinks is challenging, limiting the technical use of these composites. Furthermore, the requirement of a power generator and an inductive coil further limits the application potential of magnetically triggered systems. One alternative to inductive heating is Joule heating of electrically conductive composites by passing electric currents, which offers significant advantages over other remote heating methods, such as easy operation, long-range control, and fast response [28]. Carbon-based fillers such as carbon powder, carbon nanotubes, or carbon fibers are incorporated into SMP matrices to enable electric heating. Depending on the type, concentration and level of dispersion, these fillers have the capacity to change the mechanical properties of the composites and provide electrically induced remote heating capabilities [19]. The applicability of electric heating has been investigated for various high-tech applications such as morphing aircraft [29,30], self-deploying structures [31], and intelligent textiles [32]. However, most of these studies have focused on dual-shape effect systems, where the SMP composites were electrically triggered to transition from one shape to another [28]. Electric actuation of TSPs is rare, with few reports published on this topic [19,33]. Nevertheless, control of the two shapes was not possible, and sequential recovery of the two shapes was carried out by the application of a single voltage.

Various 3DP techniques have been used to print single or multicomponent polymer systems to produce 4D objects with enhanced properties and shape-memory capabilities [4,34,35]. For instance, vat photopolymerization (VPP) 3D printing was used to print a photocurable resin, enabling a thermally initiated TSE [34]. Material jetting 3DP has also been used to print a mixture of commercially available photosensitive resins resulting in a TSE [36]. Among different 3DP techniques, material extrusion (MEX) is the most commonly used technique to print 4D objects due to its simple operation and troubleshooting, low cost of equipment and raw materials, high speed, and the capability to print large parts [37,38]. Here, we have explored whether filament-based MEX (i.e., Fused Filament Fabrication (FFF)) can print TSPs, enabling an electrically triggered TSE.

Many thermoplastic SMPs have low stiffness, leading to filament buckling during the FFF printing process. Furthermore, in the case of reinforced composites, filler aggregation can block the nozzle; in both cases, the printing process stops. The issue of filament buckling can be avoided by adding fillers or blending with other polymers to improve filament stiffness [4]. Nevertheless, an optimal amount of filler is required to achieve 3DP by FFF and electrically triggered TSE.

We hypothesized that 4D-printable triple-shape electroactive polymers could be developed by preparing a composite with electric conductivity, multiple switching domains and elasticity. The concept pursued was the creation of an optimal balance between different domains and conductive filler to enable a suitable rigidity for FFF printing and electrically activated TSE. Our strategy involves the fabrication of a multiphase composite by incorporating electrically conductive nanoparticles into a polymer blend with heterogenous morphology containing two switching domains with separate T_{trans}. The blending of a commercially available thermoplastic polyester urethane (PEU) with poly(lactic acid) (PLA) in the presence of multi-wall carbon nanotubes (MWCNTs) was carried out to fabricate such composites. PLA is an ideal FFF material, which enables a thermally induced SME [39,40]. However, incorporating nanofillers makes PLA brittle, thus limiting its deformability [41]. Therefore, blending polyurethanes (PUs) with PLA was investigated as an effective way to obtain multiphase SMPs with improved strength and elasticity. The PEU selected was a phase-segregated PEU consisting of a crystallizable soft phase based on poly(1,4-butylene adipate) (PBA) and a 4,4′-methylenediphenyl diisocyanate (MDI)/1,4-butanediol (BD)-based hard segment [38,42,43]. The morphology of the composites was explored by using scanning electron microscopy (SEM), transmission electron microscopy (TEM) and atomic force microscopy (AFM). Differential scanning calorimetry (DSC) reveals the separate T_{trans} of both switching domains, further confirmed by dynamic mechanical analysis (DMA). Composite filaments of the targeted diameter were easily extruded as monofilaments via screw-based extrusion. Finally, the monofilament was processed in a commercial FFF 3D printer, and smart objects with electrically triggered TSE were fabricated.

2. Materials and Methods

A polyesterurethane (PEU) with the tradename Desmopan DP 2795A known for its shape memory capabilities, was received from Covestro Deutschland AG (Leverkusen, Germany). Poly(lactic acid) (PLA) with the tradename Ingeo 4032D was supplied by Nature Works LLC (Plymouth, MN, USA). The PEU and PLA pellets were dried in a vacuum oven at 60 °C overnight before melt processing. Multi-walled carbon nanotubes (MWCNTs) (Graphistrength® C100, diameter: 10–15 nm, length: 1–10 nm) [44] were procured from Arkema (Colombes, France).

Differential scanning calorimetry (DSC) experiments were performed with a Mettler Toledo DSC 3+ (Greifensee, Switzerland), using a heat-cool-heat cycle with constant heating and cooling rates of 2 K·min^{-1} under a nitrogen atmosphere. The sample granules (7–10 mg) were loaded in Netzsch DSC aluminum pans and sealed. The temperature ranges for the 1st and 2nd heating runs were from 25 °C to 200 °C and −80 °C to 200 °C, respectively. Data from the second heating and first cooling run were used.

Dynamic mechanical analysis (DMA) in tensile mode was carried out on Netzsch Gabo Eplexor 500 N DMA (Ahlden, Germany) equipped with a 25 N load cell using press molded samples with standard dimensions (ISO 527-2/1BB). The measurements were performed in temperature-sweep mode from −100 to 150 °C with a constant heating rate of 2 K·min^{-1} in air, using an oscillation frequency of 10 Hz. During the measurements, a static strain of 1% and a dynamic strain of 0.25% were used. The glass transition (T_g) was determined as the temperature at the maximum in the peak of the loss factor (tan δ) vs. temperature curve.

Scanning electron microscopy (SEM) experiments were performed using a Zeiss Supra 40VP SEM (Carl Zeiss Microscopy Deutschland GmbH, Oberkochen, Germany). For this purpose, planar block faces were prepared in an EMUC6FC6 cryo-ultramicrotome (Leica Microsystems GmbH, Wetzlar, Germany) using a diamond knife at a cutting temperature of −120 °C. Block faces were coated with 5 nm gold in a Q150 R ES sputter coater (Quorum Technologies Ltd., Laughton, UK) and imaged in a high vacuum with an accelerating voltage of 3 kV using an Everhart-Thornley backscattered electron detector. Images were obtained at 2500× to 10,000× magnification.

Transmission Electron Microscopy (TEM) was carried out to see the distribution of MWCNTs in the composite materials. For this purpose, thin films were prepared in an EMUC6FC6 cryo-ultramicrotome (Leica Microsystems GmbH, Wetzlar, Germany) using a diamond knife at a cutting temperature of −120 °C. Sections with thicknesses of 100 to 200 nm were deposited on TEM Grids (Cu, 400 mesh) and examined in a Talos™ F200X TEM (FEI Deutschland GmbH/Thermo Fisher Scientific, Dreieich, Germany) using a Gatan Cryo Transfer Holder Model 914 (AMETEK GmbH, Unterschleissheim, Germany) under cryogenic conditions (−176 °C) at an accelerating voltage of 200 kV in bright field mode. Images were acquired using a Ceta 16M CMOS camera at magnifications of 5000× to 95,000×.

Atomic force microscopy (AFM) was used to determine the phase-specific localization of MWCNTs in the PEU-PLA composites. 2 mm thick sections were cut using a razor blade from the composite and trimmed using a LEICA EM UC6 cryo-ultramicrotome. The microtomed samples were investigated using MFP-3D Infinity (Oxford Instruments,

Abingdon, UK) atomic force microscope to obtain the topography and phase contrast images. All the measurements were performed at room temperature, and a standard cantilever holder was used for operation in an air atmosphere. Images were taken with a resolution of 512 × 512 pixels at a 1.5 Hz scan rate. The analysis was performed in Amplitude Modulation-Frequency Modulation (AM−FM) mode using a silicon cantilever AC160TS (Oxford instruments) with a spring constant of about 20–30 N/m. The topography and phase-contrast images were measured at the fundamental resonance frequency of the cantilever (~300 kHz). The images were processed using the Mountains® 9 software (Digital Surf, Besancon, France) to understand the phase specific morphology of the different polymers in the bulk composite. All the measurements were acquired using the same cantilever for a single day.

Wide angle X-ray scattering (WAXS) measurements were conducted at ambient temperature and 55 °C utilizing a Bruker AXS D8 Discover x-ray diffractometer operating in transmission geometry with a two-dimensional HI-Star detector (Bruker, Karlsruhe, Germany). The samples of dimensions 2 × 0.5 cm and thickness 150 µm were fixed at both ends during characterization. The sample-detector distance was set at 150 mm, and the source wavelength was λ = 0.154 nm (Cu K$_\alpha$). A graphite monochromator and a pinhole collimator with an opening of 0.8 mm provided a parallel, monochromatic X-ray beam. The two-dimensional diffraction images were integrated to obtain plots with intensity versus diffraction angle (2θ = 5–45°). Both Bragg diffraction peaks from the crystalline phase and the broad scattering peak from the amorphous phase were fitted with Pearson VII functions. The crystallinity index (X_c) was calculated based on the sum of the areas of the fitted peaks assigned to the crystalline phase (A_{cryst}) and the amorphous phase (A_{amorph}) (Equation (1)).

$$X_c = \frac{A_{cryst}}{A_{cryst} + A_{amorp}} \times 100 \qquad (1)$$

A custom-built heating device was used to carry out the WAXS measurements of the samples at 55 °C. The sample was equilibrated at this temperature for 5 min before the measurement.

Electric heating and conductivity measurements of the printed composites were analyzed using a Series 2410 SourceMeter (Keithley, Cleveland, OH, USA) with a source voltage range between 5 µV and 100 V and a current range from 10 pA to 1.055 A. Alligator clips were used to connect the printed samples with the source meter. In contrast, the surface temperature was monitored using a VarioCAM® HiRes 384 infrared (IR) camera (InfraTec GmbH, Dresden, Germany).

Composites consisting of a PEU/PLA matrix and MWCNTs as conductive fillers were prepared in a DSM Xplore MC15 HT Vari-Batch micro compounder (Sittard, The Netherlands) in co-rotating mode with a mixing chamber volume of 15 cm³. Dried pellets of PEU and PLA were mixed in solid state at a ratio of 70:30 wt%. The content of MWCNT was varied from 5 to 20 wt%. All ingredients were pre-mixed in an aluminum weighing dish via stirring with a spatula, then fed manually into the mixing chamber preheated to 200 °C. Once all the materials were introduced into the mixing chamber, the rotational screw speed was gradually increased from 20 to 100 rpm. The material was recirculated in the mixing chamber for 5 min at 100 rpm to ensure proper mixing. The prepared composites with MWCNT contents from 5 to 20 wt% were tested for their electric heating capacity, as described above. Once it was observed that the addition of 15–20 wt% of MWCNTs was sufficient to realize Joule heating, larger batches of composites containing such levels of MWCNTs were prepared in a Thermo Scientific HAAKE Rheomex PTW16/25 OS co-rotating twin screw compounder (Karlsruhe, Germany). Barrel zone temperatures varied from 190 °C at the hopper to 220 °C at the die, while the screw speed was set to 35 rpm. The dried PEU and PLA pellets and the MWCNTs were pre-mixed manually in a glass container and fed into the hopper of the compounder slowly to avoid exceeding the torque limit. Since it was impossible to ensure the even loading of the ingredients due to granular convection during the manual feeding, it was decided to extrude the compound twice to increase the homogeneity of the mixture. After the first compounding cycle, the extrudate was pelletized using a Thermo Scientific 16 mm fixed-length strand pelletizer (Karlsruhe, Germany). After the second compounding cycle, filaments of a 2.85 mm diameter were extruded and pulled with a Schulz & Busch K-25 conveyor belt (Wülfrath, Germany) in preparation for 3D printing trials.

3D printing was performed in a 3ntr A4v3 filament-based material extrusion machine (Jdeal-Form s.r.l., Oleggio, Italy), also known as a fused filament fabrication (FFF) 3D printer. Computer-aided design (CAD) was performed using the online platform Tinkercad (Autodesk Inc, San Rafael, CA, USA) and the Gcode was prepared using Ultimaker Cura 4.10.0 (Utrecht, The Netherlands). The used FFF 3D printer has three extrusion heads.

Extrusion head 1 uses filaments with a diameter of 1.75 mm, and the other two extrusion heads support filaments with a diameter of 2.85 mm. Four types of specimens were printed with electro-active filaments (i.e., PEU70PLA30MWCNT14). The four printed demonstrators, and their dimensions are shown in Figure 1. The first object was a rectangular bar (50 × 10 × 2 mm^3) (Figure 1a), which was used for mechanical characterization and quantification of the TSE by bending experiments. A U-shape resistor (Figure 1b) and linear compression device (Figure 1c) were printed with a single electro-active composite. The fourth object (Figure 1d) was a multi-material hinge and was printed with red PLA (passive component) in extrusion head 1 and the composite (electro-active component) in extrusion head 2. Extrusion head 1 had a 0.4 mm brass nozzle, and extrusion head 2 had a 0.8 mm brass nozzle. The printing conditions used to print the different materials are shown in Table S1 (see in Supplementary Materials).

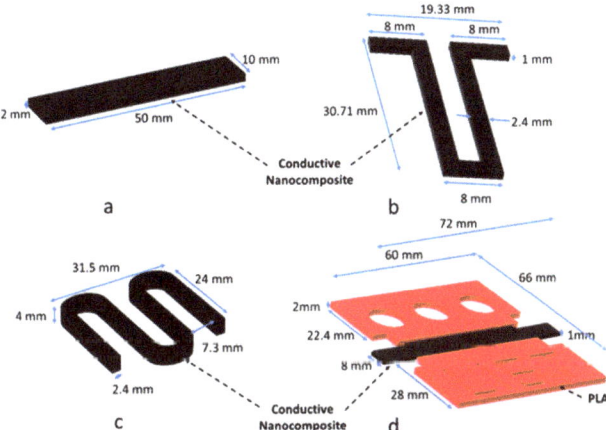

Figure 1. 3D-printed specimens with the conductive composite (PEU70PLA30MWCNT14): (**a**) rectangular bar used for mechanical characterization (**b**) U-shaped resistor (**c**) linear compression device; (**d**) hinge 3D printed with non-conductive PLA (red) and a conductive composite (black).

The triple-shape effect was quantified by measuring the recovery ratios in a triple-shape bending procedure. The 3D-printed composite bar was programmed by a two-step bending procedure. In step one, the sample (shape C) was heated to 90 °C and was bent to 90° (θ_B) and was fixed by cooling to 55 °C under constrain (shape B). After a waiting period of 15 min, the constraint was removed and a subsequent bending to 180° (θ_A) at 55 °C, followed by cooling to −10 °C (shape A), was carried out. After the removal of the constraint, the triple-shape fixation was completed.

For recovery, a stepwise reheating to T_{mid} = 55 °C by applying a low voltage (V_{low}) and to T_{high} = 90 °C by application of a higher voltage (V_{high}) was carried out. The recovery process was recorded via a video camera, and the recovery angles were recorded and evaluated using the ImageJ v1.53e software package (NIH, Bethesda, MD, USA). The ratios of different angles before and after recovery were used to calculate the shape recovery ratios R_{A-B} (shape A to shape B) or R_{A-C} (shape A to shape C) using Equations (2) and (3). Here, $\theta_B{}^{rec}$ is the angle in the partially recovered sample in shape B, and $\theta_C{}^{rec}$ is the angle in the fully recovered sample (shape C). A schematic demonstration of different angles and shapes during triple-shape bending is shown in Figure 2.

$$R_{A-B} = \frac{\theta_A - \theta_B rec}{\theta_A - \theta_B} \times 100\% \quad (2)$$

$$R_{A-C} = \frac{\theta_A - \theta_C rec}{\theta_A - \theta_C} \times 100\% \quad (3)$$

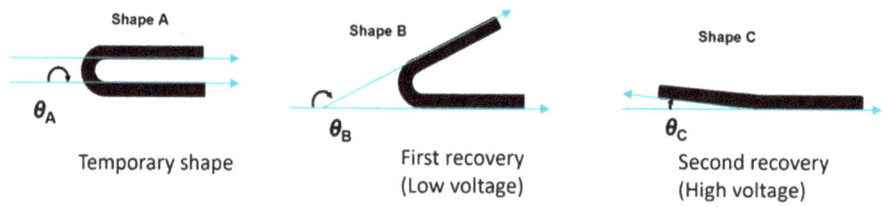

Figure 2. A schematic demonstration of the triple-shape bending procedure.

3. Results and Discussions

3.1. Selection of Composite Formulation

To retain the flexibility of the blend, PEU was selected as a continuous phase, and the mass ratio of PEU to PLA was varied between 90/10, 80/20, and 70/30. A morphological analysis of these blends confirmed the incompatibility of the PEU and PLA phases. In particular, a two-phase morphology with PLA droplets suspended in the PEU phase can be seen in all blends. Representative SEM images of the pure blends with a PEU-PLA mass ratio of 90/10 (PEU90PLA10) and 80/20 (PEU80PLA20) are shown in Figure S1a (see in Supplementary Materials). The diameter of the PLA domains in the blends was observed to be of the order of ~1–10 microns, with less variability in the case of the PEU70PLA30 formulation. To enable a TSE, the crystallizable soft segments of PEU with a melting point of T_m = 47 °C served as the first switching domain, while the amorphous segments of PLA with a glass transition of T_g = 61 °C acted as the second switching domain. The DSC thermograms of neat PEU and PLA are shown in the Supplementary Materials (Figure S1b). For the blend PEU/PLA (90/10), no T_g associated with amorphous PLA was observed, and only a melting transition related to the crystallizable soft segment of PEU was observed. For blends PEU80PLA20 and PEU70PLA30, two different transition temperatures were observed, related to the T_m of crystallizable segments of the PEU and the T_g of the amorphous PLA domains, respectively. An investigation of the mechanical properties of the blends by uniaxial tensile testing revealed a significant increase in stiffness (elastic modulus) as the PLA content was increased. The elastic modulus of pure PEU (E = 50 MPa) was increased to E = 87 MPa for PEU70PLA30. In parallel, the elongation at break (ε_b) was significantly decreased from ε_b = 860 ± 45% for pure PEU to ε_b = 319 ± 45% PEU70PLA30. The representative stress–strain curves are shown in Figure S1c (see in Supplementary Materials). To assess the 3D printability of the neat blends, filaments with a uniform diameter d = 2.85 ± 0.05 mm were extruded. One of the major concerns of 3D printing of polyurethanes-based filaments is their lack of stiffness, which can result in buckling in gear-fed 3D printing equipment due to the force applied during the feeding process and the requirement that the filament takes on the role of a piston that applies pressure on the polymer melt. Only PEU70PLA30 (with an elastic modulus of 87 MPa) provided sufficient stiffness to enable FFF printing using the equipment described. For the blends PEU90PLA10 and PEU80PLA20, in contrast, the buckling of the filaments made printing impossible. Based on these initial studies, PEU70PLA30 was chosen for the fabrication of the electro-active composite, given its attractive combination of suitable stiffness for 3D printing and two different transition temperatures for exploring TSE.

MWCNTs were used as filler to increase the electrical conductivity of the polymeric blend, thus allowing for the production of an electroactive TSE. The weight content of the MWCNT in the blends varied between 6 and 18 wt.% (3.3 and 10.5 vol.%). The nomenclature of the composite is given as PEU70PLA30MWCNTx, where "x" is the wt.% of MWCNT in the composite as determined by thermogravimetric analysis (TGA) (see Figure S2a, see in Supplementary Materials). Figure S2b (see in Supplementary Materials) shows the electrical resistance of PEU70PLA30 composites with various MWCNT contents at room temperature. The same composite dimensions and testing methodology were used for all formulations, as shown schematically in Figure S3 (see in Supplementary Materials). At filler concentrations up to 11 wt.% (6.3 vol.%), the high resistance of the composite (~1 MΩ) indicates the presence of a polymeric insulating phase. Between 11 wt.% (6.3 vol.%) and 14 wt.% (8 vol.%), a drastic decrease in the resistance to 85 ± 10 Ω was observed. This decrease in resistance was attributed to the formation of conductive pathways in the composite due to exceeding the percolation threshold. By further increasing the content of MWCNTs to 18 wt.% (10.5 vol.%), the resistance was further decreased to 45 ± 10 Ω. Compared to previous reports with a similar type of filler, the percolation threshold observed here was relatively high and implies incomplete dispersion of the MWCNT in the blend [44]. Nevertheless, as this effort focused on achieving electrical conductivity, not filler dispersion, the composites PEU70PLA30MWCNT14 and PEU70PLA30MWCNT18 were selected for further studies, given a low enough resistance to enable Joule heating. In contrast, composites with 11 wt.% of MWCNTs or lower were insufficiently conductive, and no electric heating of these composites was observed. Along with the filler content, the heating efficiency of these electroactive composites depends on the size of the sample, applied voltage, and the time for electric current exposure [45]. Therefore, to assess the heating efficiency and the maximum achievable temperature (T_{max}), composite specimens with dimensions of 50 × 10 × 2 mm^3 were printed based on PEU70PLA30MWCNT14 and PEU70PLA30MWCNT18. The current flow and the T_{max} achieved due to resistive heating in the composite specimens are shown as a function of voltage in Figure 3. A voltage of 17 V enabled a current flow of 135 mA and a T_{max} of 90 °C (well above the PLA T_g) in PEU70PLA30MWCNT14, whereas for PEU70PLA30MWCNT18, only 11 V was required to reach a similar level. Nevertheless, because of the brittle nature of the PEU70PLA30MWCNT18 compound, the PEU70PLA30MWCNT14 was selected for all further investigations of electrically activated TSE.

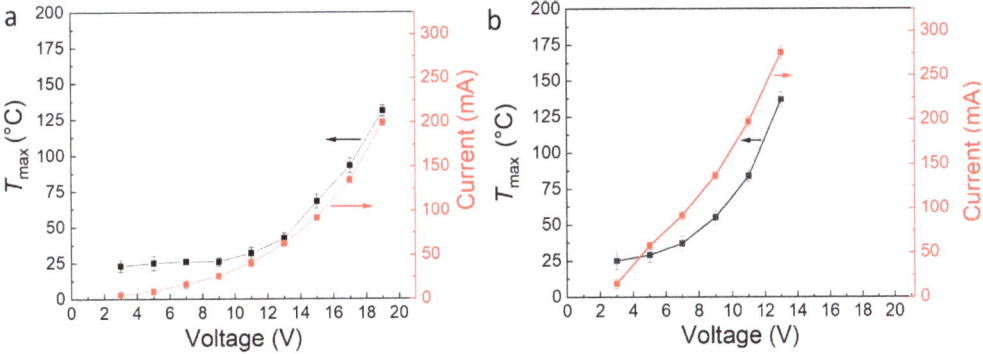

Figure 3. Maximum achievable temperatures (T_{max}) in composite samples (**a**) PEU70PLA30MWCNT14 and (**b**) PEU70PLA30MWCNT18 by passing current at variable voltages.

3.2. Morphology of the Composites

The morphologies of the composites on the micro- and nanoscale were elucidated by microscopic analysis. The backscattered SEM images of 3D-printed composite cross-sections of the sample showed irregularly shaped micron-sized aggregates of MWCNT (confirmed by EDX) that were statistically distributed within the polymer matrix. The representative SEM images of PEU70PLA30MWCNT14 are shown in Figure 4a,b. As confirmed by threshold values of MWCNTs, these aggregates enabled the formation of a conductive network in the composite that seems to originate from the high packing density of the primary agglomerates of MWCNTs. In general, due to the high concentration of MWCNTS, strong van der Waals attractions between the individual MWCNTs and poor polymer-MWCNT compatibility, the homogeneous dispersion of MWCNTs in the polymer matrix was not observed. Strategies for achieving higher levels of MWCNT dispersion would include modifying their surface and/or adding compatibilizers to enhance thermodynamic compatibility while reducing their concentration and/or increasing the level of applied shear during mixing to accelerate the kinetics of dispersion. In practice, however, achieving high levels of MWCNT dispersion is not essential for realizing Joule heating. On the contrary, a system with only well-dispersed MWCNTs would contain no MWCNT-MWCNT contacts, precluding the formation of conductive paths and leading to low levels of electrical conductivity, thus making Joule heating challenging to achieve, especially at low voltages [46]. Randomly distributed micro-level voids of different dimensions were also observed in the composite, which could be attributed to the mechanical entrapment of air or the volatilization of small molecules (e.g., water) during melt processing. Furthermore, polymer degradation (either thermal due to shear heating, or hydrolytic due to the presence of water) could also contribute to microvoid formation. In contrast to the observation of these voids, the spherical PLA domains seen in the unfilled blends were not seen in the backscattered SEM images of the composites. This observation is consistent with prior work showing that nanoparticles can cause reductions in the length scale of phase separation in immiscible blends [47].

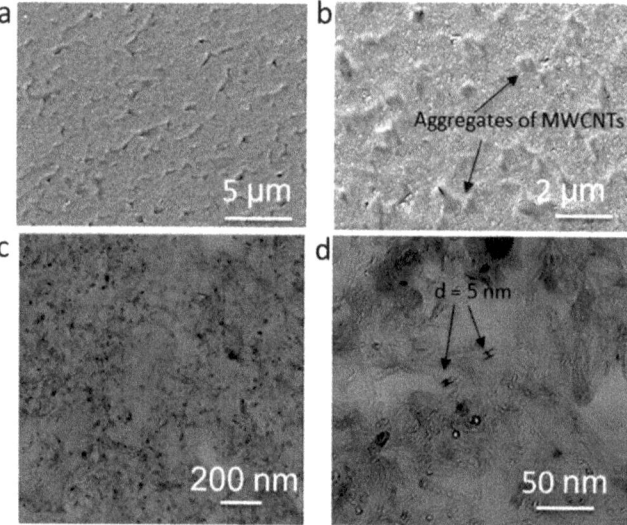

Figure 4. (**a**,**b**) Backscattered SEM images of the PEU70PLA30MWCNT14 and (**c**,**d**) TEM analysis of the composite.

To further study the distribution of MWCNTs in the polymer blend matrix, TEM analysis was carried out. Here, the difference in electron density resulted in the observation of dark cylindrical features (assigned to MWCNTs) in a lighter matrix (assigned to the

PEU/PLA blend), as shown in Figure 4c,d. While individual MWCNTs were observed via TEM imaging and their average diameter was assessed as being ~5–10 nm, no attempt was made to estimate MWCNT length given the very low probability of observing fully intact MWCNTs within a single TEM slice.

The morphology of the 3D-printed composites was further elucidated by AFM analysis. To determine the phase-specific localization of MWCNTs by AFM, four samples were analyzed: neat PEU, a PEU/MWCNT composite (PEU100MWCNT14), the neat PEU-PLA blend (PEU70PLA30), and the PEU-PLA blend/MWCNT composite (PEU70PLA30MWCNT14). AFM images of the composite PEU70PLA30MWCNT14 with sizes 3×3 µm^2 and 1.5×1.5 µm^2 are shown in Figure 5, while the images of neat PEU, PEU100MWCNT14, and PEU70PLA30 are shown in Figure S4 (see in Supplementary Materials). Furthermore, details about the micro-phase separation in pure PEU, the multi-domain architecture of pure blend PEU70PLA30 and the distribution of MWCNT in pure PEU are discussed in Supplementary Materials (Section S1). In Figure 5, images a–b represent topography, while images c–d represent phase contrast. The blend composite PEU70PLA30MWCNT14 showed a distinct two-phase morphology of PLA and PEU. The topographic images shown in Figure 5a,b reveal a distribution of MWCNTs similar to that observed in Figure S4c (see in Supplementary Materials) for PEU100MWCNT14. Similarly, the relevant phase contrast image (Figure 5c,d) shows a similar level of phase separation between the PLA and the PEU domains, as was observed in Figure S4e, for the equivalent MWCNT-free specimen. Furthermore, the phase contrast images of the composite blend indicate selective localization of MWCNTs in the PEU phase. This selective localization phenomenon of MWCNTs in an immiscible blend of thermoplastic polyurethane and PLA was also reported by Buys et al. [48]. Their investigations indicated that whenever inorganic nanoparticles are added to immiscible polymer blends, they tend to be dispersed heterogeneously, either preferentially concentrated in one of the polymer phases or localized to the interface between the two.

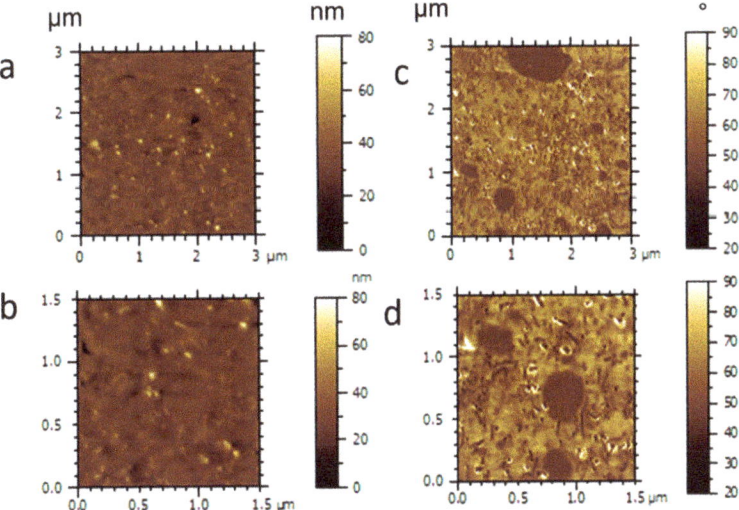

Figure 5. AFM images of the blend composite PEU70PLA30MWCNT14 (sizes 3×3 µm^2 and 1.5×1.5 µm^2; Images (**a**,**b**) are topography images, (**c**,**d**) are phase contrast images.

3.3. Thermal and Thermo-Mechanical Properties

DSC and DMA at varied temperatures were carried out to determine the transition temperatures necessary for the triple-shape programming and recovery process. The DSC thermograms of the neat PEU70PLA30 blend showed a sharp melting transition at $T_m = 49$ °C with a melting enthalpy of $\Delta H_{m,PEU} = 34$ J·g^{-1} and a weak glass transition just above the terminus of the melting transition (T_g~61 °C) (Figure 6a). The first switching domain was assigned to the melting of the crystallizable PBA soft segments of the PEU, while the second switching domain was associated with the glass transition of PLA. Furthermore, a peak at T_c~10 °C ($\Delta H_{c,PEU}$ = 28.9 J·g^{-1}), corresponding to the crystallization of the PBA soft segments of the PEU, was also observed. By the addition of MWCNTs to the blend to form the composite PEU70PLA30MWCNT14, the T_m of the PEU soft segments was increased to 50 °C (ΔH_m = 26.2 J·g^{-1}, while the T_c was decreased to 9 °C (ΔH_c = 27.5 J·g^{-1}) (Figure 6a). However, no significant change in the T_g of the amorphous PLA domains was observed. Finally, the broad cold crystallization peak (T_{cc} = 120 °C) and small melting peak (T_m = 152 °C, $\Delta H_{m,PLA}$ = 25 J·g^{-1}) assigned to crystalline PLA in the unfilled blend were observed. By adding MWCNTs, $\Delta H_{m,PLA}$ was slightly increased to 26.2 J·g^{-1}, but no significant change in the T_m of PLA in the composite was observed.

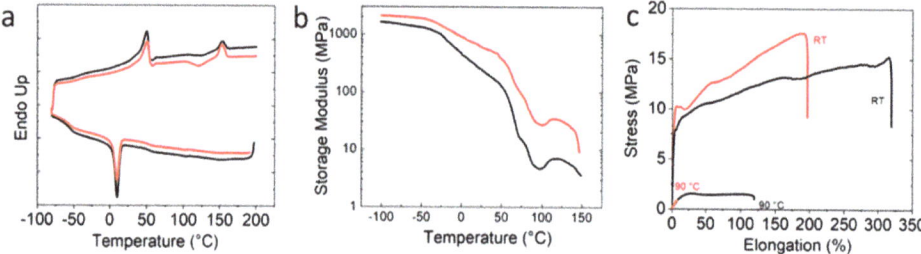

Figure 6. (a) DSC thermograms of neat blend and blend composite (b) Storage modulus at varied temperatures (c) Uniaxial tensile testing of the neat blend polymer and composite at room temperature (RT) and 90 °C. PEU70PLA30 (black line), PEU70PLA30MWCNT14 (red line).

The effects of MWCNT addition on the crystalline microstructure of the PEU-PLA blend composite were further studied by WAXS (Figure S5, see in Supplementary Materials). For the neat blend PEU70PLA30, the diffraction pattern exhibited clear peaks at 2θ = 21.7° and 24.2° and a shoulder at 22.5°, yielding χ_c = 18.2 ± 0.3%. The peaks were assigned to the PBA crystalline polymorphs (α form and β form) as indicated in the literature [49,50]. The fact that the characteristic peaks associated with crystalline PLA did not appear may be attributed to the low amount of PLA present in this system. All peaks associated with crystalline polymer disappeared at 55 °C, as expected, given that this exceeds the melting point of PBA as observed via DSC (T_m = 49 °C). For the composite PEU70PLA30MWCNT14, the addition of MWCNTs resulted in no significant changes in the position and shape of the polymer peaks. For the unfilled blend, no peaks associated with crystalline PLA were observed. The most noticeable change was a characteristic peak associated with the MWCNTs at 2θ = 26° [51]. While, as before, all crystalline polymer peaks disappeared at 55 °C, the MWCNT peak remained, as expected.

Moving from microstructure to viscoelastic response to assess the effects of MWCNTs on the thermomechanical properties of the blend, DMA was carried out as a function of temperature. A stepwise drop in the value of E' was observed in both the unfilled blend and composite as the temperature increased from −50 °C to 100 °C (Figure 6b). E' for PEU70PLA30 decreased from ~1300 MPa at −50 °C to a near-plateau of ~140 MPa at 50 °C. This drop was attributed to the glass transition temperature (T_g) and the melting point (T_m) of the semi-crystalline PBA soft segments of the PEU. A second steep decline in E' was observed between 50 °C and 100 °C, at which point a value of 4 MPa was reached; this was attributed to the T_g of PLA. Finally, the value of E' started to increase once more at temperatures around 100 °C, the result of the cold crystallization of the PLA [52].

The decrease in modulus at temperatures around 150 °C indicates further softening of the sample before the melting transition detected for the crystalline PLA phase following cold crystallization. However, the amount of PLA crystallinity is too low to explain the large change in modulus fully. It is therefore posited that, in addition to this transition, the

MDI/1,4-BDO hard segments present in the PEU are also undergoing a thermal transition. This hypothesis is supported by literature reports reporting hard segment thermal transitions over a similar temperature range [53] and showing that they become difficult to detect via DSC as the hard segment content decreases [54]. Here, given the low modulus of the pure PEU (implying a low hard segment content) coupled with its further dilution by PLA, it is not surprising that the melting transition for the PEU hard segments is not resolved by DSC but remains detectable via DMA. Finally, for the composite PEU70PLA30MWCNT14 formulation (red line in Figure 6b), a higher value of E' was observed across the entire temperature range tested, indicating an increase in rigidity in both the glassy and rubbery regions vs. the unfilled blend.

Uniaxial tensile testing of the neat blend and the composite PEU70PLA30MWCNT14 (50 × 10 × 2 mm^3) at room temperature and 90 °C was carried out (Figure 6c) to assess quasi-static mechanical properties and deformation capabilities. The elastic modulus of the neat PEU70PLA30 blend was measured as $E = 87 \pm 5$ MPa at room temperature, increasing to $E = 110 \pm 3$ MPa (+26%) in the case of the filled PEU70PLA30MWCNT14 composition. Nevertheless, the elongation (ε_b) at break was decreased from 319 ± 45% for PEU70PLA30 to $\varepsilon_b = 196 \pm 15\%$ for PEU70PLA30MWCNT14 (−39%). At an elevated temperature ($T = 90$ °C), the neat PEU70PLA30 blend displayed an elastic modulus of 0.04 MPa and an elongation at break of 140 ± 10%, while the filled PEU70PLA30MWCNT14 formulation gave an elastic modulus of 0.14 MPa (+250%) but could only be extended to 7 ± 2% (−92%) before failing. The limited deformability of the composite could be attributed to the aggregated MWCNTs, which act as stress concentrators and cause premature failure. Due to the low elongation of the printed composites, the electroactive TSE was only explored in bending mode.

3.4. 4D Printing of Electro-Active Triple-Shape Composites

The composite PEU70PLA30MWCNT14 exhibited reliable printability and was used to print 3D models of the electroactive composite alone or in combination with electrically passive subcomponents. The printing parameters and the filling density for all the models were kept constant (Table S1, see in Supplementary Materials). As explained in the experimental section, the U-shaped composite sample was programmed ($\theta_A = 180°$) by using a triple-shape bending procedure. To observe the electrically activated TSE, the two ends of the U-shaped specimen were connected using electrodes and subjected to different voltages. An IR camera monitored the surface temperature during electric heating. The application of 13 V enabled a T_{max} of 50 °C, above the T_m of the PBA crystalline domains in the PEU, and triggered a partial recovery ($\theta_B = 55°$) of the sample in 150 s. Here, it must be clarified that to avoid the involvement of the glass transition range of PLA during the first recovery step, precise heating of the composite to enable the complete melting of the PBA crystalline domain at ~50 °C was necessary. A further increase in voltage level to 17 V resulted in a quick rise of temperature to 90 °C (above the T_g of PLA) and near-complete recovery of the composite model in 25 s, as shown by thermographic imaging in Figure 7a (Video S1). The residual angle in shape C can be attributed to the mechanical constraints imposed on the sample by the electrodes connecting to the other ends of the specimen, which restrict complete recovery.

The temperature profile of the composite at two different voltages is shown in Figure 7b. Using the recovery angles during the electrical activation of TSE, the bending recovery ratios $R_{A \to B}$ and $R_{A \to C}$ were calculated. A relatively low recovery ratio $R_{A \to B} = 61\%$ for the transition from shape A to shape B was observed. However, a higher value of $R_{A \to C} = 89\%$ for the second recovery at 17 V was calculated.

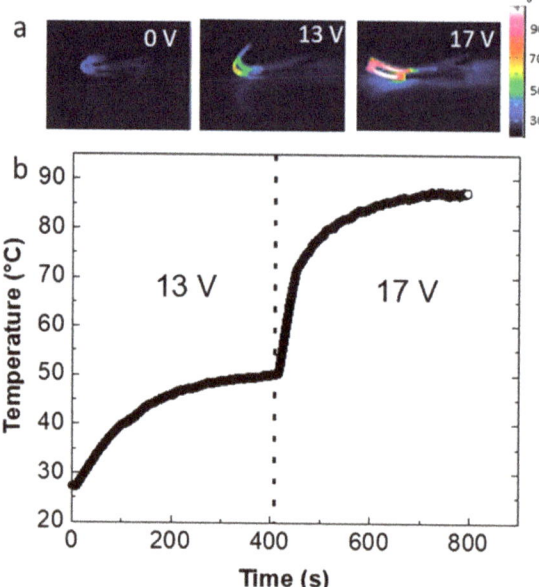

Figure 7. (a) IR thermographic images of the electrically triggered TSE of a "U" shaped composite part by a stepwise increase in voltage from 13 V to 17 V. (b) The temperature profile of the composite sample when subjected to two different voltages.

The second 3D printed demonstrator is an M-shaped linear compression device, which was printed with an internal angle $\theta_{int} = 20°$ and was programmed in a triple shape stretching mode. During the triple shape programming step, the internal angle (θ_{int}) was changed from 15° to 80°. By passing electric current through the device at two different voltages, a stepwise recovery of the device to the original shape was triggered. A voltage of 13 V enabled a change of θ_{int} from 80° to 65°, and a further increase of the voltage to 17 V resulted in full recovery of an internal angle of 20° as shown in Figure 8a (Supplementary Materials Video S2).

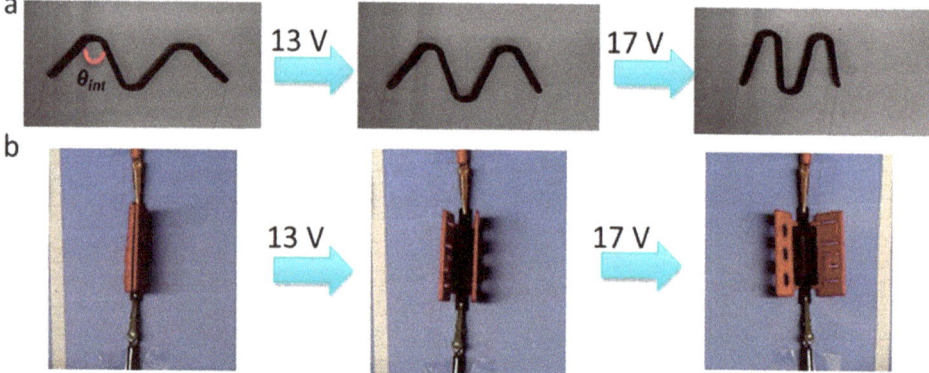

Figure 8. Electrically triggered TSE in 3D-printed composites samples (a) "M" shaped sample based the electroactive composite, (b) multi-material hinge structure consisting of electroactive composite (black) and passive PLA subcomponents (red).

The last 3D-printed object was a multi-material hinge structure. In this object, the hinge itself consists of the electroactive composite with tripe-shape capability, while the outer segments are electro-passive and are based on red-colored PLA. The multi-material 3D structure was thermally programmed using the triple-shape bending procedure described in the experimental section. Electrically induced TSE was triggered by subjecting the composite to different voltages in a stepwise manner. At 13 V, the hinge structure opened partially, increasing the voltage level to 17 V completed the process (Figure 8b). The video of the electrically activated triple-shape capability of the 3D-printed multi-material hinge is shown in the Supplementary Materials (Video S3).

4. Conclusions

In summary, 4D printing of electroactive triple-shape composites based on a PEU/PLA blend filled with MWCNTs was carried out. An optimized selection of the ratios of the three components in the blend enabled smooth FFF printability and Joule heating capability by passing electric current at different voltages. Morphological analysis revealed a phase-separated morphology with randomly distributed agglomerates of MWCNTs in the blend. The melting point of the crystallizable PBA soft segments in the PEU (T_m = 50 ± 1 °C) acted as the lower transition temperature, while the glass transition of the amorphous PLA (T_g = 61 ± 1 °C) acted as the upper transition temperature. Different single and multi-material parts were printed using a multi-nozzle FFF 3D printer and programmed using two-step bending procedures. A pronounced TSE was observed by increasing the voltage from 0 V to 13 V and 17 V to achieve well-defined temperatures below, between, and above the PEU and PLA transition temperatures, respectively. The ability to use 4D printing to fabricate composites displaying a stepwise, tunable shape change process that may be triggered remotely on-demand has clear implications for various applications in soft robotics, actuators, and smart textiles.

Supplementary Materials: The following supporting information can be downloaded at: https://www.mdpi.com/article/10.3390/polym15040832/s1, Table S1: Printing parameters of red colored PLA and conductive composite; Figure S1: (a) Back-scatter SEM images of the unfilled pure blends, (b) DSC thermograms of neat PEU, neat PLA and unfilled polymer blends, (c) Tensile testing of neat PEU and unfilled polymer blends; Figure S2: (a) Thermogravimetric analysis of pure blend and composites, (b) Electrical resistance of pure blend and composites with different MWCNT content; Figure S3: Schematic representation of the method used for measuring the resistance of the composite samples; Figure S4: AFM images of the reference samples; Figure S5: WAXS analysis of the neat blend and composite at ambient temperature and at 55 °C, Section S1: Morphology of the reference samples by atomic force microscopy; Video S1: Electrically activated triple-shape effect (TSE) in 3D printed "U" shape model; Video S2: Electric activation of the TSE in 3D printed "M" shaped compression device; Video S3: Electric activation of the TSE in a multi-material hinge structure. Refs. [55–60] are cited in Supplementary Materials file.

Author Contributions: Conceptualization, M.Y.R., S.W. and D.F.S.; methodology, M.Y.R., J.G.-G., R.D. and M.F.; software, M.Y.R., R.D. and J.G.-G.; validation, M.Y.R., J.G.-G. and M.F.; formal analysis, M.Y.R. and J.G.-G.; investigation, M.Y.R.; resources, M.Y.R. and R.D.; data curation, M.Y.R.; writing—original draft preparation, M.Y.R. and D.F.S.; writing—review and editing, M.Y.R., J.G.-G. and D.F.S.; visualization, D.F.S. and S.W.; supervision, S.W., D.R. and D.F.S.; project administration, D.F.S. and S.W.; funding acquisition, M.Y.R. and S.W. All authors have read and agreed to the published version of the manuscript.

Funding: This research received no external funding.

Institutional Review Board Statement: Not applicable.

Informed Consent Statement: Not applicable.

Data Availability Statement: Not applicable.

Acknowledgments: The authors thank Benoit Marcolini (LIST) and Yvonne Pieper (Helmholtz Zentrum Hereon, Teltow, Germany) for their support related to the thermal and microscopic analyses presented here.

Conflicts of Interest: The authors declare no conflict of interest.

References

1. Arefin, A.M.E.; Khatri, N.R.; Kulkarni, N.; Egan, P.F. Polymer 3D Printing Review: Materials, Process, and Design Strategies for Medical Applications. *Polymers* **2021**, *13*, 1499. [CrossRef]
2. Park, S.; Shou, W.; Makatura, L.; Matusik, W.; Fu, K. 3D printing of polymer composites: Materials, processes, and applications. *Matter* **2022**, *5*, 43–76. [CrossRef]
3. Arif, Z.U.; Khalid, M.Y.; Noroozi, R.; Sadeghianmaryan, A.; Jalalvand, M.; Hossain, M. Recent advances in 3D-printed polylactide and polycaprolactone-based biomaterials for tissue engineering applications. *Int. J. Biol. Macromol.* **2022**, *218*, 930–968. [CrossRef] [PubMed]
4. Razzaq, M.Y.; Gonzalez-Gutierrez, J.; Mertz, G.; Ruch, D.; Schmidt, D.F.; Westermann, S. 4D Printing of Multicomponent Shape-Memory Polymer Formulations. *Appl. Sci.* **2022**, *12*, 7880. [CrossRef]
5. Wan, X.; He, Y.; Liu, Y.; Leng, J. 4D printing of multiple shape memory polymer and nanocomposites with biocompatible, programmable and selectively actuated properties. *Addit. Manuf.* **2022**, *53*, 102689. [CrossRef]
6. Spiegel, C.A.; Hackner, M.; Bothe, V.P.; Spatz, J.P.; Blasco, E. 4D Printing of Shape Memory Polymers: From Macro to Micro. *Adv. Funct. Mater.* **2022**, *32*, 2110580. [CrossRef]
7. Lucarini, S.; Hossain, M.; Garcia-Gonzalez, D. Recent advances in hard-magnetic soft composites: Synthesis, characterisation, computational modelling, and applications. *Compos. Struct.* **2022**, *279*, 114800. [CrossRef]
8. Bastola, A.K.; Hossain, M. The shape—Morphing performance of magnetoactive soft materials. *Mater. Des.* **2021**, *211*, 110172. [CrossRef]
9. Bastola, A.K.; Paudel, M.; Li, L.; Li, W. Recent progress of magnetorheological elastomers: A review. *Smart Mater. Struct.* **2020**, *29*, 123002. [CrossRef]
10. Behl, M.; Razzaq, M.Y.; Lendlein, A. Multifunctional Shape-Memory Polymers. *Adv. Mater.* **2010**, *22*, 3388–3410. [CrossRef]
11. Xia, Y.; He, Y.; Zhang, F.; Liu, Y.; Leng, J. A Review of Shape Memory Polymers and Composites: Mechanisms, Materials, and Applications. *Adv. Mater.* **2021**, *33*, 2000713. [CrossRef] [PubMed]
12. Lendlein, A.; Kelch, S. Shape-Memory Polymers. *Angew. Chem. Int. Ed.* **2002**, *41*, 2034–2057. [CrossRef]
13. Shirole, A.; Perotto, C.U.; Balog, S.; Weder, C. Tailoring the Shape Memory Properties of Segmented Poly(ester urethanes) via Blending. *ACS Appl. Mater. Interfaces* **2018**, *10*, 24829–24839. [CrossRef]
14. Pisani, S.; Genta, I.; Modena, T.; Dorati, R.; Benazzo, M.; Conti, B. Shape-Memory Polymers Hallmarks and Their Biomedical Applications in the Form of Nanofibers. *Int. J. Mol. Sci.* **2022**, *23*, 1290. [CrossRef] [PubMed]
15. Lendlein, A.; Langer, R. Biodegradable, Elastic Shape-Memory Polymers for Potential Biomedical Applications. *Science* **2002**, *296*, 1673–1676. [CrossRef]
16. Smola-Dmochowska, A.; Śmigiel-Gac, N.; Kaczmarczyk, B.; Sobota, M.; Janeczek, H.; Karpeta-Jarząbek, P.; Kasperczyk, J.; Dobrzyński, P. Triple-Shape Memory Behavior of Modified Lactide/Glycolide Copolymers. *Polymers* **2020**, *12*, 2984. [CrossRef]
17. Ban, J.; Zhu, L.; Chen, S.; Wang, Y. The Effect of 4-Octyldecyloxybenzoic Acid on Liquid-Crystalline Polyurethane Composites with Triple-Shape Memory and Self-Healing Properties. *Materials* **2016**, *9*, 792. [CrossRef] [PubMed]
18. Tian, M.; Gao, W.; Hu, J.; Xu, X.; Ning, N.; Yu, B.; Zhang, L. Multidirectional Triple-Shape-Memory Polymer by Tunable Cross-linking and Crystallization. *ACS Appl. Mater. Interfaces* **2020**, *12*, 6426–6435. [CrossRef]
19. Wang, Z.; Zhao, J.; Chen, M.; Yang, M.; Tang, L.; Dang, Z.M.; Chen, F.; Huang, M.; Dong, X. Dually actuated triple shape memory polymers of cross-linked polycyclooctene-carbon nanotube/polyethylene nanocomposites. *ACS Appl. Mater. Interfaces* **2014**, *6*, 20051–20059. [CrossRef]
20. Ware, T.; Hearon, K.; Lonnecker, A.; Wooley, K.L.; Maitland, D.J.; Voit, W. Triple-Shape Memory Polymers Based on Self-Complementary Hydrogen Bonding. *Macromolecules* **2012**, *45*, 1062–1069. [CrossRef]
21. Yang, X.; Wang, L.; Wang, W.; Chen, H.; Yang, G.; Zhou, S. Triple shape memory effect of star-shaped polyurethane. *ACS Appl. Mater. Interfaces* **2014**, *6*, 6545–6554. [CrossRef]
22. Karasu, F.; Weder, C. Blends of poly(ester urethane)s and polyesters as a general design approach for triple-shape memory polymers. *J. Appl. Polym. Sci.* **2021**, *138*, 49935. [CrossRef]
23. Bellin, I.; Kelch, S.; Langer, R.; Lendlein, A. Polymeric triple-shape materials. *Proc. Natl. Acad. Sci. USA* **2006**, *103*, 18043–18047. [CrossRef] [PubMed]
24. Xie, T. Tunable polymer multi-shape memory effect. *Nature* **2010**, *464*, 267–270. [CrossRef]
25. Li, J.; Liu, T.; Xia, S.; Pan, Y.; Zheng, Z.; Ding, X.; Peng, Y. A versatile approach to achieve quintuple-shape memory effect by semi-interpenetrating polymer networks containing broadened glass transition and crystalline segments. *J. Mater. Chem.* **2011**, *21*, 12213–12217. [CrossRef]
26. Kumar, U.N.; Kratz, K.; Wagermaier, W.; Behl, M.; Lendlein, A. Non-contact actuation of triple-shape effect in multiphase polymer network nanocomposites in alternating magnetic field. *J. Mater. Chem.* **2010**, *20*, 3404–3415. [CrossRef]

27. Narendra Kumar, U.; Kratz, K.; Behl, M.; Lendlein, A. Shape-memory properties of magnetically active triple-shape nanocomposites based on a grafted polymer network with two crystallizable switching segments. *Express Polym. Lett.* **2012**, *6*, 26–40. [CrossRef]
28. Dong, K.; Panahi-Sarmad, M.; Cui, Z.; Huang, X.; Xiao, X. Electro-induced shape memory effect of 4D printed auxetic composite using PLA/TPU/CNT filament embedded synergistically with continuous carbon fiber: A theoretical & experimental analysis. *Compos. Part B Eng.* **2021**, *220*, 108994.
29. Huang, J.; Zhang, Q.; Scarpa, F.; Liu, Y.; Leng, J. Shape memory polymer-based hybrid honeycomb structures with zero Poisson's ratio and variable stiffness. *Compos. Struct.* **2017**, *179*, 437–443. [CrossRef]
30. Roudbarian, N.; Baniasadi, M.; Ansari, M.; Baghani, M. An experimental investigation on structural design of shape memory polymers. *Smart Mater. Struct.* **2019**, *28*, 095017. [CrossRef]
31. Zhou, J.; Li, H.; Tian, R.; Dugnani, R.; Lu, H.; Chen, Y.; Guo, Y.; Duan, H.; Liu, H. Fabricating fast triggered electro-active shape memory graphite/silver nanowires/epoxy resin composite from polymer template. *Sci. Rep.* **2017**, *7*, 5535. [CrossRef]
32. Farhan, M.; Chaudhary, D.; Nöchel, U.; Behl, M.; Kratz, K.; Lendlein, A. Electrical Actuation of Coated and Composite Fibers Based on Poly[ethylene-co-(vinyl acetate)]. *Macromol. Mater. Eng.* **2021**, *306*, 2000579. [CrossRef]
33. Sabzi, M.; Babaahmadi, M.; Rahnama, M. Thermally and Electrically Triggered Triple-Shape Memory Behavior of Poly(vinyl acetate)/Poly(lactic acid) Due to Graphene-Induced Phase Separation. *ACS Appl. Mater. Interfaces* **2017**, *9*, 24061–24070. [CrossRef]
34. Peng, B.; Yang, Y.; Gu, K.; Amis, E.J.; Cavicchi, K.A. Digital Light Processing 3D Printing of Triple Shape Memory Polymer for Sequential Shape Shifting. *ACS Mater. Lett.* **2019**, *1*, 410–417. [CrossRef]
35. Liu, H.; Wang, F.; Wu, W.; Dong, X.; Sang, L. 4D printing of mechanically robust PLA/TPU/Fe3O4 magneto-responsive shape memory polymers for smart structures. *Compos. Part B Eng.* **2023**, *248*, 110382. [CrossRef]
36. Mao, Y.; Yu, K.; Isakov, M.S.; Wu, J.; Dunn, M.L.; Jerry Qi, H. Sequential Self-Folding Structures by 3D Printed Digital Shape Memory Polymers. *Sci. Rep.* **2015**, *5*, 13616. [CrossRef] [PubMed]
37. Valvez, S.; Reis, P.N.B.; Susmel, L.; Berto, F. Fused Filament Fabrication-4D-Printed Shape Memory Polymers: A Review. *Polymers* **2021**, *13*, 701. [CrossRef] [PubMed]
38. Chalissery, D.; Pretsch, T.; Staub, S.; Andrä, H. Additive Manufacturing of Information Carriers Based on Shape Memory Polyester Urethane. *Polymers* **2019**, *11*, 1005. [CrossRef]
39. Mehrpouya, M.; Vahabi, H.; Janbaz, S.; Darafsheh, A.; Mazur, T.R.; Ramakrishna, S. 4D printing of shape memory polylactic acid (PLA). *Polymer* **2021**, *230*, 124080. [CrossRef]
40. Leist, S.K.; Gao, D.; Chiou, R.; Zhou, J. Investigating the shape memory properties of 4D printed polylactic acid (PLA) and the concept of 4D printing onto nylon fabrics for the creation of smart textiles. *Virtual Phys. Prototyp.* **2017**, *12*, 290–300. [CrossRef]
41. Fenni, S.E.; Bertella, F.; Monticelli, O.; Müller, A.J.; Hadadoui, N.; Cavallo, D. Renewable and Tough Poly(l-lactic acid)/Polyurethane Blends Prepared by Dynamic Vulcanization. *ACS Omega* **2020**, *5*, 26421–26430. [CrossRef]
42. Bothe, M.; Pretsch, T. Two-Way Shape Changes of a Shape-Memory Poly(ester urethane). *Macromol. Chem. Phys.* **2012**, *213*, 2378–2385. [CrossRef]
43. Bothe, M.; Pretsch, T. Bidirectional actuation of a thermoplastic polyurethane elastomer. *J. Mater. Chem. A* **2013**, *1*, 14491–14497. [CrossRef]
44. McAndrew, T.P.; Laurent, P.; Havel, M.; Roger, C. Arkema Graphistrength® Multi-Walled Carbon Nanotubes. In *Technical Proceedings of the 2008 NSTI Nanotechnology Conference and Trade Show, NSTI-Nanotech, Nanotechnology*; Nano Science and Technology Institute: Danville, CA, USA, 2008; Volume 1, pp. 47–50.
45. Chen, D.; Liu, Q.; Geng, P.; Tang, S.; Zhang, J.; Wen, S.; Zhou, Y.; Yan, C.; Han, Z.; Shi, Y. A 4D printing strategy and integrated design for programmable electroactive shape-color double-responsive bionic functions. *Compos. Sci. Technol.* **2021**, *208*, 108746. [CrossRef]
46. Mallakpour, S.; Zadehnazari, A. An investigation on the effects of functionalized multi-walled carbon nanotube on mechanical and thermal properties of dopamine-bearing poly(amide–imide) composite films. *J. Thermoplast. Compos. Mater.* **2015**, *28*, 1644–1661. [CrossRef]
47. Salzano de Luna, M.; Filippone, G. Effects of nanoparticles on the morphology of immiscible polymer blends—Challenges and opportunities. *Eur. Polym. J.* **2016**, *79*, 198–218. [CrossRef]
48. Buys, F.Y.; Mawardi, A.M.; Anuar, H. Selective Localization of Multi-walled CNT in Polylactic Acid/Thermoplastic Polyurethane Blends and Its Effect on Mechanical Properties. *Curr. Nanomater.* **2017**, *2*, 84–89. [CrossRef]
49. Fritzsche, N.; Pretsch, T. Programming of Temperature-Memory Onsets in a Semicrystalline Polyurethane Elastomer. *Macromolecules* **2014**, *47*, 5952–5959. [CrossRef]
50. Gan, Z.; Abe, H.; Doi, Y. Temperature-Induced Polymorphic Crystals of Poly(butylene adipate). *Macromol. Chem. Phys.* **2002**, *203*, 2369–2374. [CrossRef]
51. Sun, Y.; Teng, Y.; Kuang, Y.; Yang, S.; Yang, J.; Mao, H.; Gu, Z. Electroactive shape memory polyurethane composites reinforced with octadecyl isocyanate-functionalized multi-walled carbon nanotubes. *Front. Bioeng. Biotechnol.* **2022**, *10*, 964080. [CrossRef] [PubMed]
52. Mofokeng, J.P.; Luyt, A.S.; Tábi, T.; Kovács, J. Comparison of injection moulded, natural fibre-reinforced composites with PP and PLA as matrices. *J. Thermoplast. Compos. Mater.* **2012**, *25*, 927–948. [CrossRef]

53. Blackwell, J.; Lee, C.D. Hard-segment polymorphism in MDI/diol-based polyurethane elastomers. *J. Polym. Sci. Polym. Phys. Ed.* **1984**, *22*, 759–772. [CrossRef]
54. Gadley, J.L.; Andrade, R.J.; Maia, J.M. Effect of Soft-to-Hard Segment Ratio on Viscoelastic Behavior of Model Thermoplastic Polyurethanes during Phase Transitions. *Macromol. Mater. Eng.* **2016**, *301*, 953–963. [CrossRef]
55. Lee, C.-F.; Chen, C.-W.; Rwei, S.-P.; Chuang, F.-S. Thermal Behavior and Morphology of Thermoplastic Polyurethane Derived from Different Chain Extenders of 1,3- and 1,4-Butanediol. *Appl. Sci.* **2021**, *11*, 698. [CrossRef]
56. Sonnenschein, M.F.; Guillaudeu, S.J.; Landes, B.G.; Wendt, B.L. Comparison of adipate and succinate polyesters in thermoplastic polyurethanes. *Polymer* **2010**, *51*, 3685–3692. [CrossRef]
57. Pedrazzoli, D.; Manas-Zloczower, I. Understanding phase separation and morphology in thermoplastic polyurethanes nanocomposites. *Polymer* **2016**, *90*, 256–263. [CrossRef]
58. Nordin, N.M.; Buys, Y.F.; Anuar, H.; Ani, M.H.; Pang, M.M. Development of Conductive Polymer Composites from PLA/TPU Blends Filled with Graphene Nanoplatelets. *Mater. Today Proc.* **2019**, *17*, 500–507. [CrossRef]
59. Luo, Z.; Li, X.; Zhao, S.; Xu, L.; Liu, L. Structure and Dielectric Properties of TPU Composite Filled with CNTs@PDA Nanofibers and MXene Nanosheets. *Polymers* **2022**, *14*, 2157. [CrossRef]
60. Pötschke, P.; Häussler, L.; Pegel, S.; Steinberger, R.D.; Scholz, G.A. Thermoplastic polyurethane filled with carbon nanotubes for electrical dissipative and conductive applications. *Kgk Kautsch. Gummi Kunstst.* **2007**, *60*, 432–437.

Disclaimer/Publisher's Note: The statements, opinions and data contained in all publications are solely those of the individual author(s) and contributor(s) and not of MDPI and/or the editor(s). MDPI and/or the editor(s) disclaim responsibility for any injury to people or property resulting from any ideas, methods, instructions or products referred to in the content.

Article

Synthesis and Characterization of Orange Peel Modified Hydrogels as Efficient Adsorbents for Methylene Blue (MB)

Saedah R. Al-Mhyawi [1], Nader Abdel-Hamed Abdel-Tawab [2] and Rasha M. El Nashar [2,*]

[1] Department of Chemistry, College of Science, University of Jeddah, Jeddah 22233, Saudi Arabia
[2] Chemistry Department, Faculty of Science, Cairo University, Giza 12613, Egypt
* Correspondence: rasha.elnashar@cu.edu.eg or rashaelnashar@gmail.com

Abstract: In recent years, due to the developments in the textile industry, water contaminated with synthetic dyes such as methylene blue (MB) has become an environmental threat based on the possible impacts in terms of chemical and biochemical demand, which leads to disturbance in aquatic plants photosynthesis, besides their possible toxicity and carcinogenicity for humans. In this work, an adsorbent hydrogel is prepared via free radical polymerization comprising acrylic acid (PAA) as a monomer and orange peel (OP) as a natural modifier rich in OH and COOH present in its cellulose and pectin content. The resulting hydrogels were optimized in terms of the content of OP and the number of cross-linkers and characterized morphologically using Scanning electron microscopy. Furthermore, BET analysis was used to follow the variation in the porosity and in terms of the surface area of the modified hydrogel. The adsorption behavior was found to follow pseudo-second-order as a kinetic model, and Langmuir, Freundlich, and Temkin isotherm models. The combination of OP and PAA has sharply enhanced the adsorption percent of the hydrogel to reach 84% at the first 10 min of incubation with an adsorption capacity of more than 1.93 gm/gm. Due to its low value of pHZc, the desorption of MB was efficiently performed at pH 2 using HCl, and the desorbed OP-PAA were found to be reusable up to ten times without a decrease in their efficiency. Accordingly, OP-PAA hydrogel represents a promising efficient, cost-effective, and environmentally friendly adsorbent for MB as a model cationic dye that can be applied for the treatment of contaminated waters.

Keywords: adsorbent hydrogel; orange peel; methylene blue; poly acrylic acid hydrogel; cationic dyes

1. Introduction

Clean water availability for drinking has been a highly important health challenge in recent years. Due to the rapid expansion in industrial effluents from the textile and petrochemical industries, environmental disturbances and pollution problems have highly emerged, affecting the quality of drinking water due to their high content of dyes and toxic substances, urging the need for finding simple and cost-effective remedies to remove such contaminants. Industrial dyes represent one of the most common classes of chemicals that cause drinking water contamination [1]. Based on their charge in aqueous solutions, dyes can be classified into cationic (basic dyes), anionic (acidic dyes), and non-ionic (dispersed dyes) [2,3]. Dyes are mostly applied in the textile industry because of their bright color, water-solubility, low cost, and feasibility of application to the fabric. Yet, these dyes can be released as environmental pollutants not only during the tanning process but also as part of the waste resulting from household laundry drainage water. Due to their aromatic nature, synthetic dye degradation in the environment may lead to the production of highly toxic products that are resistant to oxidizing agents and heat or biodegradation, rendering them to be hazardous not only to humans but also to aquatic life [4].

MB is one of the common cationic industrial dyes that is highly used in cotton, silk, and paper dyeing [5,6]. It also has some applicability in biomedical and therapeutic fields [7] and in food production [8,9]. Many human health diseases were found to result from contact with MB, such as mental disorders and blindness [10,11]. The community of biological systems is also affected as a result of a decrease in the transmittance of sunlight due to MB solubility in water due to the change in its color [12,13]. MB is non-biodegradable, carcinogenic, and toxic; this requires an effective, low-cost, fast, and eco-friendly approach for its removal from water. Several methods, such as liquid–liquid extraction [14], nanofiltration [15,16], coagulation [17,18], ultrafiltration [19,20], adsorption/biosorption [21,22], etc., were reported in the literature.

Among previously reported water treatment methods, adsorption-based methods are considered highly efficient, fast, and inexpensive approaches [23]. "Natural solid bioadsorbents, especially from plant wastes, are considered to be an attractive remedy for water treatment being eco-friendly, low cost, biodegradable, and reliable in terms of green chemistry friendly materials. Their absorption efficiency depends on several parameters, including the surface area of the adsorbent, particle size, and amount, besides some other interaction conditions such as pH, temperature, and contact time with the target samples.

Many examples are reported in literature involving bio-adsorbents for the removal of toxic chemical compounds, heavy metals, and dyes, some of which include wheat and rice [24], remainings of coffee and tea [25,26], leaves and barks of dry trees [27–29], powder of saw and coir [30–32], and shell of rice [33,34] and natural polymers such as chitosan, lignin and in some cases micro-organisms that were reported to be capable of dyes degradation [25,35].

Hydrogels are materials with 10% of their cconstitutional weight or volume made of water. The existence of hydrophilic groups in the hydrogel network, such as (-NH_2, -COOH, -OH, $CONH_2$, -CONH, and -SO_3H), is the main reason for their hydrophilicity [36–38]. The hydrogel 3D structure formed by the polymeric building blocks can be tailored by different modifiers to regulate the affinity of hydrogels to various target pollutants and ionic contaminants and enhance their adsorption ability [39].

In this work, poly acrylic acid (PAA) was used as a sorbent hydrogel in the presence of orange peel (OP), a natural waste material, as a modifier to improve the removal of the MB (as a model cationic dye) from the water based on the benefit of the components comprising the cell walls of OP which are mainly hemicellulose (11%), cellulose (22), sugar (23%), and pectin (25%) [40]. These natural polymers present in OP were reported to show good adsorption properties towards cationic chemicals such as MB [41,42]. The optimum adsorption conditions, including OP content, pH, temperature, and initial dye concentration and adsorption kinetics models (pseudo-first and pseudo-second-order) and isotherm models (Langmuir, Freundlich, and Temkin), were investigated.

2. Materials and Methods

2.1. Materials

All chemicals used were of analytical grade and used without further purification. Deionized distilled water was used in preparation of reagents throughout all experiments. Acrylic acid (AA) from Merck Company, Darmstadt, Germany, Ammonium persulfate (APS), ethylene glycol dimethyl acrylate (EGDMA), and boric acid from Alfa Aesar (Kandel, Germany). Nitric acid and sodium Hydroxide from VWR Chemicals (Darmstadt, Germany). Phosphoric acid, sodium nitrate ($NaNO_3$), Acetic acid, and Methylene blue (MB) (Sigma-Aldrich company, Schnelldorf, Germany). MB Stock solutions were prepared by dissolving different amounts of powdered dye in Britton–Robinson buffer (BR-B) at different pHs.

2.2. Instrumentation

Jenway UV-Visible Spectrophotometer Model 7205 was used for MB initial and final concentration measurements (Jenway Instruments, St Neots, UK). Jenway 3510 Advanced Bench pH Meter (Jenway Instruments, St Neots, UK. Digital orbital shaker, Mini-scale, SSM1, (Stuart, London, UK), was used for shaking the tested samples during the different adsorption stages. An FTIR-Affinity-1, (Shimadzu Corporation, Kyoto, Japan) was used in the range 400–4000 cm^{-1}. Ultra-high-resolution scanning electron microscope (SEM) (Model: Leo Supra 55, Zeiss Sigma, Oberkochen, Germany) was used for surface morphology; samples were fixed on aluminum stubs and coated with gold before observation. Brunauer–Emmett–Teller (BET) measurements were used to determine the specific surface area, pore size, and volume using surface area analyzer manufacturer by Quantachrome; model of NOVA touch LX2, the sample was degassed at 150 °C for 2 h under a vacuum.

2.3. Preparation of OP Powder

The orange peel used in this study comes from local Egyptian sweet orange fruit (*Citrus sinensis*). The waste orange peels collected from fresh juice shops or household waste were washed with distilled water, then the colored outer layer of peel was removed by a scrapper to avoid any possible color overlap during the adsorption measurements. The remaining white inner layer of orange peel (OP) was cut into small parts, dried in the microwave, then ground and sieved by a 0.045 mm porous stainless-steel sieve. The resulting dried biomass was stored in a plastic cup until use without any further chemical treatment.

2.4. Adsorbent Hydrogel Preparation

In order to prepare OP-PAA absorbent hydrogel, as given in Table 1, different amounts of OP were added to deionized DW in 30 mL screw bottle and sonicated for 10 min, followed by stirring for another 10 min in an oil bath at 65 °C. After complete dispersion of OP, the AA monomer (different ratios) was added while stirring to mix, followed by EGDMA as a cross linker (different ratios). After complete homogeneity of the mixture, it was purged with nitrogen gas for 10 min to remove any entrapped oxygen; finally, 5 mL DW containing 250 mg APS was added to the mixture as an initiator. The polymerization reaction was allowed to take place for 2 h at 65 °C, and the resulting OP-PAA polymers were left to dry in a vacuum oven at 60 °C and crushed to be ready for further characterization.

Table 1. OP-PAA Adsorbent hydrogel composite component.

Polymer Name	AA (%)	AA (mL)	OP (g)	EGDMA (mL)	DW (mL)
A	10	3.0	1.0	—	22.0
B	20	6.0	1.0	—	19.0
C	30	9.0	1.0	—	16.0
D	20	6.0	0.5	—	19.0
E	20	6.0	2.0	—	19.0
F	20	6.0	—	—	19.0
G	20	6.0	0.5	0.6	18.4
H	20	6.0	0.5	1.2	17.8
I	20	6.0	0.5	1.8	17.2

2.5. Characterization

2.5.1. The Swelling Capacity Percent (SCP)

In order to study OP-PAA swelling behavior, 0.5 g of OP-PAA in a nylon tea bag was added to 50 mL BR-B (pH 4, 7, and 9) at room temperature (RT), followed by orbit shaking at 130 rpm for 3.5 h. The swelled hydrogel was then taken out from the buffer and weighted, then returned to the buffer again at definite time intervals.

The swelling capacity percent (SCP) was calculated according to the following Equation (1)

$$SCP = \frac{W_f - W_i}{W_i} * 100 \tag{1}$$

where W_f is the weight of dried OP-PAA, and W_i is the weight of swelling OP-PAA.

2.5.2. Determine the Point of Zero Charges (pHPZC)

The solid addition method was used to determine the point of zero charges (pH_{PZC}) of the polymer, according to a previous study [43]. Briefly, 25 mg of OP-PAA was added to 25 mL of 0.1 M $NaNO_3$. The pH of sodium nitrate solution was adjusted prior to the addition of OP-PAA (from 2 to 10) using either NaOH or HNO_3. The solutions were left for 2 days until equilibrium took place, and the pH of each solution was recorded. Δ pH values were calculated and plotted on the x-axis against the pH on the y-axis.

2.5.3. FTIR Spectra, SEM, and BET Characterization

Fourier transform infrared spectra in the range of 4000–400 cm^{-1} were used to characterize the chemical structures of prepared polymer with and without dye. Furthermore, SEM and BET were used to determine polymer morphology and surface area, respectively.

2.6. Dye Adsorption and Kinetic Studies

Different OP content by weight (0.5, 1.0, and 2.0 g) of OP-PAA was used to study the OP-PAA hydrogel adsorption efficiency where 250 mg OP-PAA was added to 25 mL MB dye solution (100 mg/L BR-B pH 9.0), and the mixture was agitated at 170 rpm. The dye concentration was measured at different time intervals until equilibrium took place during the kinetic studies. The initial and final concentrations of MB were determined by UV–VIS spectrophotometer and using a linear equation (y = 0.1751x − 0.1145, R^2 = 0.9998) in the range from 1 to 8 mg/L at λ_{max} 662 nm. The adsorption capacity and removal ratio for the dye can then be determined according to the following Equations (2) and (3):

$$q_e = \frac{(C_i - C_f) * V}{m} \quad (2)$$

$$RR = \frac{C_i - C_f}{C_i} * 100 \quad (3)$$

where q_e (mg/g) is the adsorption capacity, RR is the removal ratio, C_i (mg/L) is the initial dye concentration, and C_f (mg/L) is the final dye concentration. V (L) dye solution volume and m (g) represents the dried adsorbent weight.

2.7. Desorption Studies

The OP-PAA used in adsorption experiment was separated from the remaining dye by centrifugation, then washed by DW to get rid of unabsorbed MB dye. A total of 2.0 g of Separated adsorbent were then added to 25 mL DW at different pH for 1 h at RT.

3. Results and Discussion

3.1. Characterization of OP-PAA Hydrogel

3.1.1. The Swelling Capacity Percent (SCP)

The swelling behavior of a hydrogel adsorbent plays an important role in its adsorption characteristic [44]. Nine hydrogel polymers (A-I) were prepared with different amounts of AA as a monomer, OP, and EGDMA as a cross-linker, as given in Table 1. The swelling capacity percent (SCP) was determined for the prepared polymers to select the best component ratio exhibiting the highest swelling capacity.

Figure 1A shows the effect of AA percent on the swelling capacity at different pH values. It is clear that the increase in AA percent from 10% (polymer A) to 20% (polymer B) led to an increase in SCP. This increase was due to an increase in the content of carboxylic acid groups. On further increase in AA content from 20% (polymer B) to 30% (polymer C) caused a decrease in SCP. This phenomenon can be attributed to the interaction between the OH groups of OP and the COOH group in AA, rendering the polymer more rigid and, in turn, affecting its swelling tendency [45–47].

Figure 1B represents the effect of variation of OP content effect on SCP. The addition of 0.5 g OP (polymer D) causes an increase in SCP due to introducing more ionizable groups such as OH and COOH present in the cellulose and pectin content of OP, respectively. Further increase in OP to 1 or 2 g was found to cause deterioration of SCP; as a result, the viscosity increased due to the high content of OP, which hinders interaction of the adsorbents with MB [48].

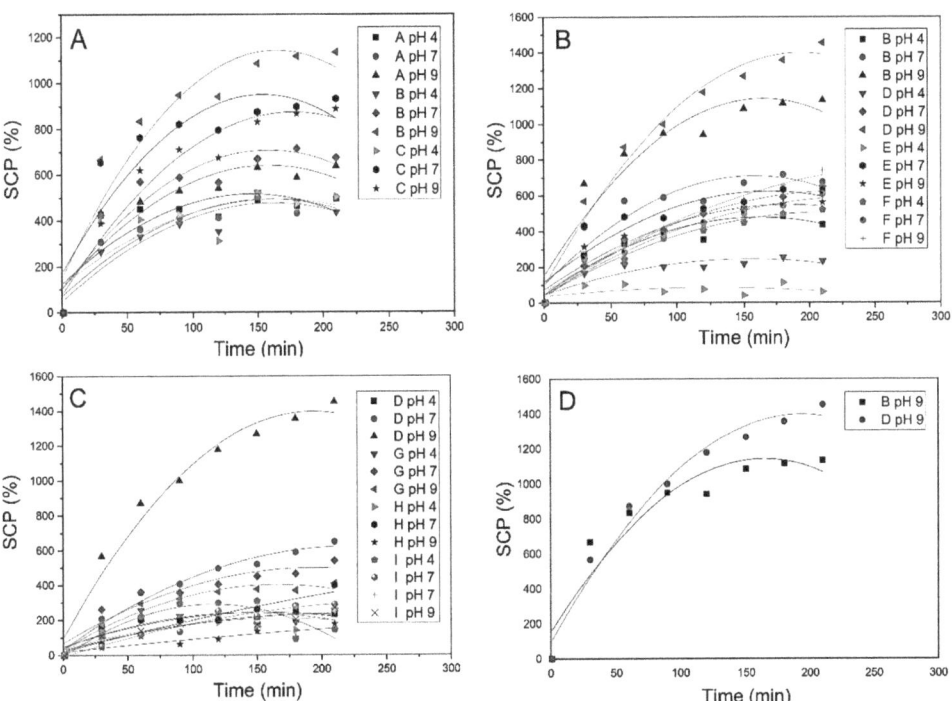

Figure 1. Effect of AA (**A**), OP (**B**) and EGDMA percent on SCP (**C**), and comparison between polymers (**B,D**) exhibiting the highest absorption efficiency (**D**).

Finally, the effect of using EGDMA as a cross-linker was studied, as given in Figure 1C. It is clear that the increase in EGDMA concentration was associated with SCP decrease because it renders the water diffusion in the compact hydrogel more difficult. The tendency of SCP to decrease with cross-linker concentration increase was found to be in agreement with the Flory theory and the results of previous studies [49].

The pH effect on SCP was also investigated, and it was found that an increase in SCP was associated with a pH increase. According to the previous result, a hydrogel adsorbent polymer composed of 6 mL AA and 0.5 g OP without using a cross-linker (polymer D) was selected to be used for further adsorption experiments.

3.1.2. Determine the Point of Zero Charges (pHPZC)

pH_{PZC} value plays an important role in the adsorption mechanism understanding where adsorption of cationic dyes such as MB is preferred to take part at pH larger than pH_{PZC} [50]. From the experimental data, it was found that pH_{PZC} values of PAA and poly E are 2.6 and 2.5, respectively, as indicated in Figure 2.

3.1.3. FT-IR Spectra

The FT-IR spectra of PAA, MB, and MB adsorbed on OP-PAA are shown in Figure 3. In PAA FT-IR spectra, a band at 2921 cm^{-1} that refers to the C-H stretching of an alkane is noticed due to complete AA polymerization. Another two bands at 1632 and 1456 cm^{-1} can be assigned to symmetric and asymmetric stretching vibrations of COO-, respectively. FT-IR spectrum of MB has many bands; the most characteristic are 1638, 1416, and 543 cm^{-1}, which correspond to C=N stretching, C–N stretching vibration in aromatic amines, and C–S skeleton vibration, respectively [51,52].

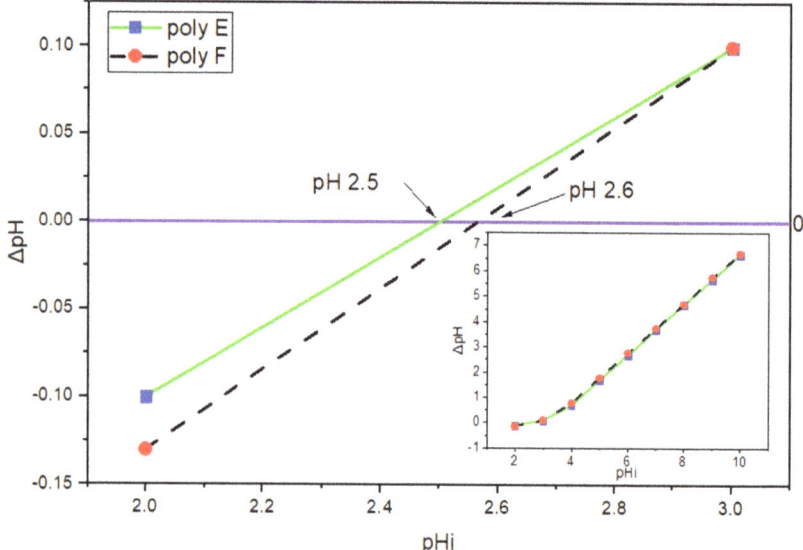

Figure 2. Relation between ΔpH and initial pH and pH$_{PZC}$ conclusion.

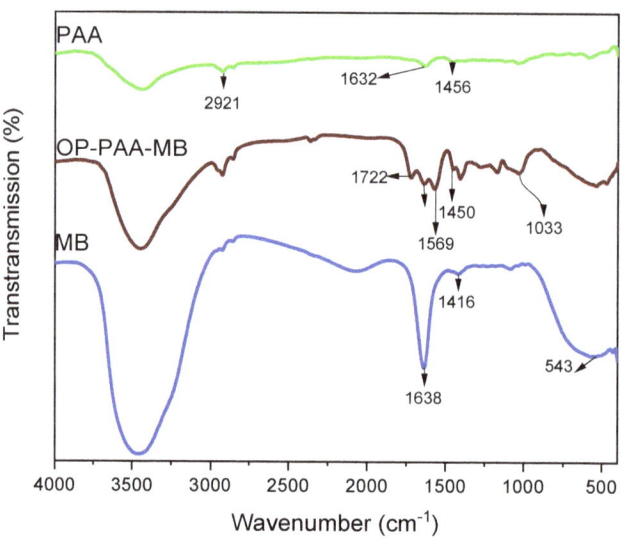

Figure 3. FT-IR spectra of PAA, MB, and MB adsorbed on polymer E hydrogel.

Due to OP's complex nature, a lot of bands were noticed in its FTIR spectrum. The bands 1722, 1033, and 1450 cm^{-1} can be associated with C=O stretching, indicating the abundant carboxylic groups that are capable of interaction and binding with MB; also, the band at 1569 cm^{-1} can be assigned to N-O stretching.

The FT-IR spectra of PAA, MB, and MB adsorbed on OP-PAA showed a clear OH band at 3500 cm^{-1}, the reduction in the intensity of the OH band indicated its participation in blending and cross-linking mechanism with the cationic MB dye, which agrees with the previously reported results for orange peel interaction with textile dyes indicating the role of OH groups in adsorption [53,54].

3.1.4. BET

The surface area (SA) of the adsorbent is one of the main influencing parameters affecting its adsorption efficiency. Multipoint Brunaur–Emmett Teller (BET) analysis was used to determine the polymer's SA using N_2 gas adsorption and desorption, as given in Figure 4A. While, Barrett–Joyner–Halenda (BJH) analysis was used to detect the pore volume of polymers F, E, and polymer E after adsorption of MB (E-MB) were found to be 0.090 cc/g, 0.111 cc/g, and 0.025 cc/g, respectively as given in Table 2 and Figure 4B,C. Polymer F was found to show a relatively higher surface area (SA) of 66.5 m^2/g compared to that of polymer E 53.1 m^2/g and polymer E-MB 22.9 m^2/g. Although the SA of polymer E is smaller [45] than that of polymer F, the adsorption capacity of polymer E was found to be the largest.

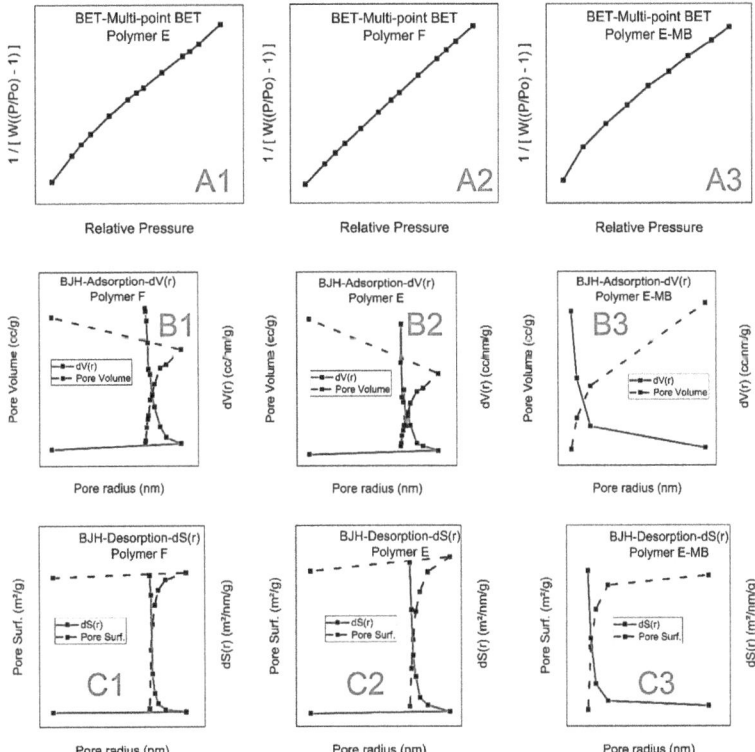

Figure 4. BET-Multi-point BET for polymer F (**A1**), polymer E (**A2**), and MB adsorbed on polymer E (**A3**). BJH Pore Size Distribution-Adsorption for polymer F (**B1**), polymer E (**B2**), and MB adsorbed on polymer E (**B3**). BJH Pore Size Distribution-Desorption for polymer F (**C1**), polymer E (**C2**), and MB adsorbed on polymer E (**C3**).

Table 2. Surface area, pore volume and radius of polymer F, polymer E, and MB adsorbed on polymer E.

	Unit	Polymer F	Polymer E	Polymer E-MB
Surface Area by multipoint BET	mc^2/g	66.51	53.11	22.86
Pore Volume by BJH	cc/g	0.090	0.111	0.025
pore radius by BJH	nm	2.087	1.734	1.684

These phenomena can be correlated to the large pore volume of polymer E and its high content of OH and COOH groups as a result of OP addition. Polymer E-MB has the smallest pore volume and SA due to the adsorption of MB dye on the polymer surface. Furthermore, the pore radius of polymer E was found to decrease from 1.734 nm to 1.684 nm in the presence of MB; this can be attributed to the high adsorbed amount of MB, which indicated the strong interaction between the OP and the PAA polymer [45,46,53].

3.1.5. SEM

Figure 5 shows the surface morphology of polymer F (5A), polymer E (5B), and polymer E after adsorption of MB (E-MB) (5C) at three different magnification power (1, 3, and 7 KX). Polymer F, comprising PAA only, showed a homogenous polymeric nature, as shown in Figure 5A, whereas the addition of OP to PAA, Figure 5B, was found to increase the surface area by increasing the pores and cavities in the polymeric matrix. The change in textural properties and roughness of polymer E after the adsorption of MB, as shown in Figure 5C, represented evidence of the accumulation of MB onto the bio-adsorbent surface.

3.2. Effect of OP Content

The effect of OP content on swelling and hydrogel adsorption capacity is presented in Figure 6. It is clear that upon the addition of 0.5 g OP to PAA, Polymer D, q_e was increased compared to polymer F comprising PAA only. This can be attributed to the introduction of OH and COOH groups from cellulose and pectin, respectively, that represented the main components in OP, where –OH reacts with AA and increases the polymeric network [55,56]. The increase in the swelling capacity also participates in adsorption capacity enhancement due to the increase in the surface area of the adsorbent [44,57,58].

Further increase in the content of OP, though, increases the adsorption capacity and velocity of adsorption, but on the other hand, results in a decrease in the swelling efficiency. This can be explained based on the increase in the number of –OH groups due to the increase in cellulose content, which may act as a cross-linker upon reaction with AA monomer increasing the polymer rigidity. However -COOH groups on the pectin surface are capable of maintaining the adsorption capacity value stable and increasing the rate of adsorption. According to OP content results, polymer E, comprising 6 mL AA and 2.0 g OP, was selected as hydrogel adsorbent to complete the next adsorption experiments as it showed the highest adsorption capacity among the prepared hydrogels.

3.3. Effect of pH

The variation of pH of the medium plays an important role in protonation and deprotonation not only of the MB but the hydrogel adsorbent itself due to its enriched content of carboxylic and hydroxyl groups; thus, the effect of pH on adsorption of MB (100 mg/L) at RT was investigated. In order to discuss the pH effect of MB adsorption, the pH$_{PZC}$ must be determined where cationic dye adsorption is preferred at a pH higher than pH$_{PZC}$ [39,50], as given in Figure 7. From experimental data, pH$_{PZC}$ values of PAA and polymer E were found to be 2.6 and 2.5, respectively.

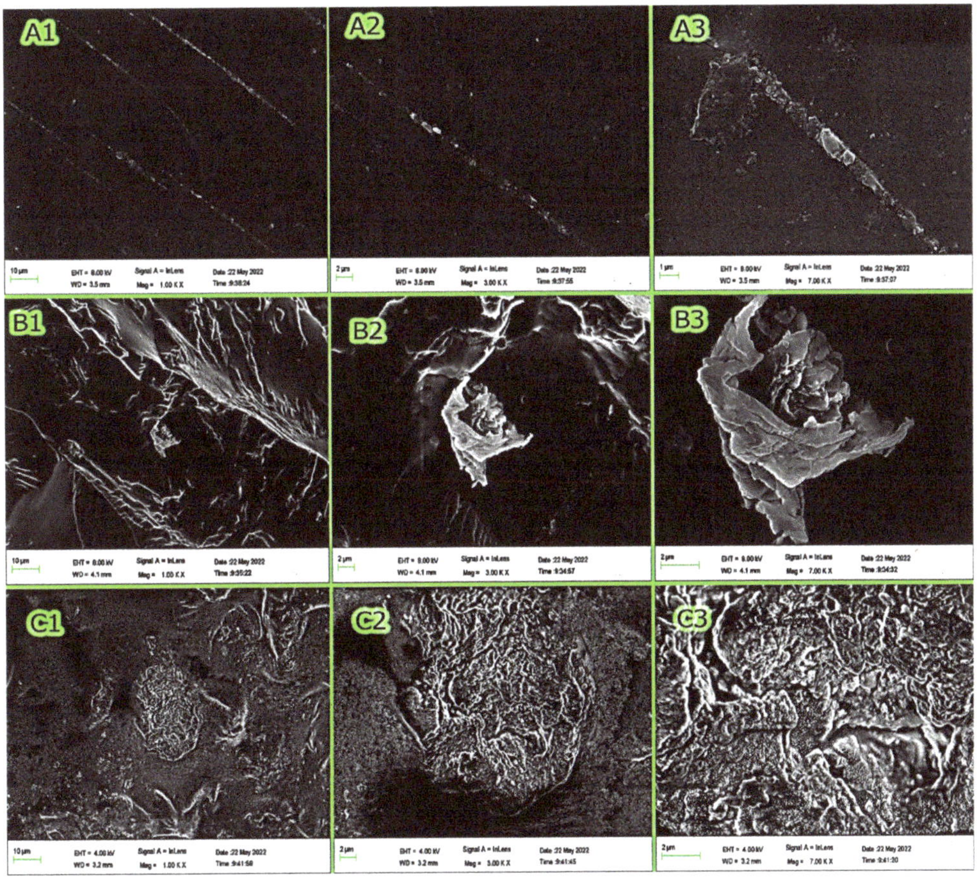

Figure 5. SEM images of polymer F (**A**), Polymer E (**B**), and MB adsorbed on polymer E (**C**).

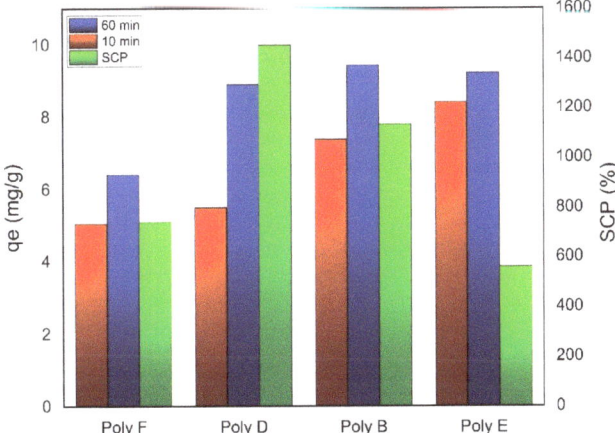

Figure 6. Effect of OP content on q_e of both PAA and OP-PAA for MB adsorption. Experimental conditions were: MB concentration 100 mg/L; adsorbent hydrogel dose 250 mg in 25 mL; pH 9 at RT.

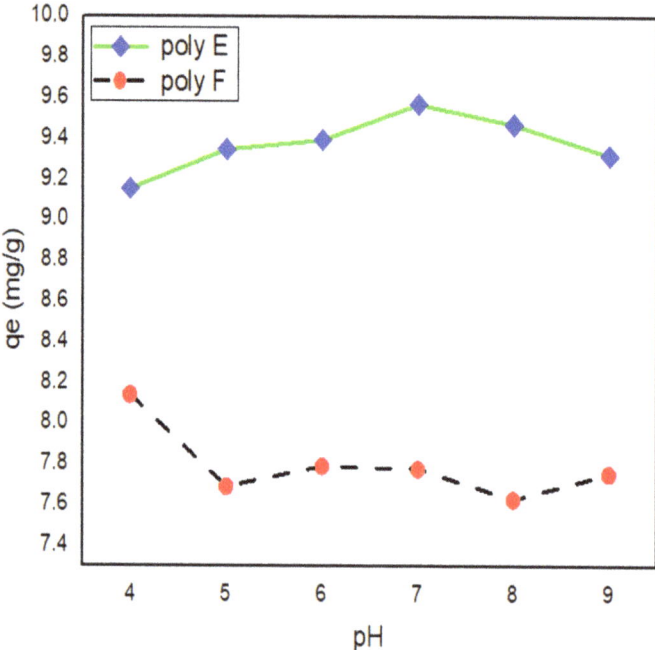

Figure 7. Effect of pH on q_e of both polymer F and polymer E for MB adsorption. Experimental conditions were: MB concentration: 100 mg/L; h adsorbents hydrogel dose: 250 mg in 25 mL and RT.

The variation in q_e was found to be relatively small at low pH and peaks at pH 7.0. This can be attributed to the protonation of -COOH at low pH, where -H from HCl is bound to the carboxylic group in acrylate. As the pH increases, COOH becomes deprotonated and more COO- groups are generated and are available for reaction with MB [59]. On the other hand, the increase in pH value results in an increase in the swelling ratio paired with increased polymer surface area and more MB penetration. At pH higher than 7.0, the q_e was found to decrease due to the charge screening effect of sodium ions and competition between sodium ions from NaOH and MB to interact with the carboxylate groups of the polymer [60,61].

3.4. Effect of Temperature

The effect of temperature on adsorption is an important factor to be investigated in order to have an insight into the adsorption thermodynamics parameters such as entropy ΔS, enthalpy ΔH, and free energy ΔG. MB adsorption on PAA and OP-PAA adsorbent hydrogel was found to decrease with a temperature increase from 30 to 60 °C, as shown in Figure 8. Although the increase in temperature commonly causes swelling of internal adsorbent construction and, in turn, increases the dye molecules penetration [62], it also causes dye ions' mobility to increase. This increase in ion mobility is associated with an adsorption capacity decrease, which agrees with the behavior previously reported [63,64].

The change in entropy ΔS, enthalpy ΔH, and free energy ΔG was calculated by using Equations (4)–(7) [65,66]:

$$K_d = \frac{[A(s)]}{[A(l)]} = \frac{q_e}{C_e} \quad (4)$$

$$\Delta G = -RT \ln K_d \quad (5)$$

$$\Delta G = \Delta H - T\Delta S \quad (6)$$

$$\ln(K_d) = \frac{\Delta S}{R} - \frac{\Delta H}{RT} \quad (7)$$

where C_f and C_i are the final and initial MB concentrations (mg/L), K_d, R, and T are the equilibrium constant, the gas constant, and temperature (K), respectively.

Figure 8 represents the Van't Hoff plot where ln K_d was plotted against 1/T. ΔS and ΔH can be calculated from the Van't Hoff plot using the intercept and slope. Table 3 shows the thermodynamic values; by increasing Temperature, ΔG values move to be a positive value, which means that the MB adsorption process on the hydrogel adsorbent is spontaneous [67,68]. ΔH and ΔS values are negative, as commonly associated with exothermic processes, indicating a randomness decrease at the solid/liquid interface [69,70].

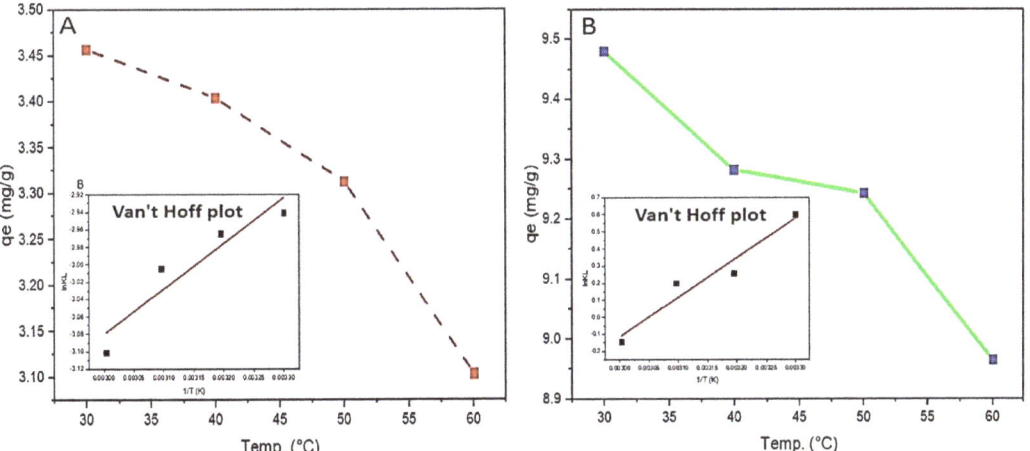

Figure 8. Effect of temperature and Van't Hoff plot on q_e of polymer F (**A**) and polymer E (**B**) for MB adsorption. Experimental conditions were: MB concentration: 100 mg/L; adsorbents hydrogel dose 250 mg in 25 mL; pH 7.

Table 3. Polymers E and F thermodynamic parameters for MB adsorption.

	Temperature	Polymer E	Polymer F
ΔG (KJmol^{-1})	30 °C	−1.52	7.41
	40 °C	−0.67	7.71
	50 °C	−0.54	8.07
	60 °C	0.40	8.59
ΔH (KJmol^{-1})		−19.23	−4.34
ΔS (KJ^{-1}mol^{-1})		−58.64	−38.62

3.5. Adsorption Kinetics

In order to investigate the adsorption kinetics of OP-PAA and MB, the time effect on adsorption capacities was studied. Figure 9 shows the relationship between time (min) and polymer E and F adsorption capacities at a definite time (min) and qt (mg/g). It can be noticed that, at the first 10 min, a rapid adsorption rate took place where about 84% of MB was adsorbed by OP-PAA. Then, the rate of adsorption became slow until equilibrium took place at 60 min with a removal ratio, RR, of about 92%). This can be attributed to the presence of a large, uncovered absorbent surface area at the beginning of the experiment, which decreased over time upon the start of MB adsorption and accumulation of the adsorbent surface. On further increase in time for another hour, the RR value was found to reach 94% and remained stable thereafter when tested up to 5 h.

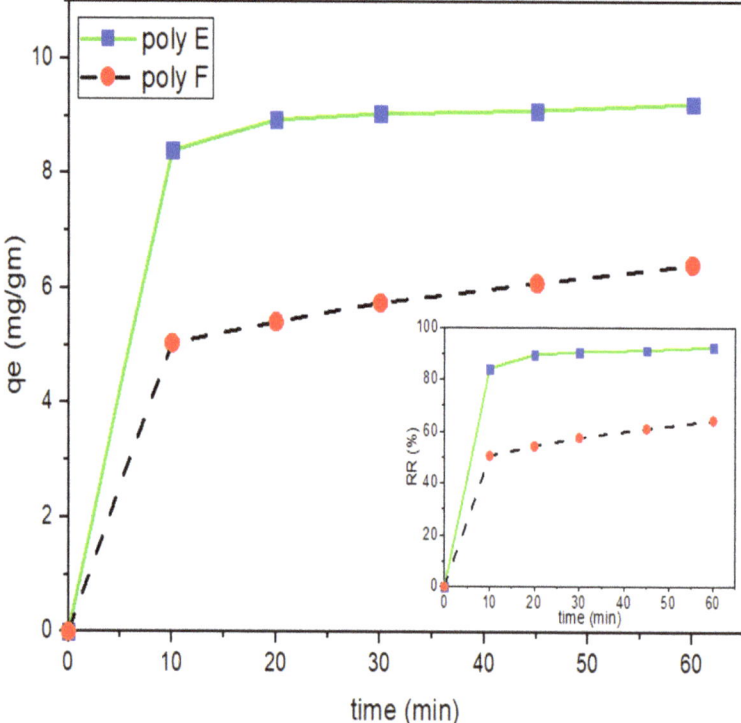

Figure 9. Effect of contact time on q_t of both PAA and OP-PAA for MB adsorption. Experimental conditions were: MB concentration 100 mg/L; adsorbents hydrogel dose 250 mg in 25 mL; pH 9 at RT.

On the other hand, on comparing with unmodified PAA as an adsorbent, RR values were found to be 50% and 64% at 10 and 60 min, respectively, with no further increase. Accordingly, it can be concluded that the addition of OP to PAA enhances the removal ratio from 50% to 84% in the first ten min and increases the RR rate to 92% after one hr.

The adsorption kinetics were tested using both pseudo-first and pseudo-second-order models [71].

The pseudo-first-order linear Equation (8) can be represented as [59]:

$$\log(q_e - q_t) = \log q_e - \frac{K_1 t}{2.303} \tag{8}$$

while the pseudo-second-order linear Equation (9) can be indicated by [72]:

$$\frac{t}{q_t} = \frac{1}{K_2 q_e^2} + \frac{t}{q_e} \tag{9}$$

where q_t (mg/g) and q_e are the adsorption capacities at a definite time (min) and equilibrium, respectively, K_1 (1/min) and K_2 (g/mg min^{-1}) are the rate constants of adsorption for pseudo-first- and second-order, respectively.

The experimental data were used to plot $\log(q_e - q_t)$ and t/q_t against t, as shown in Figure 10, in order to calculate K_1 and K_2, and q_e from the intercept and slope of the plots as given in Table 4.

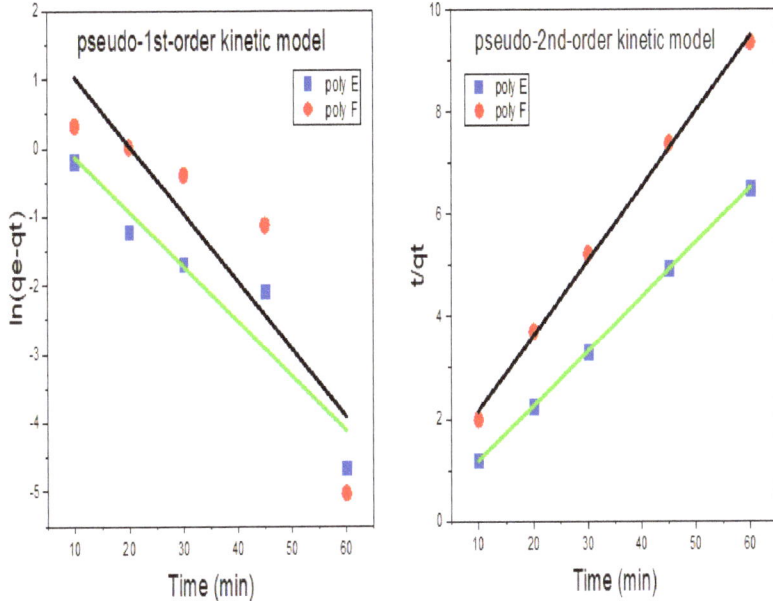

Figure 10. Kinetics curves of pseudo-first-order and pseudo-second-order model. Experimental conditions were, MB concentration 100 mg/L; adsorbents hydrogel dose 250 mg in 25 mL; pH 9; RT.

Table 4. Polymer E and F kinetic parameters for MB adsorption.

Parameters	Pseudo-First-Order				Pseudo-Second-Order			
	K_1 min^{-1}	Cal q_e mg/g	Exp q_e mg/g	R_1^2	K_2 mg^{-1} min^{-1}	Cal q_e mg/g	Exp q_e mg/g	R_2^2
Polymer E	−0.0013275	1.9	9.2	0.8692	0.09582068	9.4	9.2	0.9999
Polymer F	−0.0016448	7.5	6.4	0.7384	0.03129885	6.8	6.4	0.9969

R^2 of the pseudo-first-order model was found to be 0.739 and 0.869 for PAA and OP-PAA, respectively, while the calculated q_e were 7.5 and 1.9 mg/g for PAA and OP-PAA, respectively. The previous result indicated that the pseudo-first-order model does not fit with experimental data. In contrast, R^2 for the pseudo-second-order model was found to be 0.9969 and 0.9999 for PAA and OP-PAA, respectively, indicating better fitting to such a model. Besides the R^2 value being highly close to 1.0, the calculated q_e was close enough to the experimental q_e; thus, based on the data shown in Table 4, it can be concluded that the pseudo-second-order model is suitable to describe the adsorption of MB onto OP-PAA hydrogel.

3.6. Adsorption Isotherms

The correlation between the initial concentration of MB (C_i) and the equilibrium absorption capacity, q_e, was studied, as shown in Figure 11. It was found that the increase in C_i from 200 to 1800 mg/L led to a dramatic increase in q_e from 135 to 1309 mg/g for polymer F and from 317 to 1933 mg/g for polymer E. From this result, it can be concluded that the addition of OP to PAA improves the adsorption efficiency of the adsorbent hydrogel.

Langmuir, Freundlich, and Temkin isotherm models were used to test the interaction mechanism between MB and adsorbent hydrogel.

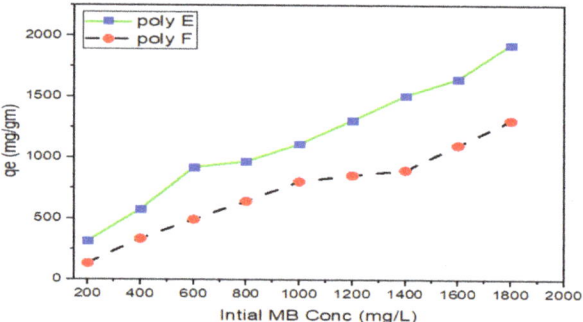

Figure 11. Effect of initial MB concentration on both polymer F and polymer E for MB adsorption. Experimental conditions were: adsorbents hydrogel dose 50 mg in 100 mL; pH 7, and RT.

The Langmuir linear Equation (10) form is [73]:

$$\frac{1}{q_e} = \frac{1}{q_m} + \frac{1}{q_m K_L} \frac{1}{C_e} \quad (10)$$

The Freundlich linear Equation (11) form is [74]:

$$\log q_e = \log K_f + \frac{1}{n} \log C_e \quad (11)$$

The Temkin linear Equation (12) form is [23]:

$$q_e = \frac{RT}{b} \ln(K_T C_e) \quad (12)$$

where q_m (mg/g) and qe represent the maximum and equilibrium adsorption capacities, respectively. C_i (mg/L) is the initial dye concentration, while K_L (L mg^1), K_F (mg$^{1-1/n}$ L$^{1/n}$ g^{-1}), and K_T Lg^{-1} are the isotherm constants of Langmuir, Freundlich, and Temkin, respectively.

Figure 12 and Table 5 show the fitting curves of the tested isotherm models and their parameters. The highest R^2 value for polymer E was found to be related to the Langmuir isotherm model (0.964) with a calculated q_m close enough to the experimental qe. Freundlich and Temkin's models showed R^2 of 0.932 and 0.873, respectively. This indicated that the adsorption of MB on polymer E could be better described by Langmuir rather than Freundlich and Temkin models.

On the other hand, R^2 of Langmuir and Freundlich for polymer F were equal (0.95) with negative sign q_m when calculated using Langmuir models, which indicated that MB adsorption on polymer F follows Freundlich rather than Langmuir's model. Furthermore, the Temkin model showed a good R^2 value (0.929) and can be used to describe the adsorption of MB on polymer F.

3.7. Desorption of MB

In order to study the regeneration and reusability of the OP-PAA, the polymers bound to MB were separated by centrifugation, then washed by DW to get rid of any unabsorbed MB dye. A total of 2.0 g of the separated adsorbent were incubated in 25 mL DW of different pH values for 1 h at RT. Desorption studies play a role in understanding the adsorption mechanism.

It is noteworthy to mention that if the MB adsorbed on the hydrogel adsorbent can be removed by DW, this means that the attraction between MB and hydrogel adsorbent is very weak. On the other hand, if the attraction between MB and hydrogel adsorbent can be destroyed by an acid such as HCl, this indicates that the mechanism can be correlated more to electrostatic attraction or ion exchange [50].

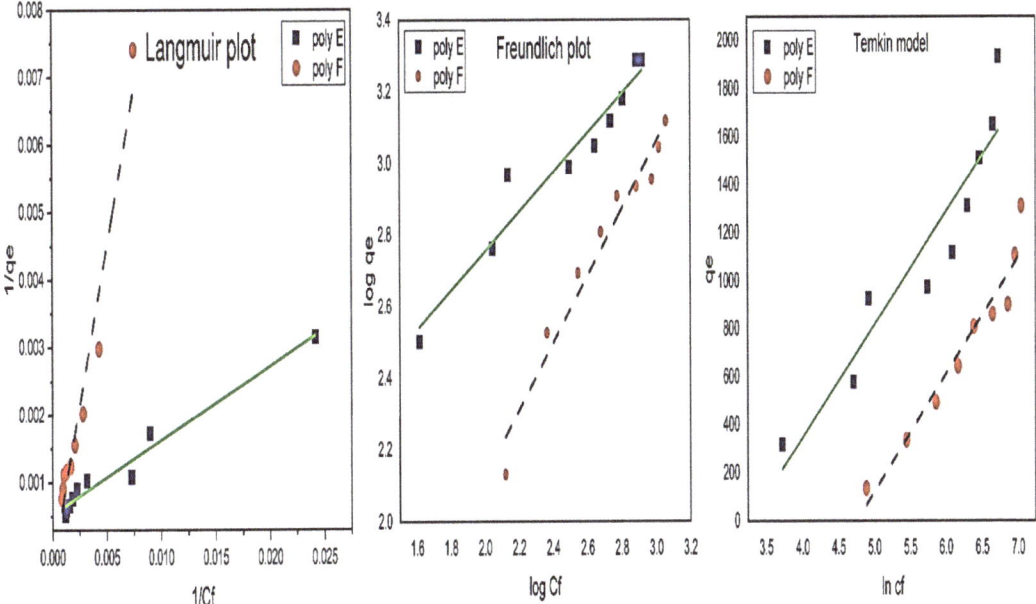

Figure 12. Isotherm model plots of both polymer F and E for MB adsorption. Experimental conditions were: adsorbents hydrogel dose 50 mg in 100 mL; pH 7, at RT.

Table 5. Polymers E and F isotherm parameters for MB adsorption at 25 °C.

	Parameters	Unit	Polymer E	Polymer F
Langmuir model	q_m K_L R^2	mg/g L/mg	1892 0.00482 0.964	−4379 −0.00024 0.953
Freundlich model	$1/n$ K_f R^2		0.5485 45.3 0.932	0.9357 1.78 0.950
Temkin model	B_T K_T R^2	J/mol L/gm	469 0.0383 0.873	489 0.0086 0.929

The experimental results showed the occurrence of no desorption of MB from the hydrogel adsorbent at pH ranging from three to nine, while efficient removal of MB was attained at pH 2, as represented in Figure 13. Desorption at pH 2 can be correlated to the low value of pH_{PZC} of the hydrogel adsorbent and indicated that the MB was attached to the hydrogel adsorbent via electrostatic attraction rather than surface adsorption.

The desorbed OP-PAA hydrogel reusability was tested, and it was found to retain its efficiency in terms of absorption capacity without any change for not less than ten times of usage and desorption, which indicated the usability of the hydrogel polymer after simple treatment with HCl.

Figure 13. Adsorption and Desorption of MB from Polymer E at different pH (2–9). Experimental conditions were: adsorbents hydrogel dose: 2.0 g in 25 mL for 1 h at RT.

4. Conclusions

In this work, a simple approach for the removal of cationic dye based on the combination of poly acrylic acid (PAA) as a sorbent hydrogel in the presence of Orange peel (OP) as a natural modifier to improve the removal of the MB (as a model cationic dye) is reported. MB adsorption by OP-PAA and PAA was found to be affected by the amount of OP, initial pH value, reaction temperature, and initial dye concentration. The addition of OP to PAA was found to enhance the adsorption capacity of PAA and accelerate the adsorption process. q_e of both polymer E and F adsorbent hydrogels were optimized at pH 7, and RT temperature increases were found to have a negative effect on q_e. The max adsorption capacity of polymer E was calculated by the Langmuir isotherm to be equal to 1892 mg/g.

Comparison of the adsorption parameters of OP-PAA hydrogel adsorbent, presented in this work and other previously reported dyes' adsorbents given in Table 6 [75–82], indicated that the presented adsorbent is not only an inexpensive and cost-effective, green and recyclable but was found to be more effective and fast acting for dye removal, where 84% of the adsorption was attained in 10 min without any need for stirring. The desorption process of MB, as a model cationic molecule, from OP-PAA hydrogel adsorbent was found to efficiently take place at pH 2.0 indicating that MB interaction with the hydrogel adsorbent took place via electrostatic attraction rather than surface adsorption which indicates the efficient ability of the hydrogel to retain the adsorbed dye particles. Furthermore, the prepared OP-PAA adsorbent hydrogel was found to be regenerable using HCl at pH 2 and retained its efficiency after several adsorption/desorption cycles, which indicated that the presented OP-PAA adsorbent hydrogel is a promising cost-effective, and efficient, reusable eco-friendly adsorbent for water treatment, from cationic moieties that can either be dyes or heavy metal ions.

Table 6. Comparison between OP-PAA and some other previously reported dyes' adsorbents.

Adsorbent	Dye	q_m mg/g	RR%	Time	Ref
Tragacanth gum and carboxyl-functionalized carbon nanotube	Methylene blue	1092	80.0	40 min	[75]
Sludge	Direct red 28	1.25	—	100 min	[76]
Haloxylon recurvum stem	Acid brown 354	6.87	81.0	50 min	[77]
Coconut Shell (Activated carbon)	Crystal violet	44.00	99.6	24 h	[78]
Orange peel	Methylene blue	—	95.7	24 h	[54]
Spent tea leave	Methylene blue	—	99.0	24 h	[54]
Rattan sawdust	Methylene blue	359.00	—	480 min	[79]
Acacia nilotica sawdust	Methylene blue	46.95	99.9	60 min	[80]
Starch-g-poly (acrylic acid)	Methylene blue	1532	—	30 min	[81]
Gum ghatti-g-poly(acrylic acid)	Methylene blue	909	99	75 min	[82]
Orange peel poly (Acrylic acid)	Methylene blue	1892	84.0	10 min	This work

Author Contributions: Conceptualization, S.R.A.-M. and R.M.E.N.; methodology, S.R.A.-M., N.A.-H.A.-T. and R.M.E.N.; validation, N.A.-H.A.-T.; investigation, N.A.-H.A.-T.; resources, S.R.A.-M. and R.M.E.N.; writing—original draft preparation, N.A.-H.A.-T.; writing—review and editing, S.R.A.-M. and N.A.-H.A.-T.; R.M.E.N. supervision, R.M.E.N. All authors have read and agreed to the published version of the manuscript.

Funding: This research received no external funding.

Institutional Review Board Statement: Not applicable.

Informed Consent Statement: Not applicable.

Data Availability Statement: All data will be available upon reasonable request from the corresponding author.

Conflicts of Interest: The authors declare no conflict of interest.

References

1. Khan, I.; Saeed, K.; Zekker, I.; Zhang, B.; Hendi, A.H.; Ahmad, A.; Ahmad, S.; Zada, N.; Ahmad, H.; Shah, L.A.; et al. Review on methylene blue: Its properties, uses, toxicity and photodegradation. *Water* **2022**, *14*, 242. [CrossRef]
2. Thomas, B.; Shilpa, E.P.; Alexander, L.K. Role of functional groups and morphology on the pH-dependent adsorption of a cationic dye using banana peel, orange peel, and neem leaf bio-adsorbents. *Emergent Mater.* **2021**, *4*, 1479–1487. [CrossRef]
3. Nirmaladevi, S.; Palanisamy, N. A comparative study of the removal of cationic and anionic dye from aqueous solutions using biochar as an adsorbent. *Desalin. Water Treat.* **2020**, *175*, 282–292. [CrossRef]
4. Pirkarami, A.; Olya, M.E.; Yousefi Limaee, N. Decolorization of azo dyes by photo electro adsorption process using polyaniline coated electrode. *Prog. Org. Coat.* **2013**, *76*, 682–688. [CrossRef]
5. Derakhshan, Z.; Baghapour, M.A.; Ranjbar, M.; Faramarzian, M. Adsorption of methylene blue dye from aqueous solutions by modified pumice stone: Kinetics and equilibrium studies. *Health Scope* **2013**, *2*, 136–144. [CrossRef]
6. Han, T.H.; Khan, M.M.; Kalathil, S.; Lee, J.; Cho, M.H. Simultaneous enhancement of methylene blue degradation and power generation in a microbial fuel cell by gold nanoparticles. *Ind. Eng. Chem. Res.* **2013**, *52*, 8174–8181. [CrossRef]
7. Dao, H.M.; Whang, C.H.; Shankar, V.K.; Wang, Y.H.; Khan, I.A.; Walker, L.A.; Husain, I.; Khan, S.I.; Murthy, S.N.; Jo, S. Methylene blue as a far-red light-mediated photocleavable multifunctional ligand. *Chem. Commun.* **2020**, *56*, 1673–1676. [CrossRef]
8. Koyuncu, H.; Kul, A.R. Removal of methylene blue dye from aqueous solution by nonliving lichen (*Pseudevernia furfuracea* (L.) Zopf.), as a novel biosorbent. *Appl. Water Sci.* **2020**, *10*, 72. [CrossRef]
9. Balarak, D.; Bazzi, M.; Shehu, Z.; Chandrika, K. Application of surfactant-modified bentonite for methylene blue adsorption from aqueous solution. *Orient. J. Chem.* **2020**, *36*, 293–299. [CrossRef]
10. Santoso, E.; Ediati, R.; Kusumawati, Y.; Bahruji, H.; Sulistiono, D.O.; Prasetyoko, D. Review on recent advances of carbon based adsorbent for methylene blue removal from waste water. *Mater. Today Chem.* **2020**, *16*, 100233. [CrossRef]

11. Abdelrahman, E.A.; Hegazey, R.M.; El-Azabawy, R.E. Efficient removal of methylene blue dye from aqueous media using Fe/Si, Cr/Si, Ni/Si, and Zn/Si amorphous novel adsorbents. *J. Mater. Res. Technol.* **2019**, *8*, 5301–5313. [CrossRef]
12. Zhou, S.; Du, Z.; Li, X.; Zhang, Y.; He, Y.; Zhang, Y. Degradation of methylene blue by natural manganese oxides: Kinetics and transformation products. *R. Soc. Open Sci.* **2019**, *6*, 190351. [CrossRef]
13. Kosswattaarachchi, A.M.; Cook, T.R. Repurposing the industrial dye methylene blue as an active component for redox flow batteries. *ChemElectroChem* **2018**, *5*, 3437–3442. [CrossRef]
14. El-Ashtoukhy, E.S.Z.; Fouad, Y.O. Liquid-liquid extraction of methylene blue dye from aqueous solutions using sodium dodecyl-benzenesulfonate as an extractant. *Alex. Eng. J.* **2015**, *54*, 77–81. [CrossRef]
15. Kong, G.; Pang, J.; Tang, Y.; Fan, L.; Sun, H.; Wang, R.; Feng, S.; Feng, Y.; Fan, W.; Kang, W.; et al. Efficient dye nanofiltration of a graphene oxide membrane: Via combination with a covalent organic framework by hot pressing. *J. Mater. Chem. A* **2019**, *7*, 24301–24310. [CrossRef]
16. Zhong, F.; Wang, P.; He, Y.; Chen, C.; Li, H.; Yu, H.; Chen, J. Preparation of stable and superior flux GO/LDH/PDA-based nanofiltration membranes through electrostatic self-assembly for dye purification. *Polym. Adv. Technol.* **2019**, *30*, 1644–1655. [CrossRef]
17. Lau, Y.Y.; Wong, Y.S.; Teng, T.T.; Morad, N.; Rafatullah, M.; Ong, S.A. Degradation of cationic and anionic dyes in coagulation-flocculation process using bi-functionalized silica hybrid with aluminum-ferric as auxiliary agent. *RSC Adv.* **2015**, *5*, 34206–34215. [CrossRef]
18. Liu, J.; Li, P.; Xiao, H.; Zhang, Y.; Shi, X.; Lü, X.; Chen, X. Understanding flocculation mechanism of graphene oxide for organic dyes from water: Experimental and molecular dynamics simulation. *AIP Adv.* **2015**, *5*, 117151. [CrossRef]
19. Zheng, L.; Su, Y.; Wang, L.; Jiang, Z. Adsorption and recovery of methylene blue from aqueous solution through ultrafiltration technique. *Sep. Purif. Technol.* **2009**, *68*, 244–249. [CrossRef]
20. Kim, S.; Yu, M.; Yoon, Y. Fouling and retention mechanisms of selected cationic and anionic dyes in a ti3c2tx mxene-ultrafiltration hybrid system. *ACS Appl. Mater. Interfaces* **2020**, *12*, 16557–16565. [CrossRef]
21. Wang, Z.; Gao, M.; Li, X.; Ning, J.; Zhou, Z.; Li, G. Efficient adsorption of methylene blue from aqueous solution by graphene oxide modified persimmon tannins. *Mater. Sci. Eng. C* **2020**, *108*, 110196. [CrossRef] [PubMed]
22. Li, H.; Liu, L.; Cui, J.; Cui, J.; Wang, F.; Zhang, F. High-efficiency adsorption and regeneration of methylene blue and aniline onto activated carbon from waste edible fungus residue and its possible mechanism. *RSC Adv.* **2020**, *10*, 14262–14273. [CrossRef] [PubMed]
23. Madrakian, T.; Afkhami, A.; Ahmadi, M. Adsorption and kinetic studies of seven different organic dyes onto magnetite nanoparticles loaded tea waste and removal of them from wastewater samples. *Spectrochim. Acta Part A Mol. Biomol. Spectrosc.* **2012**, *99*, 102–109. [CrossRef]
24. Juang, R.S.; Ju, C.Y. Equilibrium sorption of copper(ii)-ethylenediaminetetraacetic acid chelates onto cross-linked, polyaminated chitosan beads. *Ind. Eng. Chem. Res.* **1997**, *36*, 5403–5409. [CrossRef]
25. Chang, M.Y.; Juang, R.S. Adsorption of tannic acid, humic acid, and dyes from water using the composite of chitosan and activated clay. *J. Colloid Interface Sci.* **2004**, *278*, 18–25. [CrossRef] [PubMed]
26. Vieira, R.S.; Beppu, M.M. Interaction of natural and crosslinked chitosan membranes with Hg(II) ions. *Colloids Surf. A Physicochem. Eng. Asp.* **2006**, *279*, 196–207. [CrossRef]
27. Tsai, W.T.; Chang, C.Y.; Lin, M.C.; Chien, S.F.; Sun, H.F.; Hsieh, M.F. Adsorption of acid dye onto activated carbons prepared from agricultural waste bagasse by ZnCl2 activation. *Chemosphere* **2001**, *45*, 51–58. [CrossRef]
28. Mall, I.D.; Srivastava, V.C.; Agarwal, N.K.; Mishra, I.M. Removal of congo red from aqueous solution by bagasse fly ash and activated carbon: Kinetic study and equilibrium isotherm analyses. *Chemosphere* **2005**, *61*, 492–501. [CrossRef]
29. Mitchell, M.; Ernst, W.R.; Lightsey, G.R.; Rasmussen, E.T.; Bagherzadeh, P. Adsorption of textile dyes by activated carbon produced from agricultural, municipal and industrial wastes. *Bull. Environ. Contam. Toxicol.* **1978**, *19*, 307–311. [CrossRef]
30. Ghosh, D.; Bhattacharyya, K.G. Adsorption of methylene blue on kaolinite. *Appl. Clay Sci.* **2002**, *20*, 295–300. [CrossRef]
31. Muthukumar, M.; Selvakumar, N. Studies on the effect of inorganic salts on decolouration of acid dye effluents by ozonation. *Dye. Pigment.* **2004**, *62*, 221–228. [CrossRef]
32. Alinsafi, A.; Khemis, M.; Pons, M.N.; Leclerc, J.P.; Yaacoubi, A.; Benhammou, A.; Nejmeddine, A. Electro-coagulation of reactive textile dyes and textile wastewater. *Chem. Eng. Process. Process Intensif.* **2005**, *44*, 461–470. [CrossRef]
33. Han, R.; Ding, D.; Xu, Y.; Zou, W.; Wang, Y.; Li, Y.; Zou, L. Use of rice husk for the adsorption of congo red from aqueous solution in column mode. *Bioresour. Technol.* **2008**, *99*, 2938–2946. [CrossRef]
34. Chakraborty, S.; Chowdhury, S.; Das Saha, P. Adsorption of crystal violet from aqueous solution onto naoh-modified rice husk. *Carbohydr. Polym.* **2011**, *86*, 1533–1541. [CrossRef]
35. Staroń, P.; Chwastowski, J. Raphia-Microorganism Composite Biosorbent for Lead Ion Removal from Aqueous Solutions. *Materials* **2021**, *14*, 7482. [CrossRef]
36. Bekiari, V.; Sotiropoulou, M.; Bokias, G.; Lianos, P. Use of poly(N,N-dimethylacrylamide-co-sodium acrylate) hydrogel to extract cationic dyes and metals from water. *Colloids Surf. A Physicochem. Eng. Asp.* **2008**, *312*, 214–218. [CrossRef]
37. Yetimoğlu, E.K.; Kahraman, M.V.; Ercan, Ö.; Akdemir, Z.S.; Apohan, N.K. N-vinylpyrrolidone/acrylic acid/2-acrylamido-2-methylpropane sulfonic acid based hydrogels: Synthesis, characterization and their application in the removal of heavy metals. *React. Funct. Polym.* **2007**, *67*, 451–460. [CrossRef]

38. El Mansoub, A.; El Sayed, M.M.; El Nashar, R.M.; Fahmy, H.M.; Abulnour, A.M.G. Chemically/Electrically-Assisted Regeneration of Polyacrylonitrile-based Hydrogel adsorbed Heavy Metals. *Egypt. J. Chem.* **2022**, *65*, 373–384. [CrossRef]
39. Wang, L.; Zhang, J.; Wang, A. Fast removal of methylene blue from aqueous solution by adsorption onto chitosan-g-poly (acrylic acid)/attapulgite composite. *Desalination* **2011**, *266*, 33–39. [CrossRef]
40. Ayala, J.R.; Montero, G.; Coronado, M.A.; García, C.; Curiel-Alvarez, M.A.; León, J.A.; Sagaste, C.A.; Montes, D.G. Characterization of orange peel waste and valorization to obtain reducing sugars. *Molecules* **2021**, *26*, 1348. [CrossRef]
41. Sun, G.; Xu, X. Sunflower stalks as adsorbents for color removal from textile wastewater. *Ind. Eng. Chem. Res.* **1997**, *36*, 808–812. [CrossRef]
42. Panneerselvam, P.; Morad, N.; Tan, K.A. Magnetic nanoparticle (F3O4) impregnated onto tea waste for the removal of nickel(II) from aqueous solution. *J. Hazard. Mater.* **2011**, *186*, 160–168. [CrossRef] [PubMed]
43. Balistrieri, L.S.; Murray, J.W. The surface chemistry of goethite (alpha -FeOOH) in major ion seawater. *Am. J. Sci.* **1981**, *281*, 788–806. [CrossRef]
44. Tally, M.; Atassi, Y. Synthesis and characterization of pH-sensitive superabsorbent hydrogels based on sodium alginate-g-poly(acrylic acid-co-acrylamide) obtained via an anionic surfactant micelle templating under microwave irradiation. *Polym. Bull.* **2016**, *73*, 3183–3208. [CrossRef]
45. Pavithra, S.; Thandapani, G.; Sugashini, S.; Sudha, P.N.; Alkhamis, H.H.; Alrefaei, A.F.; Almutairi, M.H. Batch adsorption studies on surface tailored chitosan/orange peel hydrogel composite for the removal of Cr (VI) and Cu (II) ions from synthetic wastewater. *Chemosphere* **2021**, *271*, 129415. [CrossRef]
46. Yati, I.; Kizil, S.; Bulbul Sonmez, H. Cellulose-based hydrogels for water treatment. In *Cellulose-Based Superabsorbent Hydrogels*; Springer: Berlin/Heidelberg, Germany, 2019; pp. 1015–1037.
47. Shankar, P.; Thandapani, G.; Kumar, V.; Parappurath Narayanan, S. Evaluation of batch and packed bed adsorption column for chromium (VI) ion removal from aqueous solution using chitosan-silica–g–AM/orange peel hydrogel composite. *Biomass Convers. Biorefinery* **2022**, 1–16. [CrossRef]
48. Kowalski, G.; Kijowska, K.; Witczak, M.; Kuterasiński, L.; Lukasiewicz, M. Synthesis and effect of structure on swelling properties of hydrogels based on high methylated pectin and acrylic polymers. *Polymers* **2019**, *11*, 114. [CrossRef]
49. Flory, P.J.; Rehner, J. Statistical mechanics of cross-linked polymer networks I. rubberlike elasticity. *J. Chem. Phys.* **1943**, *11*, 512–520. [CrossRef]
50. Mall, I.D.; Srivastava, V.C.; Kumar, G.V.A.; Mishra, I.M. Characterization and utilization of mesoporous fertilizer plant waste carbon for adsorptive removal of dyes from aqueous solution. *Colloids Surf. A Physicochem. Eng. Asp.* **2006**, *278*, 175–187. [CrossRef]
51. Fernando, M.S.; de Silva, R.M.; de Silva, K.M.N. Synthesis, characterization, and application of nano hydroxyapatite and nanocomposite of hydroxyapatite with granular activated carbon for the removal of Pb2+ from aqueous solutions. *Appl. Surf. Sci.* **2015**, *351*, 95–103. [CrossRef]
52. Xie, Z.; Guan, W.; Ji, F.; Song, Z.; Zhao, Y. Production of Biologically Activated Carbon from Orange Peel and Landfill Leachate Subsequent Treatment Technology. *J. Chem.* **2014**, *2014*, 491912. [CrossRef]
53. Kallem, P.; Ouda, M.; Bharath, G.; Hasan, S.W.; Banat, F. Enhanced water permeability and fouling resistance properties of ultrafiltration membranes incorporated with hydroxyapatite decorated orange-peel-derived activated carbon nanocomposites. *Chemosphere* **2022**, *286*, 131799. [CrossRef] [PubMed]
54. Lazim, Z.M.; Mazuin, E.; Hadibarata, T.; Yusop, Z. The removal of methylene blue and remazol brilliant blue r dyes by using orange peel and spent tea leaves. *J. Teknol.* **2015**, *74*. [CrossRef]
55. Li, A.; Wang, A. Synthesis and properties of clay-based superabsorbent composite. *Eur. Polym. J.* **2005**, *41*, 1630–1637. [CrossRef]
56. Li, A.; Wang, A.; Chen, J. Studies on poly(acrylic acid)/attapulgite superabsorbent composite. I. synthesis and characterization. *J. Appl. Polym. Sci.* **2004**, *92*, 1596–1603. [CrossRef]
57. Wang, M.; Li, X.; Zhang, T.; Deng, L.; Li, P.; Wang, X.; Hsiao, B.S. Eco-friendly poly(acrylic acid)-sodium alginate nanofibrous hydrogel: A multifunctional platform for superior removal of Cu(II) and sustainable catalytic applications. *Colloids Surf. A Physicochem. Eng. Asp.* **2018**, *558*, 228–241. [CrossRef]
58. Toledo, P.V.O.; Limeira, D.P.C.; Siqueira, N.C.; Petri, D.F.S. Carboxymethyl cellulose/poly(acrylic acid) interpenetrating polymer network hydrogels as multifunctional adsorbents. *Cellulose* **2019**, *26*, 597–615. [CrossRef]
59. Paulino, A.T.; Guilherme, M.R.; Reis, A.V.; Campese, G.M.; Muniz, E.C.; Nozaki, J. Removal of methylene blue dye from an aqueous media using superabsorbent hydrogel supported on modified polysaccharide. *J. Colloid Interface Sci.* **2006**, *301*, 55–62. [CrossRef]
60. Dai, H.; Huang, H. Enhanced swelling and responsive properties of pineapple peel arboxymethyl cellulose-g-poly(acrylic acid-co-acrylamide) superabsorbent hydrogel by the introduction of carclazyte. *J. Agric. Food Chem.* **2017**, *65*, 565–574. [CrossRef]
61. Bello, K.; Sarojini, B.K.; Narayana, B.; Rao, A.; Byrappa, K. A study on adsorption behavior of newly synthesized banana pseudo-stem derived superabsorbent hydrogels for cationic and anionic dye removal from effluents. *Carbohydr. Polym.* **2018**, *181*, 605–615. [CrossRef]
62. Bhattacharyya, K.G.; Sarma, A. Adsorption characteristics of the dye, brilliant green, on neem leaf powder. *Dye. Pigment.* **2003**, *57*, 211–222. [CrossRef]

63. Wang, L.; Zhang, J.; Wang, A. Removal of methylene blue from aqueous solution using chitosan-g-poly(acrylic acid)/montmorillonite superadsorbent nanocomposite. *Colloids Surf. A Physicochem. Eng. Asp.* **2008**, *322*, 47–53. [CrossRef]
64. Lv, Q.; Hu, X.; Zhang, X.; Huang, L.; Liu, Z.; Sun, G. Highly efficient removal of trace metal ions by using poly(acrylic acid) hydrogel adsorbent. *Mater. Des.* **2019**, *181*, 107934. [CrossRef]
65. Chen, T.; Da, T.; Ma, Y. Reasonable calculation of the thermodynamic parameters from adsorption equilibrium constant. *J. Mol. Liq.* **2021**, *322*, 114980. [CrossRef]
66. Yuan, Z.; Wang, J.; Wang, Y.; Liu, Q.; Zhong, Y.; Wang, Y.; Li, L.; Lincoln, S.F.; Guo, X. Preparation of a poly(acrylic acid) based hydrogel with fast adsorption rate and high adsorption capacity for the removal of cationic dyes. *RSC Adv.* **2019**, *9*, 21075–21085. [CrossRef]
67. Ramos-Jacques, A.L.; Lujan-Montelongo, J.A.; Silva-Cuevas, C.; Cortez-Valadez, M.; Estevez, M.; Hernandez-Martínez, A.R. Lead (II) removal by poly(N,N-dimethylacrylamide-co-2-hydroxyethyl methacrylate). *Eur. Polym. J.* **2018**, *101*, 262–272. [CrossRef]
68. Raju, M.P.; Raju, K.M. Design and synthesis of superabsorbent polymers. *J. Appl. Polym. Sci.* **2001**, *80*, 2635–2639. [CrossRef]
69. Dai, J.; Yan, H.; Yang, H.; Cheng, R. Simple method for preparation of chitosan/poly(acrylic acid) blending hydrogel beads and adsorption of copper(II) from aqueous solutions. *Chem. Eng. J.* **2010**, *165*, 240–249. [CrossRef]
70. Wang, X.; Zheng, Y.; Wang, A. Fast removal of copper ions from aqueous solution by chitosan-g-poly(acrylic acid)/attapulgite composites. *J. Hazard. Mater.* **2009**, *168*, 970–977. [CrossRef]
71. Azizian, S. Kinetic models of sorption: A theoretical analysis. *J. Colloid Interface Sci.* **2004**, *276*, 47–52. [CrossRef]
72. Ho, Y.S.; McKay, G. Pseudo-second order model for sorption processes. *Process Biochem.* **1999**, *34*, 451–465. [CrossRef]
73. Osmari, T.A.; Gallon, R.; Schwaab, M.; Barbosa-Coutinho, E.; Severo, J.B.; Pinto, J.C. Statistical analysis of linear and non-linear regression for the estimation of adsorption isotherm parameters. *Adsorpt. Sci. Technol.* **2013**, *31*, 433–458. [CrossRef]
74. Freundlich, H.; Heller, W. The Adsorption of cis- and trans-Azobenzene. *J. Am. Chem. Soc.* **1939**, *61*, 2228–2230. [CrossRef]
75. Mallakpour, S.; Tabesh, F. Green and plant-based adsorbent from tragacanth gum and carboxyl-functionalized carbon nanotube hydrogel bionanocomposite for the super removal of methylene blue dye. *Int. J. Biol. Macromol.* **2021**, *166*, 722–729. [CrossRef] [PubMed]
76. Aragaw, T.A. Utilizations of electro-coagulated sludge from wastewater treatment plant data as an adsorbent for direct red 28 dye removal. *Data Br.* **2020**, *28*, 104848. [CrossRef] [PubMed]
77. Hassan, W.; Noureen, S.; Mustaqeem, M.; Saleh, T.A.; Zafar, S. Efficient adsorbent derived from Haloxylon recurvum plant for the adsorption of acid brown dye: Kinetics, isotherm and thermodynamic optimization. *Surf. Interfaces* **2020**, *20*, 100510. [CrossRef]
78. Aljeboree, A.M.; Alkaim, A.F. Role of plant wastes as an ecofriendly for pollutants (crystal violet dye) removal from aqueous solutions. *Plant Arch.* **2019**, *19*, 902–905.
79. Islam, M.A.; Ahmed, M.J.; Khanday, W.A.; Asif, M.; Hameed, B.H. Mesoporous activated carbon prepared from NaOH activation of rattan (Lacosperma secundiflorum) hydrochar for methylene blue removal. *Ecotoxicol. Environ. Saf.* **2017**, *138*, 279–285. [CrossRef]
80. Lataye, D.H. Removal of crystal violet and methylene blue dyes using Acacia nilotica sawdust activated carbon. *Indian J. Chem. Technol.* **2019**, *26*, 52–68.
81. Liu, X.; Wei, Q. Removal of methylene blue from aqueous solution using porous starch-g-poly(acrylic acid) superadsorbents. *RSC Adv.* **2016**, *6*, 79853–79858. [CrossRef]
82. Fosso-Kankeu, E.; Mittal, H.; Mishra, S.B.; Mishra, A.K. Gum ghatti and acrylic acid based biodegradable hydrogels for the effective adsorption of cationic dyes. *J. Ind. Eng. Chem.* **2015**, *22*, 171–178. [CrossRef]

Disclaimer/Publisher's Note: The statements, opinions and data contained in all publications are solely those of the individual author(s) and contributor(s) and not of MDPI and/or the editor(s). MDPI and/or the editor(s) disclaim responsibility for any injury to people or property resulting from any ideas, methods, instructions or products referred to in the content.

Article

H₂ Uptake and Diffusion Characteristics in Sulfur-Crosslinked Ethylene Propylene Diene Monomer Polymer Composites with Carbon Black and Silica Fillers after High-Pressure Hydrogen Exposure Reaching 90 MPa

Jae Kap Jung [1,*], Ji Hun Lee [1], Sang Koo Jeon [1], Un Bong Baek [1], Si Hyeon Lee [2], Chang Hoon Lee [2] and Won Jin Moon [3]

1. Hydrogen Energy Materials Research Center, Korea Research Institute of Standards and Science, Daejeon 34113, Republic of Korea
2. Department of Biochemical and Polymer Engineering, Chosun University, Gwangju 61452, Republic of Korea
3. Gwangju Center, Korea Basic Science Institute, Gwangju 61186, Republic of Korea
* Correspondence: jkjung@kriss.re.kr

Abstract: We investigated the influence of two fillers—CB (carbon black) and silica—on the H₂ permeation of EPDM polymers crosslinked with sulfur in the pressure ranges 1.2–90 MPa. H₂ uptake in the CB-blended EPDM revealed dual sorption (Henry's law and Langmuir model) when exposed to pressure. This phenomenon indicates that H₂ uptake is determined by the polymer chain and filler-surface absorption characteristics. Moreover, single sorption characteristics for neat and silica-blended EPDM specimens obey Henry's law, indicating that H₂ uptake is dominated by polymer chain absorption. The pressure-dependent diffusivity for the CB-filled EPDM is explained by Knudsen and bulk diffusion, divided at the critical pressure region. The neat and silica-blended EPDM specimens revealed that bulk diffusion behaviors decrease with decreasing pressure. The H₂ diffusivities in CB-filled EPDM composites decrease because the impermeable filler increases the tortuosity in the polymer and causes filler–polymer interactions; the linear decrease in diffusivity in silica-blended EPDM was attributed to an increase in the tortuosity. Good correlations of permeability with density and tensile strength were observed. From the investigated relationships, it is possible to select EPDM candidates with the lowest H₂-permeation properties as seal materials to prevent gas leakage under high pressure in H₂-refueling stations.

Keywords: carbon black; silica; H₂ uptake; diffusion; permeation; density; dispersion

Citation: Jung, J.K.; Lee, J.H.; Jeon, S.K.; Baek, U.B.; Lee, S.H.; Lee, C.H.; Moon, W.J. H₂ Uptake and Diffusion Characteristics in Sulfur-Crosslinked Ethylene Propylene Diene Monomer Polymer Composites with Carbon Black and Silica Fillers after High-Pressure Hydrogen Exposure Reaching 90 MPa. *Polymers* **2023**, *15*, 162. https://doi.org/10.3390/polym15010162

Academic Editor: Md Najib Alam

Received: 28 November 2022
Revised: 20 December 2022
Accepted: 23 December 2022
Published: 29 December 2022

Copyright: © 2022 by the authors. Licensee MDPI, Basel, Switzerland. This article is an open access article distributed under the terms and conditions of the Creative Commons Attribution (CC BY) license (https:// creativecommons.org/licenses/by/ 4.0/).

1. Introduction

The fundamental properties of rubbery polymers can be improved by blending fillers. Fillers in polymer composites achieve multiple purposes, the most important including reinforcement, improvement in processing, diffusing molecule impermeability, an increase in oil resistance, and a reduction in material cost [1–7]. The rubber industry uses a wide range of fillers because of their merits in rubber compounding. In particular, carbon black (CB) filler compounding with rubber enhances the mechanical properties of rubber composites, such as hardness, tear strength, tensile strength, modulus, and abrasive strength [8–10]. The size and surface area of CB filler particles are important factors affecting reinforcements in rubbers [11]. Furthermore, the reinforcement originates from filler–filler and rubber–filler interactions.

Among nonblack fillers, silica filler provides unique strength characteristics that enhance the abrasion resistance, tear strength, aging resistance, and adhesion properties of rubber [12,13]. Precipitated silica provides the highest degree of reinforcement. Silane coupling agents enhance the chemical compatibilities of silica fillers with the rubber matrix for more efficient reinforcement [14–16].

Several studies have reported the influence of filler loading on gas transport properties in polymers, which is associated with the consecutive processes of sorption, diffusion, and desorption in the polymer membrane; these processes determine the solubility, diffusion coefficient, and permeation coefficient. The diffusion characteristics of ethyl p-aminobenzoate for silicone rubber membranes containing various amounts of fumed silica filler with high surface areas were described [17]. Increased filler loading results in a decrease in the transmission rate but this apparent diffusivity decreases drastically due to the adsorption of the permeant at the filler. This result is attributed to the adsorption of permeant at the filler surface.

Ordobina et al. [18] reported that crosslinked filler particles in poly(butyl methacrylate) latex films can either enhance or delay the diffusion rate. The effect of soft filler particles on polymer diffusion is a combination of an obstacle effect, delaying polymer diffusion, and a second effect, probably a free-volume effect that promotes polymer diffusion. The change in the diffusion properties of the polymer membrane by the presence of fillers is associated with the fact that polymer chains near a filler surface have different properties than those in the bulk state. This phenomenon occurs due to the chain conformational changes attributed to the presence of a boundary and interaction between the polymer and the filler.

Moreover, the filler effect on the mechanical properties of natural rubber blended with different CB contents have been investigated. Tensile strength and hardness increased with the addition of CB. When the size of the CB particles was small, it formed a significant interaction with the natural rubber matrix [19]. The decrease in the molecular size of the CB filler generally improved the mechanical properties [20]. When the CB filler entered the rubber, the flexibility of the rubber chain decreased, resulting in a more rigid vulcanizate. This rigidity originates from filler–filler and rubber–filler interactions, occurring at various length scales due to the CB structure [21–23].

According to a previous investigation into filler influence [24], the N_2 surface area of CB slightly affected H_2 permeation in filled EPDM and obviously influenced the H_2 diffusion coefficient and solubility. The CB-filled EPDM showed a decreased diffusivity, and its solubility increased with increasing N_2 surface area because the H_2 molecules were adsorbed by the CB participate in diffusion. H_2 solubility in the filled EPDM is further influenced by the filler surface area and the interfacial structure between the filler and the polymer network. Similar studies of the filler effects on H_2 permeability were systematically conducted for nitrile butadiene rubber (NBR) [25] and EPDM [26] composites. In these studies, the contents of different fillers varied up to 60 phr. The CB and silica fillers were found to decrease H_2 diffusivity and permeation and H_2 solubility and diffusivity depended on the filler type.

Based on previous investigations, the present study contains an EPDM composite commonly used as an O-ring in gas seals under high pressure up to 90 MPa at H_2-refueling stations [27]. The study is associated with the H_2 transport properties of sulfur-crosslinked EPDM blended with CB and silica fillers. By investigating the influence of filler loadings on H_2 gas permeability in these composites, we reveal the related sorption, diffusion phenomena, and the filler-induced permeation properties of the composites. The H_2 permeation characteristics of the polymer compounds were measured precisely using a modified volumetric analysis technique (VAT) and a diffusion analysis algorithm [28,29]. The H_2 uptakes, solubilities, diffusion coefficients, and permeabilities of the EPDM composites blended with three filler types were investigated regarding exposure pressure, filler contents, and filler types. The aims of this study were to determine the pressure-dependent H_2 sorption and diffusion mechanisms, to find possible correlations between the compositions and bulk properties of the materials, and to identify the dominant effects of permeation in the filled EPDM composites.

2. Materials and Methods
2.1. Sample Composition

Table 1 shows the formulations of the EPDM compounds: one neat EPDM without filler, six EPDM specimens with CB filler, and three EPDM specimens with silica filler. Two types of CB filler were used—a high abrasion furnace (HAF) CB and a semi-reinforcing furnace (SRF) CB—with particle sizes of 32 and 65 nm, respectively, and specific surface areas of 76 and 30 m^2/g, respectively. The specific surface area of silica was 175 m^2/g. The EPDM composites blended with filler are designated EPDM HAFa, EPDM SRFb, and EPDM Sc; a, b, and c denote the parts per hundred rubber components (phr) of HAF, SRF, and silica, respectively. For instance, EPDM HAF20 represents EPDM blended with HAF CB with 20 phr. Sulfur with 1.5 phr is used as a crosslinking agent. The compounding method for EPDM rubber is detailed in the literature [26].

Table 1. Chemical compositions of sulfur-crosslinked EPDM rubber composites filled with HAF CB, SRF CB, and S fillers.

Composites	EPDM	ZnO	St/A	HAF N330	SRF N774	Silica S-175	Si-69	PEG	S	TBBS	MBT
Neat EPDM	100	3.0	1.0						1.5	1.0	0.5
EPDM HAF 20	100	3.0	1.0	20					1.5	1.0	0.5
EPDM HAF 40	100	3.0	1.0	40					1.5	1.0	0.5
EPDM HAF 60	100	3.0	1.0	60					1.5	1.0	0.5
EPDM SRF 20	100	3.0	1.0		20				1.5	1.0	0.5
EPDM SRF 40	100	3.0	1.0		40				1.5	1.0	0.5
EPDM SRF 60	100	3.0	1.0		60				1.5	1.0	0.5
EPDM S 20	100	3.0	1.0			20	1.6	0.8	1.5	1.0	0.5
EPDM S 40	100	3.0	1.0			40	3.2	1.6	1.5	1.0	0.5
EPDM S 60	100	3.0	1.0			60	4.8	2.4	1.5	1.0	0.5

2.2. Exposure to H_2 Gas

The high-pressure chamber and purge conditions used in this work are described elsewhere [29]. Cylindrical rubber samples with diameters of 13 mm and thicknesses of 3 mm were exposed to hydrogen gas for more than 20 h at pressures ranging from 1.2 to 90 MPa. After exposure to high-pressure H_2, the chamber valve was opened to emit the H_2 gas. After decompression, a specimen was loaded into a graduated cylinder. The amount of H_2 gas released during the lag time was determined by measuring the offset value using a diffusion analysis program [29].

2.3. Transmission Electron Microscopy

The microstructures of the EPDM specimens were investigated with a combination of focused ion beam (FIB) and transmission electron microscopy (TEM). Thin foil samples for TEM image observation were prepared with an FIB. The morphology, distribution, and size characteristics of the CB filler particles in EPDM were observed with a transmission electron microscope (TECNAI F20, FEI company, Hillsboro, OR, USA) operating at an accelerating voltage of 200 kV.

3. Measurement Method and Diffusion Analysis Program

3.1. Volumetric Measurement for Emitting H_2

The volumetric measurement utilized the graduated cylinders in which the emitted H_2 from the specimen was collected and measured. After exposure in the high-pressure chamber and subsequent decompression, the samples were obtained. The samples were loaded into their corresponding gas cell spaces at the top of the graduated cylinder. The details for the method are comprehensively described elsewhere [28,29].

To obtain the increased number of moles (Δn) due to the hydrogen gas released in the graduated cylinder, we measured the volume increase (ΔV); that is, we measured the reduction in the water level in the cylinder as follows [29]:

$$\Delta n = \frac{(P_o - \rho g h)\Delta V}{RT} \tag{1}$$

where ρ is the distilled water density, g is the gravitational acceleration, h is the height (vertical position) of the water level in the cylinder measured from the water level in the corresponding water container, and P_0 is the atmospheric pressure outside the cylinder.

The Δn in the cylinder is converted to the mass concentration [$C(t)$] of H_2 released from the rubber sample:

$$C(t)[\text{wtppm}] = \Delta n[\text{mol}] \times \frac{m_{H2}\left[\frac{g}{mol}\right]}{m_{sample}[g]} \times 10^6 \quad (2)$$

where m_{H2} [g/mol] is the H_2 molar mass, which is equal to 2.016 g/mol and m_{sample} is the sample mass.

3.2. Diffusion Analysis Program

Assuming that H_2 desorption is a Fickian diffusion process, the concentration $C_E(t)$ of the emitted H_2 is computed as follows [30,31]:

$$\frac{C_E(t)}{C_\infty} = 1 - \frac{32}{\pi^2} \times \left[\sum_{n=0}^{\infty} \frac{\exp\left\{\frac{-(2n+1)^2 \pi^2 Dt}{l^2}\right\}}{(2n+1)^2}\right] \times \left[\sum_{n=1}^{\infty} \frac{\exp\left\{-\frac{D\beta_n^2 t}{\rho^2}\right\}}{\beta_n^2}\right]$$

$$= 1$$

$$- \frac{32}{\pi^2} \times \left[\frac{\exp\left(-\frac{\pi^2 Dt}{l^2}\right)}{1^2} + \frac{\exp\left(-\frac{3^2 \pi^2 Dt}{l^2}\right)}{3^2} + \ldots, + \frac{\exp\left(-\frac{(2n+1)^2 \pi^2 Dt}{l^2}\right)}{(2n+1)^2} + \ldots,\right] \quad (3)$$

$$\times \left[\frac{\exp\left(-\frac{D\beta_1^2 t}{\rho^2}\right)}{\beta_1^2} + \frac{\exp\left(-\frac{D\beta_2^2 t}{\rho^2}\right)}{\beta_2^2} + \ldots, + \frac{\exp\left(-\frac{D\beta_n^2 t}{\rho^2}\right)}{\beta_n^2} + \ldots,\right]$$

where β_n is the root of the zeroth order Bessel function $J_0(\beta_n)$ with $\beta_1 = 2.40483$, $\beta_2 = 5.52008$, $\beta_3 = 8.65373, \ldots, \beta_{50} = 156.295$. Equation (3) is an infinite series expansion with two summations. This equation is a solution for Fick's second diffusion equation for a cylindrical-shaped sample. In this study, $C_E = 0$ at $t = 0$ and $C_E = C_\infty$ at $t = \infty$. C_∞ is the saturated H_2 concentration at infinity, i.e., the H_2 uptake. D is the diffusion coefficient, and l and ρ are the thickness and radius of the cylindrical specimen, respectively.

The derivative at $t = 0$, ($\frac{dC_E(t=0)}{dt}$), of two summations in Equation (3) is $-\infty$. This finding implies that the initial H_2 emission rate is extremely fast; it originates from the distribution difference of H_2 caused by the discontinuous pressure difference between the high pressure inside the specimen and the atmosphere outside the specimen after decompression. This result means that, according to Equation (3), there is a possibility of showing different evolution characteristics with time just after decompression.

Because Equation (3) has two infinite terms, we estimate that the finite number of terms (n) in the actual calculation of the two summations should be contained to obtain D and C_∞. Thus, we calculate the contributions in two summations of Equation (3), reaching $n = 50$ terms with three different times t. Figure 1 shows the normalized and calculated product of two summations versus n at three different times ($t = 1$ s, 100 s, and 10,000 s) at fixed parameters of $l = 3.0$ mm, $\rho = 6.5$ mm, and $D = 2 \times 10^{-10}$ m^2/s. With increasing n up to 50, the products of the two summations for all $t = 1$ s, 100 s, and 10,000 s converge to 1 (the value when n is infinite). The horizontal red line in Figure 1 corresponds to a 0.98 value of converged value 1. The corresponding n-term value on the x-axis exceeds 0.98 on the y-axis, for which n should be contained in two summations as nearly equal to the converged value of 1. Thus, a, b, and c, which are obtained from the intersection between the 98% line and the product of two summations (data on y-axis), indicate the minimum number of terms, n, to be included in the summations; these values correspond to 2 at $t = 10,000$ s, 8 at $t = 100$ s, and 17 at $t = 1$ s, respectively, in n. When t is sufficiently greater than 10,000 s, the two terms of the two summations in Equation (3) mainly contribute to the $C_E(t)$ value. However, if t is less than $t = 100$ s, Equation (3) cannot easily converge, and eight more n terms are needed, as shown by arrow b in Figure 1. For the calculation with an uncertainty less than 2% using Equation (3), we should include many terms that are greater than the $n = 64$ terms (8×8) in the product of the two summations at $t = 100$ s. Thus, a dedicated calculation program is needed. We developed a diffusion analysis program to calculate D and C_∞, including up to $n = 100$ terms of the first summation and $n = 50$ terms of the second summation, reaching β_{50} in Equation (3) for covering small time values, $t = 1$ s.

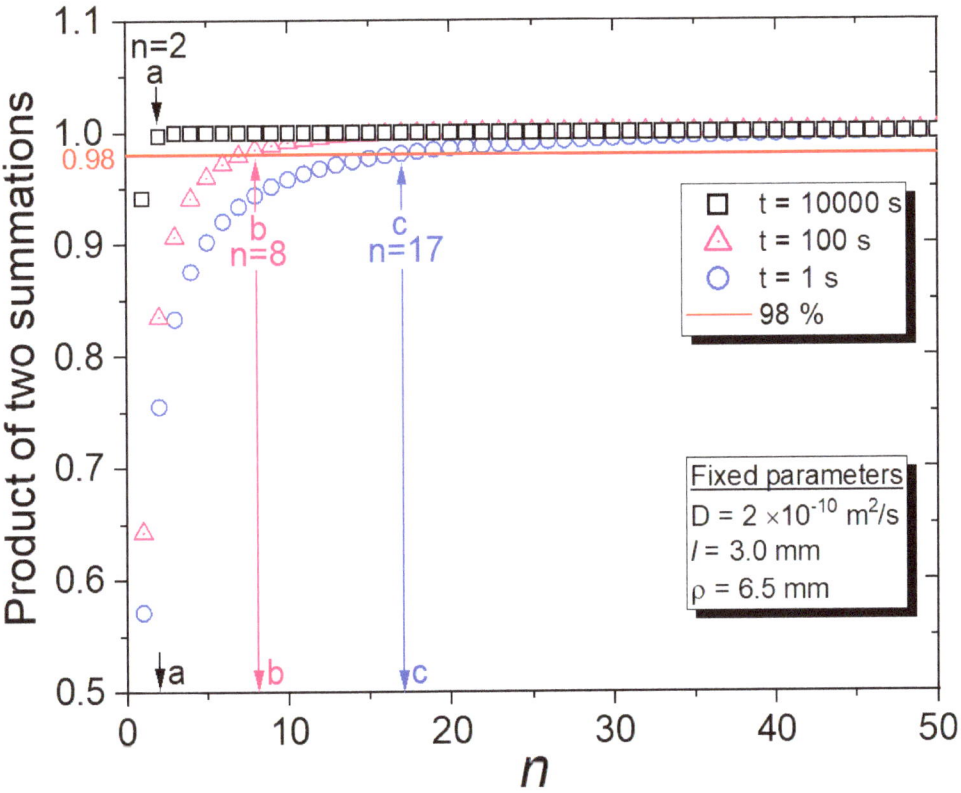

Figure 1. Product of two summations vs. n at different times.

Figure 2 shows the overall flowchart of the diffusion analysis program developed to analyze the $C_E(t)$ data using Equation (3) by a Nelder–Mead simplex nonlinear optimization algorithm [32]. As a result of the application of the developed program, the solubility, diffusivity, and permeability were finally obtained in a rubber composite system. The contents indicated by the blue color in Figure 2 are related to the selection of the solution to the diffusion equation of Equation (3); for this equation, we chose an appropriate diffusion model corresponding to the rubber shape and the number of superposition models.

The D and C_∞ for EPDM composites were determined by utilizing a diffusion analysis program based on Figures 1 and 2. An example application for the analysis program and the detailed procedure were already described in the previous literature [29]. The method of restoring H_2 content using the diffusion analysis program is regarded as a novel aspect of our research. The precise measurement of H_2 content, in particular, is possible as a result of including up to $n = 100$ terms in the summations with the help of the diffusion analysis program.

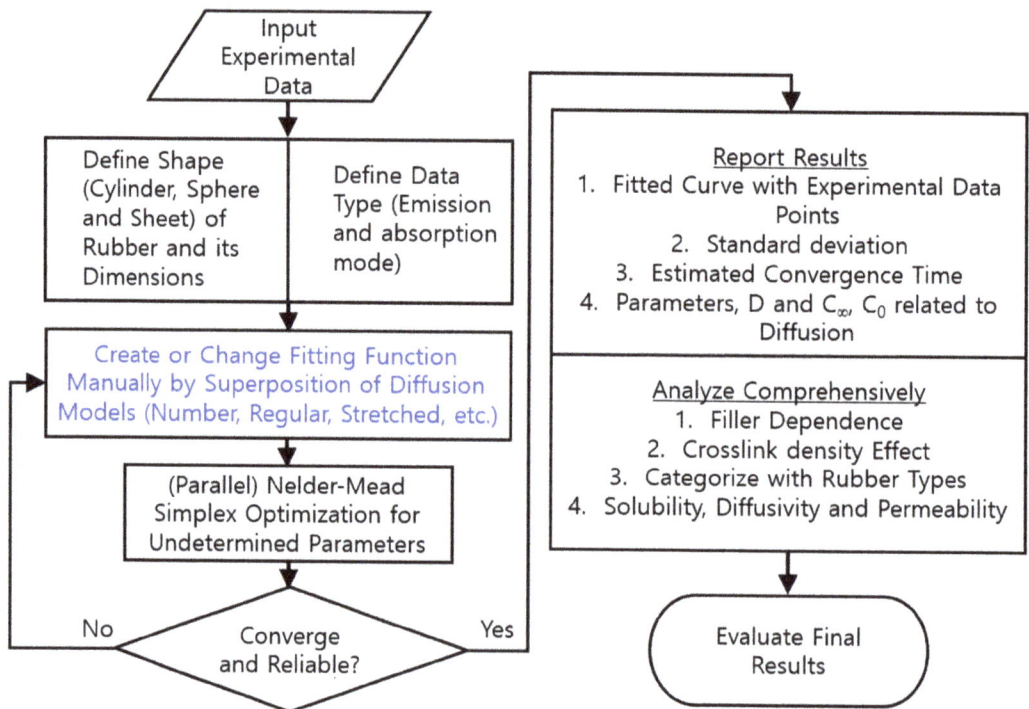

Figure 2. Flowchart for analyzing the diffusion data set for various types of rubbers via the Nelder–Mead simplex nonlinear optimization method.

4. Results and Discussion

4.1. TEM

Figure 3a–d shows the TEM micrographs of EPDM HAF20 and SRF20 filled with a CB content of 20 phr. Homogenous distributions of CB in the rubber matrix are observed in Figure 3a–c. As shown in Figure 3d, a large agglomeration of SRF CB occurs in the EPDM composite with a concentration of 20 phr. The CB filler shapes and distributions are identifiable in the visible TEM image exhibiting the CB filler particles on the rubber matrix. EPDM HAF20 and SRF20 exhibit spherical island shapes with polarized particle sizes of approximately 32 and 65 nm, respectively. The CB particles are distributed as partially condensed aggregates. The aspect ratio is defined by the ratios of the horizontal lengths to the vertical lengths of the particles. For EPDM HAF20 and SRF20 specimens with spherical shapes, the aspect ratios are 1.

Moreover, well-dispersed filler particles with small particle sizes on average, such as HAF CB filler, lead to similar results as filler particles with a high filler surface area and strong interactions with the polymer, thus affecting the gas diffusion and permeation processes. In HAF CB-filled EPDM composites crosslinked with sulfur, we measured the degree of filler dispersion according to the testing method (ASTM D7723). The measured dispersion degrees for two EPDM HAF20 and HAF60 were determined to be 98%; thus, these specimens are regarded as well-dispersed fillers in the rubber network. We did not find any remarkable differences in the dispersion degrees for samples with different filler contents.

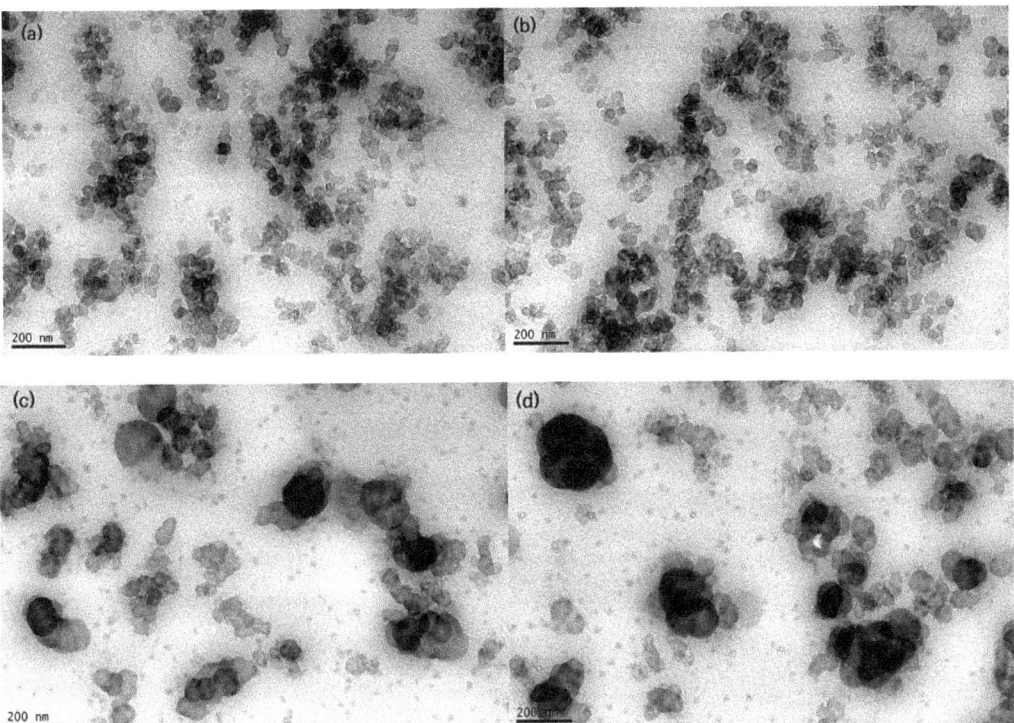

Figure 3. TEM image showing CB particles and aggregates in the CB-filled EPDM composite with (**a**,**b**) HAF20, and (**c**,**d**) SRF20. (**a**,**b**) were taken at different parts of the EPDM HAF20 specimen. (**c**,**d**) were taken at different parts of the EPDM SRF20 specimen.

4.2. Filler Effects on H_2 Uptake

The time evolution characteristics of H_2 emission after decompression at pressures ranging from 1.2 to 90 MPa were measured in ten EPDM composites blended with CB and silica, and in neat EPDM. Figure 4 shows a plot of H_2 uptake versus the elapsed time in ten EPDM rubbers after hydrogen exposure at 8.9 MPa for 20 h. The prominent characteristic is the increase in hydrogen uptake in CB-filled EPDM composites relative to that in neat EPDM. This phenomenon is attributed to H_2 adsorption due to the presence of the CB filler. Increasing the HAF CB content in the HAF CB-filled EPDM composites increased the H_2 emission content. The filler effect on the SRF CB-filled EPDM composites is similar to that of HAF CB-filled EPDM. The slight increase in H_2 uptake for HAF CB-filled EPDM might be explained by the larger specific surface areas of the HAF CB filler compared with those of the SRF CB filler. In the silica-filled EPDM composites, the variation in H_2 uptake with silica filler content is not obviously different from that of the neat EPDM polymer. This result implies that hydrogen is not adsorbed at the silica filler surface.

We measured the hydrogen emission content as a function of exposed pressure for nine EPDM composites blended with fillers and one neat EPDM. Figure 5 shows a plot of the representative hydrogen uptake data versus the pressure for four EPDM composites. Panels (a), (b), (c), and (d) of this figure display the pressure behaviors of H_2 uptake with neat EPDM, EPDM composites compounded with silica filler, EPDM HAF40, and EPDM SRF40, respectively. All EPDM composites blended with HAF CB and SRF CB fillers reveal similar uptake behaviors versus pressure. To avoid redundancy, we only present the representative hydrogen uptake data for two CB-filled EPDM composites.

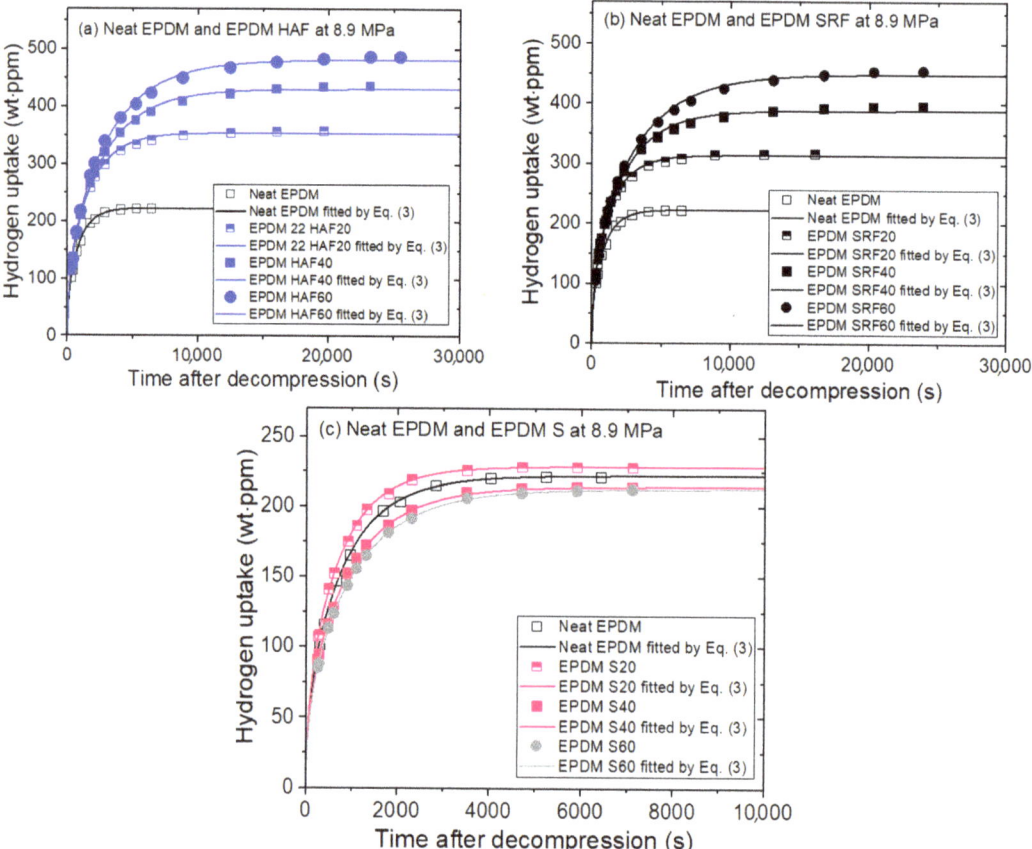

Figure 4. H_2 uptake characteristics of the (**a**) EPDM HAF, (**b**) EPDM SRF, and (**c**) EPDM S series after hydrogen exposure at 8.9 MPa for 20 h and decompression. The solid lines are the least-squares fittings of Equation (3) using the diffusion analysis program. The results of neat EPDM are included in three panels for comparison with the EPDM composites blended with fillers.

The H_2 uptakes (C_∞) of neat EPDM and EPDM S20 (Figure 5a,b) are proportional to pressures reaching 90 MPa, which is in accordance with Henry's Law [33,34]. This behavior is responsible for the absorption of H_2 into the polymer matrix. However, as shown in Figure 5c,d, the hydrogen uptakes for EPDM HAF40 and SRF40 deviate from Henry's law at pressures above 15 MPa; this phenomenon is attributed to the adsorbed hydrogen at the surface of the CB filter. Thus, dual sorption is observed for all CB-blended EPDM composites. The dual mode sorption behaviors that cover the overall pressure range reaching 90 MPa are introduced as follows:

$$C_\infty = kP + \frac{abP}{1+bP} \quad (4)$$

where C_∞ is the total H_2 gas uptake. The first term indicates Henry's law with the Henry's law coefficient k. The second term presents the Langmuir model [35,36], where a is the maximum adsorption quantity (or capacity parameter) and b is the adsorption equilibrium constant (or the Langmuir hole affinity parameter). The fitting results of the H_2 uptake characteristics according to Equation (4) are summarized in Table 2.

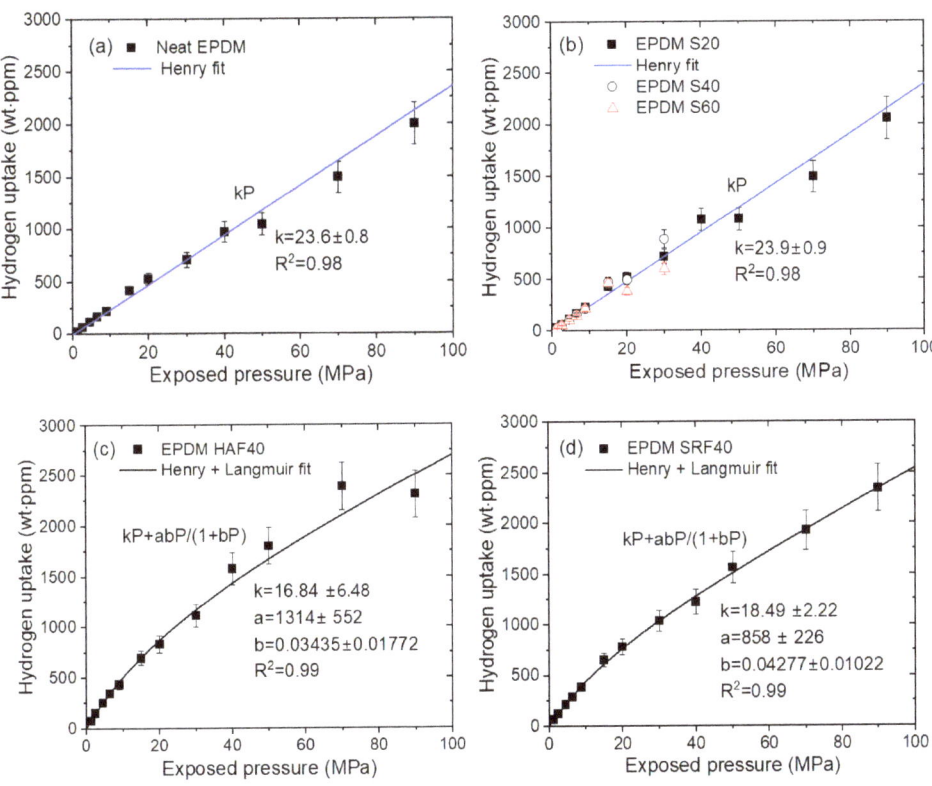

Figure 5. Relationship between H$_2$ uptake (C_∞) and exposure pressure for (**a**) neat EPDM, (**b**) EPDM S series, (**c**) EPDM HAF40, and (**d**) EPDM SRF40. The blue and black lines represent the Henry fit and the dual mode (Henry–Langmuir) fit, respectively. The legends show the linear least-squares fitting plots and their squared correlation coefficients (R^2).

Table 2. Fitting results of the sorption model for neat EPDM and EPDM rubber composites filled with HAF CB, SRF CB, and S fillers according to Equation (4).

Composites	k	a	B	R^2	Langmuir Contribution (%)
Neat EPDM	23.6	0	0	0.98	0
EPDM HAF20	17.8	909	0.0315	0.99	18
EPDM HAF40	16.8	1314	0.0344	0.99	25
EPDM HAF60	5.18	1367	0.0547	0.98	55
EPDM SRF20	18.6	502	0.0498	0.99	14
EPDM SRF40	18.5	858	0.0428	0.99	21
EPDM SRF60	18.0	1528	0.0294	0.99	32
EPDM S20	23.9	0	0	0.98	0

The Langmuir contribution is obtained with respect to total hydrogen uptake, which is the uptake sum of Henry and Langmuir contributions. The Langmuir contribution indicates that the adsorption quantity of hydrogen increases with increasing filler content, as shown in Figure 6. The deviations from linearity above 60 phr for CB-filled EPDM composites indicate an abrupt increase in hydrogen adsorption; this phenomenon may be caused by the formation of hydrogen path channels and thus lead to a percolation effect by many fillers.

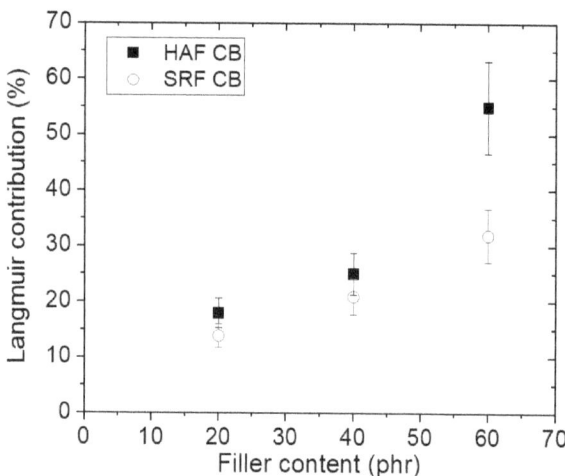

Figure 6. Langmuir contribution versus filler content for CB-filled EPDM composites.

Langmuir sorption is related to the porous solids in the gas–polymer system. The Langmuir sorption site in a glassy polymer corresponds to holes or microvoids that arise due to the nonequilibrium nature of glassy polymers. A gas sorption isotherm in a glassy polymer below the glass transition temperature (T_g) generally depends on the pressure exposure. This behavior is characteristic of dual-mode sorption with Henry's law absorption in an equilibrium state and Langmuir adsorption in a nonequilibrium state [37]. The nonequilibrium state is directly related to the excess free volume or unrelaxed free volume in a glassy polymer [38]. Bondar et al. [39] confirmed the validity of dual mode behaviors. Therefore, the dual mode sorption model for gas sorption in glassy polymers is an effective method for investigation.

However, Jung et al. [25] demonstrated that, for HAF CB-filled NBR, the experimental data at the rubbery phase polymer show dual mode sorption due to the presence of porous HAF CB filler. H_2 molecules are absorbed by rubbery NBR and are simultaneously adsorbed by porous filler, leading to dual mode sorption similar to that at the glass phase. Thus, the porous HAF CB filler in the NBR composite corresponds to the robust void structure in the glass phase polymer. The solubility result in HAF CB-filled NBR supports the dual sorption behavior.

4.3. Filler Effects on H_2 Diffusion

Similar to the pressure-dependent H_2 uptake, the H_2 diffusivities of the neat EPDM and nine filled EPDM composites were measured as functions of exposed pressure. The H_2 diffusivities of neat EPDM and the EPDM composites blended with fillers apparently depend on the exposed pressure. The pressure dependence of the diffusion coefficient is related to the decrease in the mean free path of H_2, the increased tortuosity caused by the impermeable filler in the rubber networks and the increased interactions between the filler and the rubber.

All EPDM composites blended with HAF CB and SRF CB fillers revealed similar diffusion behaviors versus pressure. To avoid redundancy, the representative pressure-dependent diffusion for EPDM HAF20 and SRF20, as shown in Figure 7a,b, respectively, can be divided into two contributions at the peak, as indicated by arrows. The contributions correspond to Knudsen diffusion for low pressure and bulk diffusion for high pressure. The pressure-dependent behavior for diffusivity is interpreted by the results of Knudsen diffusion below 7–10 MPa, and bulk diffusion above this pressure range; this combined diffusion was observed and analyzed by fractal theory in other studies [40,41]. Knudsen diffusion gradually increases with increasing pressure. Knudsen diffusion below the pressure range normally occurs when there is a large mean free path of diffusing gas molecules or a low gas density. The Knudsen diffusion coefficient ($D_{K,\,pm}$) in porous media is expressed as follows [42]:

$$D_{K,pm} = \frac{\phi}{\tau} D_K = \frac{\phi}{\tau} \frac{d_c}{3} v \qquad (5)$$

where ϕ is the pressure-dependent porosity, τ is the tortuosity caused by introducing the filler, d_c is the pore diameter, and v is the average molecular velocity derived from the kinetic theory of gases.

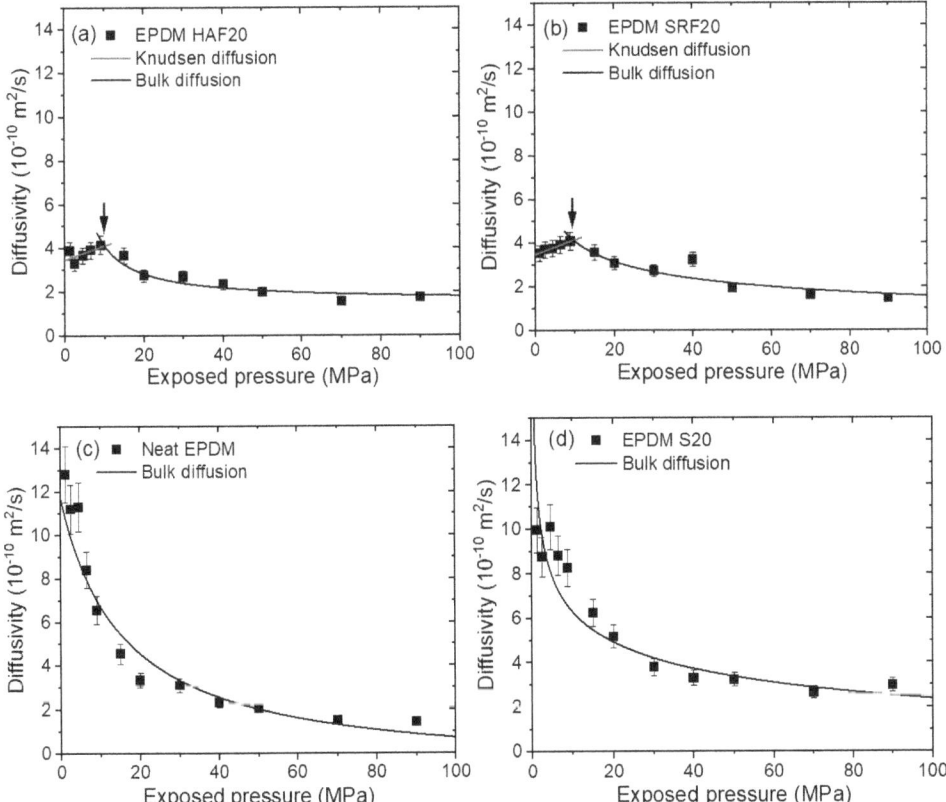

Figure 7. H_2 diffusivity versus exposed pressure in (**a**) EPDM HAF20, (**b**) EPDM SRF20, (**c**) neat EPDM, and (**d**) EPDM S20. The blue lines indicate Knudsen diffusion fitted by Equation (5). The black lines indicate bulk diffusion fitted by Equation (6). The arrows in (**a**,**b**) indicate the intersecting regions of Knudsen and bulk diffusion.

Moreover, the bulk diffusion for neat EPDM (Figure 7c) and EPDM S20 (Figure 7d), and the bulk diffusion above a critical pressure of 7–10 MPa for CB-filled EPDM composites, are inversely proportional to pressure; this phenomenon is associated with the mean free path between the H_2 molecules. Bulk diffusion is predominant when the mean free path (λ) in large pores is smaller than the pore diameters, or when high-pressure gas diffusion occurs. The bulk diffusion coefficient (D_B) is expressed as follows [43]:

$$D_B = \frac{1}{3}\lambda v = \frac{1}{3}\frac{5}{8}\frac{\mu}{P}\sqrt{\frac{RT\pi}{2M}} v \qquad (6)$$

where μ is the viscosity of the diffusion molecule in units of $kg\ m/s$ and P is the pressure. The factor 5/8 considers the Maxwell–Boltzmann distribution of molecular velocity. The experimental results of the diffusivity shown in Figure 7 are fitted by both Equations (5) and (6), as indicated by the blue and black lines, respectively. In the region of Knudsen diffusion, the diffusion coefficient is proportional to the pressure; this phenomenon may be caused by an increase in the porosity in Equation (5) due to an increase in the pressure. The decrease in the bulk diffusion coefficient is attributed to a decrease in the mean free path with increasing pressure.

Regarding the H_2 sorption and diffusion mechanism, we again justify the hydrogen sorption (diffusion) mechanism considering the role of CB. According to the hydrogen uptake data shown in Figure 5, the hydrogen sorption mechanisms in the EPDM composites blended with CBs (HAF and SRF) revealed two types of sorption (or diffusion): fast diffusion due to the hydrogen absorbed in the polymer network and slow diffusion due to the hydrogen physically adsorbed at the CB filler interface. In other words, the sorption (or diffusion) mechanism in CB-filled EPDM represents dual sorption (or diffusion) behaviors. In this study, the sorption and desorption processes of most H_2 are reversible; this finding may be attributed to physisorption rather than chemisorption by penetrated H_2.

However, a single hydrogen sorption (or diffusion) behavior in silica-filled EPDM is observed only with a fast-diffusing polymer network. The single-mode behavior is also shown in neat EPDM due to fast H_2 sorption for the polymer network. Hydrogen in the silica-filled EPDM was not adsorbed at the interface between the silica and rubber. This finding indicates that hydrogen sorption in silica or at the interface between silica and the rubber matrix did not occur. Thus, as shown in the H_2 uptake characteristics (Figure 5) for silica-filled EPDM specimens, the value (uptake) for silica-filled EPDM composites is nearly identical to that for neat EPDM. The result for CB-filled EPDM composites indicates that the fast component shows the permeation characteristics of H_2 absorbed onto the parent component of the rubber (Henry's law); the slow component shows the permeation characteristics of H_2 adsorbed by the filler (Langmuir law).

Figure 8a–c shows the variations in the diffusivity characteristics with the filler content at the three different pressures of 1.2 MPa, 8.9 MPa, and 90 MPa, respectively. At a low pressure of 1.2 MPa, all fillers extend the diffusion path due to increased tortuosity by the impermeable filler, resulting in a decrease in the diffusion rate. The diffusivity in the silica-blended EPDM is negative and it decreases linearly with increasing filler content; the diffusivity in the CB-blended EPDM decreases in the form of an asymptotic line (~1/filler content). At the low pressure of 1.2 MPa, the decrease in diffusivity of the CB-blended EPDM is larger than that of the silica-blended EPDM; this phenomenon is expected and possibly related to the additional filler–polymer interactions. However, with the increasing pressure reaching 90 MPa, the filler effect on diffusion decreases; the diffusivity characteristics for all specimens converge at values of approximately $2 \times 10^{-10}\ m^2/s$.

Two general models [18] are employed to explain the change in diffusivity by the presence of filler particles. These models differ in their descriptions of the interactions between filler particles and the polymer matrix. One model is based on the concept of free volume. Free volume ascribes the change in diffusivity to an increase or decrease in the microscopic friction coefficient of the diffusing species. This change is responsible for the influence of the filler surface on the mobilities of diffusing molecules in the vicinity of the filler particles through filler–polymer interactions. The second model is an obstacle model. Obstacles apparently decrease the diffusivity by increasing the tortuosity of the diffusion path or by creating bottlenecks without affecting the friction experienced by the diffusing species. A change in free volume can increase or decrease the polymer diffusivity; the presence of obstacles always decreases the polymer diffusivity. As shown in Figure 8a, the diffusivity for silica-filled EPDM is responsible for the tortuosity of the diffusion path by introducing filler (second model). Moreover, the diffusivity for CB-filled EPDM is attributed to both polymer–filler interactions and tortuosity (first and second model).

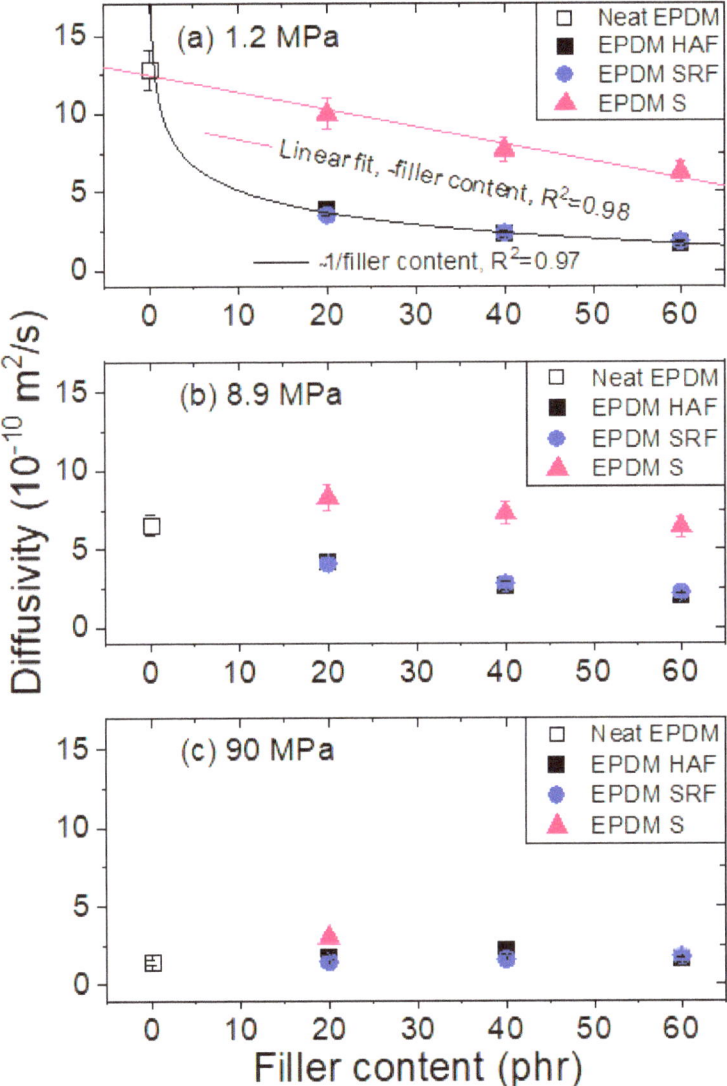

Figure 8. Diffusivity versus filler content at exposure to pressure of (**a**) 1.2 MPa, (**b**) 8.9 MPa, and (**c**) 90 MPa in EPDM composites blended with CB and silica. The pink line in (**a**) is fitted with a negative linear relationship between the diffusivity and the filler content. The black line in (**a**) is fitted with a linear relationship between the diffusivity and the reciprocal filler content. R^2 is squared correlation coefficients of fitting.

4.4. Correlations of Permeation with Density and Tensile Strength

The permeation P was determined by multiplying the solubility S by the diffusion coefficient D, i.e., $P = SD$. Figure 9a,b shows the permeability variations with density and tensile strength, respectively, for neat EPDM and blended EPDM polymer composites. The trends are similar to those of diffusivity at 1.2 MPa (Figure 8a), implying that permeation is predominantly affected by diffusivity rather than by solubility.

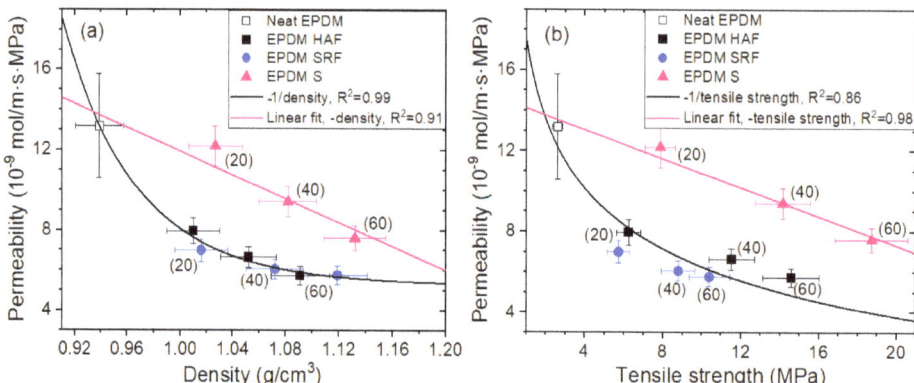

Figure 9. Correlations between permeability and (**a**) density and (**b**) tensile strength for neat EPDM and EPDM composites blended with CB/silica. The data for neat EPDM are included as pink and black curves for consistency with the fittings of the filler-blended EPDM composites. The value in parentheses indicates the phr of CB and silica. R^2 is squared correlation coefficients of fitting.

As shown in Figure 9a, the negative linear relationship (density) between permeability and density for silica-blended EPDM composites indicates a smooth decrease in permeation with increasing density, without introducing other interactions or additional parameters. However, the density effect on the permeability for CB-blended EPDM composites is inversely proportional to the filler content, i.e., ~1/density. The magnitude of the effects for the CB-blended EPDM composites is larger than that for the silica-blended EPDM composites. This result again implies an additional effect; that is, the polymer–filler interaction for CB-blended EPDM composites may originate the permeability behavior, as already shown in the pressure-dependent effect on diffusivity at 1.2 MPa. A similar behavior for the density influence on permeability was found in polyethylene gas permeability investigations with different permeants [44,45]. The decrease in permeability in polyethylene with increasing density is attributed to the volume dilution of the amorphous fraction by the relatively impermeable crystalline phase.

Moreover, the permeability changes with tensile strength shown in Figure 9b exhibit identical behaviors to the permeability changes with density, as shown in Figure 9a. The two trends may be closely related to the same origin. From the investigated relation function for physical and mechanical properties, we provide a possibility for predicting the H_2-permeation properties of compounded EPDM candidates used as seal materials under high pressure in H_2-refueling stations.

5. Conclusions

By using a volumetric analysis technique and an ungraded diffusion analysis program calculating up to a hundred summation terms in an expansion series of the concentration $C_E(t)$ of emitted H_2, we investigated the H_2-permeation characteristics of EPDM composites. The investigated results are summarized below.

The pressure-dependent H_2 uptakes for neat EPDM and silica-filled EPDM composites show single sorption models that satisfy Henry's law; this phenomenon was dominated by absorption by the polymer. The contribution from the filler was negligibly small. Moreover, H_2 uptakes for CB-filled EPDM composites followed dual sorption models that obey Henry's law and Langmuir law. The H_2 uptake in the CB-filled EPDM received contributions from absorption by the polymer networks and adsorption by the CB filler. The difference between the two CBs is attributed to the distinct specific surface areas.

The diffusivity values in all EPDMs investigated depended on pressure. The decrease in the diffusivity for silica-filled EPDM relative to that for neat EPDM was responsible for the increase in the tortuosity of the diffusion path by introducing filler. Moreover, the decrease in the diffusivity for CB-filled EPDM was attributed to both polymer–filler interactions and tortuosity by impermeable fillers.

At 1.2 MPa, the silica-blended EPDMs show negative linear correlations between diffusivity and filler content. The relationship is very similar to that of permeability with the density and tensile strength characteristics of EPDM composites. The CB-blended EPDMs exhibit reciprocal relationships between diffusivity and filler content, likely for permeability with the density and tensile strength characteristics of EPDM composites. From the investigated relationships, we predicted the H_2-permeation properties of compounded EPDM candidates when used as seal materials under high pressure in H_2-refueling stations. In the present work, it was demonstrated that EPDM HAF60 and EPDM SRF60 specimens with large density and tensile strength characteristics exhibit the lowest H_2 permeation among the specimens; these specimens are suitable candidates for high-pressure gas seals to prevent gas leakage.

Author Contributions: Conceptualization, J.K.J. and C.H.L.; Validation S.K.J. and U.B.B.; Data curation, J.K.J., J.H.L., S.H.L. and W.J.M.; Writing—original draft, J.K.J. and S.K.J.; Writing—review and editing, J.K.J.; Supervision, U.B.B. and C.H.L.; Funding acquisition, U.B.B. All authors have read and agreed to the published version of the manuscript.

Funding: This work was supported by the New Renewable Energy Core Technology Program at the Korea Institute of Energy Technology Evaluation and Planning (KETEP), granted financial resources from the Ministry of Trade, Industry and Energy, Republic of Korea, (Development of 70 MPa-class high-performance rubbers and certification standards for hydrogen refueling stations, No. 20223030040070). This research was supported by the Development of Reliability Measurement Technology for Hydrogen Fueling Station funded by the Korea Research Institute of Standards and Science (KRISS-2022-GP2022-0007).

Institutional Review Board Statement: Not applicable.

Informed Consent Statement: Not applicable.

Data Availability Statement: The data used to support the findings of this study are available from the corresponding author upon request.

Conflicts of Interest: The authors declare no conflict of interest.

References

1. Pegoretti, A.; Dorigato, A. Polymer composites: Reinforcing fillers. In *Encyclopedia of Polymer Science and Technology*; Mark, H.F., Ed.; Wiley: New York, NY, USA, 2019; pp. 1–72.
2. Mittal, G.; Rhee, K.Y.; Mišković-Stanković, V.; Hui, D. Reinforcements in multi-scale polymer composites: Processing, properties, and applications. *Compos. Part B Eng.* **2018**, *138*, 122–139. [CrossRef]
3. Liu, Q.; Paavola, J. Lightweight design of composite laminated structures with frequency constraint. *Compos. Struct.* **2016**, *156*, 356–360. [CrossRef]
4. Mittal, G.; Dhand, V.; Rhee, K.Y.; Park, S.-J.; Lee, W.R. A review on carbon nanotubes and graphene as fillers in reinforced polymer nanocomposites. *J. Ind. Eng. Chem.* **2015**, *21*, 11–25. [CrossRef]
5. Dhand, V.; Mittal, G.; Rhee, K.Y.; Park, S.-J.; Hui, D. A short review on basalt fiber reinforced polymer composites. *Compos. Part B Eng.* **2015**, *73*, 166–180. [CrossRef]
6. Al-Oqla, F.M.; Sapuan, S.M.; Anwer, T.; Jawaid, M.; Hoque, M.E. Natural fiber reinforced conductive polymer composites as functional materials: A review. *Synth. Met.* **2015**, *206*, 42–54. [CrossRef]
7. Hung, P.-Y.; Lau, K.-T.; Cheng, L.-K.; Leng, J.; Hui, D. Impact response of hybrid carbon/glass fibre reinforced polymer composites designed for engineering applications. *Compos. B Eng.* **2018**, *133*, 86–90. [CrossRef]
8. Farida, E.; Bukit, N.; Ginting, E.M.; Bukit, B.F. The effect of carbon black composition in natural rubber compound. *Case Stud. Therm. Eng.* **2019**, *16*, 100566. [CrossRef]
9. Pandey, K.; Setua, D.; Mathur, G. Material behaviour: Fracture topography of rubber surfaces: An SEM study. *Polym. Test.* **2003**, *22*, 353–359. [CrossRef]
10. Zhang, A.; Wang, L.; Zhou, Y. A study on rheological properties of carbon black extended powdered SBR using a torque rheometer. *Polym. Test.* **2003**, *22*, 133–141. [CrossRef]
11. Fröhlich, J.; Niedermeier, W.; Luginsland, H.D. The effect of filler–filler and filler–elastomer interaction on rubber reinforcement. *Compos. Part A Appl. Sci. Manuf.* **2005**, *36*, 449–460. [CrossRef]

12. Bhattacharyya, S.; Lodha, V.; Dasgupta, S.; Mukhopadhyay, R.; Guha, A.; Sarkar, P.; Saha, T.; Bhowmick, A.K. Influence of highly dispersible silica filler on the physical properties, tearing energy, and abrasion resistance of tire tread compound. *J. Appl. Polym. Sci.* **2019**, *136*, 47560. [CrossRef]
13. Mark, J.E.; Erman, B.; Eirich, F.R. Chapter 9 the science of rubber compounding. In *The Science and Technology of Rubber*; Rodgers, B., Waddell, W., Eds.; Academic Press: London, UK, 2005; pp. 401–454.
14. Dick, J.; Rader, C. *Raw Materials Supply Chain for Rubber Products*; Carl Hanser Verlag: Munich, Germany, 2014.
15. Chakraborty, S.; Sengupta, R.; Dasgupta, S.; Mukhopadhyay, R.; Bandyopadhyay, S.; Joshi, M.; Ameta, S.C. Synthesis and characterization of in situ sodium-activated and organomodified bentonite clay/styrene–butadiene rubber nanocomposites by a latex blending technique. *J. Appl. Polym. Sci.* **2009**, *113*, 1316–1329. [CrossRef]
16. Waddell, W.H.; Evans, L.R. *Rubber Technology Compounding and Testing for Performance*; Carl Hanser Verlag: Munich, Germany, 2009.
17. Most, C.F., Jr. Some filler effects on diffusion in silicone rubber. *J. Appl. Polym. Sci.* **1970**, *14*, 1019–1024. [CrossRef]
18. Odrobina, E.; Feng, J.; Pham, H.H.; Winnik, M.A. Effect of soft filler particles on polymer diffusion in poly(butyl methacrylate) latex films. *Macromolecules* **2001**, *34*, 6039–6051. [CrossRef]
19. Fu, S.-Y.; Feng, X.-Q.; Lauke, B.; Mai, Y.-W. Effects of particle size, particle/matrix interface adhesion and particle loading on mechanical properties of particulate–polymer composites. *Compos. Part B Eng.* **2008**, *39*, 933–961. [CrossRef]
20. Parkinson, D. *Reinforcement of Rubbers*; Lakeman & Co.: London, UK, 1957.
21. El-Gamal, A.A. Effect of reinforcement filler on vulcanization, diffusion, mechanical, and electrical properties of natural rubber. *J. Elastomers Plast.* **2018**, *51*, 512–526. [CrossRef]
22. Thomas, S.P.; Thomas, S.; Mathew, E.J.; Marykutty, C.V. Transport and electrical properties of natural rubber/nitrile rubber blend composites reinforced with multiwalled carbon nanotube and modified nano zinc oxide. *Polym. Compos.* **2014**, *35*, 956–963. [CrossRef]
23. Dasan, P.; Unnikrishnan, G.; Purushothaman, E. Solvent transport through carbon black filled poly(ethylene-co-vinyl acetate) composites. *Express Polym. Lett.* **2008**, *2*, 382–390. [CrossRef]
24. Yamabe, J.; Nishimura, S. Influence of carbon black on decompression failure and hydrogen permeation properties of filled ethylene-propylene–diene–methylene rubbers exposed to high-pressure hydrogen gas. *J. Appl. Polym. Sci.* **2011**, *122*, 3172–3187. [CrossRef]
25. Jung, J.K.; Lee, C.H.; Son, M.S.; Lee, J.H.; Baek, U.B.; Chung, K.S.; Choi, M.C.; Bae, J.W. Filler effects on H(2) diffusion behavior in nitrile butadiene rubber blended with carbon black and silica fillers of different concentrations. *Polymers* **2022**, *14*, 700. [CrossRef]
26. Jung, J.K.; Lee, C.H.; Baek, U.B.; Choi, M.C.; Bae, J.W. Filler influence on H(2) permeation properties in sulfur-crosslinked ethylene propylene diene monomer polymers blended with different concentrations of carbon black and silica fillers. *Polymers* **2022**, *14*, 592. [CrossRef] [PubMed]
27. Kang, H.M.; Choi, M.C.; Lee, J.H.; Yun, Y.M.; Jang, J.S.; Chung, N.K.; Jeon, S.K.; Jung, J.K.; Lee, J.H.; Lee, J.H.; et al. Effect of the high-pressure hydrogen gas exposure in the silica-filled EPDM sealing composites with different silica content. *Polymers* **2022**, *14*, 1151. [CrossRef] [PubMed]
28. Jung, J.K.; Kim, I.G.; Kim, K.T.; Ryu, K.S.; Chung, K.S. Evaluation techniques of hydrogen permeation in sealing rubber materials. *Polym. Test.* **2021**, *93*, 107016. [CrossRef]
29. Jung, J.K.; Kim, I.G.; Jeon, S.K.; Kim, K.-T.; Baek, U.B.; Nahm, S.H. Volumetric analysis technique for analyzing the transport properties of hydrogen gas in cylindrical-shaped rubbery polymers. *Polym. Test.* **2021**, *99*, 107147. [CrossRef]
30. Crank, J. *The Mathematics of Diffusion*; Oxford University Press: Oxford, UK, 1975.
31. Demarez, A.; Hock, A.G.; Meunier, F.A. Diffusion of hydrogen in mild steel. *Acta Metall.* **1954**, *2*, 214–223. [CrossRef]
32. Nelder, J.A.; Mead, R. A simplex method for function minimization. *Comput. J.* **1965**, *7*, 308–313. [CrossRef]
33. Sander, R.; Acree, W.E.; Visscher, A.D.; Schwartz, S.E.; Wallington, T.J. Henry's law constants (IUPAC Recommendations 2021). *Pure Appl. Chem.* **2022**, *94*, 71–85. [CrossRef]
34. Sander, R. Compilation of Henry's law constants (version 4.0) for water as solvent. *Atmos. Chem. Phys.* **2015**, *15*, 4399–4981. [CrossRef]
35. Kanehashi, S.; Nagai, K. Analysis of dual-mode model parameters for gas sorption in glassy polymers. *J. Membr. Sci.* **2005**, *253*, 117–138. [CrossRef]
36. Wang, J.; Kamiya, Y. Evaluation of gas sorption parameters and prediction of sorption isotherms in glassy polymers. *J. Polym. Sci. B-Polym. Phys.* **2000**, *38*, 883–888. [CrossRef]
37. Vieth, W.R.; Tam, P.M.; Michaels, A.S. Dual sorption mechanisms in glassy polystyrene. *J. Colloid Interface Sci.* **1966**, *22*, 360–370. [CrossRef]
38. Paul, D.R. Gas sorption and transport in glassy polymers. *Ber. Bunsenges. Phys. Chem.* **1979**, *83*, 294–302. [CrossRef]
39. Bondar, V.I.; Kamiya, Y.; Yampol'skii, Y.P. On pressure dependence of the parameters of the dual-mode sorption model. *J. Polym. Sci. B Polym. Phys.* **1996**, *34*, 369–378. [CrossRef]
40. Yang, Y.; Liu, S. Estimation and modeling of pressure-dependent gas diffusion coefficient for coal: A fractal theory-based approach. *Fuel* **2019**, *253*, 588–606. [CrossRef]
41. Wang, Y.; Liu, S. Estimation of pressure-dependent diffusive permeability of coal using methane diffusion coefficient: Laboratory measurements and modeling. *Energy Fuels* **2016**, *30*, 8968–8976. [CrossRef]

42. Knudsen, M. Die gesetze der molekularströmung und der inneren reibungsströmung der gase durch röhren. *Ann. Phys.* **1909**, *333*, 75–130. [CrossRef]
43. Welty, J.R.; Wicks, C.E.; Wilson, R.E. *Fundamentals of Momentum, Heat, and Mass Transfer*; Wiley: New York, NY, USA, 1984.
44. Alter, H. A critical investigation of polyethylene gas permeability. *J. Polym. Sci.* **1962**, *57*, 925–935. [CrossRef]
45. Michaels, A.S.; Parker Jr, R.B. Sorption and flow of gases in polyethylene. *J. Polym. Sci.* **1959**, *41*, 53–71. [CrossRef]

Disclaimer/Publisher's Note: The statements, opinions and data contained in all publications are solely those of the individual author(s) and contributor(s) and not of MDPI and/or the editor(s). MDPI and/or the editor(s) disclaim responsibility for any injury to people or property resulting from any ideas, methods, instructions or products referred to in the content.

Article

Anomalous Strain Recovery after Stress Removal of Graded Rubber

Quoc-Viet Do [1], Takumitsu Kida [1], Masayuki Yamaguchi [1,*], Kensuke Washizu [2], Takayuki Nagase [2] and Toshio Tada [2]

[1] School of Materials Science, Japan Advanced Institute of Science and Technology, 1-1 Asahidai, Ishikawa, Nomi 923-1292, Japan
[2] Material Research & Development HQ, Sumitomo Rubber Industries, Ltd., 1-1, 2-Chome, Tsutsui, Chuo, Hyogo, Kobe 651-0071, Japan
* Correspondence: m_yama@jaist.ac.jp

Abstract: Mechanical responses after the uniaxial deformation of graded styrene–butadiene rubber (SBR) with a gradient in the crosslink points in the thickness direction were investigated as compared with those of homogenously vulcanized SBR samples. The elongational residual strain of a graded sample was found to depend on the part with a high crosslink density. Therefore, it showed good rubber elasticity. After stress removal, moreover, the graded sample showed a marked warpage. This suggested that shrinking stress acted on the surface with a high crosslink density, which would avoid a crack growth on the surface. The sample shape was then recovered to be flat very slowly, indicating that the shrinking stress worked for a long time. This unique rubber elasticity, i.e., slow strain recovery with an excellent strain recovery, makes graded rubber highly significant.

Keywords: graded rubber; rubber elasticity; styrene–butadiene rubber

Citation: Do, Q.-V.; Kida, T.; Yamaguchi, M.; Washizu, K.; Nagase, T.; Tada, T. Anomalous Strain Recovery after Stress Removal of Graded Rubber. *Polymers* **2022**, *14*, 5477. https://doi.org/10.3390/polym14245477

Academic Editor: Md Najib Alam

Received: 1 November 2022
Accepted: 21 November 2022
Published: 14 December 2022

Copyright: © 2022 by the authors. Licensee MDPI, Basel, Switzerland. This article is an open access article distributed under the terms and conditions of the Creative Commons Attribution (CC BY) license (https://creativecommons.org/licenses/by/4.0/).

1. Introduction

A rubber has a crosslinked structure, in which crosslink points are homogeneously distributed in general. Strictly speaking, however, most rubbers have a gradient in the crosslink density, especially in a thick product, because each part in a rubber product has different thermal histories. Since polymeric materials including rubbers usually have a low thermal diffusivity, it takes a long time to be in an equilibrium temperature profile at vulcanization process [1]. Therefore, a core region in a rubber product must have a short exposure period at a high temperature compared with a skin region. This may result in the difference in the crosslink density, although its effect on the mechanical properties can be ignored for most rubber products used in industry. Such a situation is, however, pronounced and should be considered when vulcanization occurs slowly. Bellander et al. [2] vulcanized styrene–butadiene rubber (SBR) without crosslinking agents such as sulfur and found a gradient in the crosslink density. A rubber with a crosslink gradient, i.e., a graded rubber, may show poor rubber elasticity when the stress is applied in the gradient direction. This is reasonable because a layered part with no/few crosslink points will flow by applied normal force. From the viewpoint of energy absorption, however, a number of researchers reported that damping properties were improved by providing a crosslink gradient [3–6]. This must be attributed to dangling chains in the region with a low crosslink density, which showed prolonged relaxation modes with a high level of energy absorption [7–10].

A graded rubber can be obtained by different methods, including the lamination of different layers [11] and manipulated photo-curing [12–14]. For conventional rubber materials, Ikeda prepared a graded SBR by vulcanizing laminated sheets with different sulfur contents [15,16]. She found that the mechanical properties such as tensile properties were mostly decided by the area with a high crosslink density. Moreover, Glebova et al. [17] revealed that the crosslink density in SBR around zinc oxide particles was high in the nanoscale regions (ca. 200 nm). Considering that nanoparticles can show interphase transfer between different rubbers [18], this technique may provide a new idea to make a graded rubber in the future. Finally, Wang et al. [19] proposed a simple method to prepare a graded SBR by diffusing sulfur from one surface, followed by vulcanization. Since diffusion constants of curatives have been studied for a long time after the pioneering works by van Amerongen [20] and Gardiner [21], this technique must be noted. According to Wang et al., the obtained graded SBR showed high values of loss tangent in a wide temperature range due to broad distribution of glass transition temperature T_g [19]. Besides the pronounced energy absorption, however, attractive properties of a graded rubber composed of conventional materials have not been reported yet to the best of our knowledge.

Here, we prepared a graded SBR by vulcanizing under a temperature gradient and found unique strain recovery behaviors after stress removal. Dynamic mechanical properties, as well as tensile properties including stress relaxation behaviors, were also investigated in detail.

2. Materials and Methods

2.1. Materials

Styrene–butadiene rubber (SBR) with a styrene content of 25 wt.% was kindly supplied from Sumitomo Rubber Industries Ltd. (Kobe, Japan). The weight-average molecular weight as polystyrene standard was 250,000. Furthermore, carbon blacks (Diablack-H; Mitsubishi Chemical, Tokyo, Japan), stearic acid (NOF Corporation, Tokyo, Japan), zinc oxide (Fujifilm Wako Pure Chemical Corporation, Osaka, Japan), N-phenyl-N'-(1,3-dimethylbutyl)-p-phenylene diamine as an antioxidant (Nocrac 6C; Ouchi Shinko Chemical Industrial, Tokyo, Japan), aroma oil (VivaTec 500; H&R, Hamburg, Germany), sulfur (Tsurumi Chemical Industry, Tokyo, Japan), N-cyclohexyl-2-benzothiazolyl sulfenamide (CBS, Nocceler CZ-G; Ouchi Shinko Chemical Industrial), and 1,3-diphenylguanidine (DPG, Nocceler D; Ouchi Shinko Chemical Industrial) were employed. All of them were used without further purification.

2.2. Sample Preparation

The sample recipe is shown in Table 1. Mixing was performed by three steps. At the first step, all ingredients except for sulfur and accelerators, such as CBS and DPG, were added into a 1700 cc internal mixer (Mixtoron BB; Kobelco, Kobe, Japan) and mixed at 77 rpm for 3 min. The initial temperature of the mixer was controlled at 30 °C, and the final temperature was about 150 °C. After mixing, the mixture was taken out and cooled down at room temperature. Then, it was put into the mixer at 30 °C again with sulfur and accelerators and mixed at 44 rpm for 3 min as the second step. The final temperature was about 100 °C. Finally, the obtained mixture was kneaded by an 8-inch two-roll mill (Kansai Roll, Osaka, Japan) at 60 °C to prepare a sheet with a 2.5 mm thickness.

Table 1. Recipe of SBR compound.

Ingredients	Amount (phr)
SBR	100
Carbon black	50
Stearic acid	2
Zinc oxide	3
Antioxidant	1
Aroma oil	5
Sulfur	1.45
CBS *	2.3
DPG **	1.85

* N-Cyclohexyl-2-benzothiazole sulfenamide. ** Diphenyl guanidine.

Sample sheets were exposed to high temperatures in a compression molding machine under 10–20 MPa for 10 min. The temperature conditions with sample codes are shown in

Table 2. In the sample codes, "H" represents the homogeneously crosslinked samples, i.e., both plates of the compression molding machine were controlled at the same temperature, whereas "G" denotes the graded samples. The thickness of the compressed sheets was about 1.3 mm. After the vulcanization process, the surface temperature was measured again, which is also summarized in Table 2. The sample code "UV" represents the unvulcanized sample, although it was compressed at 80 °C for 30 s under 20 MPa to reduce the thickness to 1.3 mm. All samples were kept at room temperature at least 24 h before testing.

Table 2. Sample codes and temperature conditions at compression molding.

Sample Codes	Set Temperature (°C)		Temperature at the End of Vulcanization Process (°C)	
	Top Plate	Bottom Plate	Top Plate	Bottom Plate
UV	-	-	-	-
H80	80	80	80	80
H110	110	110	110	110
H170	170	170	170	170
G170-80	170	50	170	80
G170-110	170	80	170	110

2.3. Measurements

Vulcanization behaviors at various temperatures were evaluated by a rotorless curemeter (Curelastometer Type R 7; Eneos Trading Company, Tokyo, Japan) following ISO 6502.

The temperature dependence of tensile moduli was measured by a dynamic mechanical analyzer (Rheogel E4000; UBM, Muko, Japan). The frequency and heating rate were 10 Hz and 2 °C/min, respectively. The specimen had the following dimensions: 4 mm in width, 15 mm in length, and 1.3 mm in thickness.

Tensile tests were carried out by a tensile testing machine (Autograph AGS-X; Shimadzu, Kyoto, Japan) at 25 °C. The dumbbell-shaped specimens, No.7 of JIS-K6251 (corresponded to ISO37-4), were cut from the sheets using a dumbbell sample cutting machine (Super Dumbbell Cutter SDL-200; Dumbbell, Kawagoe, Japan). The crosshead speed was 100 mm/min, and the initial distance between two clamps was 20 mm. Three measurements were conducted for each sample.

Stress relaxation measurements were performed using the tensile machine. Similar to the tensile tests, the dumbbell-shaped specimens with an initial gauge length of 11 mm were stretched at 100 mm/min. The stretching was stopped at a strain of 1.0 and kept for 900 s to measure the stress relaxation. After the stress relaxation measurements, the samples were taken out from the tensile machine and kept at 25 °C to investigate the strain recovery behaviors. The sample shape was recorded by a digital camera (HDR-CX540V; Sony, Tokyo, Japan) to characterize the recovery process.

3. Results and Discussion

Figure 1 shows the torque curves versus curing time t_c at various temperatures. The torque did not increase at all in 10 min at/below 110 °C, suggesting that no/little vulcanization reaction occurred at these temperatures. Beyond 135 °C, the torque increased with the curing time beyond 300 s. Furthermore, it was found that the optimum cure time t_{90} was around 140 s at 170 °C.

The temperature dependence of tensile storage modulus E' and loss tangent $\tan \delta$ at 10 Hz is shown in Figure 2. The glassy, transition, and rubbery regions were clearly detected for all samples.

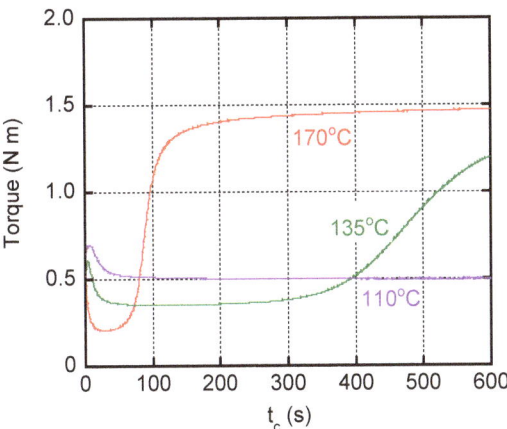

Figure 1. Torque curves plotted against curing time t_c at various temperatures.

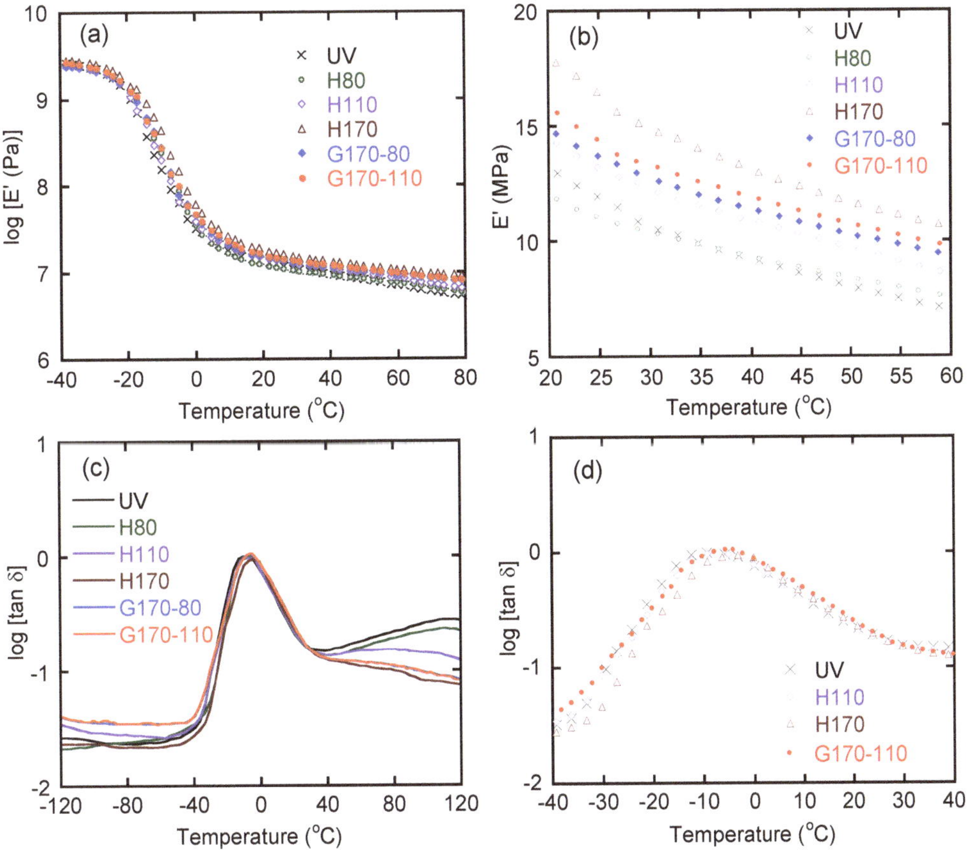

Figure 2. Temperature dependence of (**a**) and (**b**) tensile storage modulus E' and (**c**) and (**d**) loss tangent *tan δ* at 10 Hz.

Although there was no/little difference in the E' values among the samples in the glassy region, the curves in the transition region were slightly different. As seen in Figure 2a, E' decreased due to the glass-to-rubber transition that occurred at a high temperature for H170, i.e., fully crosslinked sample. Figure 2d showed that the values of $\tan \delta$ for H170 were lower than those for the others in the temperature range from −40 to −5 °C. As the vulcanization temperature increased, providing more crosslink points, segmental motion in the rubber was more restricted, which resulted in high T_g [7–10,19]. The number of crosslink points, of course, affected the modulus in the rubbery region. In Figure 2b, E' values from 20 to 60 °C were plotted. The order of E' values corresponded with the vulcanization temperature for the homogeneously crosslinked samples. E' values of the graded rubbers, i.e., G170-80 and G170-110, were found to be between those of H110 and H170. Regarding $\tan \delta$, the graded samples did not show high values in this study, which was different from some reports [8–10,12,19]. Presumably, the present samples had few dangling chains compared with those used in the previous studies. Therefore, the energy absorption was not largely expected, at least in the linear viscoelastic range. Moreover, Figure 2 indicates that H80 had almost no crosslink points because the dynamic mechanical properties were similar to those of the unvulcanized one, i.e., UV. As a result, they showed high $\tan \delta$ values in the high-temperature region (Figure 2c).

Stress–strain curves are shown in Figure 3. The stresses are engineering values, i.e., force divided by the initial cross-sectional area, while the strains are also engineering values, i.e., distance divided by the initial distance. The measurements were performed three times for each sample. Since the experimental error was not large for all samples, we showed the middle values of each in the figure. The stresses for the graded samples were between those of H170 and H110, suggesting that the part with a high crosslink density, which must be the surface region exposed to the high temperature at the vulcanization process, was responsible for the stress generation [15]. They were reasonable results and corresponded with Figure 2.

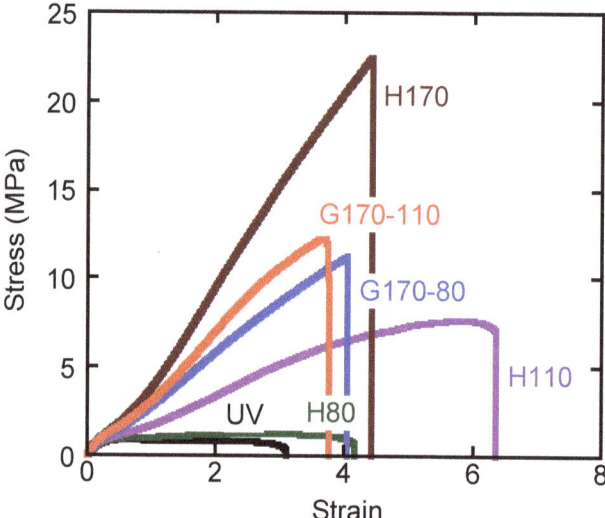

Figure 3. Stress–strain curves.

For some samples, stretching was stopped at a strain of 1.0 to measure the stress relaxation behaviors. Figure 4 shows the stress relaxation curves normalized by the stress at the cessation of stretching. The horizontal axis represents the time after the cessation of stretching. It was found that the graded samples showed high levels of normalized stress in the long-time region. The values were comparable with that of H170 and much better than that of H110, indicating that the graded samples showed good rubber elasticity. Similar to stress generation at stretching, the part with a high crosslink density was largely responsible for the rubber elasticity. It was also found from the figure that H80 showed higher values than UV, suggesting that a weak network existed in H80.

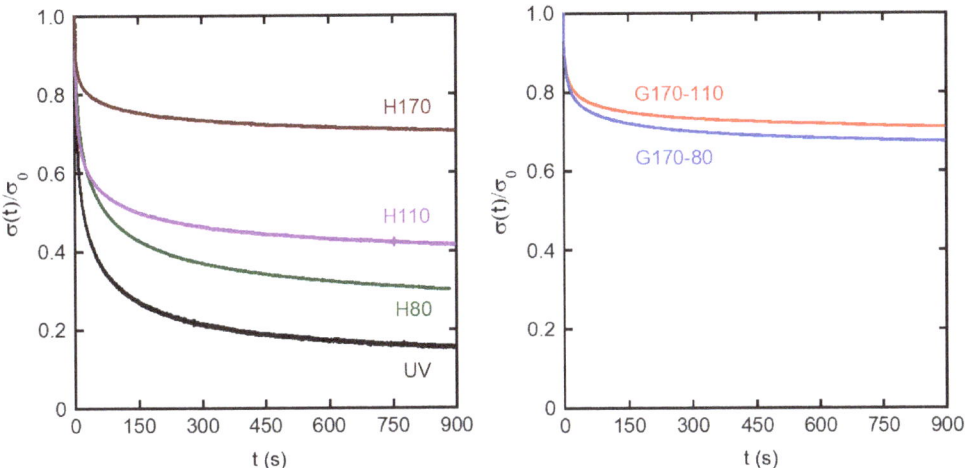

Figure 4. Relaxation curves of normalized stress, i.e., stress $\sigma(t)$ divided by that at the cessation of stretching at $\varepsilon = 1$ σ_0 for (**a**) homogenous samples and (**b**) graded samples.

After 900 s, the samples were taken out from the tensile machine and kept at room temperature to evaluate the strain recovery property. The sample shapes immediately after stress removal are exemplified in Figure 5. As shown in Figure 5a, simple shrinkage with a flat shape was detected for H170, as expected. However, the graded samples exhibited marked bending deformation (Figure 5b). The inner side was the part with a high crosslink density. The bending deformation must be attributed to the difference in the recovery stress acting on each surface of the graded samples. After the stress removal, the side with a high crosslink density was exposed to a high shrinking stress, which could avoid a crack growth on the surface. In contrast, the opposite side had no or weak shrinking stress, owing to the orientation relaxation of chain segments.

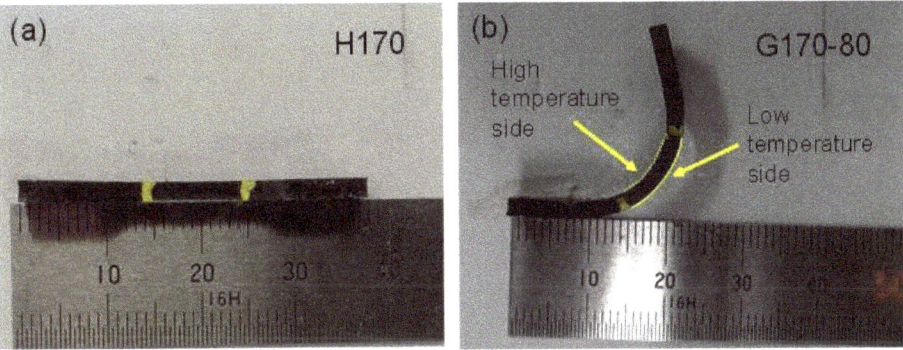

Figure 5. Sample shapes immediately after removal from the tensile machine for (**a**) H170 and (**b**) G170-80. The pictures were taken from a side view of the samples.

Eventually, the graded samples became flat, as shown in Figure 6. However, the bending deformation was still detected even after one week for both graded samples. For a better understanding of the recovery process, the recovery ratio $R(t_r)$ was calculated using the following equation as a function of the time after stress removal t_r:

$$R(t_r)\,(\%) = \left(1 - \frac{\varepsilon(t_r)}{\varepsilon_i}\right) \times 100 \tag{1}$$

where $\varepsilon(t_r)$ is the strain at time t_r and ε_i is the initial strain applied by stretching. In this experiment, ε_i is 1. For bended samples, the outer and inner lengths were evaluated by Image J software using the pictures, and the average values were used as $\varepsilon(t_r)$.

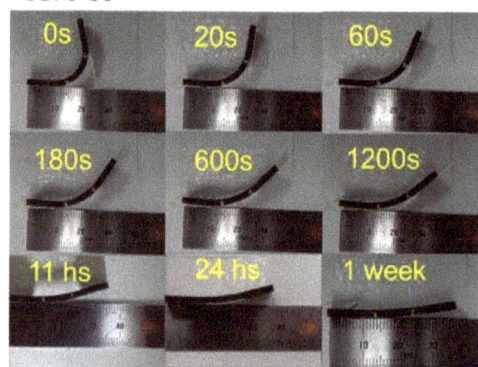
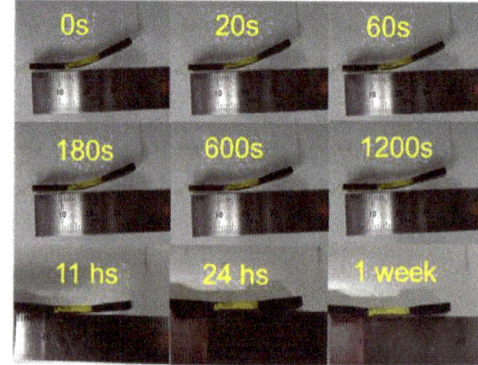

Figure 6. Recovery process after stress removal.

Figure 7 shows the recovery curves. H170 showed an immediate recovery. This is reasonable because the experiments were performed well beyond T_g. The value was 98 ± 1%, suggesting that the residual strain, $1-R(\infty)$, was around 2%. As the crosslink density for the homogenously crosslinked rubbers decreased, the time dependence became obvious, and the equilibrium values became small, i.e., poor rubber elasticity. In other words, samples showing a small residual strain exhibited a quick recovery at room temperature. As compared with the homogenously crosslinked samples, the graded ones showed a different behavior. They exhibited good strain recovery at equilibrium conditions, i.e., 96 ± 1% for G170-80 and 97 ± 1% for G170-110, which was much better than that for H110 (92%). However, it took a long time to recover; i.e., the recovery ratio slowly increased with t_r, especially for G170-80 with bending deformation. This suggests that the shrinking stress acting on the surface with a high crosslink density would work for a long time. This slow recovery cannot be predicted from the stress relaxation data because the relaxation curves were not much different from that of H170, as shown in Figure 4. In the graded rubber, segmental orientation in the side with a low crosslink density was mostly relaxed when the stress was removed. Therefore, during recovery, the shrinking stress was applied from the other side with dense crosslink points. As a result, the reorganization of segments was required in the weakly crosslinked side, which must be the origin of slow recovery.

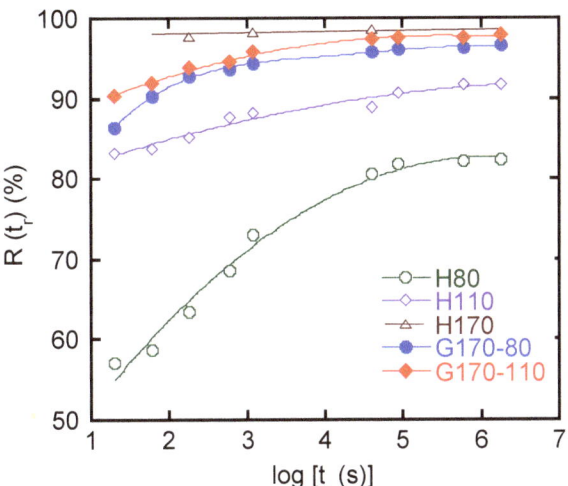

Figure 7. Recovery ratio $R(t_r)$ as a function of time after stress removal.

4. Conclusions

In this study, graded rubbers were prepared by vulcanizing a conventional SBR under a temperature gradient. The obtained samples had a gradient in crosslink density in the thickness direction. Although the dynamic mechanical properties including $\tan \delta$ were not much different from those of the homogeneously crosslinked samples, the rubber elasticity was found to be different. After the cessation of stretching, stress relaxation behavior was evaluated as compared with homogeneously crosslinked rubbers. It was found that the stress was hardly relaxed for the graded rubbers during stress relaxation measurements, which was comparable to the fully crosslinked rubber. This must be attributed to the part with a high crosslink density. Furthermore, after stress removal, the graded samples showed marked bending deformation due to the mismatch in the shrinking stress between both surfaces. Eventually, the samples showed less bending deformation and became flat, which took a long time, even well beyond T_g. Moreover, they showed good strain recovery, i.e., low residual strain, even though shrinking occurred very slowly. The segmental motion in the part with a low crosslink density must be the origin of slow strain recovery. This must be a common phenomenon for graded rubbers with a crosslink gradient in the thickness direction.

Author Contributions: Conceptualization, Q.-V.D. and M.Y.; data curation, Q.-V.D. and K.W.; formal analysis, T.K. and T.N.; investigation, Q.-V.D.; methodology, Q.-V.D. and K.W.; validation, T.T. and M.Y.; writing—original draft, Q.-V.D.; writing—review and editing, Q.-V.D. and M.Y. All authors have read and agreed to the published version of the manuscript.

Funding: This research received no external funding.

Institutional Review Board Statement: Not applicable.

Informed Consent Statement: Not applicable.

Data Availability Statement: Not available.

Conflicts of Interest: The authors declare no conflict of interest.

References

1. Tadmore, Z.; Gogos, C.G. *Principles of Polymer Processing*, 2nd ed.; Wiley: Hoboken, NJ, USA, 2006.
2. Bellander, M.; Stenberg, B.; Persson, S.J.P.E. Crosslinking of polybutadiene rubber without any vulcanization agent. *Polym. Eng. Sci.* **1998**, *38*, 1254–1260. [CrossRef]

3. Shen, M.; Bever, M.B. Gradients in polymeric materials. *J. Mater. Sci.* **1972**, *7*, 741–746. [CrossRef]
4. Varghese, H.; Johnson, T.; Bhagawan, S.S.; Joseph, S.; Thomas, S.; Groeninckx, G. Dynamic mechanical behavior of acrylonitrile butadiene rubber/poly(ethylene-co-vinyl acetate) blends. *J. Polym. Sci. B Polym. Phys. Ed.* **2002**, *40*, 1556–1570. [CrossRef]
5. Sahoo, S.K.; Mohanty, S.; Nayak, S.K. Toughened bio-based epoxy blend network modified with transesterified epoxidized soybean oil: Synthesis and characterization. *RSC Adv.* **2015**, *5*, 13674–13691. [CrossRef]
6. Sheng, Z.; Wang, J.; Song, S. Fabrication of gradient multilayer polymer with improved damping performance based on poly (butyl acrylate)-grafted polysiloxane. *J. Mater. Sci.* **2020**, *55*, 11739–11751. [CrossRef]
7. Gennes, D. *Scaling Concepts in Polymer Physics*; Cornell Univ. Press: London, UK, 1979.
8. Yamaguchi, M.; Miyata, H. Strain hardening behavior in elongational viscosity for binary blends of linear polymer and crosslinked polymer. *Polym. J.* **2000**, *32*, 164–170. [CrossRef]
9. Yamaguchi, M.; Ono, S.; Terano, M. Self-repairing property of polymer network with dangling chains. *Mater. Lett.* **2007**, *61*, 1396–1399. [CrossRef]
10. Yamaguchi, M.; Ono, S.; Okamoto, K. Interdiffusion of dangling chains in weak gel and its application to self-repairing material. *Mat. Sci. Eng. B* **2009**, *162*, 189–194. [CrossRef]
11. Lv, X.; Huang, Z.; Huang, C.; Shi, M.; Gao, G.; Gao, Q. Damping properties and the morphology analysis of the polyurethane/epoxy continuous gradient IPN materials. *Comp. Part B Eng.* **2016**, *88*, 139–149. [CrossRef]
12. Li, S.; Yang, W. Fabrication of poly(acrylamide) hydrogels with gradient crosslinking degree via photoinitiation of thick polymer system. *Polym. Adv. Technol.* **2011**, *22*, 1442–1445. [CrossRef]
13. Fan, W.; Shan, C.; Guo, H.; Sang, J.; Wang, R.; Zheng, R.; Sui, K.; Nie, Z. Dual-gradient enabled ultrafast biomimetic snapping of hydrogel materials. *Sci. Adv.* **2019**, *5*, aav7174. [CrossRef] [PubMed]
14. Fradet, C.; Lacroix, F.; Berton, G.; Méo, S.; Le Bourhis, E. Instrumented indentation of an elastomeric material, protocol and application to vulcanization gradient. *Polym. Test.* **2020**, *81*, 106278. [CrossRef]
15. Ikeda, Y. Preparation and properties of graded styrene-butadiene rubber vulcanizates. *J. Polym. Sci. B Polym. Phys. Ed.* **2002**, *40*, 358–364. [CrossRef]
16. Ikeda, Y. Graded styrene–butadiene rubber vulcanizates. *J. Appl. Polym. Sci.* **2003**, *87*, 61–67. [CrossRef]
17. Glebova, Y.; Reiter-Scherer, V.; Suvanto, S.; Korpela, T.; Pakkanen, T.T.; Severin, N.; Shershnev, V. Rabe, J.P. Nano-mechanical imaging reveals heterogeneous cross-link distribution in sulfur-vulcanized butadiene-styrene rubber comprising ZnO particles. *Polymer* **2016**, *107*, 102–107. [CrossRef]
18. Doan, V.A.; Nobukawa, S.; Ohtsubo, S.; Tada, T.; Yamaguchi, M. Selective Migration of silica particles between rubbers. *J. Polym. Res.* **2013**, *20*, 145–150. [CrossRef]
19. Wang, Y.Q.; Wang, Y.; Zhang, H.F.; Zhang, L.Q. A novel approach to prepare a gradient polymer with a wide damping temperature range by in-situ chemical modification of rubber during vulcanization. *Macromol. Rapid Commun.* **2006**, *27*, 1162–1167. [CrossRef]
20. Van Amerongen, G.J. Diffusion in elastomers. *Rub. Chem. Technol.* **1964**, *37*, 1065–1152. [CrossRef]
21. Gardiner, J.B. Curative diffusion between dissimilar elastomers and its influence on adhesion. *Rub. Chem. Technol.* **1968**, *41*, 1312–1328. [CrossRef]

Review

Fabrication of Conductive Fabrics Based on SWCNTs, MWCNTs and Graphene and Their Applications: A Review

Fahad Alhashmi Alamer * and Ghadah A. Almalki

Department of Physics, Faculty of Applied Science, Umm AL-Qura University, Al Taif Road, Makkah 24382, Saudi Arabia
* Correspondence: fahashmi@uqu.edu.sa

Abstract: In recent years, the field of conductive fabrics has been challenged by the increasing popularity of these materials in the production of conductive, flexible and lightweight textiles, so-called smart textiles, which make our lives easier. These electronic textiles can be used in a wide range of human applications, from medical devices to consumer products. Recently, several scientific results on smart textiles have been published, focusing on the key factors that affect the performance of smart textiles, such as the type of substrate, the type of conductive materials, and the manufacturing method to use them in the appropriate application. Smart textiles have already been fabricated from various fabrics and different conductive materials, such as metallic nanoparticles, conductive polymers, and carbon-based materials. In this review, we study the fabrication of conductive fabrics based on carbon materials, especially carbon nanotubes and graphene, which represent a growing class of high-performance materials for conductive textiles and provide them with superior electrical, thermal, and mechanical properties. Therefore, this paper comprehensively describes conductive fabrics based on single-walled carbon nanotubes, multi-walled carbon nanotubes, and graphene. The fabrication process, physical properties, and their increasing importance in the field of electronic devices are discussed.

Keywords: smart textile; SWCNTs; MWCNTs; graphene; applications

Citation: Alhashmi Alamer, F.; Almalki, G.A. Fabrication of Conductive Fabrics Based on SWCNTs, MWCNTs and Graphene and Their Applications: A Review. *Polymers* **2022**, *14*, 5376. https://doi.org/10.3390/polym14245376

Academic Editor: Md Najib Alam

Received: 22 November 2022
Accepted: 6 December 2022
Published: 8 December 2022

Copyright: © 2022 by the authors. Licensee MDPI, Basel, Switzerland. This article is an open access article distributed under the terms and conditions of the Creative Commons Attribution (CC BY) license (https:// creativecommons.org/licenses/by/ 4.0/).

1. Introduction

Traditional textiles were created to protect people from the elements, such as cold and rain, and to serve as covering material. The two most important qualities associated with clothing are their ability to provide protection and their aesthetics. Throughout history, advances in smart materials and electronics have contributed to a unique potential that has led to the emergence of a new field called "smart textiles". Smart textiles, also called intelligent textiles or e-textiles, are a type of intelligent materials that can detect and respond to changes in their environment [1]. The stimuli and responses can be thermal, electrical, magnetic, mechanical, chemical, or any other type of stimulus or response [2]. Smart textiles are indeed used in many applications ranging from simple to more complicated ones, for example, in military, healthcare, and wearable electronics [3–5]. They are classified into three groups based on their generation and intelligence [6,7]. First generation passive smart textiles can provide additional functions in a passive mode regardless of environmental changes. Examples of passive smart textiles include anti-odor, anti-static, anti-microbial, and bulletproof [8,9]. In the second generation, smart textiles have been developed to sense and respond to environmental stimuli. Examples include heat storage, sensors, thermoregulation, vapour-absorbing fabrics, and electrically heated suits [8]. A sophisticated smart textile consists primarily of an entity that functions similarly to the brain, with cognitive, reasoning, and activating capabilities that can sense, respond, and adapt to environmental conditions or stimuli, including health monitoring and space suits [10]. Figure 1 shows a chronology of the development of smart textiles.

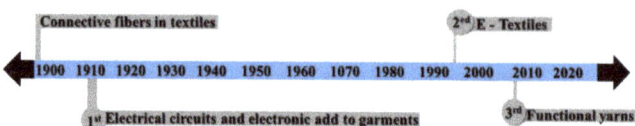

Figure 1. This timeline shows different generations of electronic textiles. Adapted [7].

Although recent advances in the field of smart textiles are extremely interesting, several obstacles still need to be overcome to make them suitable for commercial and economic use [6]. Unfortunately, the fabrication of conductive textiles is limited by several technical and non-technical features. Therefore, it is essential to develop large-scale manufacturing processes [11,12]. One of the problems in developing smart textiles is the smooth and undetectable integration of the required electronics into the fabric. Therefore, material scientists need to develop fibers with the required electrical capabilities that are strong, comfortable, and attractive at the same time [13]. In addition, temperature, perspiration, humidity, mechanical shock, continuous bending and tension, and illumination should be thoroughly investigated [14]. The autonomy of the system should be increased to reduce the burden of frequent battery charging, and the battery life also needs to be improved, which is still a research problem [15]. The garments must ensure high security against cyber threats [16]. For users to fully embrace this new technology, smart clothing must be a product that meets consumers' emotional and functional needs, and integration and connectivity tools [17].

This article is about the fabrication of smart textiles using carbon-based materials, especially SWCNTs, MWCNTs, and graphene, and is organized as follows: The smart materials, discussing the structure, physical properties, and potential applications of carbon nanotubes, SWCNTs and MWCNTs, and graphene. Then, there are three major sections focusing on the smart textiles fabricated with SWCNTs, MWCNTs and graphene, respectively. Meanwhile, the fabrication method, physical properties, especially electrical properties, factors affecting these properties, and potential applications of smart textiles are described.

2. Smart Materials

Materials that are described as "smart" or "functional" are usually part of a "smart system" that can sense and respond to its environment. If they are truly intelligent, they have a significant impact on the performance of smart textiles [18]. In this article, the focus is on single-walled carbon nanotubes, multi-walled carbon nanotubes, and graphene.

2.1. Carbon Nanotubes (CNTs)

2.1.1. Definition and Structure

Carbon nanotubes (CNTs) belong to the fullerene family, which includes carbon allotropes whose atoms are connected in cage-like configurations, such as a hollow sphere, an ellipsoid, or a cylinder [19–22], and have a thickness or diameter on the order of a few nanometers [23]. CNTs can be fabricated in a variety of ways, but the three most common methods are fabrication by electric arc, chemical vapor deposition, and laser ablation [24].

2.1.2. Types of Carbon Nanotubes

CNTs are generally classified by the number of carbon layers into single-walled (SWNTs) or multi-walled (MWNTs) carbon nanotubes, as shown in Figure 2. SWCNTs are single graphene layers wrapped in tubes. Depending on how the tube is wrapped, SWCNTs have different properties [25] and structures [26–28]. MWCNTs, on the other hand, consist of multiple graphite layers wound on top of each other [29], and the diameter between the tube walls is about 0.34 nm. The architecture of MWCNTs can be described by one of two models: the Russian doll model and the parchment model [30]. Table 1 shows the comparison between SWCNTs and MWCNTs and Table 2 shows the main physical properties of the two [31–46].

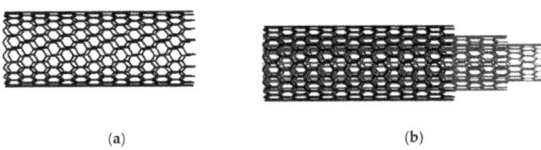

Figure 2. Structure of carbon nanotubes (**a**) SWNTs and (**b**) MWNTs.

Table 1. Comparison between SWNTs and MWNTs.

SWCNTs	MWCNTs	References
Single graphene layer	Multiple graphene layers	[40,44]
SWNTs have a diameter of 0.4 to 3.0 nm and a length of 20 to 1000 nm.	The outer diameters are between 2 and 100 nm, the inner diameters between 1–3 nm and the lengths between 1 and 50 m.	[29,45]
The synthesis of SWCNTs requires the use of a catalyst.	The synthesis of MWCNTs can in fact be made no need for a catalytic	
Bulk production is challenging because it requires precise control of growth and environmental conditions.	Bulk production is simple.	[40,44]
Purity is poor	Purity is high	
Less deposits in the body	More deposits in the body	
It is more flexible and can be twisted effortlessly.	It is complicated to twist.	
Characterization and evaluation are simple	It has a very complex structure	

Table 2. Summary of the main properties of SWCNTs and MWCNTs.

Properties	Unit	SWCNTs	MWCNTs	References
Specific gravity	g/m^3	0.8–1.3	1.8–2.6	[46]
Resistivity	μΩ/cm	5–50	5–50	
Young's modulus	TPa	~1	~1–0.3	[41,42]
Thermal conductivity	W.m^{-1}K^{-1}	3000–6000	2000–3000	[41,43]
Electrical conductivity	S/m	10^2–10^6	10^3–10^5	[43]
Thermal stability in air	°C	550–650	550–650	
Specific area	m^2/g	400–900	200–400	

2.1.3. Potential Applications of CNTs

CNTs were used in wide range of applications due to its small and lightweight, which makes them suitable for a [40]. They can be used in many fields, such as electronic and photovoltaic devices [47], solar cells [48], superconductors [49], food science [50], water purification [51], biology and medicine [52], electrical/electronic applications [53], wearable devices, and smart textiles [54]. Figure 3 shows various applications of CNTs in textiles.

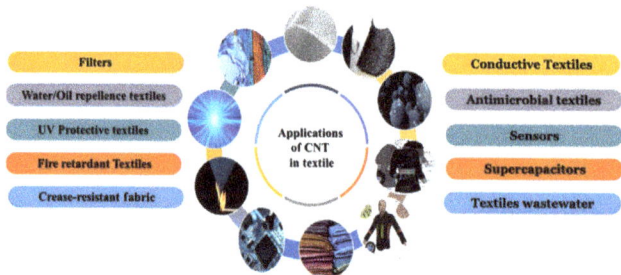

Figure 3. Potential applications for the use of carbon nanotubes in textiles.

2.2. Graphene

2.2.1. Definition and Structure

Graphene is a flat monolayer of carbon atoms densely condensed into a two-dimensional (2D) honeycomb crystal structure [55–61], as shown in Figure 4. Graphene belongs to the category of graphitic nanomaterials, which includes graphene with few layers (1–5 layers). It has numerous chemical [62,63], physical [64], electronic [65], and mechanical [66] excellent properties. In addition, graphene is said to be the thinnest known substance [67], the most hydrophobic known substance [68], possessing both brittleness and ductility [69], nontoxic, and inexpensive [70,71]. Table 3 shows a summary of the basic physical properties of graphene.

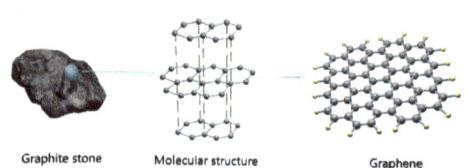

Figure 4. Structure of graphene as a honeycomb lattice of carbon atoms.

Table 3. Summary of the basic physical properties of graphene.

Quantity	Values	References
Tensile strength	130 GPa	[72]
Young's modulus	1 TPa	[73]
Weight	0.77 mg/m^2	[74]
Thermal conductivity	~3000–5000 W m^{-1} K^{-1}	[75,76]
Mobility of charge carrier	2×10^5 cm^2 V^{-1} s^{-1}	[77–79]
Electrical conductivity	~3.6 $\times 10^8$ S/m	[80]
Transmittance	≈97.7%	[81]

2.2.2. Potential Applications of Graphene

The exceptional properties of graphene can be exploited in numerous applications, including biomedicine [82], membranes [83], sensors [84], energy harvesting and storage [85], composites and coatings [86], and functional devices [87], as shown in Figure 5. In addition, graphene is a promising material for the fabrication of smart and electronic textiles, where multiple functions can be combined in a single material. The large surface area and flexibility improve conformal contact, resulting in increased sensitivity [88]. Due to the atomic structure of carbon atoms in graphene, electrons can move at incredible speeds without scattering, saving energy that would otherwise be wasted in conventional conductors. The number of graphene layers and the coupling effects with the underlying substrate affect the electronic properties of the graphene system. Seamless integration of electronics into textiles can enable various applications, including flexible, stretchable, and foldable devices [89], electrodes, and electronic textiles that can be used in various fields.

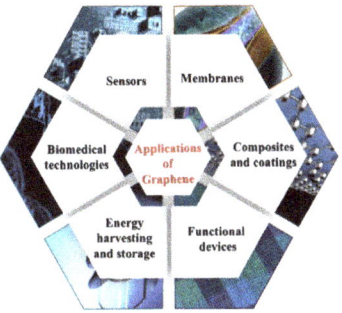

Figure 5. Potential Applications of graphene.

3. Conductive Fabrics Based on Carbon Nanotubes

Conductive fabrics are usually made from various substrates, such as cotton [90], polyester [91], wool [92], and nylon [93], using numerous techniques, such as embroidery [94], knitting [95], spinning [96], coating [90], printing [97], dipping and drying, drop casting [98], and others. To make fabrics electrically conductive, there are usually two approaches: one approach is to incorporate conductive fillers, such as metal nanoparticles and carbon-based materials [99], graphene and carbon nanotubes, into the fabric. The second approach is to coat the fabric with a conductive polymer, such as PEDOT:PSS, which contains little or no metal. This review focuses on the fabrication of conductive fabrics from carbon-based materials.

3.1. Conductive Fabrics Based on SWCNTs

Recently [100], conductive and flexible melt-blown fabrics were coated with SWCNTs by chemical vapor deposition, where the melt-blown fabrics were recycled from face masks. The results showed that the sheet resistance of the conductive fabrics depended on the deposition time and was 245, 116, and 57 $\Omega/\square$ for deposition times of 1 h, 2 h, and 3 h, respectively. It was also found that the sheet resistance decreased when the gold chloride dopant was used with values of 64, 54, and 26 $\Omega/\square$, respectively. Alamer et al. [101] fabricated a highly conductive cotton fabric impregnated with SWCNTs by using the filtration technique to produce conductive cotton. The advantage of this technique is that the residual solution that passes through the filter paper is collected in a beaker and stored for later use. This process was also safe, simple, and environmentally friendly, using renewable energy sources and using chemicals effectively. They found that the sheet resistance of the sample reached the minimum value of 0.006 $\Omega/\square$ at a concentration of 41.5 wt.%. They also ensured that the temperature behavior of the conductive cotton was consistent and reproducible for at least two months. Huang et al. [102] designed an electrode using a stretchable Lycra fabric, SWCNTs as conducting materials, and a dyeing and drying process. The impurities were first removed from the surface of the fabric using deionized water. Then, the fabric was stretched to 100% elongation and immersed in SWCNT ink, alcohol, and nitric acid, respectively. Then, the fabric was dried and stretched to allow the SWCNTs to penetrate the fabric and increase its conductivity. The resistance of the conductive electrode was stable after 5×10^2 stretching cycles with a minimum sheet resistance of 65 $\Omega/\square$ at 35% tensile load.

In another study, SWCNT ink was printed on a stretchable substrate using inkjet printing technique [103]. It was found that the sheet resistance of the conductive substrate depended on the number of coating layers and decreased as the number of layers increased.

The minimum sheet resistance was 19.08 Ω/□ for the five-layer coating, and this value changed slightly after the sample was stretched. Zhang et al. [104] fabricated conductive cotton nylon with SWCNTs by immersion drying and then modified the fabric by plasma. The results showed that the sheet resistance of the modified fabric 2.0 k Ω/□ was lower than the sheet resistance of the unmodified fabric 4.9 k Ω/□, which was attributed to the increase in the surface roughness of the fabric. SWCNTs dispersed in dodecylbenzene sulfonic acid, sodium salt, were applied to polyester fabric by the coating dry cure method [105]. Before the coating process, the fabric was treated with plasma using different working gases and different treatment times. The results showed that this treatment led to an improvement in the antistatic properties of the polyester fabric. It was found that the antistatic property increased with increasing plasma treatment and then decreased. The effect of SWCNTs as absorbers of UV light for cotton fabrics was investigated in the study published by Mahmoudifard and Safi [106] and compared with ZnO and TiO_2 absorbers. It was found that SWCNTs absorbed UV light with a high UPF value compared to ZnO and TiO2 absorbers. In another interesting study [107], a piezoresistive stretchable sensor based on SWCNTs and fabrics was fabricated, the joint movements of children were measured and compared with a rectangular sensor. It was found that the stretchable sensor had the same effect as the rectangular sensor with electrical resistance in the range of 280 Ω and 290 Ω. A flexible and stable supercapacitor with a high specific capacitance of about 70 to 80 Fg^{-1} was prepared by immersing cotton leaves in SWCNT ink [108]. The results showed that the sample exhibited high electrical conductivity with a sheet resistance of less than 1 Ω/□. In another study, SWCNTs dispersed in sodium dodecyl benzyl sulfonate and ethylene glycol were used to prepare conductive threads by the immersion drying method [109]. The results showed that the electrical conductivity was curiously dependent on the concentration of SWCNTs, with the resistance decreasing from 3.587 Ω to 0.01257 Ω as the concentration increased from 0.008049 wt% to 1.07269 wt%. In 2008, Shim et al. [110] fabricated a conductive cotton yarn by using SWNTs, MWNTs and polyelectrolytes and applying the immersion method. The cotton yarn becomes conductive after multiple immersions, with a resistivity as low as 20 $Ω.cm^{-1}$. This approach is characterized by its speed, simplicity, robustness, low cost, and easy scalability. The fabrication of conductive Lycra fabric has also been investigated using the conductive materials SWCNTs and polyaniline, using the immersion drying process [111]. The results showed that the conductive fabric had a minimum sheet resistance of 35 Ω/□. This fabric was used to manufacture antenna which worked at 2.45 GHz with reflection coefficient of about ~18.6 dB. In another study [112], conductive cotton fabrics were prepared using a composite of SWCNTs and the conductive polymer PEDOT: PSS by the technique of drop casting. The effects of applying the composite in the cotton fabrics were studied, and the results showed that a cotton fabric composed of one layer of PEDOT: PSS between two layers of SWCNTs was electrically stable for four months, with a minimum sheet resistance of 0.006 Ω at a concentration of 41.5 wt.%. The metallic conductive threads were also prepared from SWCNTs, PEDOT:PSS, and a mixture of both [113]. The results showed that the electrical resistances depend on the fabrication process. The lowest sheet resistance was obtained for the sample prepared from a mixture of SWCNTs and PEDOT:PSS with a value of 0.0072 Ω, and a lower amount of composite of 1.729 mg. Table 4 show summary list of SWCNTs-based materials with details of their manufacturing processes and electrical properties.

Table 4. Shows summary list of SWCNTs-based materials with details of their manufacturing processes and electrical properties.

Year	Substrate	Coated with	Method	Electrical Properties	References
2022	Melt blown fabrics	SWCNTs	Chemical vapor deposition	57 Ω/□	[100]
2022	Melt blown fabrics	SWCNTs + gold chloride	Chemical vapor deposition	26 Ω/□	[100]
2020	Cotton	SWNTs	Filtration	0.006 Ω/□	[101]
2017	Lycra	SWCNTs	Dyeing drying	65 Ω/□	[102]
2014	Thin films	SWCNTs	Inkjet printed	19.08 Ω/□	[103]
2012	Cotton nylon	SWCNTs	Dyeing drying	2.0 kΩ/□	[104]
2015	Polyester	SWCNTs	Coating-dry-cure	-	[105]
2012	Cotton	SWCNTs + ZnO + TiO_2	Dyeing drying	-	[106]
2020	Polyester spandex	SWCNTs + polyurethane	Dry curing machine	280–290 Ω	[107]
2010	Cotton	SWCNT ink	Dyeing drying	1 Ω/□	[108]
2020	Cotton thread	SWCNTs + SDBS	Dip coating	0.01257 Ω	[109]
2008	Cotton yarn	SWNTs + MWNTs + polyelectrolyte	Dip coating	20 Ω/cm	[110]
2017	Lycra	SWCNTs	Dyeing drying	35 Ω/□	[111]
2021	Cotton thread	SWNTs + PEDOT: PSS	drop-casting	0.0072 Ω	[112]

3.2. Conductive Fabrics Based on MWCNTs

As discussed in the previous section, conductive fabrics made of SWCNTs have excellent electrical properties; however, SWCNTs are expensive, purification is difficult, and dispersion in liquid is also difficult. Therefore, many researchers focused on fabricating conductive fabrics using MWCNTs instead of SWCNTs because they are cheaper, can be produced in large quantities, and are more stable compared to SWCNTs. Rahman et al. [113] fabricated conductive and thermal cotton fabrics with MWCNTs using the dip and dry method. The results showed that the electrical conductivity of the conductive cotton was about 0.20 S m^{-1} with a sheet resistance of 1.67 kΩ/□ after four times of immersion. In addition, the thermal conductivity of the fabric was also increased by 70%. MWCNTs dispersed in DMF were used in the recent study presented by Alamer et al. [114] to prepare conductive cotton fabrics using the drop-casting and drying method. The sheet resistance of the conductive cotton was proportional to the MWCNT loading of the fabric and reached a value of 15.92 Ω/□ at a saturation concentration of 42.20 wt.%. Moreover, the conductive fabrics exhibited semiconductor behavior as the resistance decreased with increasing temperature. The conductive cotton fabrics were also prepared by immersing the fabrics in a dispersion of MWCNTs in sodium dodecyl sulphate [115]. The amount of MWCNTs was increased up to 20 times by repeating the immersion process, and the cotton fabrics with high MWCNT concentration exhibited a minimum sheet resistance of 2.5 kΩ.cm^{-2}. The results also showed that the conductive cotton treated with HNO_3 resulted in a reduction of sheet resistance to 1.5 kΩ.cm^{-2} which was attributed to the interaction between MWCNTs and cellulose through glycosidic bonds.

The conventional dyeing method was used to deposit synthetic MWCNTs on the surface of cotton fabrics [116]. The deposition of MWCNTs was uniform and permanent, and the results showed that the sheet resistance changed in the range of 5486 MΩ/□ to 0.433 MΩ/□ due to the change of the amount of MWCNTs from 100 mg to 500 mg. In addition, the mechanical properties of the conductive fabric were also improved, and the strength was increased by increasing the amount of MWCNTs, which was attributed to the effect of van der walls force between the nanotube particles and the cotton surface. In another study [117], MWCNTs were first dispersed by grafting dimethyl phosphite and perfluorohexyl iodine, then applied to cotton fabric by the impregnation-drying method.

The conductive fabric had a sheet resistance of 225.6 kΩ/□ and exhibited UV resistance, with the UPF value reaching the maximum value of 121. Costa et al. [118] fabricated electrodes for supercapacitors based on cotton fabric and MWCNTs dispersed in sodium dodecylbenzene sulfonate. The MWCNT dispersion was applied to the surface of the conductive cotton using the dip-pad drying method. By repeating this method eight times, the resistance of the fabric electrodes reached the minimum value of 2.62 $\Omega \cdot cm^{-2}$, had a specific capacitance of 8.01 F g^{-1}, a high energy density of 6.30 Wh kg^{-1}, and a cyclability of 5000. In the study presented by Nafeiea et al. [119], conductive wool fabrics were prepared using MWCNTs and carboxylated MWCNTs, both dispersed in water, and the effect of sodium dodecyl sulfate as an anionic surfactant and cetyltrimethylammonium bromide as a cationic surfactant was investigated. The results showed that the use of a cationic surfactant improved the dispersion of MWCNTs in water, while the dispersion of carboxylated MWCNTs in water was better without the use of a surfactant. The electrical conductivity of the wool fabric prepared with 5 g/L carboxylated MWCNTs reached a maximum value of 2×10^{-3} S cm^{-1}, which is ten times higher than the conductivity of the wool fabric treated with MWCNTs. In the study presented by Kowalczyk et al. [120], MWCNTs were also dispersed in sodium dodecyl sulfate, then applied to polyester/cotton fabrics using the padding-drying method. The resistance depended on the number of pads and changed from 5.79 kΩ to 1.07 kΩ when the number of pads was increased from one to three, which was attributed to the formation of the MWCNT networks. Polyester fabric treated with MWCNTs dispersed in enzymes was used as an electrode in dye-sensitised solar cells [121], where the MWCNT dispersion was applied to the surface of the fabric by the tape-casting method. The sheet resistance of the treated fabric depended on the thickness of the coating and changed from 38 Ω/□ to 12 Ω/□ when the coating thickness increased from 5 µm to 28 µm. It was also found that the sheet resistance depended on the size of MWCNTs. The energy conversion efficiency of the conductive electrode reached about 5.69%. Hao et al. [122] fabricated flexible conductive cotton electrodes for supercapacitors using carboxyl MWCNTs. The carboxyl MWCNTs were deposited on the cotton fabric at high temperature and pressure by immersion method. The electrical resistance of the composite reached a value of 2.606 Ω with a high specific capacitance of 94.3 F g^{-1}, and the sample exhibited good stability up to 3000 cycles. The conductive yarns based on MWCNTs were fabricated in the study presented by Abbas et al. [123,124], in which the spin-dry method was used for the fabrication process. The results showed that the resistance of the conductive yarn depended on the diameter of the yarn. It was 2.55 kΩ and 120 Ω for the yarns with diameters of 12 µm and 100 µm, respectively. In addition, the absorption coefficients of the conductive yarns were measured in the range of 50 MHz to 20 GHz and were found to depend on the diameter of the conductive yarns. Table 5 shows summary list of MWCNTs-based materials with details of their manufacturing processes and electrical properties.

Table 5. Summary list of MWCNTs-based materials with details of their manufacturing processes and electrical properties.

Year	Substrate	Coated with	Method	Electrical Properties	References
2015	Cotton	MWCNTs	Dyeing drying	1.67 kΩ/□ 0.20 S m^{-1}	[115]
2022	Cotton	MWCNTs	Dyeing drying	15.92 Ω/□	[116]
2015	Cotton	MWCNTs	Dyeing drying	1.5 Ωk cm^{-2}	[117]
2019	Cotton	MWCNTs	Dyeing	0.433 MΩ/□	[118]
2020	Cotton	MWCNTs	Dipping drying	225.6 kΩ/□	[119]
2020	Cotton	MWCNTs	Dip-pad-dry	2.625 Ω cm^{-2}	[120]
2016	Wool	MWCNTs	-	2×10^{-3} S cm^{-1}	[121]
2015	Polyester/Cotton	MWCNTs	Padding machine	1.03×10^3 Ω/□	[122]
2015	Polyester	MWCNTs	Tape casting	15 Ω/□	[123]
2018	Cotton	MWCNTs-COOH	Dyeing drying	2.606 Ω	[124]

3.3. Conductive Fabrics Based on Graphene

Incorporating graphene into textiles not only imparts conductivity to the textiles, but also enables the production of multifunctional textiles due to the excellent physical properties of graphene, as we discussed in Section 2. Gan et al. [125] fabricated conductive cotton fabrics using graphene nanoribbons by wet coating method. The mechanical and electrical properties of the fabrics were improved after repeating the wet coating method. The achieved low resistance was about 80 Ω with an increase in tensile stress and elastic modulus of 58.9% and 64.1%, respectively. In another study [126], the trapping method was used to fabricate conductive PET graphene-based fabrics. The main feature of this method is to reduce the insolubility of graphene so that it can easily penetrate the fabrics. The sheet resistance of the fabrics was strongly dependent on the graphene loading and changed from 77.9 MΩ/□ to 2.5 kΩ/□ when the graphene loading increased from 2.5 wt.% to 10.7 wt.%. Sahito et al. [127] developed a flexible and conductive cotton fabric coated with graphene nanosheets. Briefly, the charge of the surface of the cotton fabric was modified by cationization, which resulted in a positive charge that enabled strong bonding between the graphene oxide nanosheets and the cotton fabric and formed a uniform layer on the surface of the fabric, then the chemical reduction method was used to convert the graphene oxide nanosheets into graphene nanosheets. This conductive flexible cotton fabric with sheet resistance of 7 Ω/□ was used as a counter electrode for a dye-sensitive solar cell, and the calculated photovoltaic conversion efficiency was 6.93%. Ren et al. [128] fabricated conductive cotton fabrics with graphene oxide, where the graphene oxide was synthesized from graphite flakes, dispersed in DI water, applied to the cotton fabrics by a vacuum filtration method, and then reduced by a hot-pressing method. The sheet resistance was about 0.9 kΩ/□ and increased to about 1.2 kΩ/□ after 10 washing cycles. This conductive cotton fabric was used as a strain sensor and showed good stability up to 400 bending cycles. In an interesting method, Atta et al. [129] immersed cotton yarns in a graphene oxide dispersion, then reduced them with gamma rays. The resulting cotton yarns were used as portable supercapacitors and the specific capacitance reached a maximum value of 97 F/g. It was also found that the series resistance and charge transfer resistance depended on the graphene oxide concentration and reached a minimum value of 34 Ω and 22 Ω for the series resistance and charge transfer resistance, respectively. Maneval et al. [130] prepared conductive cotton yarns by using two methods: cationization to improve electrostatic interactions, and dip coating to coat the surface of cotton yarns with a graphene dispersion (see Figure 6). Before the yarn breaks, the electrical conductivity of the yarn reached a maximum of 1.1 S cm^{-1} at a graphene concentration of 14% by weight and under continuous mechanical stress.

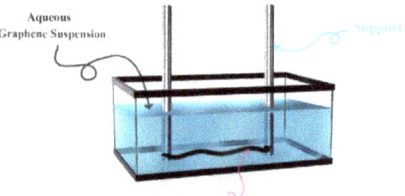

Figure 6. Schematic representation of the device used for the coating of cotton yarns with graphene sheets [130].

Lu et al. [131] fabricated silk fabric with a high conductivity of a single fiber of 3595 S m^{-1} by using graphene oxide nanosheets as a conductive material and a coating reduction method. Briefly, the untreated silk fabric was immersed in bovine serum albumin, which generates a positive charge on the surface of the fabric and increases the absorption of the conductive material when the fabric is immersed in the graphene oxide nanosheet solution. Then, a hydrazine vapor reduction method was used to reduce the graphene oxide on the fabric. In the study presented by Zulan et al. [132], the conductive silk fabric was also prepared with graphene oxide after the fabric was modified. The modification of the silk fabric was performed as follows: the fabric was immersed in a solution containing regenerated silk fibroin as an electrostatic adhesive, deionized water, and bovine serum albumin. The modified fabric was coated with graphene oxide, then thermally reduced to convert the graphene oxide into graphene. The results showed that the conductive silk fabric was thermally stable and exhibited an electrical conductivity of 3.06×10^{-6} S cm^{-1}. In another study [133], a flexible, stable, conductive cotton yarn with an electrical conductivity of about 1.0 S cm^{-1} was prepared using reduced graphene oxide and a dip coating and reduction method. The results also showed that the conductive cotton yarn exhibited mechanical stability up to 1000 cycles and absorbed UV irradiation of about 1.0 mA/W under bending deformation. Yarns from Calotropis gigantean [134], which have a unique structure, excellent hydrophilicity, and lower natural longitudinal crimp, were used to produce conductive yarns on a large scale by dyeing graphene oxide onto the surface of the yarn and applying a reduction process (see Figure 7). The obtained conductivity of the treated yarn depended on the concentration of graphene oxide and reached a maximum value of 6.9 S m^{-1} at high concentration and was shown to be resistant to washing, which was due to the hydrogen bonding formed between the fiber and graphene during the dyeing process.

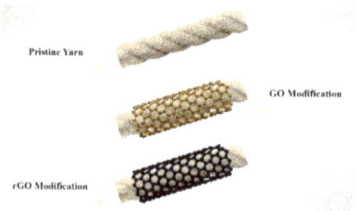

Figure 7. Images of the pristine, GO-modified, and rGO-modified CGYs [134].

Molina et al. [135] fabricated conductive fabrics by chemical reduction of graphene oxide on polyester fabric. The resistance of the fabric decreased from 10^{11} Ω. cm^2 for the untreated fabric to 23.15 Ω. cm^2 for the fabric coated with three layers of reduced graphene oxide. In another study [136], the knitted fabric was also immersed in graphene oxide solution, then subjected to a reduction process. The sheet resistance of the resulting promoted fabric was dependent on the amount of reduced graphene oxide in the fabric and the number of immersion cycles. It reached the value of 0.19 MΩ/□ after 15 dipping cycles. The graphene/polyurethane composite material and the dip coating method [137] were used to fabricate a conductive para-aramid fabric, as shown in Figure 8. It was found that the sheet resistance and electrical capacitance of the conductive fabric decreased as the number of dip coating cycles increased due to the increase in the amount of composite materials. The minimum sheet resistance and electrical capacitance of the conductive fabric reached 75 kΩ/□ and 89.4 pF, respectively, after 5 dip coating cycles. It was also found that this sample can be used for a heat-resistant para-aramid knitted glove with a phone touch screen when hot-pressed at 140 degrees.

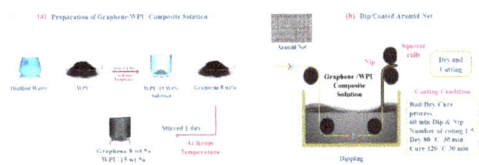

Figure 8. Illustration of the fabrication process for graphene/WPU with dip coating on para-aramid knitted fabric, which consists of two steps: (**a**) preparation of graphene/WPU composite solution, (**b**) dip coating of para-aramid knitted fabric with different coating cycles [137].

A conductive stretch-sensitive fabric was fabricated using graphene oxide nanosheets and a reducing deposition method [138]. Briefly, graphene oxide nanosheets were deposited on nylon/polyurethane fabric, then reduced with sodium borohydride. The results showed that the electrical resistivity of the conductive fabric with a value of 112 kΩ m^{-2} was four times lower than that of the untreated nylon/polyurethane fabric. In addition, the electrical resistivity of the conductive fabric increased from 112 kΩ m^{-2} to 154 kΩ m^{-2} after eight washes. The conductive fabric was also used to fabricate a strain sensor in the strain range of 0 to 30 percent, and the strain sensor exhibited good sensitivity and stability. Ba et al. [139] found a method to improve the bonding between the graphene and the functional group on the cotton fabric using karaya gum as a bioinspired exfoliating agent, in which the synthesized graphene solution was applied to the surface of the cotton fabric by dip coating or brush coating. The electrical conductivity of the conductive cotton fabric reached a maximum value of 13,000 S m^{-1} at a graphene concentration of 6 wt.%. Another interesting method to improve the bonding between graphene and cotton fabric and to fabricate a scalable conductive fabric with a length of 150 m and a sheet resistance of about 11.9 Ω/□ was presented in the study by Afroj et al. [140]. The conductive graphene dispersion was prepared using the microfluidization technique for natural graphite flakes, then applied to the cotton fabric using the pad dry curing method. Another important observation was that the conductivity of the conductive cotton did not change even after washing ten times. The screen printing method [141] was used to print graphene ink on the surface of the textile after the textile was modified using heat transfer technology. The sheet resistance of the conductive textile reached a minimum value of 100 Ω/□ after three printing cycles, and this textile was used to fabricate a conductive electrode for electrocardiogram monitoring. It was found that the efficiency of the graphene electrode was comparable to the conventional electrode. In another study, the screen printing method with graphene ink was also used, but for the two sides of the cotton fabric in the study presented by Zhang et al. [142], and the fabric produced was used as a portable heater (see Figure 9). The small voltage difference of 3 V applied to the conductive fabric resulted in a high heating temperature, 52.6 °C, which confirmed that this conductive fabric could be used as a wearable heater. The conductive cotton fabric also exhibited a high electrical conductivity of 1.18×10^4 S m^{-1}. The biocompatible conductive fabric sensor was fabricated using graphene nanoplatelets dispersed in a water-based ink and screen-printed onto the fabric surface [143]. The results showed that the fabric sensor was stable and sensitive, that the stiffness of the fabric increased with the amount of material applied, and that the electrical conductivity reached the maximum value of about 10.26 S m^{-1} at a graphene concentration of 3.8 wt%.

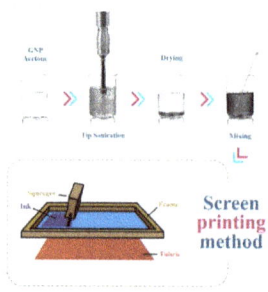

Figure 9. Screen printing process for the production of conductive cotton with graphene ink [143].

In addition, Yapici et al. [144] also fabricated electrocardiogram electrodes based on nylon fabric coated with reduced graphene oxide using the immersion drying method. The resulting electrode had an electrical conductivity of 4.5 S cm^{-1}, and this value was stable up to five washing cycles. In the study by Sahito et al. [145], the surface of a cotton fabric was modified with bovine serum albumin, which resulted in a positive charge on the surface. In their study, the electrical properties of cotton and modified cotton were compared after they were immersed in graphene oxide. The results showed that the amount of graphene oxide in the modified cotton was greater than the amount of graphene oxide in the cotton. Then, the graphene oxide was reduced and converted to graphene by the chemical vapor reduction method. The minimum sheet resistances obtained were 40 Ω/□ and 510 Ω/□ for the conductive cotton and the conductive modified cotton, respectively. The conductive nylon 6 fabric was prepared by depositing reduced graphene oxide on the surface of the fabric, which was presented in the study by Yun et al. [146]. A new method was used in this process: electrostatic self-assembly of graphene oxide and bovine serum albumin to improve the adhesion of graphene sheets on the fabric was used to deposit them on the surface of the fabric, followed by a low-temperature reduction process. The conductive fabric exhibited high electrical conductivity of 1000 S cm^{-1}, which is not affected by bending and washing cycles. In another process [147], UV light was used to reduce graphene oxide on cotton and wool fabrics without using a reducing agent or high annealing temperature. Briefly, graphene oxide was first applied to the surface of the fabric using the brush coating drying method, the process was repeated to increase the concentration of the materials, then the fabric was irradiated with UV light to reduce the graphene oxide and convert it to graphene. It was found that the sheet resistance reached a minimum value of 100.80 kΩ/□ and 45 kΩ/□ when the conductive cotton and wool fabrics were irradiated with UV light, respectively. Table 6 shows summary list of graphene-based materials with details of their manufacturing processes and electrical properties.

Table 6. Summary list of graphene-based materials with details of their manufacturing processes and electrical properties.

Year	Substrate	Coated with	Method	Electrical Properties	References
2015	Cotton	Graphene nanoribbon	Wet coating	80 Ω	[125]
2014	Cotton yarn	Graphene/graphite	Trapping	2.5 kΩ/□	[126]
2016	Cotton	Graphene nanosheets	Dyeing drying	7 Ω	[127]
2017	Cotton	Graphene oxide	Vacuum filtration	0.9 kΩ/□	[128]
2021	Cotton	Graphene oxide	Immersing	22–34Ω	[129]
2021	Cotton	Graphene	Dyeing drying	1.1 S cm^{-1}	[130]
2015	Silk	Graphene oxide	Immersing	3595 S m^{-1}	[131]
2019	Silk	Graphene Oxide	Dip-pad	3.06×10^{-4} S cm^{-1}	[132]
2019	Cotton thread	Graphene oxide	Dip-coating chemical reduction	~1.0 S cm^{-1}	[133]
2022	Gigantea yarn	Graphene oxide	Pad dyeing	6.9 S m^{-1}	[134]
2013	Polyester	Graphene oxide	Chemical reduction	23.15 Ω·cm^2	[135]
2017	knitted	Graphene oxide	Dyeing drying	0.19 MΩ/□	[136]
2019	Para-aramid	Graphene/waterborne/polyurethane	Dip coating	7.5×10^4 Ω/□	[137]
2017	Nylon	Graphene oxide	Dip coating	112 KΩ/m^2	[138]
2020	Cotton	Graphene	Dip- and brush-coated	13000 S m^{-1}	[139]
2020	Poly-cotton	Graphene	Pad dry	11.9 Ω/□	[140]
2020	Cotton	Graphene	Screen-printing	100 Ω/□	[141]
2021	Cotton	Graphene ink	Screen-printing	1.18×10^4 S m^{-1}	[142]
2021	Polyester elastane	Graphene nanoplatelets	Screen-printing	10.26 S m^{-1}	[143]
2015	Nylon	Graphene oxide	Dip coating	4.5 S cm^{-1}	[144]
2015	Cotton	Graphene oxide	Dip coating	40 Ω/□	[145]
2015	Modified cotton	Graphene oxide	Dip coating	510 Ω/□	[145]
2013	Nylon yarns	Graphene oxide	Electrostatic self-assembly and low temperature reduction	1000 S/m	[146]
2014	Cotton	Graphene oxide	Brush coating drying	100.8 kΩ/□	[147]
2014	Wool	Graphene oxide	Brush coating drying	45 kΩ/□	[147]

4. Summary

This review article summarizes the method of designing and fabricating electrical, flexible, and lightweight conductive fabrics with embedded SWCNTs, MWCNTs, and graphene, and their applications in the field of smart textiles. The development of smart textiles was carried out in three stages: the first stage is to impart conductivity to the textiles, the second stage is to fabricate the smart textiles, and the final stage is to functionalize the conductive yarns. Carbon-based materials, particularly SWCNTs, MWCNTs, and graphene, are discussed from their structure, physical properties, and potential applications to their use in the design and fabrication of conductive fabrics with a wide range of electrical conductivity and interesting physical properties that make them suitable for various wearable electronic applications. We come to the following conclusions:

1. Conductive fabrics based on SWCNTs have been prepared by various fabrication methods: chemical vapor deposition, filtration technique, dyeing and drying method, inkjet printing method, dipping and drying method, and drop-casting method. It has

been shown that the electrical conductivity of conductive fabrics and sheet resistance has a wide range from low to high values, and depends on various factors, such as: deposition times, dopants, content of SWCNTs in the fabric, stretching cycles, number of coating layers, treatment of the fabric with plasma, and mixing of SWCNTs with other carbon-based materials, such as MWCNTS and graphene, or with conductive polymers, such as polyaniline and PEDOT:PSS. These SWCNTs based fabrics have been used in various applications such as: UV light shielding, piezoresistive sensor, supercapacitor, antenna, and metal thread.

2. Conductive fabrics based on MWCNTs were prepared by various fabrication methods: Dipping and drying, drop casting and drying, dipping, impregnation and drying, dipping and drying, and tape casting. The fabricated conductive fabrics exhibited a wide range of electrical conductivity, which was influenced by several factors: Size of MWCNTs, number of dipping operations, type of organic solvents, content of MWCNTs, temperature, dopants, repetition of the fabrication process, use of anionic and cationic surfactants, and use of enzymes. These conductive fabrics based on MWCNTs have been used as electrodes in supercapacitors, and as electrodes in dye-sensitised solar cells and block UV light.

3. Conductive graphene-based fabrics have been fabricated using graphene, graphene nanoribbons, graphene oxide, graphene nanosheets, natural graphite flakes, and graphene ink. Various methods were used in the fabrication process: wet coating, trapping method, chemical reduction method, vacuum filtration—hot press method, dipping and gamma ray reduction method, cationization—dip coating, coating and reduction method, dyeing and reduction method, dipping and reduction method, reducing deposition method, brush coating, pad dry curing method, screen printing method, chemical vapor reduction method, electrostatic self-assembly—low temperature reduction process, UV reduction method. The electrical properties of graphene-based materials are influenced by several factors: repetition of the manufacturing process, graphene content, graphene oxide concentration, tensile stress, cationization process, number of coating layers, number of dipping cycles, exfoliation agent, and number of printing cycles. Thus, the conductive fabrics produced have been used in various applications: electrode for a dye-sensitive solar cell, strain sensor, portable supercapacitors, UV blockers, heat-resistant gloves, electrocardiogram electrodes, and portable heaters. Finally, the field of smart textile fabrication from carbon-based materials is developing rapidly but needs further development to bring these applications from small scales (in research laboratories) to large scales (industrial applications). Therefore, more basic research is needed to enable the next wave of smart textile products.

Author Contributions: F.A.A.: Conceptualization, Investigation, Writing—review & editing; G.A.A.: Writing—original draft, Resources. All authors have read and agreed to the published version of the manuscript.

Funding: This research received no external funding.

Institutional Review Board Statement: Not applicable.

Data Availability Statement: Not applicable.

Conflicts of Interest: The authors declare that they have no conflict of interest.

References

1. Van Langenhove, L.; Hertleer, C. Smart clothing: A new life. *Int. J. Cloth. Sci. Technol.* **2004**, *16*, 63–72. [CrossRef]
2. Tao, X. *Smart Technology for Textiles and Clothing–Introduction and Overview*; Woodhead Publishing Ltd.: Cambridge, UK, 2001.
3. Shi, J.; Liu, S.; Zhang, L.; Yang, B.; Shu, L.; Yang, Y.; Ren, M.; Wang, Y.; Chen, J.; Chen, W.; et al. Smart Textile-Integrated Microelectronic Systems for Wearable Applications. *Adv. Mater.* **2019**, *32*, e1901958. [CrossRef] [PubMed]
4. Massaroni, C.; Saccomandi, P.; Schena, E. Medical Smart Textiles Based on Fiber Optic Technology: An Overview. *J. Funct. Biomater.* **2015**, *6*, 204–221. [CrossRef] [PubMed]
5. Dang, T.; Zhao, M. The application of smart fibers and smart textiles. *J. Phys. Conf. Ser.* **2021**, *1790*, 012084. [CrossRef]

6. Kongahage, D.; Foroughi, J. Actuator Materials: Review on Recent Advances and Future Outlook for Smart Textiles. *Fibers* **2019**, *7*, 21. [CrossRef]
7. Hughes-Riley, T.; Dias, T.; Cork, C. A Historical Review of the Development of Electronic Textiles. *Fibers* **2018**, *6*, 34. [CrossRef]
8. Sarif Ullah Patwary, M.S. Smart Textiles and Nano-Technology: A General Overview. *J. Text. Sci. Eng.* **2015**, *5*, 181. [CrossRef]
9. Kandi, S.G.; Tehran, M.A.; Rahmati, M. New method for obtaining proper initial clusters to perform FCM algorithm for colour image clustering. *J. Text. Inst.* **2009**, *100*, 237–244. [CrossRef]
10. Stoppa, M.; Chiolerio, A. Wearable Electronics and Smart Textiles: A Critical Review. *Sensors* **2014**, *14*, 11957–11992. [CrossRef] [PubMed]
11. Lymberis, A.; Paradiso, R. Smart fabrics and interactive textile enabling wearable personal applications: R&D state of the art and future challenges. In Proceedings of the 2008 30th Annual International Conference of the IEEE Engineering in Medicine and Biology Society, Vancouver, BC, Canada, 20–25 August 2008; IEEE: New York, NY, USA, 2008; pp. 5270–5273.
12. Wang, L.; Fu, X.; He, J.; Shi, X.; Chen, T.; Chen, P.; Wang, B.; Peng, H. Application Challenges in Fiber and Textile Electronics. *Adv. Mater.* **2019**, *32*, e1901971. [CrossRef] [PubMed]
13. Foroughi, J.; Mitew, T.; Ogunbona, P.; Raad, R.; Safaei, F. Smart Fabrics and Networked Clothing: Recent developments in CNT-based fibers and their continual refinement. *IEEE Consum. Electron. Mag.* **2016**, *5*, 105–111. [CrossRef]
14. Fernández-Caramés, T.M.; Fraga-Lamas, P. Towards The Internet of Smart Clothing: A Review on IoT Wearables and Garments for Creating Intelligent. *Electronics* **2018**, *7*, 405. [CrossRef]
15. Pu, X.; Hu, W.; Wang, Z.L. Toward Wearable Self-Charging Power Systems: The Integration of Energy-Harvesting and Storage Devices. *Small* **2017**, *14*, 1702817. [CrossRef]
16. Fraga-Lamas, P.; Fernandez-Carames, T.M. Reverse engineering the communications protocol of an RFID public transportation card. In Proceedings of the 2017 IEEE International Conference on RFID, Phoenix, AZ, USA, 9–11 May 2017; IEEE: New York, NY, USA, 2017; pp. 30–35. [CrossRef]
17. Cho, G. *Smart Clothing Technology and Applications*, 1st ed.; CRC Press: Boca Raton, FL, USA, 2009.
18. Schwarz, A.; van Langenhove, L.; Guermonprez, P.; Deguillemont, D. A roadmap on smart textiles. *Text. Prog.* **2010**, *42*, 99–180. [CrossRef]
19. Poole, C.P., Jr.; Owens, F.J. *Introduction to Nanotechnology*; John Wiley: New York, NY, USA, 2003.
20. Iijima, S. Helical microtubules of graphitic carbon. *Nature* **1991**, *354*, 56–58. [CrossRef]
21. Zhang, M.; Li, J. Carbon nanotube in different shapes. *Mater. Today* **2009**, *12*, 12–18. [CrossRef]
22. Dresselhaus, M.; Dresselhaus, G. Nanotechnology in carbon materials. *Nanostruct. Mater.* **1997**, *9*, 33–42. [CrossRef]
23. Nasir, S.; Hussein, M.Z.; Zainal, Z.; Yusof, N.A. Carbon-based nanomaterials/allotropes: A glimpse of their synthesis, properties and some applications. *Materials* **2018**, *11*, 295. [CrossRef]
24. Journet, C.; Bernier, P. Production of Carbon Nanotubes. *Appl. Phys. A* **1998**, *67*, 1–9. [CrossRef]
25. Sahoo, N.G.; Rana, S.; Cho, J.W.; Li, L.; Chan, S.H. Polymer nanocomposites based on functionalized carbon nanotubes. *Prog. Polym. Sci.* **2010**, *35*, 837–867. [CrossRef]
26. Danailov, D.; Keblinski, P.; Nayak, S.; Ajayan, P.M. Bending Properties of Carbon Nanotubes Encapsulating Solid Nanowires. *J. Nanosci. Nanotechnol.* **2002**, *2*, 503–507. [CrossRef] [PubMed]
27. Malik, S.R.; Maqbool, M.A.; Hussain, S.; Irfan, H. Carbon nanotubes: Description, properties and applications. *J. Pak. Mater. Soc.* **2008**, *2*, 21–26.
28. Mamedov, A.A.; Kotov, N.; Prato, M.; Guldi, D.M.; Wicksted, J.P.; Hirsch, A. Molecular design of strong single-wall carbon nanotube/polyelectrolyte multilayer composites. *Nat. Mater.* **2002**, *1*, 190–194. [CrossRef] [PubMed]
29. Purohit, R.; Purohit, K.; Rana, S.; Rana, R.; Patel, V. Carbon Nanotubes and Their Growth Methods. *Procedia Mater. Sci.* **2014**, *6*, 716–728. [CrossRef]
30. Varshney, K. Carbon Nanotubes: A Review on Synthesis, Properties and Applications. *Int. J. Eng. Res. Gen. Sci.* **1991**, *2*, 660–677.
31. Saifuddin, N.; Raziah, A.Z.; Junizah, A.R. Carbon Nanotubes: A Review on Structure and Their Interaction with Proteins. *J. Chem.* **2012**, *2013*, 1–18. [CrossRef]
32. Yakobson, B.I.; Avouris, P. Mechanical properties of carbon nanotubes. In *Carbon Nanotubes*; Springer: Berlin/Heidelberg, Germany, 2001.
33. Tessonnier, J.-P.; Rosenthal, D.; Hansen, T.W.; Hess, C.; Schuster, M.E.; Blume, R.; Girgsdies, F.; Pfänder, N.; Timpe, O.; Su, D.S.; et al. Analysis of the structure and chemical properties of some commercial carbon nanostructures. *Carbon* **2009**, *47*, 1779–1798. [CrossRef]
34. Wei, W.; Sethuraman, A.; Jin, C.; Narayan, R.J. Biological Properties of Carbon Nanotubes Biological Properties of Carbon Nanotubes. *Nanosci. Nanotechnol.* **2007**, *18*, 035201.
35. Hone, J. Carbon Nanotubes: Thermal Properties. *Dekker Encycl. Nanosci. Nanotechnol.* **2004**, *7*, 603–610.
36. Wan, X.; Dong, J. Optical properties of carbon nanotubes. *Phys. Rev.* **1998**, *58*, 6756. [CrossRef]
37. Zhu, J.; Holmen, A.; Chen, D. Carbon Nanomaterials in Catalysis: Proton Affinity, Chemical and Electronic Properties, and their Catalytic Consequences. *ChemCatChem* **2013**, *5*, 378–401. [CrossRef]
38. Collins, P.G.; Avouris, P. Nanotubes for electronics. *Sci. Am.* **2000**, *283*, 62–69. [CrossRef] [PubMed]
39. Abdalla, S.; Al-Marzouki, F.; Al-Ghamdi, A.A.; Abdel-Daiem, A. Different Technical Applications of Carbon Nanotubes. *Nanoscale Res. Lett.* **2015**, *10*, 1–10. [CrossRef] [PubMed]

40. Pitroda, J.; Jethwa, B.; Dave, S.K. A Critical Review on Carbon Nanotubes. *Int. J. Constr. Res. Civ. Eng.* **2016**, *2*, 36–42.
41. Harris, P.J.; Harris, P.J.F. *Carbon Nanotube Science: Synthesis, Properties and Applications*; Cambridge University Press: Cambridge, UK, 2009.
42. Part, C.; Determination, I.; Moduli, E. Transverse Mechanical Properties of Single-Walled Carbon Nanotube. *Compos. Sci. Technol.* **2003**, *63*, 1543–1550.
43. Cao, Q.; Yu, Q.; Connell, D.W. Titania/carbon nanotube composite (TiO_2/CNT) and its application for removal of organic pollutants. *Clean Technol. Environ. Policy* **2013**, *15*, 871–880. [CrossRef]
44. Iijima, S.; Ichihashi, T. Single-shell carbon nanotubes of 1-nm diameter. *Nature* **1993**, *363*, 603–605. [CrossRef]
45. Kostarelos, K.; Lacerda, L.; Partidos, C.; Prato, M.; Bianco, A. Carbon nanotube-mediated delivery of peptides and genes to cells: Translating nanobiotechnology to therapeutics. *J. Drug Deliv. Sci. Technol.* **2005**, *15*, 41–47. [CrossRef]
46. Tiwari, S.K.; Kumar, V.; Huczko, A.; Oraon, R.; de Adhikari, A.; Nayak, G.C. Magical Allotropes of Carbon: Prospects and Applications. *Crit. Rev. Solid State Mater. Sci.* **2016**, *41*, 257–317. [CrossRef]
47. Ago, H.; Petritsch, K.; Shaffer, M.S.; Windle, A.H.; Friend, R.H. Composites of Carbon Nanotubes and Conjugated Polymers for Photovoltaic Devices. *Adv. Mater.* **1994**, *11*, 1281–1285. [CrossRef]
48. Yan, J.; Uddin, M.; Dickens, T.; Okoli, O. Carbon nanotubes (CNTs) enrich the solar cells. *Sol. Energy* **2013**, *96*, 239–252. [CrossRef]
49. Kasumov, A.Y.; Deblock, R.; Kociak, M.; Reulet, B.; Bouchiat, H.; Khodos, I.I.; Gorbatov, Y.B.; Volkov, V.T.; Journet, C.; Burghard, M. Supercurrents through single-wailed carbon nanotubes. *Science* **1999**, *284*, 1508–1511. [CrossRef] [PubMed]
50. Kumar, M.; Ando, Y. A simple method of producing aligned carbon nanotubes from an unconventional precursor–Camphor. *Chem. Phys. Lett.* **2003**, *374*, 521–526. [CrossRef]
51. Long, R.Q.; Yang, R.T. Carbon Nanotubes as Superior Sorbent for Dioxin Removal Richard. *J. Am. Chem. Soc.* **2001**, *123*, 2058–2059. [CrossRef]
52. Wang, X.; Liu, Z. Carbon nanotubes in biology and medicine: An overview. *Chin. Sci. Bull.* **2012**, *57*, 167–180. [CrossRef]
53. Beckett, P. Exploiting Multiple Functionality for Nano-Scale Reconfigurable Systems. In Proceedings of the 13th ACM Great Lakes symposium on VLSI, New York, NY, USA, 28 April 2003.
54. Di, J.; Zhang, X.; Yong, Z.; Zhang, Y.; Li, D.; Li, R.; Li, Q. Carbon-Nanotube Fibers for Wearable Devices and Smart Textiles. *Adv. Mater.* **2016**, *28*, 10529–10538. [CrossRef]
55. Zhang, Y.; Yin, Q. Carbon and other light element contents in the Earth's core based on fi rst-principles molecular dynamics. *Proc. Natl. Acad. Sci. USA* **2012**, *109*, 19579–19583. [CrossRef]
56. Ebbesen, T.W. Carbon nanotubes. *Annu. Rev. Mater. Sci.* **1996**, *24*, 235–264. [CrossRef]
57. Liu, J.; Rinzler, A.G.; Dai, H.; Hafner, J.H.; Bradley, R.K.; Boul, P.J.; Lu, A.; Iverson, T.; Shelimov, K.; Huffman, C.B.; et al. Fullerene pipes. *Science* **1998**, *280*, 1253–1256. [CrossRef]
58. Geim, A.K.; Novoselov, K.S. The rise of graphene. In *Nanoscience and Technology: A Collection of Reviews from Nature Journals*; Macmillan Publishers Ltd., 4-6 Crinan Street: London, UK, 2009.
59. Novoselov, K.S.; Geim, A.K.; Morozov, S.V.; Jiang, D.; Zhang, Y.; Dubonos, S.V.; Grigorieva, I.V.; Firsov, A.A. Electric field effect in atomically thin carbon films. *Science* **2004**, *306*, 666–669. [CrossRef]
60. Shiraishi, M. Spin Injection into Graphene at Room Temperature. *Hyomen Kagaku* **2008**, *29*, 310–314. [CrossRef]
61. Feng, L.; Liu, Z. Graphene in biomedicine: Opportunities and challenges. *Nanomedicine* **2011**, *6*, 317–324. [CrossRef] [PubMed]
62. Liu, L.; Ryu, S.; Tomasik, M.R.; Stolyarova, E.; Jung, N.; Hybertsen, M.S.; Steigerwald, M.L.; Brus, L.E.; Flynn, G.W. Graphene Oxidation: Thickness-Dependent Etching and Strong Chemical Doping. *Nano Lett.* **2008**, *8*, 1965–1970. [CrossRef] [PubMed]
63. Elias, D.C.; Nair, R.R.; Mohiuddin, T.M.G.; Morozov, S.V.; Blake, P.; Halsall, M.P.; Ferrari, A.C.; Boukhvalov, D.W.; Katsnelson, M.I.; Geim, A.K.; et al. Control of Graphene's Properties by Reversible Hydrogenation: Evidence for Graphane. *Science* **2009**, *323*, 610–613. [CrossRef] [PubMed]
64. Mark, F.; Goerbig, O.; Notes, L. Introduction to the Physical Properties of Graphene. *Lect. Notes* **2008**, *10*, 11–12.
65. Neto, A.H.C.; Guinea, F.; Peres, N.M.R.; Novoselov, K.S.; Geim, A.K. The electronic properties of graphene. *Rev. Mod. Phys.* **2009**, *81*, 109. [CrossRef]
66. Papageorgiou, D.G.; Kinloch, I.A.; Young, R.J. Progress in Materials Science Mechanical properties of graphene and graphene-based nanocomposites. *Prog. Mater. Sci.* **2017**, *90*, 75–127. [CrossRef]
67. Si, Y.; Samulski, E.T.; Hill, C.; Carolina, N. Synthesis of Water Soluble Graphene. *Nano Lett.* **2008**, *8*, 1679–1682. [CrossRef]
68. Jishnu, A.; Jayan, J.S.; Saritha, A.; Sethulekshmi, A.S.; Venu, G. Superhydrophobic graphene-based materials with self-cleaning and anticorrosion performance: An appraisal of neoteric advancement and future perspectives. *Colloids Surf. A Physicochem. Eng. Asp.* **2020**, *606*, 125395. [CrossRef]
69. Geim, A.K. Graphene: Status and Prospects. *Science* **2009**, *324*, 1530–1534. [CrossRef]
70. Tu, Y.; Lv, M.; Xiu, P.; Huynh, T.; Zhang, M.; Castelli, M.; Liu, Z.; Huang, Q.; Fan, C.; Fang, H.; et al. Destructive extraction of phospholipids from Escherichia coli membranes by graphene nanosheets. *Nat. Nanotechnol.* **2013**, *8*, 594–601. [CrossRef]
71. Katsnelson, M.I. Graphene: Carbon in two dimensions. *Mater. Today* **2007**, *10*, 20–27. [CrossRef]
72. Adetayo, A.; Runsewe, D. Synthesis and Fabrication of Graphene and Graphene Oxide: A Review. *Open J. Compos. Mater.* **2019**, *9*, 207–229. [CrossRef]
73. Lee, C.; Wei, X.; Kysar, J.W.; Hone, J. Measurement of the elastic properties and intrinsic strength of monolayer graphene. *Science* **2008**, *321*, 385–388. [CrossRef] [PubMed]

74. Radadiya, T. A properties of graphene. *Mater. Sci.* **2015**, *2*, 6–18.
75. Seol, J.H.; Jo, I.; Moore, A.L.; Lindsay, L.; Aitken, Z.H.; Pettes, M.T.; Li, X.; Yao, Z.; Huang, R.; Broido, D.; et al. Two-Dimensional Phonon Transport in Supported Graphene. *Science* **2010**, *328*, 213–216. [CrossRef] [PubMed]
76. Huang, X.; Qi, X.; Zhang, H. Graphene-based composites. *Chem. Soc. Rev.* **2012**, *2*, 666–686. [CrossRef]
77. Pop, E.; Mann, D.; Wang, Q.; Goodson, K.; Dai, H. Thermal Conductance of an Individual Single-Wall Carbon Nanotube above Room Temperature. *Nano Lett.* **2006**, *6*, 96–100. [CrossRef]
78. Kim, P.; Shi, L.; Majumdar, A.; McEuen, P.L. Thermal Transport Measurements of Individual Multiwalled Nanotubes. *Phys. Rev. Lett.* **2001**, *87*, 215502. [CrossRef]
79. Morozov, S.V.; Novoselov, K.S.; Katsnelson, M.I.; Schedin, F.; Elias, D.C.; Jaszczak, J.A.; Geim, A.K. Giant Intrinsic Carrier Mobilities in Graphene and Its Bilayer. *Phys. Rev. Lett.* **2008**, *100*, 016602. [CrossRef]
80. Chandrasekhar, P. *Conducting Polymers, Fundamentals and Applications Including Carbon Nanotubes and Graphene*; Springer: Berlin/Heidelberg, Germany, 2018.
81. Nair, R.R.; Blake, P.; Grigorenko, A.N.; Novoselov, K.S.; Booth, T.J.; Stauber, T.; Peres, N.M.R.; Geim, A.K. Fine Structure Constant Defines Visual Transparency of Graphene. *Science* **2008**, *320*, 1308. [CrossRef] [PubMed]
82. Syama, S.; Mohanan, P.V. *Comprehensive Application of Graphene: Emphasis on Biomedical Concerns*; Springer: Berlin/Heidelberg, Germany, 2019.
83. Song, N.; Gao, X.; Ma, Z.; Wang, X.; Wei, Y.; Gao, C. A review of graphene-based separation membrane: Materials, characteristics, preparation and applications. *Desalination* **2018**, *437*, 59–72. [CrossRef]
84. Nag, A.; Mitra, A.; Chandra, S. Graphene and its sensor-based applications: A review. *Sens. Actuators A Phys.* **2018**, *270*, 177–194. [CrossRef]
85. Li, X.; Wang, Y.; Zhao, Y.; Zhang, J.; Qu, L. Graphene Materials for Miniaturized Energy Harvest and Storage Devices. *Small Struct.* **2022**, *3*, 2100124. [CrossRef]
86. Cui, G.; Bi, Z.; Zhang, R.; Liu, J.; Yu, X.; Li, Z. A comprehensive review on graphene-based anti-corrosive coatings. *Chem. Eng. J.* **2019**, *373*, 104–121. [CrossRef]
87. Wang, R.; Ren, X.-G.; Yan, Z.; Jiang, L.-J.; Sha, W.E.I.; Shan, G.-C. Graphene based functional devices: A short review. *Front. Phys.* **2018**, *14*, 13603. [CrossRef]
88. Xu, W.; Kwok, K.S.; Gracias, D.H. Ultrathin Shape Change Smart Materials. *Accounts Chem. Res.* **2018**, *51*, 436–444. [CrossRef]
89. Kim, S.J.; Choi, K.; Lee, B.; Kim, Y. Materials for Flexible, Stretchable Electronics: Graphene and 2D Materials. *Annu. Rev. Mater. Res.* **2015**, *45*, 63–84. [CrossRef]
90. Hu, X.; Tian, M.; Qu, L.; Zhu, S.; Han, G. Multifunctional cotton fabrics with graphene/polyurethane coatings with far-infrared emission, electrical conductivity, and ultraviolet-blocking properties. *Carbon* **2015**, *95*, 625–633. [CrossRef]
91. Çetiner, S.; Köse, H. A Systematic Study on Morphological, Electrical and Electromagnetic Shielding Performance of Polypyrrole Coated Polyester Fabrics. *Text. Appar.* **2021**, *31*, 111–121. [CrossRef]
92. Jia, Y.; Xin, B. Preparation and Characterization of Polypyrrole-coated Wool Fabric for High Electrical Conductivity. *J. Phys. Conf. Ser.* **2021**, *1790*, 012080. [CrossRef]
93. Zhao, S.Q.; Zheng, P.X.; Cong, H.L.; Wan, A.L. Facile fabrication of flexible strain sensors with AgNPs-decorated CNTs based on nylon/PU fabrics through polydopamine templates. *Appl. Surf. Sci.* **2021**, *558*, 149931. [CrossRef]
94. Mikkonen, J.; Pouta, E. Flexible Wire-Component for Weaving Electronic Textiles. In Proceedings of the 2016 IEEE 66th Electronic Components and Technology Conference (ECTC), Las Vegas, NV, USA, 31 May 2016; IEEE: New York, NY, USA, 2016; pp. 1656–1663.
95. Jia, X.; Tennant, A.; Langley, R.J.; Hurley, W.; Dias, T. A knitted textile waveguide. In Proceedings of the 2014 Loughborough Antennas and Propagation Conference (LAPC), Loughborough, UK, 10 November 2014; IEEE: New York, NY, USA, 2014; pp. 679–682.
96. Zheng, T.; Xu, N.; Kan, Q.; Li, H.; Lu, C.; Zhang, P.; Li, X.; Zhang, D.; Wang, X. Wet-Spinning Assembly of Continuous, Highly Stable Hyaluronic/Multiwalled Carbon Nanotube Hybrid Microfibers. *Polymers* **2019**, *11*, 867. [CrossRef]
97. Tseghai, G.B.; Malengier, B.; Fante, K.A.; Nigusse, A.B.; Van Langenhove, L. Development of a Flex and Stretchy Conductive Cotton Fabric Via Flat Screen Printing of PEDOT:PSS/PDMS Conductive Polymer Composite. *Sensors* **2020**, *20*, 1742. [CrossRef] [PubMed]
98. Alamer, F.A. A simple method for fabricating highly electrically conductive cotton fabric without metals or nanoparticles, using PEDOT:PSS. *J. Alloys Compd.* **2017**, *702*, 266–273. [CrossRef]
99. Wang, F.; Zhao, S.; Jiang, Q.; Li, R.; Zhao, Y.; Huang, Y.; Wu, X.; Wang, B.; Zhang, R. Advanced functional carbon nanotube fibers from preparation to application. *Cell Rep. Phys. Sci.* **2022**, *3*, 100989. [CrossRef]
100. Cao, J.; Zhang, Z.; Dong, H.; Ding, Y.; Chen, R.; Liao, Y. Dry and Binder-Free Deposition of Single-Walled Carbon Nanotubes on Fabrics for Thermal Regulation and Electromagnetic Interference Shielding. *ACS Appl. Nano Mater.* **2022**, *9*, 13373–13383. [CrossRef]
101. Alamer, F.A.; Badawi, N.M.; Alsalmi, O. Preparation and Characterization of Conductive Cotton Fabric Impregnated with Single-Walled Carbon Nanotubes. *J. Electron. Mater.* **2020**, *49*, 6582–6589. [CrossRef]
102. Huang, Y.; Wang, Y.; Gao, L.; He, X.; Liu, P.; Liu, C. Characterization of stretchable SWCNTs/Lycra fabric electrode with dyeing process. *Mater. Electron.* **2017**, *28*, 4279–4287. [CrossRef]

103. Kim, T.; Song, H.; Ha, J.; Kim, S.; Kim, D.; Chung, S.; Lee, J.; Hong, Y. Inkjet-printed stretchable single-walled carbon nanotube electrodes with excellent mechanical properties. *Appl. Phys. Lett.* **2014**, *104*, 113103. [CrossRef]
104. Zhang, W.; Johnson, L.; Silva, S.R.P.; Lei, M. The effect of plasma modification on the sheet resistance of nylon fabrics coated with carbon nanotubes. *Appl. Surf. Sci.* **2012**, *258*, 8209–8213. [CrossRef]
105. Wang, C.; Lv, J.; Ren, Y.; Zhi, T.; Chen, J.; Zhou, Q.; Lu, Z.; Gao, D.; Jin, L. Surface modification of polyester fabric with plasma pretreatment and carbon nanotube coating for antistatic property improvement. *Appl. Surf. Sci.* **2015**, *359*, 196–203. [CrossRef]
106. Mahmoudifard, M.; Safi, M. Novel study of carbon nanotubes as UV absorbers for the modification of cotton fabric. *J. Text. Inst.* **2012**, *103*, 893–899. [CrossRef]
107. Cho, H.-S.; Yang, J.-H.; Lee, J.-H.; Lee, J.-H. Evaluation of Joint Motion Sensing Efficiency According to the Implementation Method of SWCNT-Coated Fabric Motion Sensor. *Sensors* **2020**, *20*, 284. [CrossRef] [PubMed]
108. Pasta, M.; La Mantia, F.; Hu, L.; Deshazer, H.D.; Cui, Y. Aqueous supercapacitors on conductive cotton. *Nano Res.* **2010**, *3*, 452–458. [CrossRef]
109. Badawi, N.M.; Batoo, K.M. Conductive Nanocomposite Cotton Thread Strands for Wire and Industrial Applications. *J. Electron. Mater.* **2020**, *49*, 6483–6491. [CrossRef]
110. Shim, B.S.; Chen, W.; Doty, C.; Xu, C.; Kotov, N.A. Smart Electronic Yarns and Wearable Fabrics for Human Biomonitoring made by Carbon Nanotube Coating with Polyelectrolytes. *Nano Lett.* **2008**, *8*, 4151–4157. [CrossRef]
111. Guo, X.; Huang, Y.; Wu, C.; Mao, L.; Wang, Y.; Xie, Z.; Liu, C.; Zhang, Y. Flexible and reversibly deformable radio-frequency antenna based on stretchable SWCNTs/PANI/Lycra conductive fabric. *Smart Mater. Struct.* **2017**, *26*, 105036. [CrossRef]
112. Alamer, F.A.; Badawi, N.M. Fully flexible, highly conductive Threads based on SWCNTs and PEDOT:PSS. *Adv. Eng. Mater.* **2021**, *23*, 2100448. [CrossRef]
113. Rahman, M.J.; Mieno, T. Conductive Cotton Textile from Safely Functionalized Carbon Nanotubes. *J. Nanomater.* **2015**, *2015*, 1–10. [CrossRef]
114. Alamer, F.A.; Alnefaie, M.A.; Salam, M.A. Preparation and characterization of multi-walled carbon nanotubes-filled cotton fabrics. *Results Phys.* **2022**, *33*, 105205. [CrossRef]
115. Bharath, S.P.; Manjanna, J.; Javeed, A.; Yallappa, S. Multi-walled carbon nanotube-coated cotton fabric for possible energy. *Bull. Mater. Sci.* **2015**, *38*, 169–172. [CrossRef]
116. You, A.; Be, M.A.Y.; In, I. Coating of multi-walled carbon nanotubes on cotton fabric via conventional dyeing for enhanced electrical and mechanical properties. *AIP Conf. Proc.* **2019**, *2142*, 140019.
117. Xu, J.; Zhang, J.Y.; Xu, J.; Chang, Y.; Shi, F.; Zhang, Z.; Zhang, H. Design of functional cotton fabric via modified carbon nanotubes. *Pigment. Resin Technol.* **2020**, *49*, 71–78. [CrossRef]
118. Costa, R.S.; Guedes, A.; Pereira, A.M.; Pereira, C. Fabrication of all-solid-state textile supercapacitors based on industrial-grade multi-walled carbon nanotubes for enhanced energy storage. *J. Mater. Sci.* **2020**, *55*, 10121–10141. [CrossRef]
119. Nafeie, N.; Montazer, M.; Hemmati, N.; Harifi, T. Electrical conductivity of different carbon nanotubes on wool fabric: An investigation on the effects of different dispersing agents and pretreatments. *Colloids Surf. A Physicochem. Eng. Asp.* **2016**, *497*, 81–89. [CrossRef]
120. Kowalczyk, D.; Brzeziński, S.; Makowski, T.; Fortuniak, W. Conductive hydrophobic hybrid textiles modified with carbon nanotubes. *Appl. Surf. Sci.* **2015**, *357*, 1007–1014. [CrossRef]
121. Arbab, A.A.; Sun, K.C.; Sahito, I.A.; Qadir, M.B.; Jeong, S.H. Multiwalled carbon nanotube coated polyester fabric as textile based flexible counter electrode for dye sensitized solar cell. *Phys. Chem. Chem. Phys.* **2015**, *17*, 12957–12969. [CrossRef]
122. Hao, T.; Sun, J.; Wang, W.; Yu, D. MWCNTs-COOH/cotton flexible supercapacitor electrode prepared by improvement one-time dipping and carbonization method. *Cellulose* **2018**, *25*, 4031–4041. [CrossRef]
123. Abbas, S.M.; Sevimli, O.; Heimlich, M.C.; Esselle, K.P.; Kimiaghalam, B.; Foroughi, J.; Safaei, F. Microwave Characterization of Carbon Nanotube Yarns For UWB Medical Wireless Body Area Networks. *IEEE Trans. Microw. Theory Tech.* **2013**, *61*, 3625–3631. [CrossRef]
124. Abbas, S.M.; Foroughi, J.; Ranga, Y.; Matekovits, L.; Esselle, K.; Hay, S.; Heimlich, M.; Safaei, F. Stretchable and Highly Conductive Carbon Nanotube-Graphene Hybrid Yarns for Wearable Systems. *EAI Endorsed Trans. Internet Things* **2015**, *2*, 12–14. [CrossRef]
125. Gan, L.; Shang, S.; Yuen, C.W.M.; Jiang, S.-X. Graphene nanoribbon coated flexible and conductive cotton fabric. *Compos. Sci. Technol.* **2015**, *117*, 208–214. [CrossRef]
126. Woltornist, S.J.; Alhashmi, F.; Mcdannald, A.; Jain, M.; Sotzing, G.A.; Adamson, D.H. Preparation of conductive graphene/graphite infused fabrics using an interface trapping method. *Carbon* **2014**, *81*, 38–42. [CrossRef]
127. Ali, I.; Chul, K.; Ayoub, A.; Bilal, M.; Seon, Y.; Hoon, S. Flexible and conductive cotton fabric counter electrode coated with graphene nanosheets for high ef fi ciency dye sensitized solar cell. *J. Power Sources* **2016**, *319*, 90–98. [CrossRef]
128. Ren, J.; Wang, C.; Zhang, X.; Carey, T.; Chen, K.; Yin, Y.; Torrisi, F. Environmentally-friendly conductive cotton fabric as flexible strain sensor based on hot press reduced graphene oxide. *Carbon* **2017**, *111*, 622–630. [CrossRef]
129. Atta, M.M.; Maksoud, M.I.A.A.; Sallam, O.I.; Awed, A.S. Gamma irradiation synthesis of wearable supercapacitor based on reduced graphene oxide/cotton yarn electrode. *J. Mater. Sci. Mater. Electron.* **2021**, *32*, 3688–3698. [CrossRef]
130. Maneval, L.; Atawa, B.; Serghei, A.; Sintes-Zydowicz, N.; Beyou, E. In situcoupled electrical/mechanical investigations of graphene coated cationized cotton yarns with enhanced conductivity upon mechanical stretching. *J. Mater. Chem. C* **2021**, *9*, 14247–14255. [CrossRef]

131. Lu, Z.; Mao, C.; Zhang, H. Highly conductive graphene-coated silk fabricated via a repeated coating-reduction approach. *Mater. Chem.* **2015**, *3*, 4265–4268. [CrossRef]
132. Zulan, L.; Zhi, L.; Lan, C.; Sihao, C.; Dayang, W.; Fangyin, D. Reduced Graphene Oxide Coated Silk Fabrics with Conductive Property for Wearable Electronic Textiles Application. *Adv. Electron. Mater.* **2019**, *5*, 1–9. [CrossRef]
133. Yang, H.Y.; Jun, Y.; Yun, Y.J. Ultraviolet response of reduced graphene oxide/natural cellulose yarns with high flexibility. *Compos. Part B Eng.* **2019**, *163*, 710–715. [CrossRef]
134. Zhang, J.; Liu, J.; Zhao, Z.; Huang, D.; Chen, C.; Zheng, Z.; Fu, C.; Wang, X.; Ma, Y.; Li, Y.; et al. A facile scalable conductive graphene-coated Calotropis gigantea yarn. *Cellulose* **2022**, *29*, 3545–3556. [CrossRef]
135. Molina, J.; Fernández, J.; Inés, J.C.; Río, A.I.; Bonastre, J.; Cases, F. Electrochemical characterization of reduced graphene oxide-coated polyester fabrics. *Electrochim. Acta* **2013**, *93*, 44–52. [CrossRef]
136. Chatterjee, A.; Kumar, M.N.; Maity, S. Influence of graphene oxide concentration and dipping cycles on electrical conductivity of coated cotton textiles. *J. Text. Inst.* **2017**, *108*, 1910–1916. [CrossRef]
137. Kim, H.; Lee, S.; Kim, H. Electrical Heating Performance of Electro-Conductive Para-aramid Knit Manufactured by Dip-Coating in a Graphene/Waterborne Polyurethane Composite. *Sci. Rep.* **2019**, *9*, 1511. [CrossRef] [PubMed]
138. Cai, G.; Yang, M.; Xu, Z.; Liu, J.; Tang, B.; Wang, X. Flexible and wearable strain sensing fabrics. *Chem. Eng. J.* **2017**, *325*, 396–403. [CrossRef]
139. Ba, H.; Truong-Phuoc, L.; Papaefthimiou, V.; Sutter, C.; Pronkin, S.; Bahouka, A.; Lafue, Y.; Nguyen-Dinh, L.; Giambastiani, G.; Pham-Huu, C. Cotton Fabrics Coated with Few-Layer Graphene as Highly Responsive Surface Heaters and Integrated Lightweight Electronic-Textile Circuits. *ACS Appl. Nano Mater.* **2020**, *3*, 9771–9783. [CrossRef]
140. Afroj, S.; Tan, S.; Abdelkader, A.M.; Novoselov, K.S.; Karim, N. Highly Conductive, Scalable, and Machine Washable Graphene-Based E-Textiles for Multifunctional Wearable Electronic Applications. *Adv. Funct. Mater.* **2020**, *30*, 2000293. [CrossRef]
141. Xu, X.; Luo, M.; He, P.; Yang, J. Washable and Flexible Screen Printed Graphene Electrode on Textile for Wearable Healthcare Monitoring. *J. Phys. D Appl. Phys.* **2020**, *53*, 125402. [CrossRef]
142. Zhang, Y.; Ren, H.; Chen, H.; Chen, Q.; Jin, L.; Peng, W.; Xin, S.; Bai, Y. Cotton Fabrics Decorated with Conductive Graphene Nanosheet Inks for Flexible Wearable Heaters and Strain Sensors. *ACS Appl. Nano Mater.* **2021**, *4*, 9709–9720. [CrossRef]
143. Marra, F.; Minutillo, S.; Tamburrano, A.; Sarto, M.S. Production and characterization of Graphene Nanoplatelet-based ink for smart textile strain sensors via screen printing technique. *Mater. Des.* **2021**, *198*, 109306. [CrossRef]
144. Yapici, M.K.; Alkhidir, T.; Samad, Y.A.; Liao, K. Graphene-clad textile electrodes for electrocardiogram monitoring. *Sens. Actuators B Chem.* **2015**, *221*, 1469–1474. [CrossRef]
145. Ali, I.; Chul, K.; Ayoub, A.; Bilal, M.; Hoon, S. Integrating high electrical conductivity and photocatalytic activity in cotton fabric by cationizing for enriched coating of negatively charged graphene oxide. *Carbohydr. Polym.* **2015**, *130*, 299–306.
146. Yun, Y.J.; Hong, W.G.; Kim, W.-J.; Jun, Y.; Kim, B.H. A Novel Method for Applying Reduced Graphene Oxide Directly to Electronic Textiles from Yarns to Fabrics. *Adv. Mater.* **2013**, *25*, 5701–5705. [CrossRef]
147. Javed, K.; Galib, C.; Yang, F.; Chen, C.-M.; Wang, C. A new approach to fabricate graphene electro-conductive networks on natural fibers by ultraviolet curing method. *Synth. Met.* **2014**, *193*, 41–47. [CrossRef]

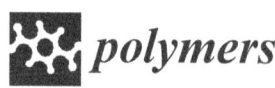

Article

Advances in Rubber Compounds Using ZnO and MgO as Co-Cure Activators

Md Najib Alam, Vineet Kumar and Sang-Shin Park *

School of Mechanical Engineering, Yeungnam University, 280, Daehak-ro, Gyeongsan 38541, Republic of Korea
* Correspondence: pss@ynu.ac.kr

Abstract: Zinc oxide performs as the best cure activator in sulfur-based vulcanization of rubber, but it is regarded as a highly toxic material for aquatic organisms. Hence, the toxic cure activator should be replaced by a non-toxic one. Still, there is no suitable alternative industrially. However, binary activators combining ZnO and another metal oxide such as MgO can largely reduce the level of ZnO with some improved benefits in the vulcanization of rubber as investigated in this research. Curing, mechanical, and thermal characteristics were investigated to find out the suitability of MgO in the vulcanization of rubber. Curing studies reveal that significant reductions in the optimum curing times are found by using MgO as a co-cure activator. Especially, the rate of vulcanization with conventional 5 phr (per hundred grams) ZnO can be enhanced by more than double, going from 0.3 Nm/min to 0.85 Nm/min by the use of a 3:2 ratio of MgO to ZnO cure activator system that should have high industrial importance. Mechanical and thermal properties investigations suggest that MgO as a co-cure activator used at 60% can provide 7.5% higher M100 (modulus at 100% strain) (0.58 MPa from 0.54 MPa), 20% higher tensile strength (23.7 MPa from 19.5 MPa), 15% higher elongation at break (1455% from 1270%), 68% higher fracture toughness (126 MJ/m^3 from 75 MJ/m^3), and comparable thermal stability than conventionally using 100 % ZnO. Especially, MgO as a co-cure activator could be very useful for improving the fracture toughness in rubber compounds compared to ZnO as a single-site curing activator. The significant improvements in the curing and mechanical properties suggest that MgO and ZnO undergo chemical interactions during vulcanization. Such rubber compounds can be useful in advanced tough and stretchable applications.

Keywords: rubber; cure activators; cross-linking; synergism; tensile properties; thermal properties

---

Citation: Alam, M.N.; Kumar, V.; Park, S.-S. Advances in Rubber Compounds Using ZnO and MgO as Co-Cure Activators. *Polymers* **2022**, *14*, 5289. https://doi.org/10.3390/polym14235289

Academic Editor: Changwoon Nah

Received: 10 November 2022
Accepted: 2 December 2022
Published: 3 December 2022

Copyright: © 2022 by the authors. Licensee MDPI, Basel, Switzerland. This article is an open access article distributed under the terms and conditions of the Creative Commons Attribution (CC BY) license (https://creativecommons.org/licenses/by/4.0/).

1. Introduction

The discovery of rubber mastication in the year 1821 by Hancock and the rubber vulcanization by Charles Goodyear in the year 1839 revolutionized the industrial utility of rubbers. Without vulcanization, rubbers remain stiff in cold weather and sticky in hot weather, which restricts their industrial applications. Charles Goodyear first started the vulcanization of rubber by simply heating rubber with sulfur. With this process, vulcanization takes a longer time and is currently uneconomic for industrial applications. In modern vulcanization systems, many ingredients have been used for the vulcanization of rubber. Among them, sulfur, accelerator, and activator are the basic ingredients. With these vulcanizing ingredients, cure activators play an important role in regenerating precursors that can effectively cross-link the rubber. A combination of metallic oxide and fatty acid acts as a cure activator. Currently, the combination of zinc oxide and stearic acid is the most successful cure activator system in the rubber industries. Generally, zinc oxide at 2 to 5 phr (per hundred grams of rubber) and stearic acid at 0.5 to 3 phr are used as cure activators in the vulcanization of rubber [1,2]. Mostly, 5 phr zinc oxide along with 2 phr stearic acid is the conventional amount in the tire industries to achieve a better modulus, low heat build-up, and good abrasion resistance properties [3]. Zinc oxide in rubber compound also acts as an adhesion promoter between the interfaces of brass-coated steel cords and the rubber in radial tires [4]. About 10^5 tons of zinc oxide are produced annually, of which 50–60% is used in rubber industries [5]. Zinc oxide is well known as a high carcinogen for aquatic organisms and thus an environmental pollutant. Importantly, soluble zinc compounds are toxic to aquatic species [6]. The amount of zinc oxide can be reduced by using higher surface-active zinc oxide such as nano zinc oxide. However, recent toxicological studies suggest that even nano zinc oxide is more toxic directly or by dissolution than zinc ion or some combination thereof [7,8]. When rubber products are thrown into the environment after the end of their use, zinc oxide releases into the environment during degradation or by leaching from the landfill sites. Zinc oxide release to the environment by leaching should enhance the ecosystem exposure, even though it is difficult to measure. Modeling efforts suggest that the zinc oxide content presently is high in wastewater treatment plant effluent and can cause toxicological risk to aquatic species [9]. Another major source of zinc oxide in the environment was detected from tire wear during the service life [10]. Hence, to relieve the environmental pollution, either an alternative cure activator that is less toxic or at least the amount of zinc oxide should be reduced in the rubber formulation.

To solve this environmental issue, the reduction of zinc oxide amount was first considered through zinc-based materials [5,11–16], because zinc ion is almost necessary for the formation of a zinc-accelerator complex that can effectively cross-link the rubber chains. For example, layered double hydroxides or zinc-containing clays can effectively reduce the level of zinc oxide [5,11–13], but they are less dispersible in the pristine form to the non-polar rubber and result in reduced physical properties which are economically unfavorable [17]. Some researchers found some zinc complexes [18–21] might be the alternative, but in some selective rubbers. Nono zinc oxide or nano zinc hydroxide [22–30] could be the alternative but are relatively expensive. Wu et al. investigated carbon nanodots as an alternative eco-friendly cure activator for sulfur-based rubber vulcanization and found some promising results in diene rubbers [31]. Other metal oxides such as CaO, MgO, CdO, CuO, PbO, and NiO can be used as cure activators. Among different metal oxides, MgO is the most promising candidate [32–35] because of its non-toxicity. Unlike other basic metal oxides, the hydrolyzed form of MgO is also solid and has a negligible effect on rubber plasticity. Some attempts have already been made using nano magnesium oxide-based cure activator in the vulcanization of rubber [36–39], and the improvements in the properties are mainly due to nanoparticle reinforcement. Generally, nano cure activators are expensive due to complicated synthetic procedures, and also according to Ding et al. metal oxide nanoparticles are more toxic for aquatic organisms than conventional microparticles [40]. However, according to Kuschner et al. [41], micro magnesium oxide has very low environmental toxicity compared to zinc oxide. Tire industries always have a high demand for the simplest, easiest, and most economical way to reduce production costs with enhanced properties. Although magnesium oxide did not provide similar properties to zinc oxide, a significant improvement was possible by using binary accelerators instead of single-site curing accelerators [42]. In particular, binary accelerators, one of which contains zinc ions, can show synergistic effects on the vulcanization properties [42]. Hence, it is believed that MgO-only can, with some difficulty, replace the ZnO by using a single accelerator for the vulcanization of rubber. It was also concluded that MgO can undergo a reaction similar to ZnO with the vulcanizing accelerators. However, the lower cross-linking capacity of MgO could be due to a lack of active sulfurating complex, as was evident in the ZnO-based cure activator [42]. From different studies [11–16,31,32,38,42], it was revealed that partial or complete replacement of ZnO could be possible depending upon the purpose of application. For tire application, complete replacement of ZnO is quite impossible because the tires need the higher modulus and other advantages that cannot be achieved without ZnO. It was quite familiar that binary accelerators comprising thiuram and thiazole functional groups undergo mutual activity to produce higher vulcanization properties such as improved cross-link density, mechanical modulus, etc., than single accelerator systems in the presence of a zinc oxide-based cure activator [43–47]. Magnesium oxide as a cure activator also provided synergism on the vulcanization properties, but it was quite low due to the formation of a lower amount of cross-linking precursors [42]. However, in the presence of the zinc ion-containing accelerator, the synergistic activity was higher. Hence, it was believed that zinc compounds either formed in-situ or externally added may undergo interactions with MgO and can deliver the synergistic effect. Generally, lower zinc oxide can be useful in practice for a binary accelerators system where the accelerators undergo mutual interactions to produce higher cross-link density compared to single accelerator systems [42]. Thus, Guzmán et al. [33–35] studied the efficiency of single MgO and Zn/Mg oxide nanoparticles in reducing the amount of ZnO in the vulcanization of rubber. Interestingly, they found that mixed metal oxide nanoparticles improved the cross-link density and the rate of vulcanization compared to single activator systems [33–35]. However, instead of binary accelerators, they considered single accelerators for the vulcanization [33–35]. Recently, Alam et al. [42] found that, similar to ZnO, MgO could also promote the mutual interactions between thiuram- and thiazole-based accelerator systems to enhance the cross-link densities. Hence MgO as a co-cure activator with ZnO could be more effective in the binary accelerator systems than in single accelerator systems to reduce the amount of ZnO from the vulcanization. Moreover, MgO is quite cheap, nontoxic, and abundant. While most studies were done to reduce ZnO levels by nano cure activators or modified zinc compounds that could have additional toxicity, here we use conventional micro MgO, which is non-toxic. In this way, the rubber compounds are expected to have much less environmental toxicity. Since vulcanization is almost necessary for all practical applications with better properties, we use binary accelerators rather than single accelerators to obtain higher vulcanization properties followed by synergism with MgO.

In this article, we investigate MgO as a co-cure activator along with ZnO in the vulcanization of natural rubber. Low sulfur and a binary accelerators system are chosen to understand the mutual interactions between the accelerators and the activators. Moreover, a high accelerator to low sulfur ratio that is known as an efficient vulcanization system (EV), which can produce higher mono and disulfide cross-links with better thermal stability than other vulcanizing systems such as conventional vulcanization (CV) and semi-efficient vulcanization (SEV), is considered. Detailed curing, mechanical and thermal properties are investigated to establish the utility of MgO as a co-cure activator. Special attention is given to the fracture toughness of the rubber compounds, since this property is highly important for stretchable mechanical and electronic devices. Possible mechanisms of chemical interactions between the cure activators causing the synergistic activities on the properties are proposed and discussed in detail.

2. Materials and Methods
2.1. Materials

Zinc oxide, stearic acid, sulfur, and natural rubber (NR, RSS-3, density = 0.96 g/cm^3) were supplied by the Thai Rubber Research Institute. Cure accelerators such as tetramethyl thiuram disulfide (TMTD) and dibenzothiazyl disulfide (MBTS) were purchased from Tokyo Chemical Industry Co., Ltd., Japan. Magnesium oxide light of fine powder was purchased from AppliChem PanReac, Thailand. X-ray studies of ZnO and MgO confirmed their excellent purities with hexagonal and cubic crystalline structures respectively. From the X-ray studies and using the Scherrer equation, it is confirmed that MgO bears a lower crystalline size compared to ZnO. The XRD plots of ZnO and MgO are given in Figure 1a,b.

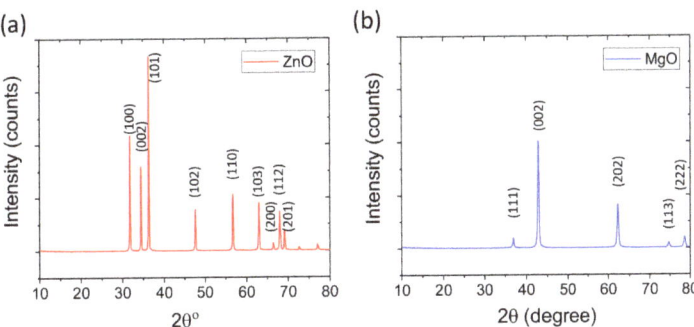

Figure 1. XRD plots of used cure activators; (**a**) ZnO and (**b**) MgO.

2.2. Rubber Compounding

Natural rubber was first masticated in a laboratory size two-roll mill for 5 min to promote additives dispersion. After that, cure activator(s) and stearic acid were mixed for another 5 min. Finally, accelerators and sulfur were mixed for the last 5 min. After complete mixing, the compounded rubbers were cut as sheets. The friction ratio of the front and rear roller was maintained at a 1.2:1 ratio with roller speeds of 24 and 20 rpm, respectively. The nip gap between the two rollers was kept at 1 mm during mastication and the rest at 0.5 mm. The different mixing ingredients with the mixing formulations are provided in Table 1.

Table 1. Mixing composition of different ingredients in phr (per hundred gram of rubber).

Formulation	Mixing Composition				
	NR/5-MgO	NR/4-MgO/1-ZnO	NR/3-MgO/2-ZnO	NR/2-ZnO	NR/5-ZnO
NR	100	100	100	100	100
MgO	5	4	3	0	0
ZnO		1	2	2	5
Stearic Acid	2	2	2	2	2
TMTD	0.72	0.72	0.72	0.72	0.72
MBTS	2	2	2	2	2
Sulfur	0.5	0.5	0.5	0.5	0.5

2.3. Measurements of Cure Characteristics

About 5 g of compounded rubber was placed in the cavity of a Moving Die Rheometer (MDR) to measure the curing characteristics at 140 °C. Moderate vulcanization temperature was used because some curing reactions were very fast. The MDR provided rheographs (torque vs. time curve) from which the different curing parameters were obtained. The different curing parameters such as lowest torque (M_L), highest torque (M_H), torque difference ($M_H - M_L = \Delta$ torque), scorch safety time (t_2), optimum curing time (t_{90}), cure rate index (CRI = $100/(t_{90} - t_2)$), and rate of vulcanization ($R_v = (Mt_{90} - Mt_2)/(t_{90} - t_2)$) were obtained from the rheographs.

2.4. Measurement of Cross-link Density of Rubber Vulcanizates

The cross-link densities of the vulcanized rubbers were determined by the swelling method and by applying the Flory-Rehner equation [48]. The rubber specimens were kept for swelling in toluene for 7 days to reach equilibrium swelling, and the cross-link densities were obtained as follows

$$V_c = -\{\ln(1 - V_r) + V_r + \chi V_r^2\}/\{V_s d_r (V_r^{1/3} - V_r/2)\}$$

where V_c is the cross-link density, V_r is the volume fraction of rubber in the equilibrium-swollen specimen, V_s is the molar volume of solvent (toluene), d_r is the density of the rubber (0.96 g/cm^3), and χ is the solvent-rubber interaction parameter.

The volume fraction of rubber in the equilibrium swelling stage was determined according to the formula

$$V_r = (W_r/d_r)/\{(W_r/d_r) + (W_s/d_s)\}$$

where W_r is the weight of dry rubber, W_s is the weight of solvent swelled, and d_s is the density of solvent. In this experiment $\chi = 0.3795$, $V_s = 106.2$ cm^3/mol, and $d_s = 0.87$ g/cm^3 were considered.

2.5. Mechanical Properties of Rubber Vulcanizates

The rubber compounds were cured corresponding to their t_{90} values in a hot press molding machine at 100 psi as sheets of 2 mm thickness. The cured rubber sheets were placed in a refrigerator to control the aging effects [45] due to the drastic changes in environmental temperature. Before measuring the tensile properties, the sheets were placed at ambient temperature for 24 h. Dumbbell-shaped test pieces were cut from the sheets according to the standard (ISO 37, type 2) to measure the tensile properties. The tensile tests were performed in a tensile testing machine (UTM, LLOYD LR 100 K, Lloyd Instruments, Hampshire, UK) using a 1 kN load cell and a cross-head speed of 500 mm/min. The gauge length of the dumbbell-shaped specimen was fixed at 25 mm. Tensile properties such as modulus at 100% elongation (M100), modulus at 300% elongation (M300), tensile strength (T.S), and elongation at break (E.B) were obtained from the stress-strain curves.

2.6. Scanning Electron Microscopic Analysis and Elemental Mapping

Scanning electron microscopic (SEM) analyses were performed on the tensile fractured surface of rubber samples by field emission scanning electron microscope (FE-SEM, S-4800, Hitachi, Tokyo, Japan). Elemental mapping to understand the dispersion of the curatives was achieved through the energy-dispersive X-ray spectroscopic technique. Before SEM analyses, the samples were pre-coated with platinum by a sputter coater.

2.7. Thermo Gravimetric Analysis

Thermal analyses were performed using a Thermo gravimetric analyzer (TGA, NETZSCH TG 209F3 TGA209F3A-0364-L) and heating the samples from 35 to 800 °C in a nitrogen environment at a heating rate of 10 °C/min in a crucible made of alumina.

3. Results and Discussion

3.1. Curing Characteristics

Cure curves (rheographs) of different rubber vulcanizates are provided in Figure 2a. From the rheographs, it is found that the nature of all the cure curves is more or less similar except for the vulcanizate with an MgO single activator system. After reaching the highest torque, MgO-only-based vulcanizate showed substantial reversion in the curing process. This suggests that MgO itself acts as a poor cure activator compared to ZnO alone. It is recognized that MgO may break the polysulfide linkages but is unable to reform as stable sulfur cross-links. However, the reversion can be completely removed by using binary curing activator systems. It is known that ZnO in the presence of a curing accelerator and sulfur produces an active sulfurating complex [49], whereas MgO does not produce such a type of active complex but it can decompose accelerators quickly and start the vulcanization early, as seen in Figure 2a. Moreover, MgO can decompose the poly-sulfidic bridges and suppress the total number of cross-links with increasing cure time. In the presence of ZnO and MgO combined cure activators, the vulcanization starts earlier, as does higher torque without reversion. Hence, it can be assumed that MgO as a co-cure activator may help to produce a zinc-accelerators complex more effectively than ZnO alone and that the degradation of sulfur cross-links caused by MgO can be suppressed completely. Since MgO can degrade higher-ranked sulfur bridges, a high accelerator-to-sulfur ratio, i.e., efficient vulcanization, should be the proper choice, rather than semi-efficient and conventional vulcanization systems [34,35,42] to get the benefits of MgO as a co-cure activator.

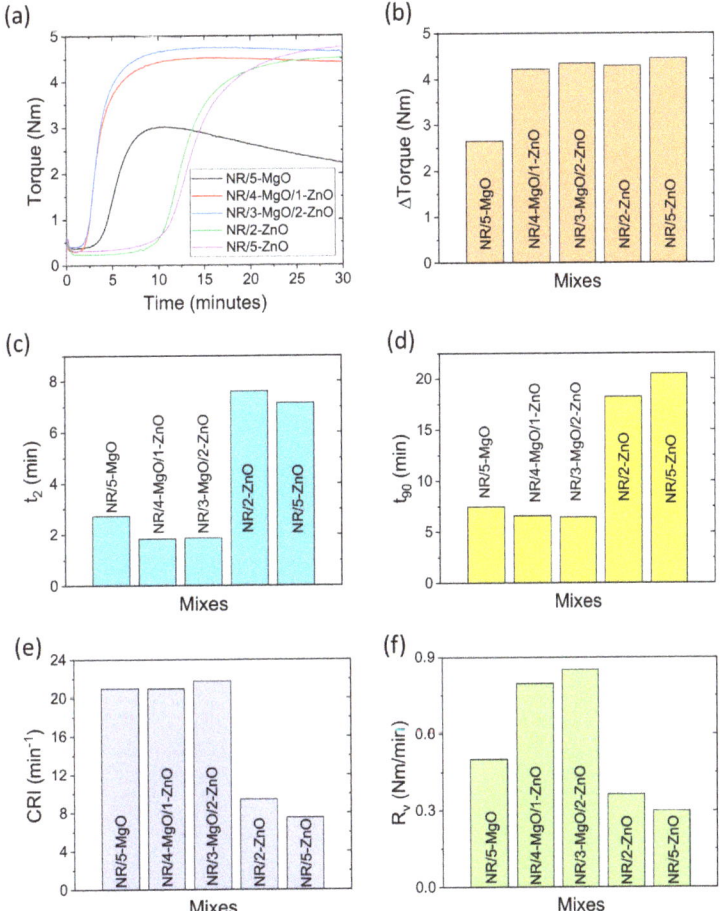

Figure 2. Curing characteristics of rubber vulcanizates; (**a**) rheographs, (**b**) Δ torque, (**c**) t_2, (**d**) t_{90}, (**e**) CRI, and (**f**) R_v.

It seems that the M_L values are a little higher for MgO-based vulcanizates, which may be due to the faster rate of vulcanization. However, such little increases in the M_L values did not affect the flowing properties of compounded rubber during molding. The M_H values are increased with ZnO content in binary activator systems. The NR/3-MgO/2-ZnO shows an M_H value very near to that of NR/5-ZnO. This result suggests that, concerning M_H value, 60% ZnO can be replaced by MgO easily.

The torque differences for different vulcanizates are plotted in Figure 2b. From this figure, it can be found that the 3:2 ratio of MgO to ZnO in the binary activator system (NR/3-MgO/2-ZnO) provides similar Δ torque compared with 5 phr ZnO containing vulcanizate

(NR/5-ZnO). A considerable increase in Δ torque from NR/5-MgO to NR/4-MgO/1-ZnO vulcanizate indicates that 80% ZnO can also be replaced by compromising a little lower value than that of the NR/5-ZnO vulcanizate. If we compare the Δ torque values of binary activator systems with single activator systems, prominent synergisms can be seen, which suggests that MgO may further activate the effect of ZnO on the cross-linking reactions.

Scorch safety is an important parameter for thick vulcanizate. The different scorch safety times are shown in Figure 2c. From this figure, it can be found that MgO-based vulcanizate bears a lower scorch safety value than ZnO-based vulcanizate. Actually, in the presence of MgO, the accelerators began to decompose and started cross-linking reactions much faster. On the other hand, ZnO reacts quite slowly with accelerators to form the active sulfurating complex, and then the cross-linking reaction starts at a faster rate. Hence, ZnO-only shows a higher scorch safety value than other compounds. Although MgO as a co-activator provides lower scorch safety values, these values can be improved by controlling the rate of vulcanization using a lower vulcanization temperature.

Optimum cure time is an important curing parameter, of which values below and above can result in lower vulcanizate properties. The different optimum cure times are shown in Figure 2d. From this figure, it can be seen that optimum cure times for MgO-based vulcanizates are lower compared to ZnO-only-based vulcanizates. Interestingly, the optimum cure time for 3 phr MgO as a co-cure activator is about 3 times lower than 5 phr ZnO containing vulcanizate, keeping similar Δ torque values. This implies that MgO as a co-cure activator with ZnO will be much more economical for rubber vulcanization than using ZnO-only as a single-site curing activator.

The cure rate index (CRI) values are plotted in Figure 2e. From this figure, it can be seen that MgO-based compounds have higher CRI values than ZnO-based single activator systems. It is also to be noted that a slight decrease in the CRI value can be found with increasing 2 phr to 5 phr ZnO content. From the equation, it can be seen that the CRI value depends on both scorch safety and optimum cure times. Since ZnO as a single-site curing activator poses a higher optimum cure time, the composites showed lower CRI values than MgO-based vulcanizates. The actual rate of vulcanization (R_v) can be found in Figure 2f. Interestingly, although the MgO-only compound has a CRI value similar to that of the MgO/ZnO binary activators system, huge differences have been found in vulcanization rates among them. Moreover, the R_v value of the MgO-only-based activator system has a higher value than ZnO-only-based activator systems. These results suggest that MgO can improve the vulcanization rate, but the total cross-linking level is quite low, as is evident from the lower Δ torque. However, in the cases of MgO/ZnO binary curing activator systems, an increase in ZnO content increases R_v as well as Δ torque values. It is believed that in the presence of MgO, the accelerators decompose at faster rates, which increases the rate of vulcanization, and the decomposition of accelerators helps to produce a higher amount of zinc-accelerator complex, which efficiently vulcanizes the rubber.

The swelling index and cross-link density data are plotted in Figure 3a,b. From Figure 3a,b, it can be seen that the MgO/ZnO binary and ZnO-only cure activator systems provide a similar swelling index and cross-link density values. Moreover, MgO/ZnO binary activators provide much better curing efficiency than ZnO-only as a single-site curing activator in the vulcanization of rubber. MgO-only provides the highest swelling index and lowest cross-link density. The highest swelling index and low cross-link density suggest that MgO-only has poor efficiency in cross-linking.

The vital step in vulcanization is the formation of a metal complex combining the activator and the accelerator [50,51]. It is well known that zinc-dithiocarbamate is an ultrafast accelerator compared to corresponding thiuram disulfide [42]. It is well accepted that in the presence of ZnO, the thiuram-type accelerator forms a complex like zinc-dithiocarbamate [42,49–51]. This zinc-dithiocarbamate is then processed to an active sulfurating complex in the presence of sulfur and ultimately undergoes cross-linking. The detailed mechanistic aspects of sulfur vulcanization can be found in Figure 4 (steps 1–7) in the presence of ZnO and cure accelerators. Steps 1–5 (Figure 4) regard the formation of thiocarbamic acid, which undergoes an acid-base type reaction with ZnO. Since thiocarbamic acid is regarded as a weak acid and ZnO is a weak base, the formation of the zinc-dithiocarbamate type complex is very favorable. This dithiocarbamate undergoes an ionic exchange reaction with MBTS and again forms thiuram disulfide and produces the cross-links in steps 4 and 5 (Figure 4). It is believed that in the presence of ZnO and thiuram disulfide, a catalytic-type complex is formed. In the presence of sulfur, this active complex produces more and more cross-links between the rubber chains. It is believed that after cross-linking, the complexes return to dithiocarbamate, which itself has low cross-linking efficiency [44]. However, the presence of an oxidizing reagent, such as a secondary accelerator, can convert this dithiocarbamate to a more active in-situ thiuram disulfide [44]. It was observed that the rate of vulcanization was greater when thiuram disulfide was formed in-situ rather than being added externally [44]. This result suggests that, to improve the kinetics of vulcanization, in-situ conversion of dithiocarbamate to thiuram disulfide should be preferable.

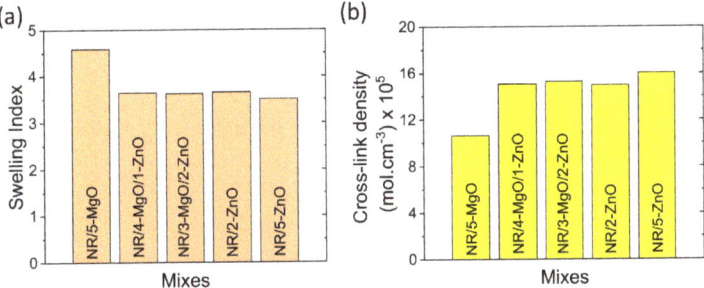

Figure 3. (a) Swelling index and (b) cross-link density of rubber vulcanizates.

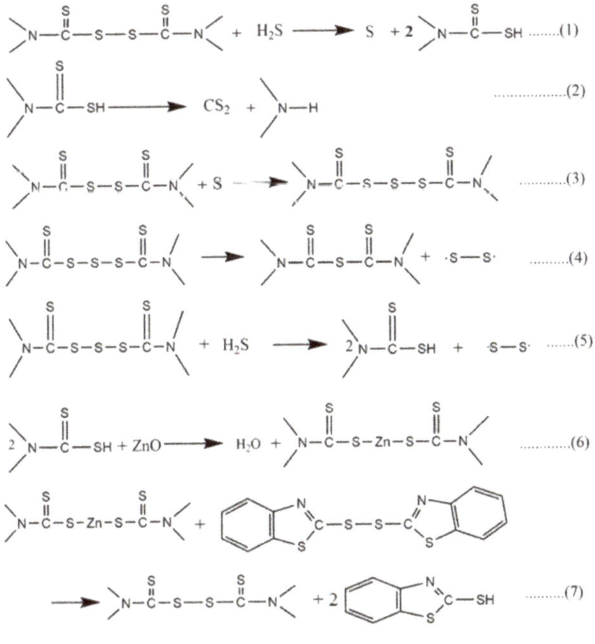

Figure 4. Mutual interaction of thiuram- and thiazole-based accelerators in presence of ZnO cure activator.

The different possible steps in the presence of an MgO/ZnO binary curing activator are given in Figure 5. A similar type of zinc-dithiocarbamate, magnesium-dithiocarbamate can be formed in step 1 (Figure 5). This dithiocarbamate can undergo synergism with MBTS to produce higher cross-link density compared to single accelerator systems [42]. However, the number of cross-links achieved based on MgO as a single-site curing activator is much lower than based on ZnO as a single-site curing activator. It is believed that in the presence of MgO as a cure activator, a higher amount of magnesium-dithiocarbamate is formed; however, the resulting compound has no catalytic activity regarding the final cross-links. However, in the presence of ZnO, the magnesium-dithiocarbamate readily converts the zinc-accelerator complex of higher reactivity. Depending upon the amount of ZnO, the amount of conversion is also varied. The beauty of this conversion is that it produces a higher amount of zinc-accelerator complex at a higher rate than ZnO-only. In the presence of MgO, due to the higher basicity of MgO, the dithiocarbamate may produce at a faster rate but it has no catalytic activity with sulfur to form an active sulfurating complex, and hence it only enhances the rate of vulcanization. In this perspective, MgO can be assumed as a simple base that stabilizes the thiocarbamic acid and protects it from thermal decomposition (Figure 4, step 2). In the presence of the MgO/ZnO binary curing activator, the magnesium-dithiocarbamate finally converts to zinc-dithiocarbamate (Figure 5, step 3) which improves the cross-linking efficiency as well as the rate of vulcanization. From the concept of the acid-base theory of weak acids and weak bases, the reaction in step 3 (Figure 5) is highly feasible.

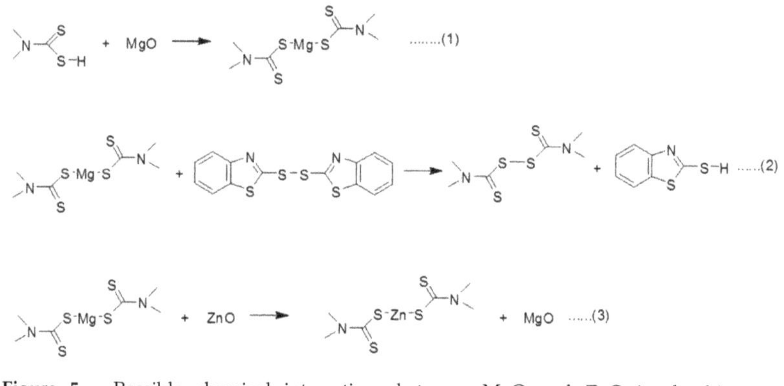

Figure 5. Possible chemical interactions between MgO and ZnO in the binary curing activators systems.

3.2. Tensile Mechanical Properties

The different tensile mechanical properties are plotted in Figure 6a–f. The most representative of the average stress-strain curves is provided in Figure 6a for different vulcanizates. From this figure, we can roughly say that MgO-only as a cure activator provides a lower overall modulus and higher elongation at break values compared to other vulcanizates. The specific modulus such as M100 and M300 in Figure 6b,c have similar trends for all the vulcanizates. It can be noted that the MgO/ZnO binary activator at a 3:2 ratio of MgO to ZnO provides the best modulus values (0.58 MPa in M100 and 1.36 MPa in M300) among the compounds. Regarding tensile strength in Figure 6d, the MgO/ZnO binary activator at a 4:1 ratio of MgO to ZnO provides the highest tensile strength (24.8 MPa). Interestingly, it can be noted that MgO-only as a cure activator can achieve a similar tensile strength value compared to ZnO as a single-site curing activator system. The elongation at break values for different compounds is shown in Figure 6e. From this figure, it can be seen that elongation at break values decreased with the increase of ZnO content in the binary activator systems. Fracture toughness is an important mechanical property that is necessarily useful for stretchable electronic applications [52,53]. Regarding the toughness value, MgO-only and binary activator systems provide better toughness values compared to ZnO-only as a cure activator. The highest toughness value (132 MJ/m^3) was obtained for the MgO/ZnO binary activator system at a 4:1 ratio of MgO to ZnO content. The better toughness values of the binary activator systems might be due to a better modulus and elongation at break values. From the above discussion, it can be concluded that a complete ZnO-free vulcanizate can be possible where high toughness is necessary, sacrificing the modulus values.

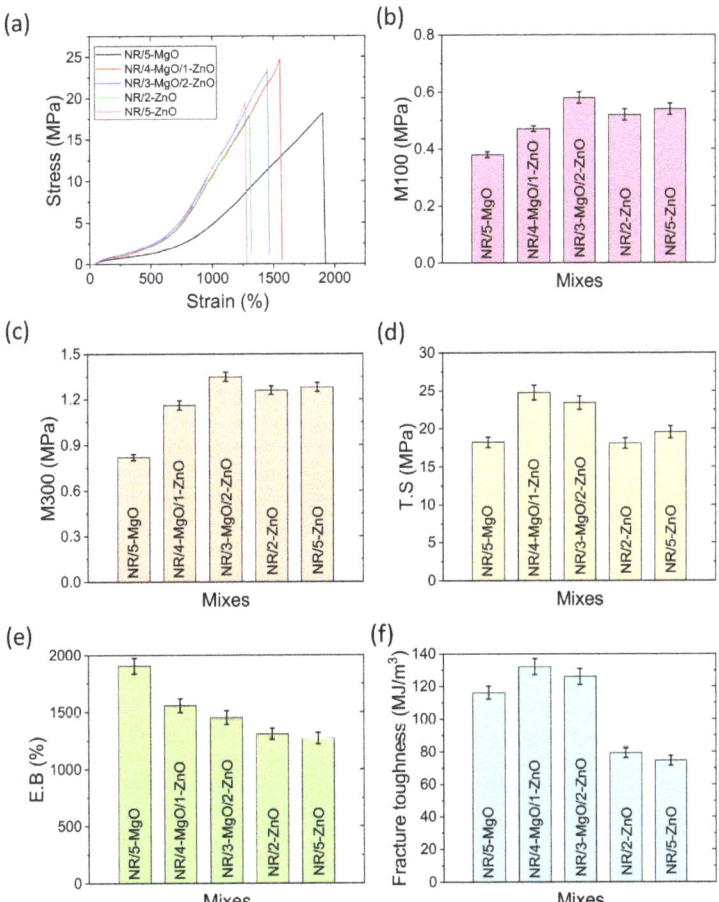

Figure 6. Tensile mechanical properties of vulcanized rubber; (**a**) stress-strain, (**b**) M100, (**c**) M300, (**d**) T.S, (**e**) E.B, and (**f**) fracture toughness.

Boonkerd et al. described the relationship between tensile properties with the number of sulfur cross-links and the sulfur ranks per cross-link [54]. A greater number of cross-links with a lower sulfur rank indicate low elongation and a higher modulus. On the other hand, at similar cross-link density, a higher sulfur rank (i.e., polysulfide cross-links) enhanced the tensile strength and elongation at break values. It is believed that elongation at break values is highly dependent on the dispersion of the curatives as well as cross-link density and sulfur ranks among the cross-links. From the cross-link density measurements and tensile properties, it can be predicted that higher MgO content provides higher sulfur-ranked cross-links and excellent dispersion of the curatives. Good dispersion of curatives not only improves the cross-link density but also enhances the physical bonding between the remaining unreacted or vulcanization byproducts and the rubber molecules that have some contribution to the enhanced mechanical properties. Since the toughness is related to the number of total linked bonds and their strengths, it can be assumed that the highest number of linked bonds exist in the MgO-only-activated system reacting with the most sulfur elements in the system. On the other hand, binary activator systems provide a higher number of linked bonds as well as a higher number of stronger C-S bonds that assure excellent toughness and modulus values compared to single-cure activator systems. According to Borros et al. [33], when both zinc and magnesium existed in the vulcanization, a higher amount of disulfide linkages were obtained than mono and polysulfide linkages [33]. For the existence of disulfide linkages in higher amounts in the vulcanized compound [54], excellent dispersion of curatives (Figure 7), a smaller particle size of MgO, a better modulus and tensile strength were obtained in MgO/ZnO binary systems. Thus, magnesium oxide plays an important role in improving the tensile properties, especially the tensile strength and fracture toughness of the rubber vulcanizates.

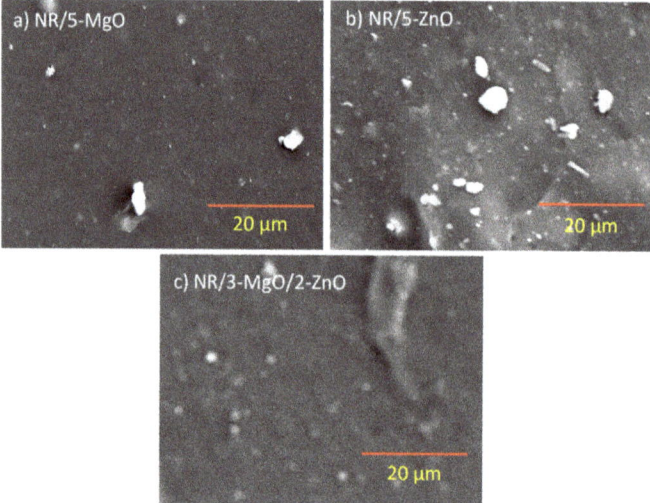

Figure 7. SEM micrographs of vulcanized rubber; (**a**) NR/5-MgO, (**b**) NR/5-ZnO, and (**c**) NR/3-MgO/2-ZnO.

3.3. SEM and EDS-Mapping

The distribution of curatives in the rubber matrix is an important factor that ensures homogeneity in the cross-link density throughout the rubber compounds. It is believed that more homogeneity in the cross-links throughout the rubber matrix provides better tensile properties. Figure 7a–c represents the SEM images of single and binary curing activator systems. From Figure 7a, it is evident that some curatives remain separated from the matrix because of their incomplete chemical interactions with the rubber matrix in the presence of MgO as a cure activator. Similarly, in the presence of a ZnO-only as a cure activator more curatives remain unreacted and are separated from the rubber matrix (Figure 7b). On the other hand, the binary curing activator ensures excellent distribution and reactivity of all curatives, and only a few remain unreacted (Figure 7c). These results are highly correlated with the improved toughness in the MgO/ZnO binary curing activator systems. To confirm the homogeneous distribution of curatives in the MgO/ZnO binary curing activator system, EDS mapping for different elements is performed and is presented in Figure 8. EDS-mapping for different elements indicates the homogeneous distribution of curatives for the 3:2 ratio of MgO to ZnO-based binary curing activator system that ensures better tensile mechanical properties than ZnO-only as a single-site curing activator in the vulcanization of rubber.

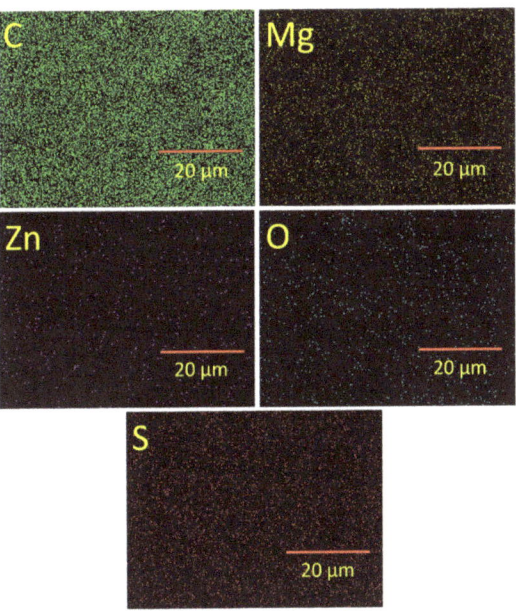

Figure 8. Particle distribution through EDS-mapping in NR/3-MgO/2-ZnO compound.

3.4. Thermo Gravimetric Properties of Rubber Vulcanizates

The thermal properties of conventional 5 phr ZnO-based vulcanizate were compared with 5 phr MgO-based and binary curing activators containing 3 phr MgO and 2 phr ZnO-based vulcanizates. Literature showed that the main products of thermal degradation of NR are isoprene, dipentene, and *p*-menthene [55]. The weight losses from ~150 to ~327 °C in Figure 9a,b, are mainly due to decomposition of the un-reacted crosslink precursors, sulfur cross-links, stearic acid and accelerator, and partial breakage of the rubber backbone [56]. The rapid degradation region from ~350 to ~450 °C in the thermal gravimetric analysis (TGA) curves in Figure 9a represents the main chain breakdown of rubber polymer. This behavior is clearer on the derivative thermo-gravimetric analysis (DTA) curves in Figure 9b. From these figures, it is clear that there are no significant differences in the thermal stabilities of the compounds cured by single MgO and MgO/ZnO binary curing activators compared to 5 phr ZnO as a single-site curing activator. The comparable thermal stabilities may be due to better homogeneity in the cross-links and reduced thermal motion of rubber chains in the presence of MgO [38,57] compared to ZnO-only.

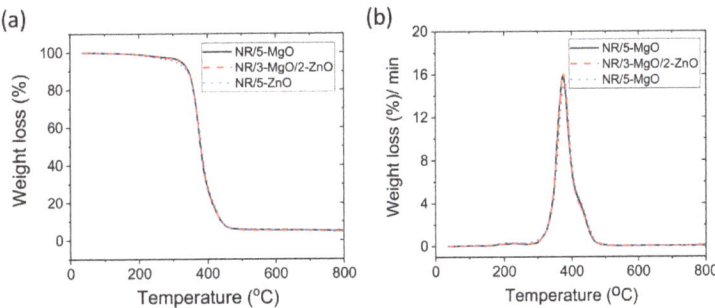

Figure 9. (**a**) TGA and (**b**) DTA results of vulcanized rubber.

4. Conclusions

In this research non-toxic MgO was tested as a cure activator for the sulfur vulcanization of rubber. We aimed to reduce the amount of environmentally hazardous ZnO cure activator without compromising curing and mechanical properties in the vulcanization of rubber. The curing, mechanical, and thermal properties were investigated to find out the suitability of the MgO by itself or in combination with ZnO in the vulcanization of rubber. Results revealed that considering some advancements in properties like curing time, elongation at break, and fracture toughness, MgO can compete with the conventionally used ZnO as a cure activator. However, to achieve the present industrial level of crosslink density, mechanical modulus, and tensile strength, the binary combination of MgO and ZnO should be used as curing activators. Considerable reductions (60–80%) of ZnO from the conventional amount could be possible by utilizing an MgO/ZnO-based binary curing activator system with improved efficiencies in the curing kinetics and mechanical properties with negligible reduction in the thermal stability. For example, 60% MgO in MgO/ZnO binary activators system provides 0.58 MPa of M100, 23.7 MPa of tensile strength, 1455% of elongation at break, 126 MJ/m^3 of fracture toughness, and 0.85 Nm/min of vulcanization rate values, which are 7.5%, 20%, 15%, 68%, and 184%, respectively, higher than the 100% ZnO-based activator system. Investigations on the curing properties suggest that certain chemical interactions might have happened between the two cure activators following the paths of accelerators-activators complex formation reactions. Instead of a single-site curing activator, the proposed binary curing activators could be very useful in the rubber industries for considerable improvements in the vulcanization kinetics and tensile properties.

Author Contributions: Conceptualization, M.N.A.; methodology, M.N.A. and V.K.; validation, M.N.A., V.K. and S.-S.P.; formal analysis, M.N.A.; investigation, M.N.A.; data curation, M.N.A. and V.K.; writing—original draft preparation, M.N.A.; writing—review and editing, M.N.A., V.K. and S.-S.P.; visualization, M.N.A.; supervision, S.-S.P.; project administration, S.-S.P.; funding acquisition, S.-S.P. All authors have read and agreed to the published version of the manuscript.

Funding: This research received no external funding.

Institutional Review Board Statement: Not applicable.

Informed Consent Statement: Not applicable.

Data Availability Statement: The data presented in this study are available on request from the corresponding author.

Conflicts of Interest: The authors declare no conflict of interest.

References

1. Chapman, A.V. Safe rubber chemicals: Reduction of zinc levels in rubber compounds. In Proceedings of the International Rubber Conference, Tun Abdul Razak Research Centre (TARRC)/Malaysian Rubber Producers' Research Association (MRPRA), Kuala Lumpur, Malaysia, 6–9 October 1997.
2. Duchacek, V. Effect of zinc oxide concentration on the course of thiuram-accelerated sulfur vulcanization. *J. Appl. Polym. Sci.* **1976**, *20*, 71–78. [CrossRef]
3. Heideman, G.; Datta, R.N.; Noordermeer, J.W.M.; van Baarle, B. Influence of zinc oxide during different stages of sulfur vulcanization. Elucidated by model compound studies. *J. Appl. Polym. Sci.* **2005**, *95*, 1388–1404. [CrossRef]
4. Paulthangam, K.M.; Som, A.; Ahuja, T.; Srikrishnarka, P.; Nair, A.S.; Pradeep, T. Role of Zinc Oxide in the Compounding Formulation on the Growth of Nonstoichiometric Copper Sulfide Nanostructures at the Brass–Rubber Interface. *ACS Omega* **2022**, *7*, 9573–9581. [CrossRef]
5. Das, A.; Wang, D.Y.; Leuteritz, A.; Subramaniam, K.; Greenwell, H.C.; Wagenknecht, U.; Heinrich, G. Preparation of zinc oxide free, transparent rubber nanocomposites using a layered double hydroxide filler. *J. Mater. Chem.* **2011**, *21*, 7194–7200. [CrossRef]
6. Fosmire, G.J. Zinc toxicity. *Am. J. Clin. Nutr.* **1990**, *51*, 225–227. [CrossRef] [PubMed]
7. Poynton, H.C.; Lazorchak, J.M.; Impellitteri, C.A.; Smith, M.E.; Rogers, K.; Patra, M.; Hammer, K.A.; Allen, H.J.; Vulpe, C.D. Differential gene expression in Daphnia magna suggests distinct modes of action and bioavailability for ZnO nanoparticles and Zn ions. *Environ. Sci. Technol.* **2011**, *45*, 762–768. [CrossRef] [PubMed]

8. Reed, R.B.; Ladner, D.A.; Higgins, C.P.; Westerhoff, P.; Ranville, J.F. Solubility of nano-zinc oxide in environmentally and biologically important matrices. *Environ. Toxicol. Chem.* **2012**, *31*, 93–99. [CrossRef] [PubMed]
9. Gottschalk, F.; Sonderer, T.; Scholz, R.W.; Nowack, B. Modeled environmental concentrations of engineered nanomaterials (TiO$_2$, ZnO, Ag, CNT, Fullerenes) for different regions. *Environ. Sci. Technol.* **2009**, *43*, 9216–9222. [CrossRef]
10. Kole, P.J.; Löhr, A.J.; Van Belleghem, F.G.; Ragas, A.M. Wear and tear of tyres: A stealthy source of microplastics in the environment. *Int. J. Environ. Res. Public Health* **2017**, *14*, 1265. [CrossRef]
11. Heideman, G.; Datta, R.N.; Noordermeer, J.W.M.; van Baarle, B. Effect of zinc complexes as activator for sulfur vulcanization in various rubbers. *Rubber Chem. Technol.* **2005**, *78*, 245–257. [CrossRef]
12. Heideman, G.; Datta, R.N.; Noordermeer, J.W.M.; van Baarle, B. Zinc loaded clay as activator in sulfur vulcanization: A new route for zinc oxide reduction in rubber compounds. *Rubber Chem. Technol.* **2004**, *77*, 336–355. [CrossRef]
13. Heideman, G.; Datta, R.N.; Noordermeer, J.W.M.; van Baarle, B. Multifunctional additives as zinc-free curatives for sulfur vulcanization. *Rubber Chem. Technol.* **2006**, *79*, 561–588. [CrossRef]
14. Mostoni, S.; Milana, P.; Di Credico, B.; D'Arienzo, M.; Scotti, R. Zinc-based curing activators: New trends for reducing zinc content in rubber vulcanization process. *Catalysts* **2019**, *9*, 664. [CrossRef]
15. Qin, X.; Xu, H.; Zhang, G.; Wang, J.; Wang, Z.; Zhao, Y.; Wang, Z.; Tan, T.; Bockstaller, M.R.; Zhang, L.; et al. Enhancing the performance of rubber with nano ZnO as activators. *ACS Appl. Mater. Interfaces* **2020**, *12*, 48007–48015. [CrossRef]
16. Utrera-Barrios, S.; Verdugo Manzanares, R.; Araujo-Morera, J.; González, S.; Verdejo, R.; López-Manchado, M.Á.; Hernández Santana, M. Understanding the molecular dynamics of dual crosslinked networks by dielectric spectroscopy. *Polymers* **2021**, *13*, 3234. [CrossRef]
17. Basu, D.; Das, A.; Stöckelhuber, K.W.; Wagenknecht, U.; Heinrich, G. Advances in layered double hydroxide (LDH)-based elastomer composites. *Prog. Polym. Sci.* **2014**, *39*, 594–626. [CrossRef]
18. Pysklo, L.; Pawlowski, P.; Parasiewicz, W.; Slusarski, L. Study on reduction of zinc oxide level in rubber compounds. Part I influence of zinc oxide specific surface area and the interphase transfer catalyst 18-crown-6. *Kautsch. Gummi Kunstst.* **2007**, *60*, 548–553.
19. Pysklo, L.; Pawlowski, P.; Nicinski, K.; Slusarski, L.; Wlodarska, M.; Bak, G. Study on reduction of zinc oxide level in rubber compounds: Part II mechanism of activation of sulphur vulcanisation by means of the interphase transfer catalyst 18-crown-6-ether. *Kautsch. Gummi Kunstst.* **2008**, *61*, 442–446.
20. Henning, S.K. Reduced Zinc Loading: Using Zinc Monomethacrylate to Activate Accelerated Sulfur Vulcanization. In Proceedings of the 172nd Technical Meeting of the Rubber Division, American Chemical Society and International Rubber Conference 2007, Cleveland, OH, USA, 16–18 October 2007.
21. Przybyszewska, M.; Zaborski, M.; Jakubowski, B.; Zawadiak, J. Zinc chelates as new activators for sulphur vulcanization of acrylonitrile-butadiene elastomer. *Express Polym. Lett.* **2009**, *3*, 256–266. [CrossRef]
22. Przybyszewska, M.; Zaborski, M. The effect of zinc oxide nanoparticle morphology on activity in crosslinking of carboxylated nitrile elastomer. *Express Polym. Lett.* **2009**, *3*, 542–552. [CrossRef]
23. Sahoo, S.; Bhowmick, A.K. Influence of ZnO nanoparticles on the cure characteristics and mechanical properties of carboxylated nitrile rubber. *J. Appl. Polym. Sci.* **2007**, *106*, 3077–3083. [CrossRef]
24. Xiong, M.; Gu, G.; You, B.; Wu, L. Preparation and characterization of poly(styrene butylacrylate) latex/nano-ZnO nanocomposites. *J. Appl. Polym. Sci.* **2003**, *90*, 1923–1931. [CrossRef]
25. Sahoo, S.; Maiti, M.; Ganguly, A.; George, J.J.; Bhowmick, A.K. Effect of zinc oxide nanoparticles as cure activator on the properties of natural rubber and nitrile rubber. *J. Appl. Polym. Sci.* **2007**, *105*, 2407–2415. [CrossRef]
26. Jincheng, W.; Yuehui, C. Application of nano-zinc oxide master batch in polybutadiene styrene rubber system. *J. Appl. Polym. Sci.* **2006**, *101*, 922–930. [CrossRef]
27. Wang, Z.; Lu, Y.; Liu, J.; Dang, Z.; Zhang, L.; Wang, W. Preparation of nano-zinc oxide/EPDM composites with both good thermal conductivity and mechanical properties. *J. Appl. Polym. Sci.* **2011**, *119*, 1144–1155. [CrossRef]
28. Roy, K.; Alam, M.N.; Mandal, S.K.; Debnath, S.C. Sol-Gel derived nano zinc oxide for the reduction of zinc oxide level in natural rubber compounds. *J. Sol-Gel Sci. Technol.* **2014**, *70*, 378–384. [CrossRef]
29. Alam, M.N.; Potiyaraj, P. Synthesis of nano zinc hydroxide via sol-gel method on silica surface and its potential application in the reduction of cure activator level in the vulcanization of natural rubber. *J. Sol-Gel Sci. Technol.* **2017**, *81*, 903–911. [CrossRef]
30. Alam, M.N.; Potiyaraj, P. Precipitated nano zinc hydroxide on the silica surface as an alternative cure activator in the vulcanization of natural rubber. *Rubber Chem. Technol.* **2017**, *90*, 714–727. [CrossRef]
31. Wu, S.; Xiao, C.; Kong, S.; Li, B.; Yang, Z.; Tang, Z.; Liu, F.; Guo, B. Carbon nanodots as an eco-friendly activator of sulphur vulcanization in diene-rubber composites. *Compos. Commun.* **2021**, *25*, 100755. [CrossRef]
32. da Silva, A.A.; da Rocha, E.B.; Linhares, F.N.; de Sousa, A.M.F.; Carvalho, N.M.; Furtado, C.R. Replacement of ZnO by ecofriendly synthesized MgO in the NBR vulcanization. *Polym. Bull.* **2022**, *79*, 8535–8549. [CrossRef]
33. Guzmán, M.; Reyes, G.; Agulló, N.; Borrós, S. Synthesis of Zn/Mg oxide nanoparticles and its influence on sulfur vulcanization. *J. Appl. Polym. Sci.* **2011**, *119*, 2048–2057. [CrossRef]
34. Guzmán, M.; Vega, B.; Agulló, N.; Giese, U.; Borrós, S. Zinc oxide versus magnesium oxide revisited. Part 1. *Rubber Chem. Technol.* **2012**, *85*, 38–55. [CrossRef]

35. Guzmán, M.; Vega, B.; Agulló, N.; Borrós, S. Zinc oxide versus magnesium oxide revisited. Part 2. *Rubber Chem. Technol.* **2012**, *85*, 56–67. [CrossRef]
36. Kar, S.; Bhowmick, A.K. Nanostructured magnesium oxide as cure activator for polychloroprene rubber. *J. Nanosci. Nanotechnol.* **2009**, *9*, 3144–3153. [CrossRef] [PubMed]
37. Siti, N.Q.M.; Kawahara, S. Evaluating performance of magnesium oxide at different sizes as activator for natural rubber vulcanization. *Int. J. Adv. Chem. Eng. Biol. Sci.* **2016**, *3*, 97–101.
38. Roy, K.; Alam, M.N.; Mandal, S.K.; Debnath, S.C. Preparation of zinc-oxide-free natural rubber nanocomposites using nanostructured magnesium oxide as cure activator. *J. Appl. Polym. Sci.* **2015**, *132*, 42705. [CrossRef]
39. Roy, K.; Alam, M.N.; Mandal, S.K.; Debnath, S.C. Development of a suitable nanostructured cure activator system for polychloroprene rubber nanocomposites with enhanced curing, mechanical and thermal properties. *Polym. Bull.* **2016**, *73*, 191–207. [CrossRef]
40. Zhang, L.; Jiang, Y.; Ding, Y.; Povey, M.; David, Y. Investigation into the antibacterial behaviour of suspensions of ZnO nanoparticles (ZnO nanofluids). *J. Nanopart. Res.* **2007**, *9*, 479–489. [CrossRef]
41. Kuschner, W.G.; Wong, H.; D'Alessandro, A.; Quinlan, P.; Blanc, P.D. Human pulmonary responses to experimental inhalation of high concentration fine and ultrafine magnesium oxide particles. *Environ. Health Perspect.* **1997**, *105*, 1234–1237. [CrossRef] [PubMed]
42. Alam, M.N.; Kumar, V.; Potiyaraj, P.; Lee, D.-J.; Choi, J. Synergistic activities of binary accelerators in presence of magnesium oxide as a cure activator in the vulcanization of natural rubber. *J. Elastomers Plast.* **2022**, *54*, 123–144. [CrossRef]
43. Alam, M.N.; Mandal, S.K.; Debnath, S.C. Bis (N-benzyl piperazino) thiuram disulfide and dibenzothiazyl disulfide as synergistic safe accelerators in the vulcanization of natural rubber. *J. Appl. Polym. Sci.* **2012**, *126*, 1830–1836. [CrossRef]
44. Alam, M.N.; Mandal, S.K.; Debnath, S.C. Effect of zinc dithiocarbamates and thiazole-based accelerators on the vulcanization of natural rubber. *Rubber Chem. Technol.* **2012**, *85*, 120–131. [CrossRef]
45. Alam, M.N.; Mandal, S.K.; Roy, K.; Debnath, S.C. Synergism of novel thiuram disulfide and dibenzothiazyl disulfide in the vulcanization of natural rubber: Curing, mechanical and aging resistance properties. *Int. J. Ind. Chem.* **2014**, *5*, 8. [CrossRef]
46. Craig, D.; Davidson, W.L.; Juve, A.E. Tetramethylthiuram disulfide vulcanization of extracted rubber, V: Low molecular products and the mechanism of zinc oxide activation. *J. Polym. Sci.* **1951**, *6*, 177–187. [CrossRef]
47. Scheele, W. Kinetic studies of the vulcanization of natural and synthetic rubbers. *Rubber Chem. Technol.* **1961**, *34*, 1306–1401. [CrossRef]
48. Flory, P.J.; John, R., Jr. Statistical mechanics of cross-linked polymer networks II. Swelling. *J. Chem. Phys.* **1943**, *11*, 521–526. [CrossRef]
49. Akiba, M.; Hashim, A.S. Vulcanization and crosslinking in elastomers. *Prog. Polym. Sci.* **1997**, *22*, 475–521. [CrossRef]
50. Kruger, F.W.H.; McGill, J. A DSC study of curative interactions. I. The interaction of ZnO, sulfur, and stearic acid. *J. Appl. Polym. Sci.* **1991**, *42*, 2643–2649. [CrossRef]
51. Sahoo, N.G.; Das, C.K.; Panda, A.B.; Pramanik, P. Nanofiller as vulcanizing aid for styrene-butadiene elastomer. *Macromol. Res.* **2002**, *10*, 369–372. [CrossRef]
52. Wang, Z.; Xiang, C.; Yao, X.; Le Floch, P.; Mendez, J.; Suo, Z. Stretchable materials of high toughness and low hysteresis. *Proc. Natl. Acad. Sci. USA* **2019**, *116*, 5967–5972. [CrossRef]
53. Ghosh, G.; Meeseepong, M.; Bag, A.; Hanif, A.; Chinnamani, M.V.; Beigtan, M.; Kim, Y.; Lee, N.-E. Tough, transparent, biocompatible and stretchable thermoplastic copolymer with high stability and processability for soft electronics. *Mater. Today* **2022**, *57*, 43–56. [CrossRef]
54. Boonkerd, K.; Deeprasertkul, C.; Boonsomwong, K. Effect of sulfur to accelerator ratio on crosslink structure, reversion, and strength in natural rubber. *Rubber Chem. Technol.* **2016**, *89*, 450–464. [CrossRef]
55. Brandrup, J.; Immergut, E.H.; Mc Dowell, W. *Polymer Handbook*, 2nd ed.; Wiley: New York, NY, USA, 1975.
56. De, D.; De, D.; Singharoy, G.M. Reclaiming of ground rubber tire by a novel reclaiming agent. I. virgin natural rubber/reclaimed GRT vulcanizates. *Polym. Eng. Sci.* **2007**, *7*, 1091–1100. [CrossRef]
57. Gilman, J.W.; Jackson, C.L.; Morgan, A.B.; Harris, R.; Manias, E.; Giannelis, E.P.; Wuthenow, M.; Hilton, D.; Phillips, S.H. Flammability properties of polymer— layered-silicate nanocomposites. Polypropylene and polystyrene nanocomposites. *Chem. Mater.* **2000**, *12*, 1866–1873. [CrossRef]

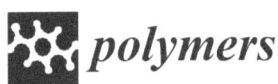

Article

New Insight into Rubber Composites Based on Graphene Nanoplatelets, Electrolyte Iron Particles, and Their Hybrid for Stretchable Magnetic Materials

Vineet Kumar, Md Najib Alam , Sang-Shin Park * and Dong-Joo Lee *

School of Mechanical Engineering, Yeungnam University, 280 Daehak-ro, Gyeongsan 38541, Korea
* Correspondence: pss@ynu.ac.kr (S.-S.P.); djlee@yu.ac.kr (D.-J.L.)

Abstract: New and soft composites with good mechanical stretchability are constantly addressed in the literature due to their use in various industrial applications such as soft robotics. The stretchable magnetic materials presented in this work show a promising magnetic effect of up to 28% and improved magnetic sensitivity. The composites are soft in nature and possess hardness below 65. These composites were prepared by mixing silicone rubber with fillers such as graphene nanoplatelets (GNP), electrolyte-iron particles (EIP), and their hybrid via solution mixing. The final composites were cured at room temperature for 24 h and their isotropic and anisotropic properties were studied and presented. The mechanical properties under compressive and tensile strain were studied in detail. The results show that the compressive modulus was 1.73 MPa (control) and increased to 3.7 MPa (GNP) at 15 per hundred parts of rubber (phr), 3.2 MPa (EIP), and 4.3 MPa (hybrid) at 80 phr. Similarly, the mechanical stretchability was 112% (control) and increased to 186% (GNP) at 15 phr, 134% (EIP), and 136% (hybrid) at 60 phr. Thus, GNP emerges as a superior reinforcing filler with high stiffness, a high compressive modulus, and high mechanical stretchability. However, the GNP did not show mechanical sensitivity under a magnetic field. Therefore, the hybrids containing GNP and EIP were considered and an improved mechanical performance with magnetic sensitivity was noticed and reported. The mechanism involves the orientation of EIP under a magnetic field causing a magnetic effect, which is 28% for EIP and 5% for hybrid.

Keywords: mechanical stretchability; silicone rubber; graphene nanoplatelets; electrolyte iron particles; compressive modulus; anisotropy

Citation: Kumar, V.; Alam, M.N.; Park, S.-S.; Lee, D.-J. New Insight into Rubber Composites Based on Graphene Nanoplatelets, Electrolyte Iron Particles, and Their Hybrid for Stretchable Magnetic Materials. Polymers 2022, 14, 4826. https://doi.org/10.3390/polym14224826

Academic Editor: Mikhail Shamonin

Received: 7 October 2022
Accepted: 7 November 2022
Published: 9 November 2022

Copyright: © 2022 by the authors. Licensee MDPI, Basel, Switzerland. This article is an open access article distributed under the terms and conditions of the Creative Commons Attribution (CC BY) license (https://creativecommons.org/licenses/by/4.0/).

1. Introduction

Stretchable magnetic materials (SMM) consist of composites reinforced with different types of iron particles and polymers with a mainly elastomeric matrix [1]. SMM have a stretchable behavior when strained [2]. The mechanical properties of SMM are influenced by the type of strain [3]. The type of strain can be tensile or compressive in nature [4]. The alternative way of affecting their mechanical properties is through the influence of a magnetic field [5]. The iron particles present in SMM tend to orient in the direction of the magnetic field, thereby influencing the mechanical properties of SMM [6].

Such an orientation of the iron particles is also called an anisotropic effect, while the samples without such an orientation are called isotropic samples [7] (Scheme 1). However, (a) the influence of the mechanical properties orienting the iron particles or (b) the addition of iron particles as a source of reinforcement constitute insufficient pathways for obtaining devices with industrial value [8]. So, other reinforcing fillers must be added to achieve optimum mechanical properties [9]. These fillers can be carbon black [10], carbon nanotubes [11], or graphene [12]. Among them, carbon black is traditionally used as a filler that improves mechanical properties but is used at a high content [13]. This high content alters the viscoelastic properties of the samples [13]. Thus, nanofillers such as carbon nanotubes or graphene are employed to improve and obtain the desired mechanical properties at a low filler content [14,15].

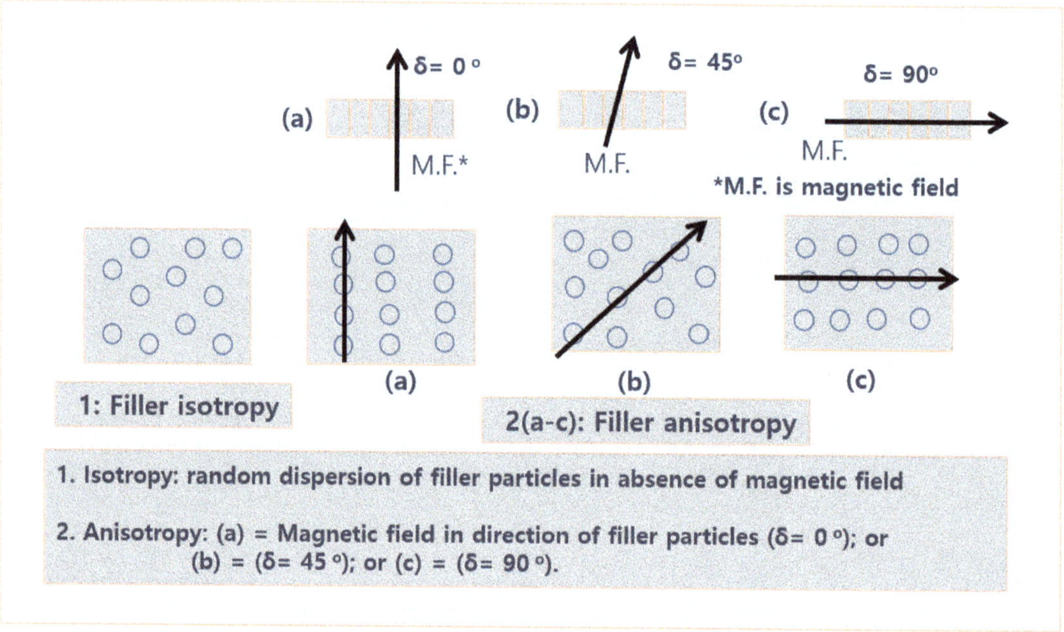

Scheme 1. Concept of isotropy and anisotropy in magneto-rheological elastomers.

The elastomeric matrix family is quite large, including rubber as one member [16]. The rubber matrix can be synthetic [17] or natural in origin [18]. Rubber with a natural origin is known as "natural rubber latex", and is obtained from trees [19]. Synthetic rubber is dynamic, with vast classes ranging from diene rubber [20] to silicone rubber [21]. Among them, silicone rubber is more promising than diene rubber in terms of hardness, easy processing, and easy curing [22]. Silicone rubber is categorized based on the type of vulcanization, which can be room temperature or high temperature [23]. Among them, room temperature-vulcanized silicone rubber is more promising due to its versatile behavior, softness, and curability without the use of sophisticated machines [24].

Nanofillers based on carbon allotropes added to silicone rubber lead to drastic improvements in the composites' mechanical, electrical, or thermal properties [25]. A review study by Kumar et al. showed that these improved properties may be useful for a range of soft industrial applications such as strain sensors [26]. The review study by Kumar et al. showed that among the different ranges of carbon-based nanofillers, CNT and graphene emerge as the best candidates for reinforcement [26]. These improved properties are due to (a) the high aspect ratio of these nanofillers [27]; (b) the favorable morphology of these nanofillers, which allows for their uniform dispersion [28]; and (c) the high interfacial area of these nanofillers, which enables high stress-transfer from polymers to these nanofillers in composites [29].

Moreover, a silicone rubber matrix filled with iron particles may be useful for "magneto-rheological elastomers" [30] or SMM, in which mechanical properties can be influenced by switching magnetic fields, as performed in this work [1,30]. The mechanism behind such an increase is the orientation of the iron particles in the direction of a magnetic field, thereby forming a chain-like structure and influencing the mechanical properties such as the modulus [31]. Another way to improve these mechanical properties is to add a reinforcing filler along with iron particles to obtain high-performance SMM [32,33]. Thus, graphene nanoplatelets were used in the present work along with electrolyte iron particles in a silicone rubber matrix.

Various studies have been reported that show the use of binary carbon-based reinforcing fillers such as carbon nanotubes [34], graphene [35], or carbon black [36] along with different types of iron particles in the rubber matrix to obtain SMM [1,34–36]. These studies show that the incorporation of secondary fillers based on carbon not only improves mechanical performance but also does not influence the magnetic sensitivity exhibited by the iron particles [37]. This study is an advancement from the previously reported studies because it studies the anisotropic effects of fillers at a high magnetic field of 1 Tesla, wherein the effect of the orientation of EIP on the dispersion of GNPs was correlated. Moreover, hybrid composites were also prepared in this study that investigates the composites' possible synergistic effects and their relation to the improvement of various mechanical properties. We hypothesize that the hybrid composite possesses the advantages of a higher compressive modulus, increased reinforcing effects, and optimum magnetic sensitivity, while EIP possesses the advantages of an increased magnetic effect, which was supported experimentally in this work.

2. Materials and Methods

2.1. Materials

The RTV-silicone rubber was used as a rubber matrix in the present work and was purchased from Shin-Etsu Chemical Corporation Limited, Tokyo, Japan. Its commercial name is "KE-441-KT", and it is transparent in nature. The vulcanizing agent used in the present work was "CAT-RM", which was purchased from Shin-Etsu Chemical Corporation Limited, Tokyo, Japan. The graphene nanoplatelets were used as a reinforcing nanofiller in the present work. Their commercial name is "XG C750" and they were purchased from XG Science, Lansing, MI, USA. The nanoplatelets had a total surface area of around 750 m^2/g, lateral dimensions from 500 nm–1 μm, and thickness of 1–2 nm. The micron-size electrolyte iron particles (EIP) with the commercial name "Fe#400" were obtained from Aometal Corporation Limited, Gomin-si, Korea. The average particle size of each EIP is >10 μm, with each particle possessing an irregular shape, being light greyish in color, possessing a density of 2–3 g cm^{-3}, and a purity of 98.8% iron, while other traces of carbon, oxygen, and nitrogen were also found. All the materials were used in a pristine state without any further purification. The mold-releasing agent was purchased from Nabakem, Pyeongtaek-si, Korea.

2.2. Fabrication of Rubber Composites

The steps of composites' preparation were optimized and reported in previous studies [38]. The procedure involved spraying the molds with mold-releasing agents and then drying them at room temperature for 3 h. In the next step, the liquid RTV-SR rubber was poured into a beaker and a known amount of filler (Table 1) was mixed in. Rubber–filler mixing was performed for around 10 min. Next, the known amount of vulcanizing agent was added to the sample and mixed for nearly 1 min. Then, the composite was added to the molds and kept for 24 h at ambient conditions before the vulcanized composite was ready (Scheme 2) for testing of mechanical and anisotropic properties.

Table 1. Fabrication of the different rubber composites.

Formulation	RTV-SR (phr)	GNP (phr)	EIP (phr)	Vulcanizing Solution (phr)
Control	100	-	-	2
RTV-SR/GNP	100	5, 10, 15	-	2
RTV-SR/EIP	100	-	40, 60, 80	2
RTV-SR/Hybrid	100	5, 10, 15	35, 50, 65	2

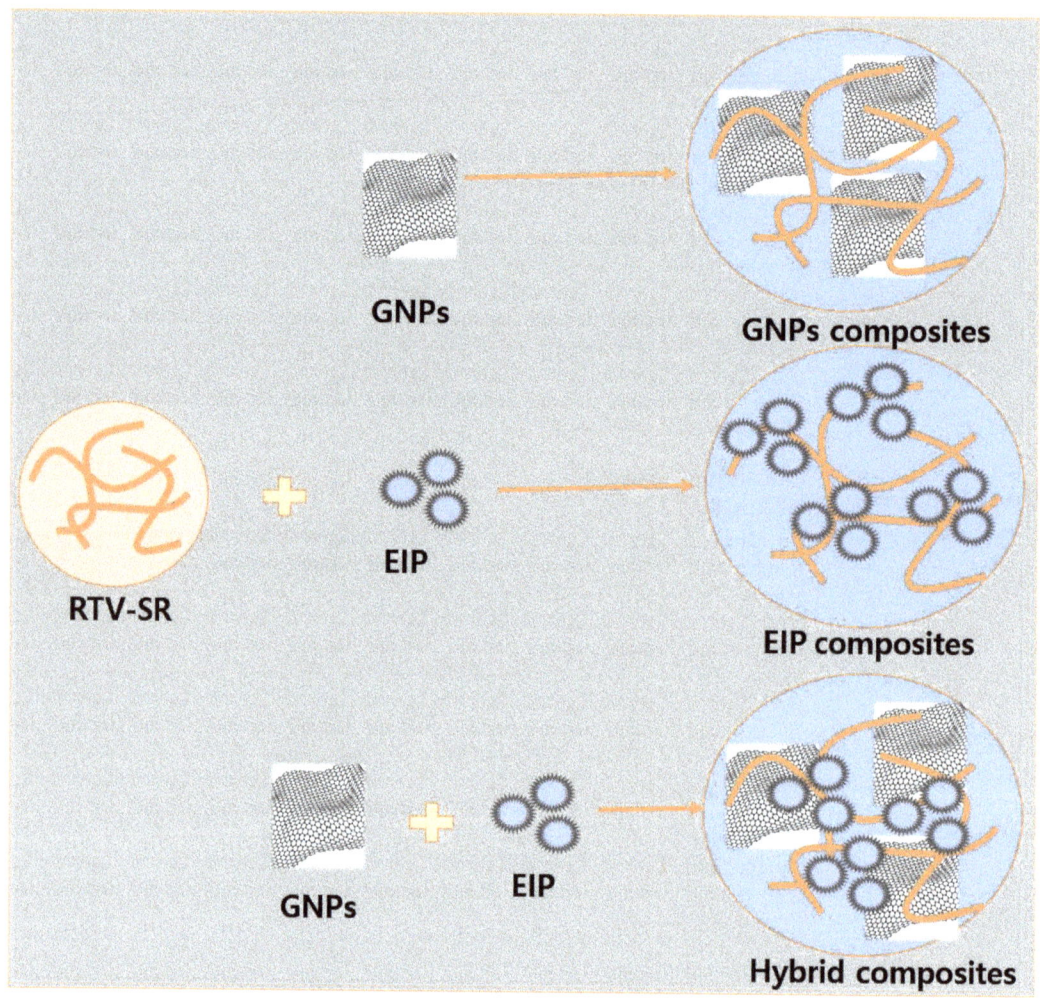

Scheme 2. Schematic of the formulation of different types of composites.

2.3. Characterization Technique

The nanofiller's morphology and its dispersion in the rubber matrix were studied by SEM (S-4800, Hitachi, Tokyo, Japan). The composite specimen was sectioned to a thickness of 0.5 mm using a surgical blade and then placed on the SEM stub before the coating process was initiated. The SEM samples were coated with conductive platinum for 2 min to make the surface of the samples electrically conductive. The mechanical properties under compressive and tensile strain were studied using a universal testing machine (UTS, Lloyd Instruments, Bognor Regis, UK). The mechanical properties under compressive strain were measured at a 4 mm/min strain rate and under a load of 0.5 kN using cylindrical samples. These cylindrical samples were 10 mm in thickness and 20 mm in diameter. The maximum strain of 35% was applied to these cylindrical samples as higher strain leads to fracture of the sample. The tests of the mechanical properties under tensile strain were performed at a strain rate of 100 mm/min on a dumbbell-shaped sample with a gauge length of 25 mm and thickness of 2 mm. The tensile specimens were strained until fracture and their mechanical parameters such as their moduli, tensile strength, or fracture strain were estimated. These mechanical properties were obtained following DIN 53 504 standards. The anisotropic magnetic properties were studied at 1 T by placing the specimen inside a magnetic field for 90 min.

3. Results and Discussion

3.1. Morphologies of Nanofillers

The morphology of the nanofillers used as a reinforcing agent in the rubber matrix is known to affect the properties of composites [39]. Therefore, the morphology of the fillers was studied and is presented in Figure 1. Figure 1a shows the typical platelet-like morphology of GNPs. The platelets in the graphene nanoparticles were three-dimensional in nature [40]. The particle size in the lateral dimension was in the range from 500 nm–1 μm, and the thickness of 1–2 nm led to a very high aspect ratio. This high aspect ratio provides very high mechanical, electrical, and thermal properties when added as filler in the composite [41,42]. It is also expected that the high aspect ratio of GNPs leads to the formation of long-range and connective filler networks throughout the rubber matrix [43]. These filler networks with filler–filler and polymer–filler interactions within the composite significantly improve mechanical properties [44]. GNPs have a strong lubricating effect and are known to improve the fracture strain of the composites [45]. The GNPs' structure includes the 3-D arrangement of 2-D graphene sheets held together by weak Vander Waals forces. On the other hand, EIP are micron-sized particles—namely, with sizes in the range of 10–12 μm—possessing a 3-dimensional morphology. These particles are magnetically active and algin themselves in the presence of a magnetic field in the composite. This behavior of these iron particles makes them promising in terms of their anisotropic effects and potential magnetic sensitivity applications [46]. The micron-sized EIP are rough, have an irregular shape, and can be easily aligned under a magnetic field. This anisotropic property of EIP leads to an improvement in the mechanical properties of the composites and will be discussed in the coming sections.

Figure 1. SEM images: (**a**) GNPs; (**b**) EIP.

3.2. Filler Dispersion Analyzed through SEM Microscopy

The effect of the filler dispersion on the properties of the composites is well-known. It is also known that composites with a uniform filler dispersion exhibit more optimum properties than those with aggregated fillers or non-uniform dispersion. Thus, the study of the filler dispersion in composites is a key aspect for ascertaining their properties. In this study, the filler dispersion was studied through SEM images. A number of images were studied, and their representative images are presented in Figure 2. Figure 2a–c show SEM images of the control sample. It is evident that there are no filler particles, as expected. Then, different types of fillers such as GNP, EIP, or their hybrids were added, and their dispersion was studied. From Figure 2d–f, it was found that the GNP particles are dispersed uniformly while very few aggregates can be noticed at a high resolution, as in Figure 2e. However, the influence of these aggregates is not severe enough to affect the composites' properties. So, the properties of the GNP-filled composites were expected to be higher, and were studied, as shown in Figures 3–6. Figure 2g–i shows the dispersion of EIP in the rubber matrix. It can be observed from the SEM images that the EIP are also uniformly distributed. However, due to the large particle size of the EIP, the surface roughness was higher and there were fewer EIP when compared to the GNPs' particle distribution in the composites. Similarly, the distribution of the hybrid filler was studied in Figure 2j–l. It was found that, in general, the GNP particles are found in the vicinity of the EIP, and the interfacial interaction of the EIP in the composites could be improved by the GNP particles. So, a sort of synergistic aspect was generated in their dispersion and this led to better properties in the hybrid composites.

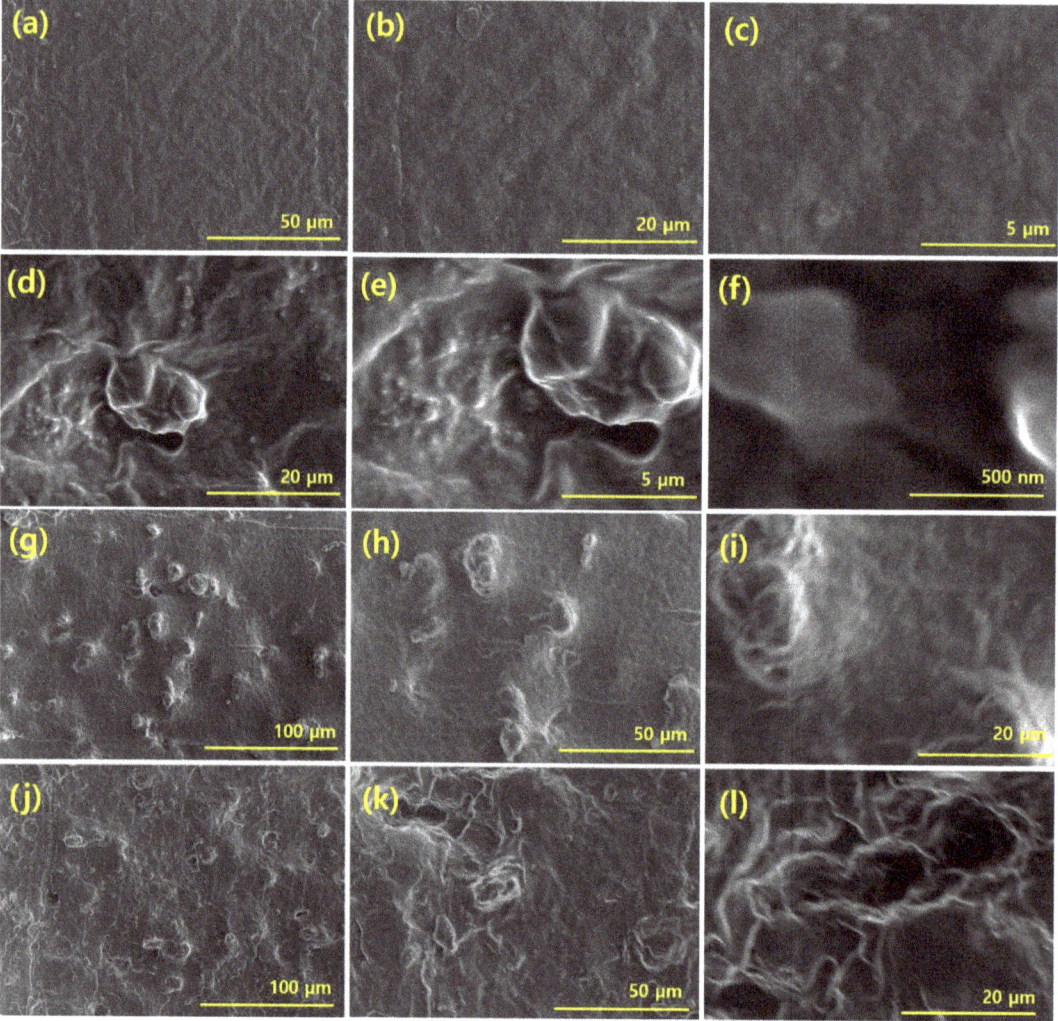

Figure 2. SEM of composites: (**a**–**c**) control; (**d**–**f**) 10 phr of GNPs (**g**–**i**) and 60 phr of EIP; (**j**–**l**) 60 phr of hybrid.

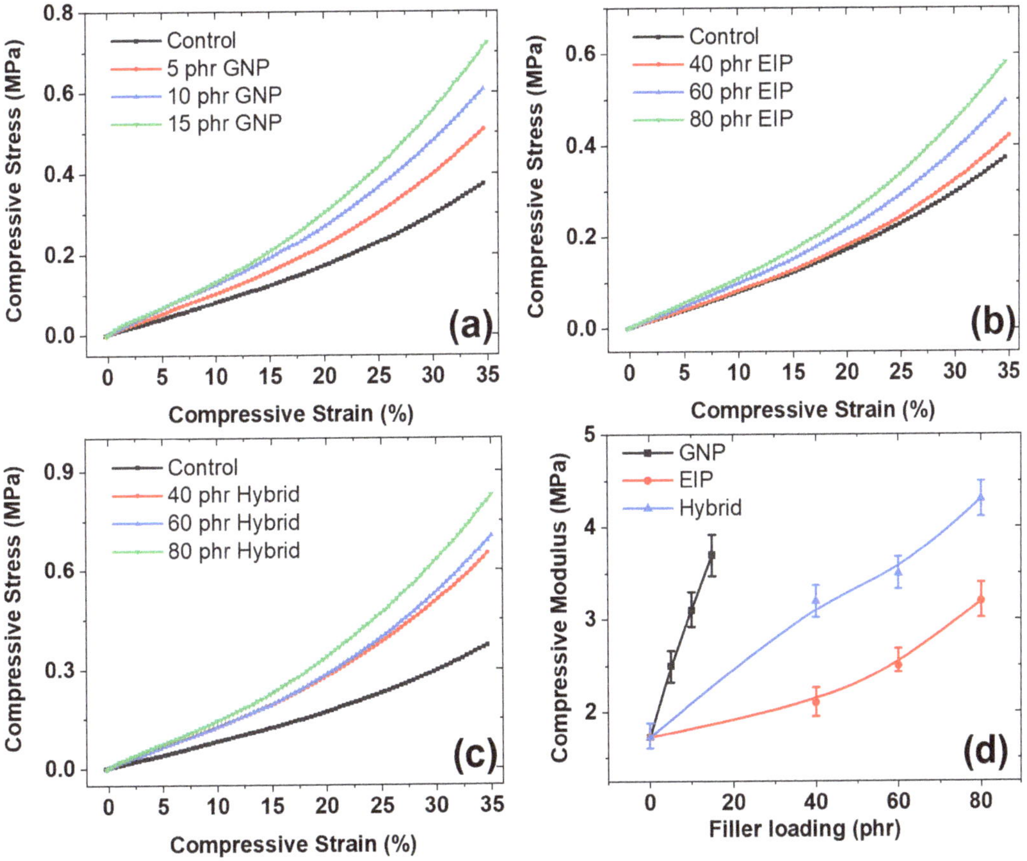

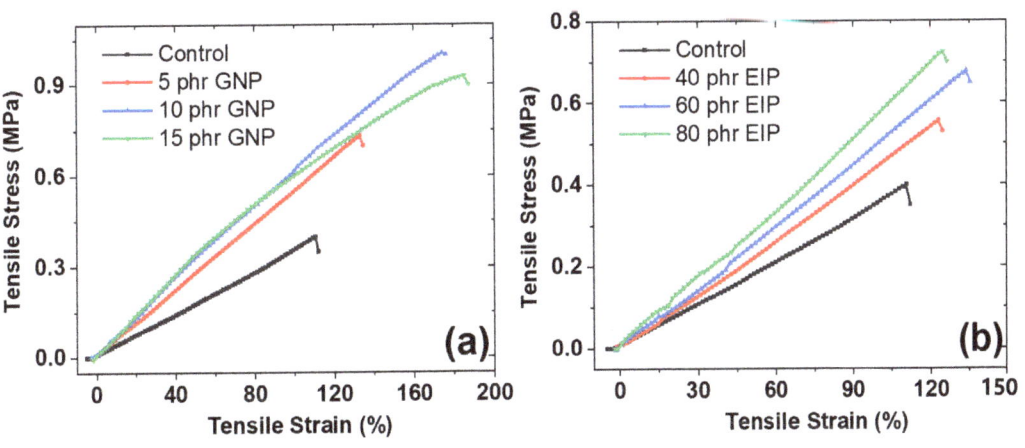

Figure 3. Compressive stress–strain profiles: (**a**) GNP; (**b**) EIP; (**c**) hybrid; (**d**) compressive modulus for different fillers.

Figure 4. *Cont.*

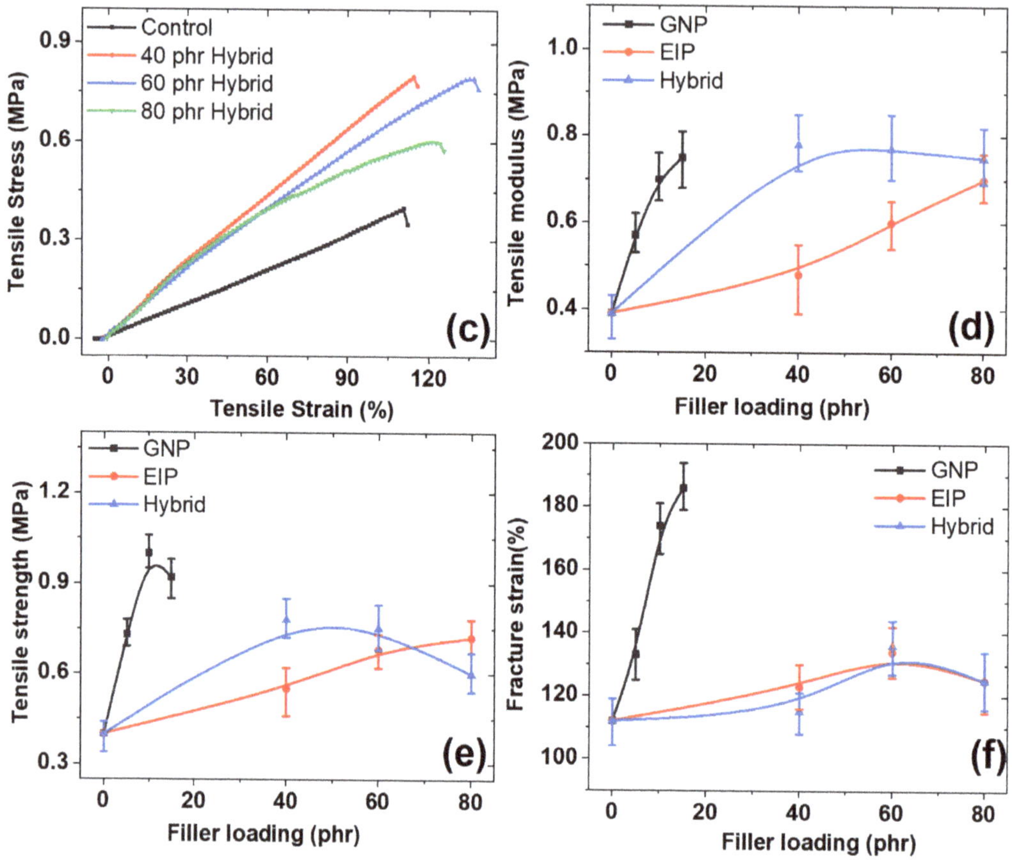

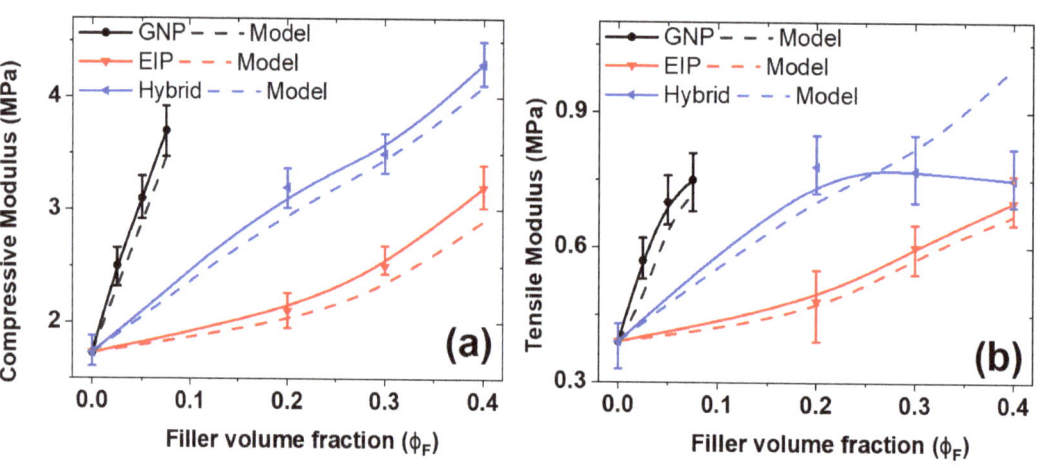

Figure 4. Stress–strain profiles under tensile strain: (**a**) GNP composites; (**b**) EIP composites; (**c**) hybrid composites; (**d**) tensile modulus of different composites; (**e**) tensile strength of different composites; (**f**) fracture strain of composites.

Figure 5. *Cont.*

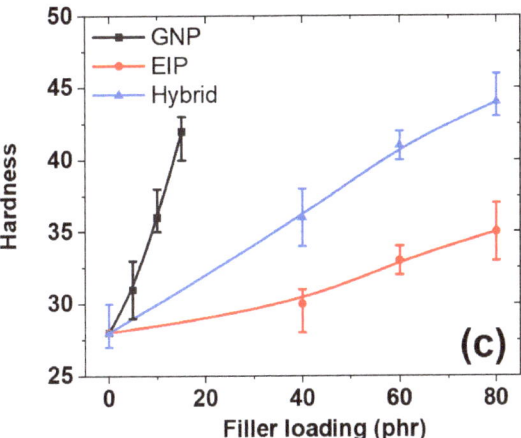

Figure 5. The experimental values compared to the theoretical modeling: (**a**) compressive modulus and prediction through models; (**b**) tensile moduli and their prediction through models; (**c**) hardness of different composites.

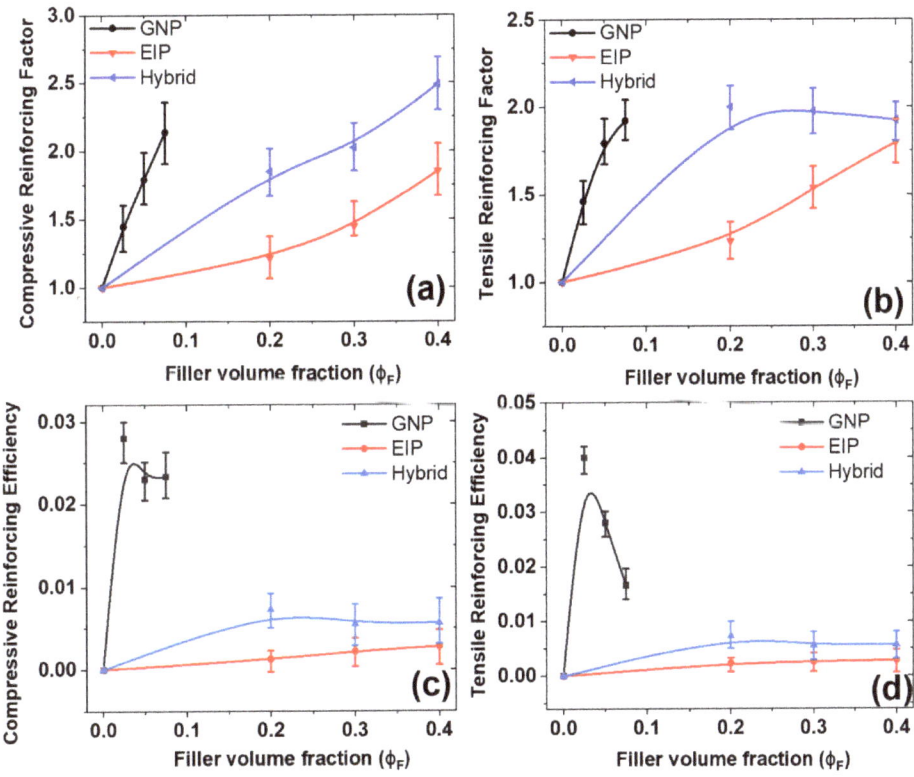

Figure 6. (**a**) Compressive-reinforcing factor of composites; (**b**) tensile-reinforcing factor of composites; (**c**) compressive-reinforcing efficiency of composites; (**d**) tensile-reinforcing efficiency of composites.

3.3. Mechanical Properties under Compressive Strain

Composites' mechanical properties depend on the distribution of the filler [47]; the filler's characteristics [48], such as the type of filler, the nature of the filler, the shape of the filler, the size of the filler, or the filler's aspect ratio; the cross-linking density of the curatives [49]; the type of rubber matrix [50]; and the type of applied strain under which the mechanical properties are tested [51]. Herein, fillers with different characteristics were used in silicone rubber and their effects on the rubber's mechanical properties were tested. Figure 3a–c provide the compressive stress–strain behavior of the composites with different types of fillers and their hybrids. All the stress–strain profiles show that the stress increases with an increase in the compressive strain. This trend is attributed to an increase in the packing fraction of the polymer chains and filler particles with the increasing compressive strain [51]. This process increases the stiffness in the composite, thereby leading to higher compressive stress at a higher level of compressive strain. It is also interesting to note that the GNP-filled composites show higher compressive strain values at all strain levels, even at lower filler loadings. These results agree with the results obtained for GNPs in the literature [52]. This phenomenon is due to the higher aspect ratio of the GNPs, which induces improved filler networking of the GNPs in the composites leading to higher compressive stress [52]. It is also interesting to note that the hybrid composite shows improved stress resistance, which could be due to the higher filler loading and synergism among the binary filler particles in the composite [53].

The behavior of the compressive modulus at different filler loadings is presented in Figure 3d. The GNPs show excellent modulus values—even at a lower loading—than the EIP. This is attributed to (a) the high aspect ratio of the GNPs, which led to the formation of continuous filler networks throughout the rubber matrix even at a lower filler loading [52]; (b) the high surface area of the GNPs, which leads to the availability of a higher interfacial area that allows for better stress transfer from the polymer to the filler particles [54]; and (c) the high levels of filler–filler and polymer–filler interactions due to the presence of the large interfacial area and small particle size of the GNPs [55]. Moreover, the hybrid specimen shows a significantly higher compressive modulus than the EIP at the same filler loading. This can be attributed to (a) the higher reinforcing ability of the GNPs in the hybrid filler that leads to a remarkable increase in the compressive modulus, and (b) the favorable positive synergism among the binary fillers that leads to a remarkable increase in the compressive modulus [53]. In the end, the EIP show poor reinforcing properties in all the composites. This is attributed to (a) the micron-sized particles that lead to poor stress transfer even at high filler loadings and their small aspect ratio that leads to the formation of non-efficient and discontinuous filler networks.

3.4. Mechanical Properties under Tensile Strain

The stress–strain behavior of the different composites was studied and presented in Figure 4a–c. For all the composites, it was witnessed that the stress increases with the increasing strain until fracture. This behavior was attributed to improved interfacial interaction between the filler and rubber matrix [56]. The increase in tensile strain leads to filler particles and polymer chains oriented against the direction of the applied strain [57]. This causes mechanical resistance against the applied strain and leads to an increase in stress with the increasing strain [58].

It is also interesting to note that the GNP-based composites show higher tensile stress and fracture strain at the applied strain than EIP and the hybrid composites. The improved stress is due to the efficient filler networking that allows for better stress transfer within the composites [59]. The improved fracture strain is due to the lubricating effect of the GNPs, which is higher with a higher number of GNPs [60].

The behavior of other mechanical properties such as the moduli (Figure 4d), tensile strength (Figure 4e), and fracture strain (Figure 4f) are presented. Among all the experimental data, the control sample shows the lowest mechanical strength and stiffness. This is due to the absence of reinforcing fillers in virgin rubber. Besides this, it was found that

GNP exhibited outstanding performance among all the fillers studied irrespective of the smaller loadings, showing values of up to 15 phr greater than EIP and their hybrid at 80 phr. The tensile modulus was higher for the GNP-filled composites followed by the hybrid composites and was lowest for the EIP-filled composites. This result is attributed to the high reinforcing effect of the GNPs due to their favorable morphological features such as a high aspect ratio [61].

The EIP show poor properties due to their poor reinforcing effects resulting from their poor morphological features, such as their large particle size and small aspect ratio. The hybrid specimen shows medium modulus values, which are higher than the EIP and lower than the GNPs. This is due to the semi-effect of the GNPs in the hybrid composite that forms synergism among the GNPs and EIP in the hybrid specimen. Moreover, as expected, the tensile strength and fracture strain were also superior for the GNP-based filler. This behavior agrees with the mechanical performance obtained in Figures 3 and 4a–d. The higher tensile strength is due to the improved filler network formation of the GNPs, its interaction with the rubber matrix, and its ability to achieve good stress transfer from the rubber matrix to GNP particles [62]. The improved fracture strain of the GNPs is due to their lubricating effect and favorable platelet morphology that form three-dimensional networks in the rubber matrix, leading to higher fracture strain.

3.5. Theoretical Models and Hardness of Composites

The prediction of mechanical properties is a well-known subject of research in the literature [63]. These models help to estimate the deviation of experimental results from theoretical models. The existing theoretical models such as the Guth–Gold Smallwood equations [64,65] were used in the present work to understand their deviation from the experimental findings. These models are known to depend on the aspect ratio of fillers such as GNP or EIP for the one component and the hybrid for the two-component system, the filler volume fraction of the fillers, and their interactive factors, especially for the hybrid two-component system. The Guth–Gold Smallwood equations for a one and two-component system [66] are as follows:

$$E_{GNP} = E_o (1 + 0.67 f_{GNP} \phi_{GNP}) \quad (1)$$

$$E_{EIP} = E_o (1 + 0.67 f_{EIP} \phi_{EIP}) \quad (2)$$

$$E_{GNP+EIP} = E_o [(1 + 0.67 f_{GNP} \phi_{GNP}) + (1 + 0.67 f_{EIP} \phi_{EIP})] \times a \quad (3)$$

where E_o is the modulus of the control composite, E_{GNP} is the predicted modulus for GNP, E_{EIP} is the predicted modulus for EIP, $E_{GNP+EIP}$ is the predicted modulus for the hybrid filler, f_{GNP} is the aspect ratio of GNP, f_{EIP} is the aspect ratio of EIP, ϕ_{GNP} is the volume fraction of GNP, ϕ_{EIP} is the volume fraction of EIP, and a is the interacting factor of the binary fillers in the composite.

From Figure 5a,b, it is evident that our experimental data are in good agreement with the theoretical models, except for the hybrid data of the tensile modulus in Figure 4b. The agreement of the experiments with the theoretical model confirms that the experimental data are reliable and is an important contribution to the composite field. However, the deviation of the hybrid modulus with the predicted modulus in Figure 5b after 60 phr could be due to the aggregation of the binary fillers that leads to a decrease in the modulus and the deviation of the results. In principle, the model assumes perfect dispersion and perfect filler–polymer interaction, which are hard to establish experimentally. So, few results deviate from the predicted model [66,67].

The hardness of the composites was studied to determine whether each filled composite was hard or soft in nature. In many cases, if the hardness of the composites falls below 65, it is termed a soft composite [68]. These soft composites can be useful for industrial applications such as soft robotics [69]. In the present work, the hardness of different composites was studied and is presented in Figure 5c. It was found that the GNP composite shows a more robust hardness increase with the addition of GNPs compared to the hybrid and

EIP. The higher hardness in the GNP composite can be attributed to the higher reinforcing effect of GNPs, which is due to favorable features such as their high aspect ratio. These results agree with the mechanical data presented in Figures 3 and 4 in the mechanical properties section.

3.6. Reinforcement by Particulate Nanofillers

The reinforcing factor and reinforcing efficiency of the different composites under compressive and tensile strain were studied and are presented in Figure 6. The reinforcing factor was calculated by

$$\text{R.F.} = \frac{EF}{Eo} \qquad (4)$$

where R.F. is the reinforcing factor, EF is the modulus of the filled composites, and Eo is the modulus of the control specimens. Similarly, the reinforcing efficiency [70] was calculated:

$$\text{R.E. at compressive strain} = \frac{\sigma\,(35\%)\text{filled} - \sigma\,(35\%)\text{unfilled}}{\text{wt\% of filler}} \qquad (5)$$

$$\text{R.E. at tensile strain} = \frac{\sigma\,(100\%)\text{filled} - \sigma\,(100\%)\text{unfilled}}{\text{wt\% of filler}} \qquad (6)$$

where R.E. is the reinforcing efficiency; σ is the stress at a particular strain, which is 35% for compressive strain and 100% for tensile strain; and wt% is the weight of the filler. The compressive and tensile R.F. (Figure 6a,b) show that the GNPs have a significant reinforcing effect over the EIP-filled composites. This can be attributed to the small particle size of the GNPs that disperse in the rubber matrix in a highly exfoliated state [71].

Besides this, the EIP have a poor reinforcing effect due to their large particle size, which is dispersed unevenly in the rubber matrix. The filler networks are also more efficient in the GNP-filled composites than in the EIP-reinforced ones. These results agree with the mechanical properties studied in Figures 3–5. Moreover, the compressive and tensile R.E. (Figure 6c,d) also show a superior reinforcing effect of the GNPs in the rubber matrix. These results agree with the results presented in the mechanical properties section in Figures 3–5. It is also interesting to note that the R.E. decreased with an increase in the filler loading. This trend can be explained by Equations (5) and (6), wherein the R.E. is inversely proportional to the filler loading [70]. So, when the filler loading increases, the R.E. decreases for all the filled composites. It is noteworthy that the hybrid shows higher R.E. than the EIP-filled composites at all filler loadings. This can be attributed to the synergism between the binary fillers in the hybrid composite. The EIP-filled composites show poor R.E. at all loadings due to their poor aspect ratio and large particle size that is unevenly distributed in the rubber matrix even at large filler loadings of 60–80 phr.

3.7. Anisotropic Magnetic Effect in Mechanical Properties

The iso-anisotropic stress–strain curves are presented for the EIP-filled composites (Figure 7a) and hybrid composites (Figure 7b). Two types of effects were noticed from the stress–strain profiles: (a) a stress change due to the effect of the magnetic field in the composites, and (b) a stress change due to a change in the compressive strain [72,73]. The first change in stress under a magnetic field is due to the orientation of EIP in the direction of the magnetic field, which leads to a change in compressive stress. The second change in stress under compressive strain is due to the packing of filler particles, which increases with the increases in stress, as described previously in Figure 3.

Figure 7c shows the behavior of the compressive modulus in the presence and absence of a magnetic field. It can be seen that when the composites containing EIP or the hybrid were subjected to a magnetic field, the compressive modulus increased. This increase in the compressive modulus could be due to the orientation of the iron particles forming chain-like structures in both the EIP and hybrid composites [74]. Such an effect is also called the anisotropic effect, as described in Figure 7d. The anisotropic effect is defined as an effect in which a change in mechanical stiffness occurs when isotropic composites are subjected to a magnetic field [75].

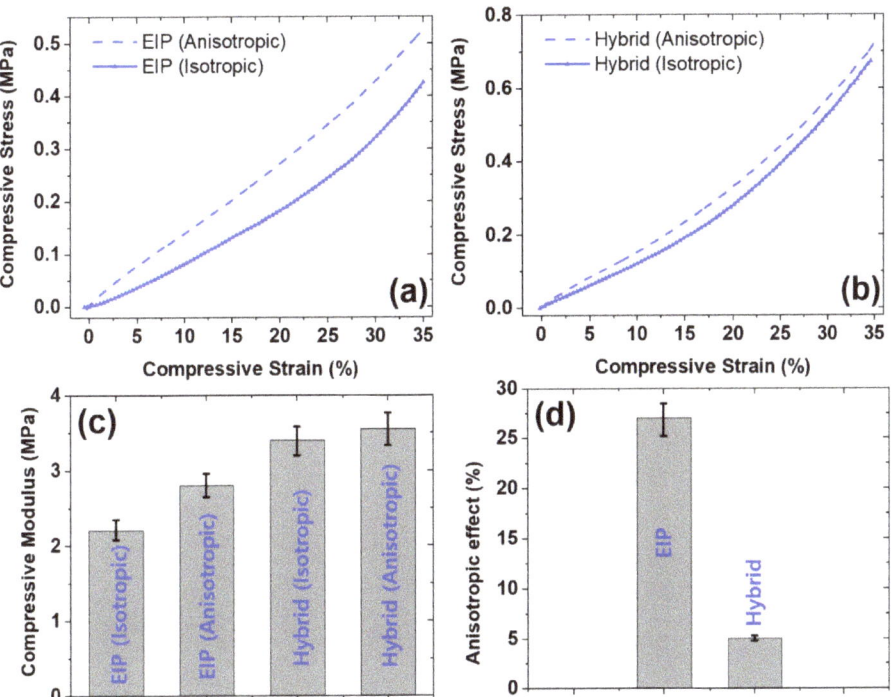

Figure 7. Iso-anisotropic compressive stress–strain behavior at 60 phr and 1 T magnetic field of (a) EIP filled composites; (b) hybrid-filled composites; (c) compressive modulus for different composites; (d) anisotropic effect of different composites.

It is interesting to note that the anisotropic effect was higher for the EIP-filled composites than for the hybrid composites, even though the modulus was higher in the hybrid composites. The higher anisotropic effect in the EIP is due to the greater number of vacancies in the composite containing EIP, which allows them to orient freely thereby causing a higher degree of uniform orientation. On the other hand, the GNPs in the hybrid occupy almost all the vacancies, thereby making it difficult for the EIP to align. Due to this reason, the anisotropy was higher in the EIP than in the hybrid composites.

3.8. Mechanism of the Anisotropic Magnetic Effect

Different types of filler–polymer microstructures are formed in the presence and absence of a magnetic field. In the present work, the magnetic field was applied in the pre-curing state of the composite. The magnitude of the magnetic field was 1 T, and the duration of the field was 90 min. It was found from the tests that the EIP are oriented in the direction of the magnetic field, thereby influencing their mechanical properties such as their moduli, as presented in Figure 7.

The mechanism involves the presence of a restorative force that opposes the field-aligned orientation of the magnetic particles in the composites. Therefore, the magnetic field should be higher than this restorative force to orient the magnetic EIP. It is clear from Figure 8 that the restorative force in the case of the hybrid filler is higher than in the EIP-filled composites. So, under the same magnetic field, the degree of orientation of the EIP is lower in the hybrid composites than in the EIP composites.

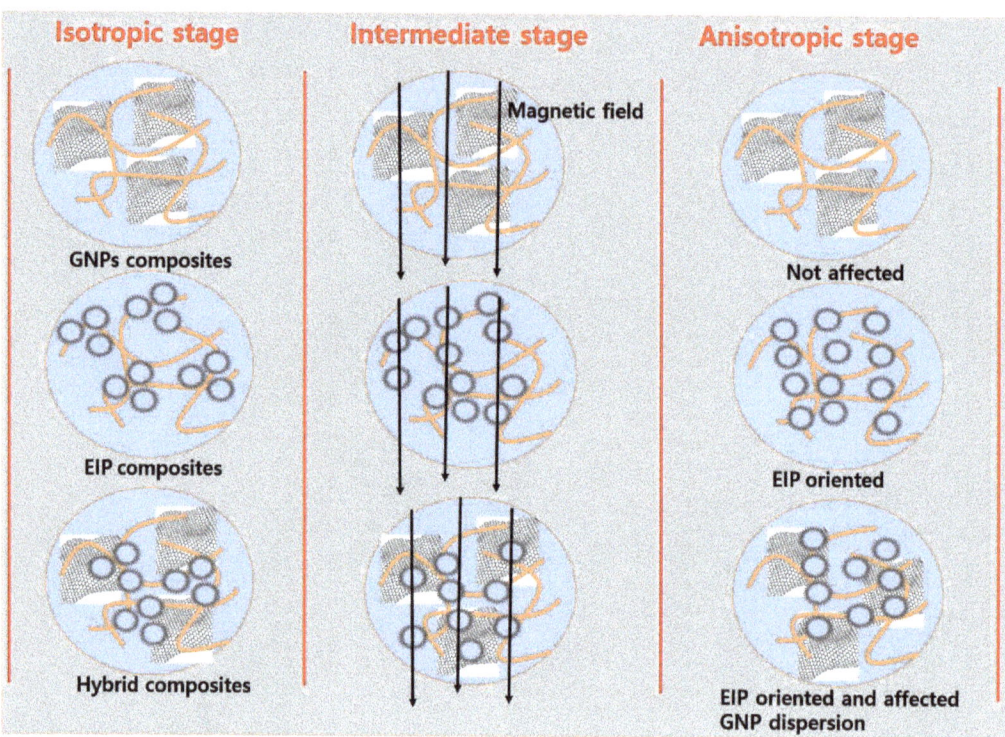

Figure 8. Mechanism of isotropic and anisotropic effects.

There are three stages of the microstructure formation of the filler and polymer chains in the composites during the curing process. These processes are (a) the isotropic stage, in which the filler particles are randomly distributed in all the composites investigated (0 T of magnetic field); (b) the intermediate stage, in which the EIP or magnetic particles are partially oriented at 1 T of the magnetic field; and (c) the mature or anisotropic stage, in which the EIP orientation attains maturity and the magnetic exposure of 90 min—as in the present work—is achieved. The isotropic and anisotropic effects on the mechanical properties are presented in Figure 7. The alignment of the EIP under a magnetic field could be due to several reasons. However, the key factors that influence their orientation could be (a) the morphology, shape, size, and polymer–filler interactions in the composite; (b) the occurrence of a field-induced dipole among the magnetic filler particles [76]; or (c) the filler networking density, which significantly affects the orientation of the magnetic filler particles. These features of MREs are useful for various applications such as electromagnetic absorbers and various other functions as reported in the literature [77–80].

4. Conclusions

The robust performance of the stretchable magnetic materials is presented in this work. After the successful preparation of the composites by solution mixing, the mechanical properties of the composites were studied in the presence and absence of a magnetic field. The mechanical properties were investigated under compressive and tensile strain. The results obtained from these mechanical properties show that both the compressive modulus and mechanical stretchability increase with an increasing filler content, wherein GNPs emerge as an outstanding filler material. For example, the compressive modulus was 1.73 MPa (control) and increased to 3.7 MPa (GNP) at 15 per hundred parts of rubber (phr), 3.2 MPa (EIP), and 4.3 MPa (hybrid) at 80 phr. Similarly, the mechanical stretchability was

112% (control) and increased to 186% (GNP) at 15 phr, 134% (EIP), and 136% (hybrid) at 60 phr. Moreover, the GNPs show a higher reinforcing factor and reinforcing efficiency; however, the GNPs lack magnetic sensitivity due to a lack of iron particles and are not suitable for stretchable magnetic materials. Therefore, EIP and a GNP hybrid were prepared and studied. The mechanism involves the orientation of EIP under a magnetic field causing a magnetic effect, which is 28% for EIP and 5% for the hybrid. Thus, EIP and the hybrid-filled composites emerge as promising candidates for stretchable magnetic materials, which is the main theme of this work.

Author Contributions: Conceptualization, V.K.; Methodology, V.K. and M.N.A.; software, V.K.; validation, V.K. and M.N.A.; formal analysis, V.K.; investigation, V.K. and M.N.A.; resources, S.-S.P. and D.-J.L.; data curation, V.K.; writing—original draft preparation, V.K.; writing—review and editing, V.K.; visualization, V.K.; supervision, S.-S.P. and D.-J.L.; project administration, S.-S.P. and D.-J.L. All authors have read and agreed to the published version of the manuscript.

Funding: This work received no external funding.

Institutional Review Board Statement: Not applicable.

Informed Consent Statement: Not applicable.

Data Availability Statement: Not applicable.

Conflicts of Interest: The authors declare no conflict of interest.

References

1. Melzer, M.; Makarov, D.; Schmidt, O.G. A review on stretchable magnetic field sensorics. *J. Phys. D Appl. Phys.* **2019**, *53*, 083002. [CrossRef]
2. Lazarus, N.; Bedair, S.S. Improved power transfer to wearable systems through stretchable magnetic composites. *Appl. Phys. A* **2016**, *122*, 1–7. [CrossRef]
3. Abramchuk, S.; Kramarenko, E.; Stepanov, G.; Nikitin, L.V.; Filipcsei, G.; Khokhlov, A.R.; Zrínyi, M. Novel highly elastic magnetic materials for dampers and seals: Part I. Preparation and characterization of the elastic materials. *Polym. Adv. Technol.* **2007**, *18*, 883–890. [CrossRef]
4. Eldin, N.N.; Senouci, A.B. Rubber-Tire Particles as Concrete Aggregate. *J. Mater. Civ. Eng.* **1993**, *5*, 478–496. [CrossRef]
5. Li, T.; El-Aty, A.A.; Cheng, C.; Shen, Y.; Wu, C.; Yang, Q.; Hu, S.; Xu, Y.; Tao, J.; Guo, X. Investigate the Effect of the Magnetic Field on the Mechanical Properties of Silicone Rubber-Based Anisotropic Magnetorheological Elastomer during Curing Process. *J. Renew. Mater.* **2020**, *8*, 1411–1427. [CrossRef]
6. Khimi, S.R.; Pickering, K. The effect of silane coupling agent on the dynamic mechanical properties of iron sand/natural rubber magnetorheological elastomers. *Compos. Part B Eng.* **2016**, *90*, 115–125. [CrossRef]
7. Bica, I. The influence of the magnetic field on the elastic properties of anisotropic magnetorheological elastomers. *J. Ind. Eng. Chem.* **2012**, *18*, 1666–1669. [CrossRef]
8. Kumar, V.; Alam, N.; Park, S.S. Robust magneto-rheological elastomers performance for composites based on iron oxide and carbon black in silicone rubber. *J. Polym. Res.* **2022**, *29*, 1–11. [CrossRef]
9. Aloui, S.; Klüppel, M. Magneto-rheological response of elastomer composites with hybrid-magnetic fillers. *Smart Mater. Struct.* **2014**, *24*, 025016. [CrossRef]
10. Robertson, C.G.; Hardman, N.J. Nature of Carbon Black Reinforcement of Rubber: Perspective on the Original Polymer Nanocomposite. *Polymers* **2021**, *13*, 538. [CrossRef]
11. Coleman, J.N.; Khan, U.; Blau, W.J.; Gun'Ko, Y.K. Small but strong: A review of the mechanical properties of carbon nanotube–polymer composites. *Carbon* **2006**, *44*, 1624–1652. [CrossRef]
12. Young, R.J.; Kinloch, I.A.; Gong, L.; Novoselov, K.S. The mechanics of graphene nanocomposites: A review. *Compos. Sci. Technol.* **2012**, *72*, 1459–1476. [CrossRef]
13. Fan, Y.; Fowler, G.D.; Zhao, M. The past, present and future of carbon black as a rubber reinforcing filler—A review. *J. Clean. Prod.* **2020**, *247*, 119115. [CrossRef]
14. Bokobza, L. Multiwall carbon nanotube elastomeric composites: A review. *Polymer* **2007**, *48*, 4907–4920. [CrossRef]
15. Zhu, Q.; Wang, Z.; Zeng, H.; Yang, T.; Wang, X. Effects of graphene on various properties and applications of silicone rubber and silicone resin. *Compos. Part A Appl. Sci. Manuf.* **2020**, *142*, 106240. [CrossRef]
16. Coran, A.Y. Chemistry of the vulcanization and protection of elastomers: A review of the achievements. *J. Appl. Polym. Sci.* **2002**, *87*, 24–30. [CrossRef]
17. Wood, L.A. Synthetic Rubbers: A Review of Their Compositions, Properties, and Uses. *Rubber Chem. Technol.* **1940**, *13*, 861–885. [CrossRef]
18. Bokobza, L. Natural Rubber Nanocomposites: A Review. *Nanomaterials* **2019**, *9*, 12. [CrossRef]

19. Ranta, P.M.; Ownby, D.R. A Review of Natural-Rubber Latex Allergy in Health Care Workers. *Clin. Infect. Dis.* **2004**, *38*, 252–256. [CrossRef]
20. Singha, N.K.; Bhattacharjee, S.; Sivaram, S.J.R.C. Hydrogenation of diene elastomers, their properties and applications: A critical review. *Rubber Chem. Technol.* **1997**, *70*, 309–367. [CrossRef]
21. Shit, S.C.; Shah, P. A review on silicone rubber. *Natl. Acad. Sci. Lett.* **2013**, *36*, 355–365. [CrossRef]
22. Lewis, F.M. The Science and Technology of Silicone Rubber. *Rubber Chem. Technol.* **1962**, *35*, 1222–1275. [CrossRef]
23. El-Hag, A.; Jayaram, S.; Cherney, E. Fundamental and low frequency harmonic components of leakage current as a diagnostic tool to study aging of RTV and HTV silicone rubber in salt-fog. *IEEE Trans. Dielectr. Electr. Insul.* **2003**, *10*, 128–136. [CrossRef]
24. Kumar, V.; Tang, X.-W.; Liu, S.-C.; Lee, D.-J. Studies on nanocomposites reinforced with CNTs in different types of dielectric rubber. *Sensors Actuators A Phys.* **2017**, *267*, 310–317. [CrossRef]
25. Kumar, V.; Lee, G.; Singh, K.; Choi, J.; Lee, D.-J. Structure-property relationship in silicone rubber nanocomposites reinforced with carbon nanomaterials for sensors and actuators. *Sens. Actuators A Phys.* **2019**, *303*, 111712. [CrossRef]
26. Kumar, V.; Alam, N.; Manikkavel, A.; Song, M.; Lee, D.-J.; Park, S.-S. Silicone Rubber Composites Reinforced by Carbon Nanofillers and Their Hybrids for Various Applications: A Review. *Polymers* **2021**, *13*, 2322. [CrossRef]
27. George, J.; Ishida, H. A review on the very high nanofiller-content nanocomposites: Their preparation methods and properties with high aspect ratio fillers. *Prog. Polym. Sci.* **2018**, *86*, 1–39. [CrossRef]
28. Bhattacharya, M.; Bhowmick, A.K. Synergy in carbon black-filled natural rubber nanocomposites. Part I: Mechanical, dynamic mechanical properties, and morphology. *J. Mater. Sci.* **2010**, *45*, 6126–6138. [CrossRef]
29. Kozlov, G.V.; Dolbin, I.V. Transfer of Mechanical Stress from Polymer Matrix to Nanofiller in Dispersion-Filled Nanocomposites. *Inorg. Mater. Appl. Res.* **2019**, *10*, 226–230. [CrossRef]
30. Samal, S.; Škodová, M.; Abate, L.; Blanco, I. Magneto-rheological elastomer composites. A review. *Appl. Sci.* **2020**, *10*, 4899. [CrossRef]
31. Kim, Y.; Zhao, X. Magnetic Soft Materials and Robots. *Chem. Rev.* **2022**, *122*, 5317–5364. [CrossRef]
32. Arumugam, A.B.; Subramani, M.; Dalakoti, M.; Jindal, P.; Selvaraj, R.; Khalife, E. Dynamic characteristics of laminated composite CNT reinforced MRE cylindrical sandwich shells using HSDT. *Mech. Based Des. Struct. Mach.* **2021**, 1–17. [CrossRef]
33. Li, R.; Wang, D.; Yang, P.A.; Tang, X.; Liu, J.; Li, X. Improved magneto-sensitive adhesion property of magnetorheological elastomers modified using graphene nanoplatelets. *Ind. Eng. Chem. Res.* **2020**, *59*, 9143–9151. [CrossRef]
34. Selvaraj, R.; Ramamoorthy, M. Experimental and finite element vibration analysis of CNT reinforced MR elastomer sandwich beam. *Mech. Based Des. Struct. Mach.* **2020**, *50*, 2414–2426. [CrossRef]
35. Arani, A.G.; Shahraki, M.E.; Haghparast, E. Instability analysis of axially moving sandwich plates with a magnetorheological elastomer core and GNP-reinforced face sheets. *J. Braz. Soc. Mech. Sci. Eng.* **2022**, *44*, 1–22. [CrossRef]
36. Fan, L.; Wang, G.; Wang, W.; Lu, H.; Yang, F.; Rui, X. Size effect of carbon black on the structure and mechanical properties of magnetorheological elastomers. *J. Mater. Sci.* **2018**, *54*, 1326–1340. [CrossRef]
37. Rahaman, M.; Chaki, T.K.; Khastgir, D. Development of high performance EMI shielding material from EVA, NBR, and their blends: Effect of carbon black structure. *J. Mater. Sci.* **2011**, *46*, 3989–3999. [CrossRef]
38. Kumar, V.; Alam, N.; Park, S.S. Soft Composites Filled with Iron Oxide and Graphite Nanoplatelets under Static and Cyclic Strain for Different Industrial Applications. *Polymers* **2022**, *14*, 2393. [CrossRef]
39. Xie, L.; Zhu, Y. Tune the phase morphology to design conductive polymer composites: A review. *Polym. Compos.* **2017**, *39*, 2985–2996. [CrossRef]
40. Cataldi, P.; Bayer, I.S.; Nanni, G.; Athanassiou, A.; Bonaccorso, F.; Pellegrini, V.; Castillo, A.E.D.R.; Ricciardella, F.; Artyukhin, S.; Tronche, M.-A.; et al. Effect of graphene nano-platelet morphology on the elastic modulus of soft and hard biopolymers. *Carbon* **2016**, *109*, 331–339. [CrossRef]
41. Li, J.; Kim, J.-K. Percolation threshold of conducting polymer composites containing 3D randomly distributed graphite nanoplatelets. *Compos. Sci. Technol.* **2007**, *67*, 2114–2120. [CrossRef]
42. Chong, H.M.; Hinder, S.J.; Taylor, A.C. Graphene nanoplatelet-modified epoxy: Effect of aspect ratio and surface functionality on mechanical properties and toughening mechanisms. *J. Mater. Sci.* **2016**, *51*, 8764–8790. [CrossRef]
43. Safdari, M.; Al-Haik, M.S. Synergistic electrical and thermal transport properties of hybrid polymeric nanocomposites based on carbon nanotubes and graphite nanoplatelets. *Carbon* **2013**, *64*, 111–121. [CrossRef]
44. Yu, L.-M.; Huang, H.-X. Temperature and shear dependence of rheological behavior for thermoplastic polyurethane nanocomposites with carbon nanofillers. *Polymer* **2022**, *247*, 124791. [CrossRef]
45. Nasser, K.I.; del Río, J.M.L.; Mariño, F.; López, E.R.; Fernández, J. Double hybrid lubricant additives consisting of a phosphonium ionic liquid and graphene nanoplatelets/hexagonal boron nitride nanoparticles. *Tribol. Int.* **2021**, *163*, 107189. [CrossRef]
46. Wang, Z.; Wang, X.; Liu, M.; Gao, Y.; Hu, Z.; Nan, T.; Liang, X.; Chen, H.; Yang, J.; Cash, S.; et al. Highly Sensitive Flexible Magnetic Sensor Based on Anisotropic Magnetoresistance Effect. *Adv. Mater.* **2016**, *28*, 9370–9377. [CrossRef]
47. Hussain, M.; Oku, Y.; Nakahira, A.; Niihara, K. Effects of wet ball-milling on particle dispersion and mechanical properties of particulate epoxy composites. *Mater. Lett.* **1996**, *26*, 177–184. [CrossRef]
48. Tan, J.C.; Cheetham, A.K. Mechanical properties of hybrid inorganic–organic framework materials: Establishing fundamental structure–property relationships. *Chem. Soc. Rev.* **2011**, *40*, 1059–1080. [CrossRef]

49. Zhao, F.; Bi, W.; Zhao, S. Influence of Crosslink Density on Mechanical Properties of Natural Rubber Vulcanizates. *J. Macromol. Sci. Part B* **2011**, *50*, 1460–1469. [CrossRef]
50. McMillin, C.R. Biomedical Applications of Rubbers and Elastomers. *Rubber Chem. Technol.* **2006**, *79*, 500–519. [CrossRef]
51. Budzien, J.; McCoy, J.D.; Adolf, D.B. Solute mobility and packing fraction: A new look at the Doolittle equation for the polymer glass transition. *J. Chem. Phys.* **2003**, *119*, 9269–9273. [CrossRef]
52. Kumar, V.; Kumar, A.; Alam, N.; Park, S. Effect of graphite nanoplatelets surface area on mechanical properties of room-temperature vulcanized silicone rubber nanocomposites. *J. Appl. Polym. Sci.* **2022**, *139*, e52503. [CrossRef]
53. Sagalianov, I.; Vovchenko, L.; Matzui, L.; Lazarenko, O. Synergistic Enhancement of the Percolation Threshold in Hybrid Polymeric Nanocomposites Based on Carbon Nanotubes and Graphite Nanoplatelets. *Nanoscale Res. Lett.* **2017**, *12*, 140. [CrossRef]
54. Cilento, F.; Martone, A.; Carbone, M.G.P.; Galiotis, C.; Giordano, M. Nacre-like GNP/Epoxy composites: Reinforcement efficiency vis-a-vis graphene content. *Compos. Sci. Technol.* **2021**, *211*, 108873. [CrossRef]
55. Wang, M.-J. Effect of polymer-filler and filler-filler interactions on dynamic properties of filled vulcanizates. *Rubber Chem. Technol.* **1998**, *71*, 520–589. [CrossRef]
56. Fröhlich, J.; Niedermeier, W.; Luginsland, H.D. The effect of filler–filler and filler–elastomer interaction on rubber reinforcement. *Compos. Part A Appl. Sci. Manuf.* **2005**, *36*, 449–460. [CrossRef]
57. Scotti, R.; Wahba, L.; Crippa, M.; D'Arienzo, M.; Donetti, R.; Santo, N.; Morazzoni, F. Rubber–silica nanocomposites obtained by in situ sol–gel method: Particle shape influence on the filler–filler and filler–rubber interactions. *Soft Matter* **2012**, *8*, 2131–2143. [CrossRef]
58. Kim, D.B.; Lee, J.W.; Cho, Y.S. Anisotropic In Situ Strain-Engineered Halide Perovskites for High Mechanical Flexibility. *Adv. Funct. Mater.* **2021**, *31*, 2007131. [CrossRef]
59. Galimberti, M.; Kumar, V.; Coombs, M.; Cipolletti, V.; Agnelli, S.; Pandini, S.; Conzatti, L. Filler networking of a nanographite with a high shape anisotropy and synergism with carbon black in poly (1, 4-cis-isoprene)–based nanocomposites. *Rubber Chem. Technol.* **2014**, *87*, 197–218. [CrossRef]
60. Srivyas, P.D.; Charoo, M. Tribological behavior of hybrid aluminum self-lubricating composites under dry sliding conditions at elevated temperature. *Tribol. Mater. Surfaces Interfaces* **2021**, *16*, 1–15. [CrossRef]
61. Rohm, K.; Solouki Bonab, V.; Manas-Zloczower, I. In situ TPU/graphene nanocomposites: Correlation between filler aspect ratio and phase morphology. *Polym. Eng. Sci.* **2021**, *61*, 1018–1027. [CrossRef]
62. Rahman, O.A.; Sribalaji, M.; Mukherjee, B.; Laha, T.; Keshri, A.K. Synergistic effect of hybrid carbon nanotube and graphene nanoplatelets reinforcement on processing, microstructure, interfacial stress and mechanical properties of Al_2O_3 nanocomposites. *Ceram. Int.* **2018**, *44*, 2109–2122. [CrossRef]
63. Yu, J.; Kil Choi, H.; Kim, H.S.; Kim, S.Y. Synergistic effect of hybrid graphene nanoplatelet and multi-walled carbon nanotube fillers on the thermal conductivity of polymer composites and theoretical modeling of the synergistic effect. *Compos. Part A Appl. Sci. Manuf.* **2016**, *88*, 79–85. [CrossRef]
64. Fukahori, Y.; Hon, A.A.; Jha, V.; Busfield, J.J.C. Modified guth–gold equation for carbon black–filled rubbers. *Rubber Chem. Technol.* **2013**, *86*, 218–232. [CrossRef]
65. Wolff, S.; Donnet, J.-B. Characterization of Fillers in Vulcanizates According to the Einstein-Guth-Gold Equation. *Rubber Chem. Technol.* **1990**, *63*, 32–45. [CrossRef]
66. Lee, J.-Y.; Kumar, V.; Tang, X.-W.; Lee, D.-J. Mechanical and electrical behavior of rubber nanocomposites under static and cyclic strain. *Compos. Sci. Technol.* **2017**, *142*, 1–9. [CrossRef]
67. Domurath, J.; Saphiannikova, M.; Heinrich, G. The concept of hydrodynamic amplification in filled elastomers. *Kautsch. Gummi Kunstst.* **2017**, *70*, 40.
68. Wang, Y.-X.; Wu, Y.-P.; Li, W.-J.; Zhang, L.-Q. Influence of filler type on wet skid resistance of SSBR/BR composites: Effects from roughness and micro-hardness of rubber surface. *Appl. Surf. Sci.* **2010**, *257*, 2058–2065. [CrossRef]
69. Whitesides, G.M. Soft robotics. *Angew. Chem. Int. Ed.* **2018**, *57*, 4258–4273. [CrossRef]
70. Das, C.; Bansod, N.D.; Kapgate, B.P.; Reuter, U.; Heinrich, G.; Das, A. Development of highly reinforced acrylonitrile butadiene rubber composites via controlled loading of sol-gel titania. *Polymer* **2017**, *109*, 25–37. [CrossRef]
71. Sharma, A.; Morisada, Y.; Fujii, H. Bending induced mechanical exfoliation of graphene interlayers in a through thickness Al-GNP functionally graded composite fabricated via novel single-step FSP approach. *Carbon* **2022**, *186*, 475–491. [CrossRef]
72. Périgo, E.A.; Weidenfeller, B.; Kollár, P.; Füzer, J. Past, present, and future of soft magnetic composites. *Appl. Phys. Rev.* **2018**, *5*, 031301. [CrossRef]
73. Filipcsei, G.; Csetneki, I.; Szilágyi, A.; Zrínyi, M. Magnetic Field-Responsive Smart Polymer Composites. In *Oligomers—Polymer Composites Molecular Imprinting*; Springer: Berlin/Heidelberg, Germany, 2007; pp. 137–189. [CrossRef]
74. Erb, R.M.; Libanori, R.; Rothfuchs, N.; Studart, A.R. Composites Reinforced in Three Dimensions by Using Low Magnetic Fields. *Science* **2012**, *335*, 199–204. [CrossRef] [PubMed]
75. Shen, X.-J.; Dang, C.-Y.; Tang, B.-L.; Yang, X.-H.; Nie, H.-J.; Lu, J.-J.; Zhang, T.-T.; Friedrich, K. The reinforcing effect of oriented graphene on the interlaminar shear strength of carbon fabric/epoxy composites. *Mater. Des.* **2020**, *185*, 108257. [CrossRef]
76. Nielsen, L.E. Models for the Permeability of Filled Polymer Systems. *J. Macromol. Sci. Part A Chem.* **1967**, *1*, 929–942. [CrossRef]
77. Yang, Z.; Gao, R.; Hu, N.; Chai, J.; Cheng, Y.; Zhang, L.; Wei, H.; Kong, E.S.-W.; Zhang, Y. The Prospective Two-Dimensional Graphene Nanosheets: Preparation, Functionalization and Applications. *Nano-Micro Lett.* **2012**, *4*, 1–9. [CrossRef]

78. Lou, Z.; Wang, Q.; Zhou, X.; Kara, U.I.; Mamtani, R.S.; Lv, H.; Zhang, M.; Yang, Z.; Li, Y.; Wang, C.; et al. An angle-insensitive electromagnetic absorber enabling a wideband absorption. *J. Mater. Sci. Technol.* **2022**, *113*, 33–39. [CrossRef]
79. Lou, Z.; Wang, Q.; Sun, W.; Liu, J.; Yan, H.; Han, H.; Bian, H.; Li, Y. Regulating lignin content to obtain excellent bamboo-derived electromagnetic wave absorber with thermal stability. *Chem. Eng. J.* **2022**, *430*, 133178. [CrossRef]
80. Lou, Z.; Han, X.; Liu, J.; Ma, Q.; Yan, H.; Yuan, C.; Yang, L.; Han, H.; Weng, F.; Li, Y. Nano-Fe_3O_4/bamboo bundles/phenolic resin oriented recombination ternary composite with enhanced multiple functions. *Compos. Part B Eng.* **2021**, *226*, 109335. [CrossRef]

Article

One-Shot Synthesis of Thermoplastic Polyurethane Based on Bio-Polyol (Polytrimethylene Ether Glycol) and Characterization of Micro-Phase Separation

Yang-Sook Jung [1,2], Sunhee Lee [3], Jaehyeung Park [2,*] and Eun-Joo Shin [1,*]

1. Department of Organic Materials and Polymer Engineering, Dong-A University, Busan 49315, Korea
2. Department of Bio-Fibers and Materials Science, Kyungpook National University, Daegu 41566, Korea
3. Department of Fashion Design, Dong-A University, Busan 49315, Korea
* Correspondence: parkj@knu.ac.kr (J.P.); sejoo6313@dau.ac.kr (E.-J.S.); Tel.: +82-539505738 (J.P.); +82-512007343 (E.-J.S.)

Abstract: In this study, a series of bio-based thermoplastic polyurethane (TPU) was synthesized via the solvent-free one-shot method using 100% bio-based polyether polyol, prepared from fermented corn, and 1,4-butanediol (BDO) as a chain extender. The average molecular weight, degree of phase separation, thermal and mechanical properties of the TPU-based aromatic (4,4-methylene diphenyl diisocyanate: MDI), and aliphatic (bis(4-isocyanatocyclohexyl) methane: $H_{12}MDI$) isocyanates were investigated by gel permeation chromatography, Fourier transform infrared spectroscopy, atomic force microscopy, X-ray Diffraction, differential scanning calorimetry, dynamic mechanical thermal analysis, and thermogravimetric analysis. Four types of micro-phase separation forms of a hard segment (HS) and soft segment (SS) were suggested according to the [NCO]/[OH] molar ratio and isocyanate type. The results showed (a) phase-mixed disassociated structure between HS and SS, (b) hydrogen-bonded structure of phase-separated between HS and SS forming one-sided hard domains, (c) hydrogen-bonded structure of phase-mixed between HS, and SS and (d) hydrogen-bonded structure of phase-separated between HS and SS forming dispersed hard domains. These phase micro-structure models could be matched with each bio-based TPU sample. Accordingly, H-BDO-2.0, M-BDO-2.0, H-BDO-2.5, and M-BDO-3.0 could be related to the (a)—form, (b)—form, (c)—form, and (d)—form, respectively.

Keywords: thermoplastic polymer; bio-based polyurethane; polymerization; biomaterials; micro-phase separation

1. Introduction

Nowadays, thermoplastic polyurethanes (TPUs) are one of the most consumed families of polymers worldwide. These unique polymeric materials with a wide range of physical and chemical properties are broadly used in paints, coatings, synthetic rubbers, foams, fibers, adhesives, and packaging and in numerous fields such as the automotive industry, consumer or domestic equipment, construction engineering and biomedical applications [1–3]. The performances and properties of TPUs depend on the chemical nature of the reacting components and the utilized processes. Thus, by tailoring these factors, the TPUs can exhibit many useful properties, including modulable flexibility, elasticity, strength, good abrasion resistance, and high transparency [4]. TPUs are characterized by a segmented-block structure that is composed of a hard segment (HS; adduct of isocyanates and chain extender) and a soft segment (SS). Aliphatic and aromatic isocyanates are used to synthesize TPUs [5]. In addition, the choice of chain extender determines the characteristics of the HS and predominantly the physical properties of the TPU, and a low-molecular-weight diol or diamine is mainly used. The most important chain extender for the TPU is linear glycols such as 1,3-propandiol (PDO) and 1,4-butanediol (BDO). A SS introduces flexibility consisting of polyols (polyether or polyester polyols). Polyols constitute more than half of the total composition of the TPU and poly(ethylene glycol), poly(propylene glycol), poly(tetramethylene glycol) and poly(tetrahydrofuran) are commonly used polyether polyols, which are petrochemical-based. The synthesis of TPUs, including isocyanates, polyols, catalysts, and additives, and their application require abundant volatile organic compounds and emit hazardous pollutants that cause severe environmental pollution [6,7]. Since the application of TPU is increasingly extensive, decreasing the petrochemical monomers and energy required for polymerization and processing is becoming significant. The chemical industry is exploring new solutions based on natural raw materials, and, recently, the growing interest in applying bio-based substrates as a primary component in TPU synthesis is rapidly increasing. The new trend involves the replacement of petrochemical polyols and chain extenders used in synthesizing TPUs with materials obtained from biomass, including polysaccharides [8], sugars (such as xylose, mannose, glucose, galactose, and idose) [9,10], lignin [11], and vegetable oils [12,13] (such as soybean oil [14], castor oil [15], rubber seed oil [16], and palm oil [17]. Although vegetable oils are triglycerides, the esterification product of glycerol with three fatty acids, they do not possess suitable hydroxyl groups for polyurethane manufacture in many cases [4]. Moreover, after chemical modification, most vegetable oil-based polyols contain more than two hydroxyl groups or free long-chain fatty acids, which limits their application as the starting material for TPU elastomers [18]. Currently, bio-based polyols are used often for polymer production. This study concerns bio-based TPUs obtained using 100% bio-based polyether polyol, PO3G (Poly trimethylene ether glycol, SK Chemical). This polymer is linear with a molecular weight of 1000 g/mol and has excellent biodegradability and thermal and oxidation stability [19].

Using a diisocyanate, polyol, and low-molecular chain extender, the TPUs can be synthesized using two methods, particularly a one-shot and a pre-polymer process [20]. In the pre-polymer process, polyols and isocyanates first react to form an isocyanate-terminated pre-polymer, followed by the second step of chain extension by adding the chain extender. Numerous TPU elastomers and all polyurethane-ureas are prepared via NCO pre-polymer intermediates. This method allows the complete reaction (even of low-reactivity polyether diols) in the absence of catalysts and the intentional preparation of a segmented structure. Conversely, in the one-shot process, the starting materials are mixed in a reactor, and polymerization is performed. This process, which is conducted without solvents, is generally rapid, particularly in the presence of catalysts. Therefore, formed materials, for example, are prepared to utilize the one-shot process by mixing the reactants directly with additives [21]. The reaction is exothermic and is substantially terminated within 2–30 min, depending on the catalyst applied.

TPUs are block copolymers with a specific organization of HSs and SSs, and this segmented structure is the key feature of the TPU molecular chain [22,23]. The HS (may be glassy or semicrystalline) is typically composed of a rigid diisocyanate and a chain extender (for example, a short polyol) [24,25]. However, even for isocyanate compounds with the same NCO, each compound has a different molecular shape and property. Hence, the fields wherein each compound is applied are also different. Due to the thermodynamic incompatibility between the two structural units, the polymers undergo microphase separation resulting in HS domains dispersed in the SS matrix [26,27]. Moreover, the incompatibility results in a characteristic microphase-separated structure of TPU, with hard domains acting as the tie points for the flexible SS phase. The microphase separation of these two chemically distinct components generates unusual and useful physical and mechanical properties of TPUs. This phase separation between HSs and SSs occurs rapidly and simultaneously with polymerization and results from the thermodynamic immiscibility between the rigid and soft phases, forming microphases bound together by hydrogen bonding [28,29]. Since a wide range of monomeric materials is now commercially available, extensive investigations have been devoted to the structure-property relationships of TPUs, and tailor-made properties can be obtained from well-designed combinations of monomeric materials. The lengths and chemical structures of HS and SS, including the soft/hard segment [30,31], diisocyanate symmetry [25,32], ability hydrogen bonding ability, and crystallinity of soft/hard phase domains, and the synthesis and processing methods [33,34] are important features that determine the phase separation and physical properties of TPUs. In TPUs, the observed domain morphology and microphase separation are strongly related to the hydrogen bonding of HSs and their crystallization kinetics [3,35]. The structures and properties of TPUs are known to change as a function of temperature. These changes have been extensively investigated by thermal analysis [36,37], spectroscopy [38,39], dynamic mechanical analysis [21,25,32], wide-angle X-ray diffraction [30,40], and small-angle X-ray scattering [41]. Numerous studies have reported the relationship between the mechanical and thermal properties and the microphase structure of TPUs [42,43]. Although the combination of HS and SS of the TPU can have various morphologies depending on the synthesis conditions, studies that clearly define the difference in the morphology according to the synthesis conditions are insufficient. Therefore, in this study, we presented various models according to the [NCO]/[OH] molar ratio and isocyanate type. The present contribution is aimed at TPU synthesis based on bio-based polyol using the one-shot process and controlled microphase separation structure according to the HS type (diisocyanate symmetry) and HS contents. Thus, this study focuses on: (1) the effect of increased bio-content on the synthesis and (2) the effect of SSs or HSs (with different [NCO]/[OH] molar ratios, isocyanate type (MDI, H_{12}MDI)) on the surface (morphological behavior and hydrogen bonding), crystallinity, molecular weight, and thermodynamical and mechanical properties. Through this analysis, (3) several types of micro-phase-separated in a two-phase system as the HS and SS matrix were modeled, depending on the [NCO]/[OH] molar ratio and the isocyanate type when synthesizing TPU. The combination of spectroscopy, X-ray scattering, microscopy, and thermal and mechanical analyses was used to characterize these complex materials in detail.

2. Materials and Methods

2.1. Synthesis of Bio-Based Thermoplastic Polyurethanes

We prepared different segmented TPUs by changing the [NCO]/[OH] molar ratio and content of the HS. The HS consisted of 4,4-methylene diphenyl diisocyanate (MDI) and dicyclohexylmethane diisocyanate (H_{12}MDI) as diisocyanate and BDO as the chain extender; the SS consisted of PO3G. TPUs were synthesized using a solvent-free one-shot polymerization procedure (Figure 1). The molar ratio of the OH groups of the polyol, NCO groups of the MDI or H_{12}MDI, and OH groups BDO was maintained at 1:2.0:1, 1:2.5:1, and 1:3.0:1, respectively, and the resulting TPUs were identified (Table 1). The MDI that had been frozen was melted in the oven at 80 °C for 4 h. In the one-shot method, the PO3G polyols (M_n = 1000) and the chain extender BDO were thoroughly mixed in polypropylene beakers using a mechanical stirrer, which was then placed in the oven to maintain the temperature of the reaction mixture at 80 °C. Next, diisocyanate H_{12}MDI or MDI and the catalyst dibutylin dilaurate (0.03 wt%) were added to the reaction mixture and mechanically stirred at room temperature (20–25 °C) for 1–2 min. As the mixture was stirred and polymerized, the transparent liquid became opaque, and the TPU was subsequently obtained. The reaction mixture was then poured into a Teflon-coated pan, cured in an oven at 100 °C for 24 h, and then kept to complete polymerization.

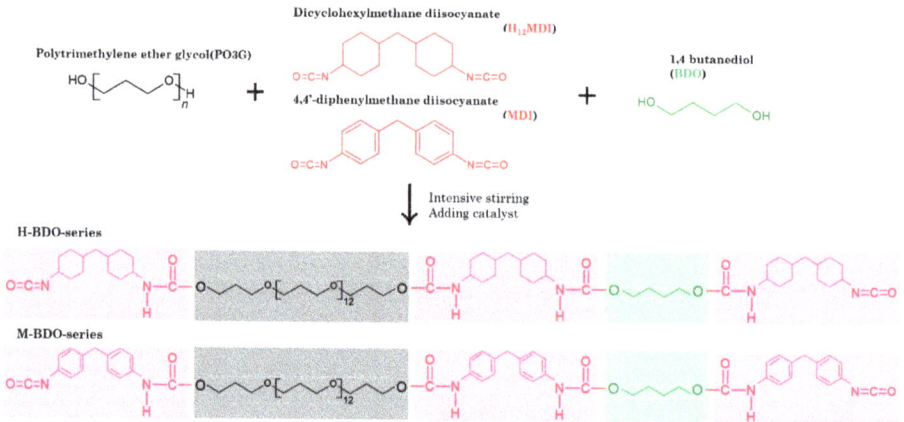

Figure 1. A schematic representation of the bio-based thermoplastic polyurethane (TPU) synthesis.

Table 1. The formulation design of bio-based TPU samples.

Sample	Diisocyanate	OH/NCO/OH	Hard Segment (HS) Content (wt%) [a]	Content of Bio-Based Sources (wt%) [b]
H-BDO-2.0	H_{12}MDI	1:2.0:1	36.9	63
H-BDO-2.5	H_{12}MDI	1:2.5:1	41.5	59
H-BDO-3.0	H_{12}MDI	1:3.0:1	45.5	55
M-BDO-2.0	MDI	1:2.0:1	33.3	67
M-BDO-2.5	MDI	1:2.5:1	37.6	62
M-BDO-3.0	MDI	1:3.0:1	41.4	59

[a] Hard segment concentration is defined as the ratio of the mass of non-polyol components to the total mass;
[b] Bio-content is defined as the ratio of the mass of bio-based components to the total mass.

The final product was pressed at 180 °C for 5 min to obtain a film. All bio-based TPU films with 0.5–1 mm thickness were shown transparent white, and the light transmittance values of the H_BDO- and M_BDO-series ranged from 60–73% and 35–68%, respectively. After cooling to room temperature, the sheet was removed from the mold and used for further structural, thermal, and mechanical testing and characterization. All the reagents were of analytical grade and used without further purification. Detailed information on the materials is listed in Table 2.

Table 2. Characteristics, structures, and molecular weights of the pure materials used in bio-based TPU synthesis.

Reagent Name	Supplier	Description	Molecular Structure
Polytrimethylene ether glycol (PO3G)	SK chemical, Korea	100% bio-based polyether polyol (1,3-propanediol based) by corn oil M_w 1000 g/mol, hydroxyl number = 53.4–59.0, T_m = 16–18 °C	HO–[–O–]$_n$–H

Table 2. *Cont.*

Reagent Name	Supplier	Description	Molecular Structure
Dicyclohexylmethane diisocyanate (H_{12}MDI)	Sigma Aldrich, Germany	Aliphatic diisocyanate, M_w 262.35 g/mol	
4,4'-Diphenylmethane diisocyanate (MDI)	Sigma Aldrich, Germany	Aromatic diisocyanate, M_w 250.25 g/mol	
1,4-Butanediol (BDO)	Sigma Aldrich, Germany	Chain extender, M_w 90.122 g/mol	
Dibutylin dilaurate (DBTDL)	Sigma Aldrich, Germany	Catalyst, M_w 631.56 g/mol	

2.2. Characterization

2.2.1. Molecular Characteristics

The number average (M_n) and weight average (M_w) molecular weights and the polydispersity index (PDI) were measured by gel permeation chromatography (GPC) using a Viscotek GPCmax (VE-2001 system, Malvern, Worcestershire, UK). To test the solubility of the synthesized bio-based TPU in organic solvents, THF was used and applied to a mobile phase solvent. The column was 300 × 810 mm and the flow rate was 1 mL min^{-1} at 60 °C.

2.2.2. Fourier Transform Infrared Spectroscopy (FT-IR)

Fourier transform infrared (FT-IR) analyses were performed using a Nicolet Nexus FT-IR spectrometer (PerkinElmer, Shelton, CT, USA) over the wavelength range of 400–4000 cm^{-1}, equipped with an attenuated total reflectance accessory.

2.2.3. Atomic Force Microscopy (AFM)

The surface morphology of TPU films was investigated using an Inova system (Bruker, Billerica, MA, USA), equipped with a standard silicon nitride probe, SuperSharpSilicon™-SPM-Probe (NanoSensors™, Zurich, Switzerland; spring constant 42 N and resonant frequency 320 kHz). The analyses were performed under ambient conditions using the tapping mode atomic force microscopy (AFM) technique, and the surface images were taken in the sizes of 20 × 20 µm. The bulk morphology was evaluated by imaging the fracture area after the previous freeze-fracturing of sheets at a temperature of −80 °C. The AFM images were processed using NanoScope analysis software.

2.2.4. X-ray Diffraction (XRD)

X-ray diffractograms were collected using a Shimadzu diffractometer (XRD-6000) with mono-chromatic CuKα radiation (λ = 0.15418 nm) and a generator working at 40 kV and 30 mA. Intensities were measured in the range of 5 < 2θ < 40°, typically with scan steps of 0.05° and 2 s/step (1.5° min^{-1}). Peak separations were performed by Gaussian deconvolution.

2.2.5. Dynamic Mechanical Analysis (DMA)

Dynamic mechanical properties of the TPU films were determined using a DMA Q800 analyzer (TA instruments, New Castle, DE, USA) in the tensile mode at a frequency of 1 Hz. The samples were initially cooled to −100 °C and subsequently heated to 150 °C at a heating rate of 4 °C/min.

2.2.6. Differential Scanning Calorimetry (DSC)

The thermal properties of the obtained samples were measured on a DSC 8500 thermal analyzer (TA Instrument, New Castle, DE, USA). All the samples were weighed between 2 and 10 mg. The measurement was conducted from −70 °C to 250 °C at a heating rate of 20 °C/min under a nitrogen purge.

2.2.7. Thermogravimetric Analysis (TGA)

The thermal properties of the obtained samples were also determined using TGA Q500 (TA Instrument, New Castle, DE, USA), which were measured at a temperature range of 40–650 °C with a ramp heating rate of 10 °C/min in the presence of a nitrogen atmosphere. The weight of each sample was approximately 5 mg. The weight loss of 5% and 50%, the maximum degradation rate, and ash residue at 600 °C were registered.

2.2.8. Shore A Hardness

Hardness was measured at room temperature using a Zwick Roell GS-706N analogical hard-ness testing apparatus (Teclock Co., Tokyo, Japan) using the "UNE-EN ISO 868:1998: Plastics and ebonite—Determination of indentation hardness by means of a durometer (Shore hardness)" standard procedure at (23 ± 2) °C and 50% relative humidity.

2.2.9. Mechanical Properties

Bio-based TPUs films were tested on an Instron 4201 autograph tester (Shimadzu, Tokyo, Japan) to measure the stress-strain behavior of the samples in tension. The length × width × thickness of the specimens was 10 × 2 × 0.5 mm.

3. Results and Discussion

3.1. Molecular Weight of Synthesized Thermoplastic Polyurethanes

The molecular weight of the synthesized TPUs was characterized by GPC analysis, and detailed measurements are summarized in Table 3. The M_n and M_w of the TPUs were in the range of 37,723–112,117 and 70,953–236,689, respectively. The M_w of TPUs strongly increased as the [NCO]/[OH] molar ratio increased. In addition, the number of urethane units in the hard domains increased with the [NCO]/[OH] molar ratio, resulting in higher molecular weights [3,44,45]. The M-BDO-series exhibited a higher molecular weight than the H-BDO-series, although it had a similar HS mole ratio, as shown in Table 3. Furthermore, the melting temperature increased as the M_w of TPUs was increased (the [NCO]/[OH] molar ratio was increased) (see Section 3.5). The HS melting temperature was found to depend on the HS–SS interaction and the number of hydrogen bonds in the HS [30,46]. Moreover, the M-BDO-series exhibited a high melting temperature because of its high M_w. The M_n of the resulting bio-based TPUs was more than 30,000 g/mol, which is sufficiently high to satisfy the industrial application requirements and be the strong hydrogen-bonding character of TPUs.

Table 3. Average molecular weight (M_n and M_w) and polydispersity index (PDI) of bio-based TPUs.

Sample	M_n	M_w	PDI
H-BDO-2.0	37,723	70,953	1.88
H-BDO-2.5	40,375	72,013	1.78
H-BDO-3.0	112,117	236,689	2.11
M-BDO-2.0	57,476	112,220	1.95
M-BDO-2.5	60,829	153,285	2.51
M-BDO-3.0	n.m. [a]	n.m.	n.m.

[a]: n.m. = not measured.

The PDI values of the TPUs were in the range of 1.78–2.51, which is consistent with the molecular weights. The low PDI values indicated a narrow molecular weight distribution of the prepared TPU samples. However, the PDI values were approximately 2 (the theoretical value for linear step-growth polymers is 2 according to the Flory's theory [47]) suggesting that the conversions of polymerizations were sufficiently high for the one-shot bulk reaction. In the case of M-BDO-2.0, the PDI was 1.95; thus, this condition is better for the one-shot bulk polymerization reaction with a short polymerization time because of the good control of the polymerization process. This is true for most TPUs, and it is nevertheless sufficient for polymer processing (injection and extrusion) and various applications [48]. These increasing values of M_n and M_w with increasing HS and using aromatic diisocyanate lead to differences in the TPU detected by FT-IR spectra, X-ray diffraction (XRD), AFM, DSC thermograms, and tensile strength and hardness, which are discussed below.

3.2. Chemical Structure Characterization (FT-IR)

The chemical structure of the synthesized TPUs was confirmed through FT-IR spectroscopy. In Figure 2, the excess isocyanate can be detected using IR spectroscopy. The NCO band at 2270 cm^{-1} is one of the most intense bands and is practically undisturbed through the absorption of other groups. Therefore, the reaction is confirmed to be completed owing to the absence of absorption bands related to the NCO groups and at 3470 cm^{-1} corresponding to the OH groups of the polyol end group for all samples [49]. The N-H bond stretching vibration of urethane groups appeared at 3330 cm^{-1} due to hydrogen bonding. It is commonly known that the N-H bond can be observed in two separate bands, that is, the hydrogen-bonded N-H at 3275–3300 cm^{-1} and the free N-H bond at 3500 cm^{-1} [50]. The C-H asymmetric and symmetric stretching vibrations of the -CH$_2$ groups were observed as bimodal bands with the maxima at 2850 and 2950 cm^{-1}, respectively, which are assigned to the SS of the TPU matrix [32]. Bands 2900 cm^{-1} are assigned to the CH groups, particularly at 2950 cm^{-1} corresponding to the asymmetric CH$_2$ stretching and the 2850 cm^{-1} band that is associated with the symmetric CH$_2$ stretching [51]. Thus, TPU formulations with more content of SS exhibited bands with more intensity in this zone. The double peak observed in the 1680–1740 cm^{-1} range corresponds to the carbonyl group (C=O) stretching vibrations [52,53]. Further characteristic bands at 1600 cm^{-1} and 1819 cm^{-1} correspond to the C=C aromatic stretching vibration of the M-BDO-series. The band observed at 1530 cm^{-1} is associated with the stretching vibration of the –CN bond of the urethane groups. [35] The strong absorption band at 1104 cm^{-1} is ascribed to the free ether bond (C-O-C) of the used polyether polyol. The band maximum associated with the asymmetric stretching vibrations of the non-associated ether group is marked by the 1104 cm^{-1} band, while the 1063 cm^{-1} band is related to hydrogen bond interaction between N-H and C-O-C groups [33].

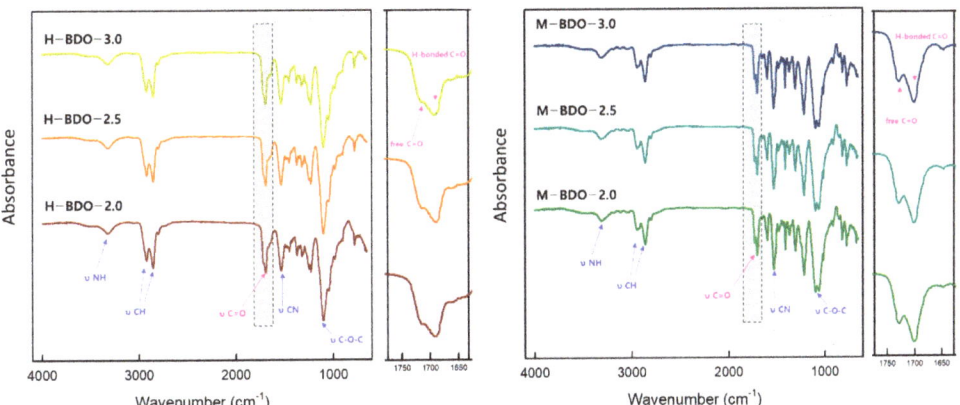

Figure 2. Fourier transform infrared (FT-IR) spectra of TPU-based H-BDO and M-BDO.

FT-IR analysis is a useful method for characterizing the band intensity and shape of the localized vibrations associated with specific functional groups, N-H or C=O, which are involved in specific hydrogen bonding in various domains, as shown in Figure 3. The position and intensity of these vibrations are known to be susceptible to the strength and specificity of the formed hydrogen bond [22]. Thus, the phase separation in TPUs can be characterized by measuring the intensity and position of the hydrogen-bonded N-H stretching vibration. It is usually interpreted that extensive phase separation has occurred when there is significant N-H—O=C hydrogen bonding since both units are associated with the HS. It has also been suggested that N-H can form a strong hydrogen bond with the oxygen of the ether groups from polyol associated with the SS when available. The N-H bond vibration appeared in all synthesized TPUs regardless of the [NCO]/[OH] molar ratio. This region also exhibited a small shoulder at 3500 cm^{-1} in all the curves, corresponding to the non-hydrogen-bonded N-H group [23]. Furthermore, the band intensities in the hydrogen-bonding association of the carbonyl group (C=O) region can be utilized to characterize the phase separation between the HS and SS. The specific method to calculate the degree of phase separation (DPS) [24] is to analyze the spectra in the 1750–1680 cm^{-1} region by deconvolution of the carbonyl bands using Origin software (Origin 2018) in Table 4. In particular, the H-bonded C=O only existed in the HS, whereas free C=O was merely scattered in the SS. Therefore, the DPS was calculated based on the amounts of free carbonyl and hydrogen-bonded –C=O in the amorphous and ordered regions.

$$\text{Degree of phase separation (DPS)} = R/(R+1) \tag{1}$$

$$R = \frac{\text{Ab(Absorption intensity of hydrogen bonded C}=\text{O)}}{\text{Af(Absorption intensity for free C}=\text{O)}} \tag{2}$$

$$\text{Degree of phase mixing (DPM)} = 1 - \text{DPS} \tag{3}$$

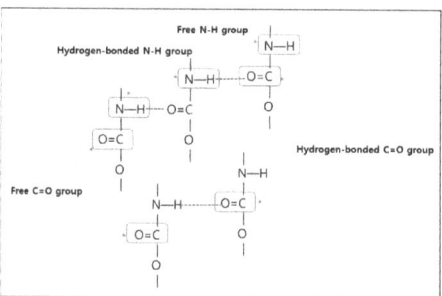

Figure 3. Specific hydrogen bonding in the TPUs.

The exact band positions indicated only some differences among the TPUs containing different diisocyanates and [NCO]/[OH] molar ratios. These results indicate that, in the case of H-BDO and M-BDO, the DPS slightly increased with the [NCO]/[OH] molar ratio in correlation with the increasing HS content. The H-BDO-2.0 (R = 1.67) sample contained 55.5% of HS, which were connected with hydrogen bonds. For the H-BDO-3.0 sample (R = 1.40), slightly more hydrogen bonds were generated, with some limitations of the phase separation function depending on the diisocyanate type. Furthermore, increasing the amount of isocyanate to create the HS phase favored the elevation of the hydrogen bonding, as indicated by the corresponding increase in the R index. Compared to the H-BDO-3.0 sample, the M-BDO-3.0 sample exhibited high DPS despite having lower HS content. In the M-BDO-2.0 sample, it could be concluded that more than 57% of the HS was microphase separated, while only 42% of the HS was mixed within the polyether polyol matrix. Therefore, the amount of hydrogen-bonded C=O groups was affected by the HS content and diisocyanate type. In conclusion, DPS slightly increased for TPUs prepared with the MDI as diisocyanate. Factors influencing DPS in TPU materials include hydrogen bonding between polymer chains, segment length, polarity and crystallizability, overall composition, and mechanical and thermal history [25]. In the following section, the reason for phase separation between the HS and SS is discussed, and the experimental evidence supporting the presence of microdomains is presented.

Table 4. Deconvolution of the FT-IR absorbance bands in the 1750–1680 cm^{-1} range that occurred in the prepared TPUs.

Sample	Hard Segment (HS)	Absorption Intensity of C=O Band		R	Degree of Phase Separation (DPS)	Degree of Phase Mixing (DPM)
		Free C=O 1730 cm^{-1}	H-Bonded C=O 1700 cm^{-1}			
H-BDO-2.0	36.9	29.37	35.88	1.22	0.550	0.450
H-BDO-2.5	41.5	27.87	35.53	1.24	0.553	0.447
H-BDO-3.0	45.5	29.80	37.16	1.25	0.555	0.445
M-BDO-2.0	33.3	30.12	40.53	1.35	0.574	0.426
M-BDO-2.5	37.6	30.55	41.89	1.37	0.578	0.422
M-BDO-3.0	41.4	30.08	42.00	1.40	0.583	0.417

3.3. Atomic Force Microscopy (AFM) Analysis

The size and shape of the hard domains in the TPU were evaluated using AFM. Although AFM is typically used to analyze the surface physical structure and quantify the surface roughness [54–56], it is also a useful tool for the investigation of the internal structure of heterogeneous materials [32,57], particularly when other methods are inefficient due to low contrast or when a comparison with other analytical methods is desirable. The freeze-fractured cross-sections of the prepared bio-based TPU samples were analyzed to investigate whether the HS/SS morphology of the surface patterns of the films was irregular and relatively rough. Figure 4a–f shows the height and 3D AFM images of the TPU sample using the tapping mode, and Table 5 illustrates the supporting data of the AFM images. The samples exhibited two types of phase contrast: a dark, featureless matrix corresponding to the SS and bright elements of different sizes dispersed in this matrix [57]. The roughness increased when the HS content increased with the [NCO]/[OH] molar ratio, and the phase separation and hard domain structure became more pronounced in the sample morphology. The shape of the hilly was significantly different when the H-BDO- and M-BDO-series were compared. The H-BDO-series showed a sharp hilly, whereas a more rounded hilly was observed in the M-BDO-series. These characteristics are affected by the HS content and isocyanate type.

The samples containing a high SS content (H-BDO-2.0 in Figure 4a) exhibited smooth surfaces that were separated from each other in the samples. Conversely, the TPU with high hard domain content showed an irregular "hilly" break surface (Figure 4b–f). Dramatic changes were observed in the phase images when comparing H-BDO-2.5 to M-BDO-3.0 in Figure 4b,f, although the HS content was similar in both samples (41% of HS content). The M-BDO-3.0 sample showed the largest hilly surface with large globules and the highest roughness value. The depth and height of the globules in the z direction on the TPU surface were examined using Nanotec Electronica WSM software (2019). The roughness in the Z-profile of the TPU contributes to the co-continuous network morphology [58]. Various microdomains observed on the TPU surface could have significantly contributed to the roughness enhancement. The non-uniform distribution of the hard domain increased the surface roughness of the H-BDO-series from 23 to 228 nm. In addition, the surface roughness in the M-BDO-series was observed in a highly spiked region (R_{max} = 1872–4243 nm) from 178 to 347 nm. Consequently, variation in the size and shape of the hard domains depending on the isocyanate type can be confirmed. When the HS content increases by varying the [NCO]/[OH] molar ratio, the HS domains become larger (up to multi-μm size); the roughness and the extent of phase separation increase, and the hard domain structures become more pronounced and visible in the sample morphology. Many factors such as the [NCO]/[OH] molar ratio, HS content, and isocyanate type exhibiting diverse morphological structures are related to the AFM topographic images. Therefore, these factors support the DPS between the HS and SS and hydrogen bonding obtained by FT-IR analysis. When the HS content increases by varying the [NCO]/[OH] molar ratio, the HS domains become larger (up to multi-μm size); the roughness and the extent of phase separation increase, and the hard domain structures become more pronounced and visible in the sample morphology.

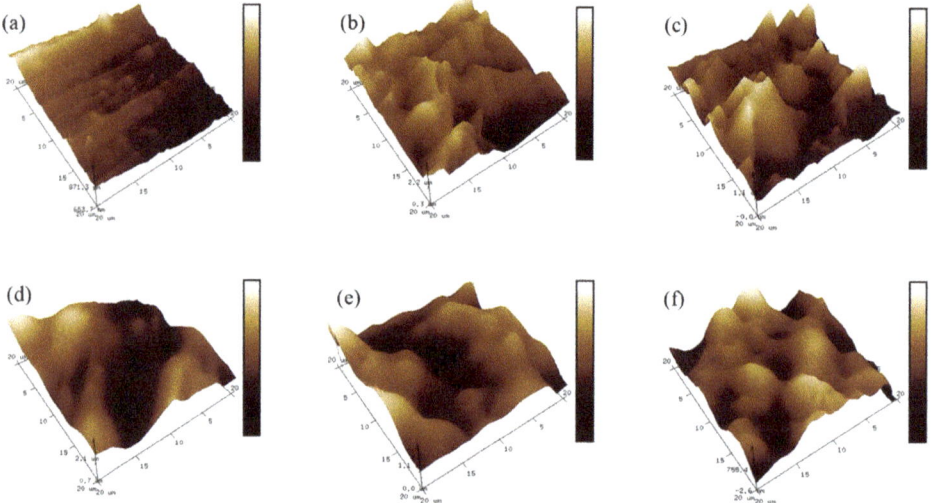

Figure 4. Atomic force microscopy phase images of the freeze-fractured surfaces of the bio-based TPU: (**a**) H−BDO−2.0, (**b**) H−BDO−2.5 (**c**) H−BDO−3.0, (**d**) M−BDO−2.0, (**e**) M−BDO−2.5, and (**f**) M−BDO−3.0.

Table 5. Characterization of phase images of TPU films.

Sample	Surface Area (μm²)	R_q (nm) [a]	R_a (nm) [b]	R_{max} (nm) [c]
H-BDO-2.0	401	29	23	277
H-BDO-2.5	437	257	177	1604
H-BDO-3.0	441	294	228	2177
M-BDO-2.0	414	227	178	1872
M-BDO-2.5	415	305	248	2328
M-BDO-3.0	408	417	347	4243

Surface area: total area of the examined sample surface. Mean: average of all the Z values within the enclosed area. [a] R_q (rms): standard deviation of the Z values within the given area. [b] R_a (mean roughness): mean value of the surface relative to the center place. [c] R_{max} (maximum height): difference in height between the highest and the lowest points on the surface relative to the mean plane.

3.4. X-ray Diffraction (XRD) Analysis

The degree of crystallinity in the prepared bio-based TPU was investigated by wide-angle XRD in the region of wider angles. The changes in the crystalline structure due to varying [NCO]/[OH] molar ratio and isocyanate type were determined by XRD analysis. A comparison of the XRD patterns with different HS contents is illustrated in Figure 5. The peaks in the TPU samples at 2θ = 19° generally correspond to the hard domain related to the hydrogen bonds between urethane groups in 2θ = 19–23° [59].

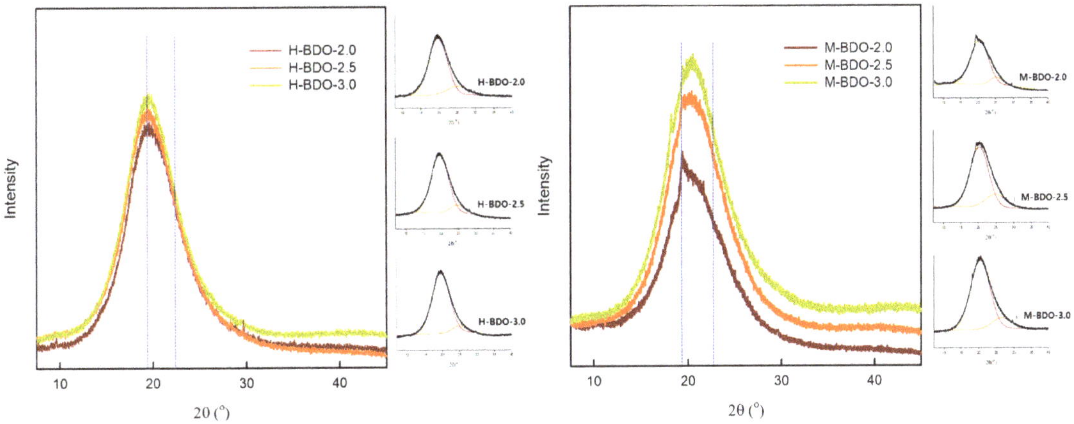

Figure 5. X-ray diffractograms of the bio-based TPU.

The observed diffraction patterns exhibited significant broadening of peaks, which resulted in two main hard domain regions with a maximum at approximately 19.4° and 23.5° by fitting with Gaussian distributions using Origin software (2018). The HSs demonstrated a diffraction peak at 19.4° with higher intensity when present in higher concentration. Furthermore, as the [NCO]/[OH] molar ratio was increased, the intensity of the peak localized at 2θ = 19.4° increased from 4858 to 5047 and 3355 to 5292 for the H-BDO- and M-BDO-series, respectively, with a sharper full width at half maximum (FWHM) in Table 6. This peak indicated a hard domain based on a higher-ordered arrangement of the HS with hydrogen bonding [41]. The phenomenon of the higher hard domain rate of the HS in M-BDO-3.0 attributed to the globular morphology of the hard phases, which was also confirmed by the AFM image in Figure 4f.

Table 6. Peak analysis data for X-ray diffraction peaks.

Sample	Peak 1 (θ = 19.40)		Peak 2 (θ = 23.46)	
	Intensity	FWHM [a]	Intensity	FWHM
H-BDO-2.0	4858.01	6.00	752.15	4.44
H-BDO-2.5	4664.34	5.85	713.71	4.98
H-BDO-3.0	5046.76	5.96	609.42	6.53
M-BDO-2.0	3354.74	6.55	914.39	4.07
M-BDO-2.5	5069.32	6.23	820.04	6.16
M-BDO-3.0	5291.63	5.97	591.92	6.36

[a] full width at half maximum (FWHM): peak width.

In contrast, as the HS content increased, the diffraction peak at 23.5° exhibited a lower diffraction peak height and broader FWHM, indicating a smaller hard domain that decreased the peak intensity, which is also supported by the AFM images illustrating the decreased height of the surface in Figure 4a. Although a discrepancy was observed between the XRD and AFM results, AFM showed agglomeration, while a single aggregate containing a number of hard domains was observed in XRD. In segmented TPU, the phase separation of the SSs and HSs can occur depending on their relative contents, structural regularity, and thermodynamics incompatibility. The XRD studies revealed that the hard domain depends on the structure of diisocyanates and the [NCO]/[OH] molar ratio in the

TPU hard domain. Moreover, Figure 4 shows that hard domain contents increased from aliphatic to aromatic characters of the diisocyanates utilized in the bio-based TPU.

3.5. Thermal Analysis (DSC)

The thermal characteristics of the bio-based TPUs and their segmented structure were investigated using DSC. Regarding the thermal properties of TPUs, four types of thermal effects (the glass transition temperature of SS, melting temperature of SS, glass transition temperature of the amorphous part of the HS, and the melting of the HS) could be distinguished from the DSC runs (Figure 6), and the results are summarized in Table 7. The TPUs are segmented or block copolymers consisting of alternating HS and SS. The microphase separation of these two chemically distinct components gives thermodynamic incompatibility, generating separated peaks in T_g and T_m of the SS and T_g and T_m of the HS [60]. Generally, the HS and SS of a TPU are incompatible because they have a positive heat mixing. Thus, there is a tendency toward phase separation of the two components; however, the topology of the block copolymer molecules imposes restrictions on segregation, thereby forming a microdomain [61].

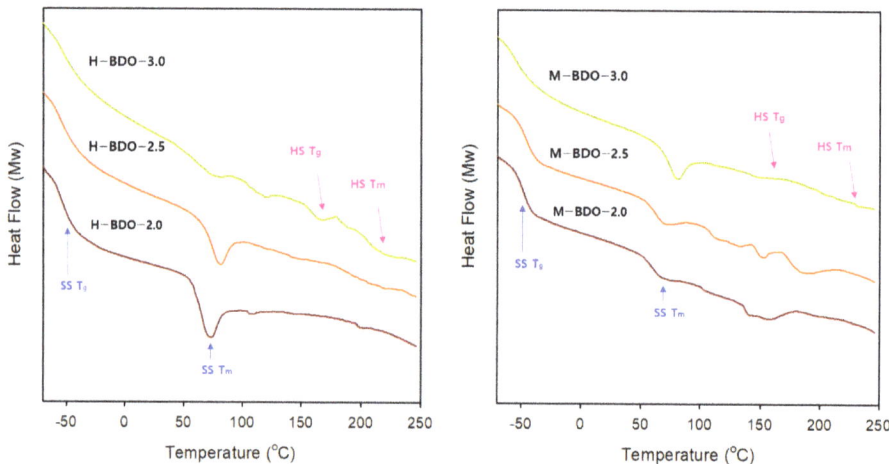

Figure 6. Differential scanning calorimetry (DSC) of the first heating ramp measured at 20 °C/min from −100 °C to 200 °C for the TPU-based sample series.

Table 7. A summary of DSC results of the bio-based TPUs.

Sample	SS T_g (°C)	SS T_m (°C)	HS T_g (°C)	HS T_m (°C)
H-BDO-2.0	−55.4	71.6	-	195.8
H-BDO-2.5	−55.0	80.3	145.6	217.4
H-BDO-3.0	−52.4	83.5	166.0	219.6
M-BDO-2.0	−47.7	67.1	136.3	160.1
M-BDO-2.5	−46.0	69.3	151.8	183.8
M-BDO-3.0	−44.3	80.4	157.3	231.6

Up to four transitions were observed for the bio-based TPU in Figure 6, including the glass transition temperature (between −55 °C to −44 °C), the melting temperature of SS (67 °C to 80 °C), and the melting temperature of HS (160 °C to 231 °C). The SS T_g in the H-BDO- and M-BDO-series was detected at approximately −54 °C and −45 °C, respectively. H-BDO-2.0, H-BDO-2.5, and H-BDO-3.0 exhibited the melting of the SS at 72, 80, and 84 °C, respectively. The HS T_g in the M-BDO-series was detected at 136, 151, and 157 °C, while the melting temperature of the HS was observed at 160, 183, and 231 °C. When considering the [NCO]/[OH] molar ratio, the SS T_g of the H-BDO- and M-BDO-series increased with the [NCO]/[OH] molar ratio. The result of the H-BDO-series is explicated because of the increasing miscibility of the HS and SS, that is, the partial mixing of the HS within the SS matrix. The sample of the M-BDO-series demonstrated higher T_g of the SS as higher contents of the HS were well phase-separated, which increased the DPS as confirmed by the FT-IR analysis; this phenomenon may be an indicator of kinetically favorable and stable phase separation. Moreover, the T_m of SS slightly increased as the [NCO]/[OH] molar ratio was increased, while the heat capacity increased with increasing SS contents, which corresponded to the melting peak of the SS. Therefore, a TPU sample showing insignificant T_m of SS with higher melting enthalpy could be obtained. Since the T_m of SS is the deciding factor of the shape memory transition temperature, this result also suggested that the shape memory TPU can be adjusted to various temperature ranges by carefully selecting the [NCO]/[OH] molar ratio and isocyanate type, thus possibly expanding the application field. As an example of a shape memory polymer, H-BDO-2.0 exhibited a relatively sharp endothermic peak at approximately 72 °C, which resulted from the T_m of the SS-associated thermal transition temperature (T_{trans}). When the temperature increase is higher than 72 °C (T_{trans}) of the switching segments, the segments are flexible, and the polymer can be deformed elastically. The temporary shape is fixed by cooling below 72 °C, and the permanent shape is recovered when the polymer is heated again.

The length of the HS blocks forms the upper limit to the size of the HS crystals in the chain direction, which determines the melting point and thermal stability [62]. Figure 6 illustrates the temperature range wherein most HS melting temperatures shift towards higher temperatures as the HS content is increased. The T_m at higher temperatures corresponded to the dissociation of a great order structure, which was related to the mixing between the HS and SS. The lower temperatures could be attributed to the melting of the less-ordered structure or suitable SS [63,64]. In particular, the samples of the M-BDO-series exhibited increased high-temperature T_m peaks and their shift to higher temperatures, which resulted from better HS ordering and the formation of stronger and more stable HS domains. From a morphological viewpoint, the HS formed globular domains in a continuous SS matrix in this sample. The degree of order within the HS of the TPU depends on the chemistry, rigidity, and hydrogen-bonding within the HS [65]. In addition, the phase separation between the HS and SS depends on their respective lengths and affinity for each other, which is closely related to the ability of the HS and SS to establish hydrogen-bonding interactions. Therefore, the phase separation is affected by the chemical composition and HS content in the synthesized TPUs. The effect of the isocyanate type on the T_m of the TPUs was determined at a higher temperature using MDI. The materials obtained from aromatic isocyanates exhibited a higher glass transition temperature than those obtained from aliphatic isocyanates. The lower glass transition temperatures observed for TPUs obtained using aliphatic isocyanates are attributed to the better separations of the phases [34]. Conversely, the presence of aromatic isocyanate in the HS produces a stiffer polymer chain with a higher melting point.

3.6. Dynamic Mechanical Analysis (DMA)

The DMA results are shown in Figure 7 as a functional relationship of the storage modulus (G′) and tan δ, revealing the thermal transitions and indicating the applicable temperature ranges. The storage modulus defines the energy stored elastically by the materials at deformation supplying information about the polymer stiffness [36]. The highest values of G′ were observed between 1085 and 2850 MPa for the H-BDO-series and 2918 and 3030 MPa for M-BDO-series. The G′ values below −50 °C remained nearly constant due to the restriction of the molecular motions to vibrations and short-range rotations of the SS. The gradual decrease in G′ at approximately −50 °C corresponded to the T_g of SS. The α-relaxation process, which is indicated by the low-temperature peak in the tan δ plots and represented the glass transition of the SS, was broad, suggesting only fair phase separation between SS and HS. In addition, this peak was sufficiently defined for the M-BDO-series, which allowed the assignment of the T_g based on the position of the peak maximum. The difference in the shape of the SS T_g peak is attributed to higher DPM in the H-BDO-series with high DPM, as indicated by the FT-IR analysis. In tan δ curves, the double peaks, which suggest the existence of the two phases of the SSs related to the SS T_g (approximately −54.77 °C to −40.56 °C) and SS T_m (approximately 1.35 °C to 48.23 °C), appeared in H-BDO-2.5 and H-BDO-3.0. The temperature difference between the above double peaks decreased as the [NCO]/[OH] ratio increased, particularly for the H-BDO-series. This result indicates that as the ratio of the HS increases, a sharp increase in tan δ appears due to the relaxation of the hard domain of the TPU molecules related to the presence of more HS in the soft domains that restrict the mobility of the SS. The T_g of TPU can be detected by DSC and DMA, and the obtained T_g values were similar. Since the HS content increased, the modulus, T_{flow} due to the melting of the HS, and the T_g increased.

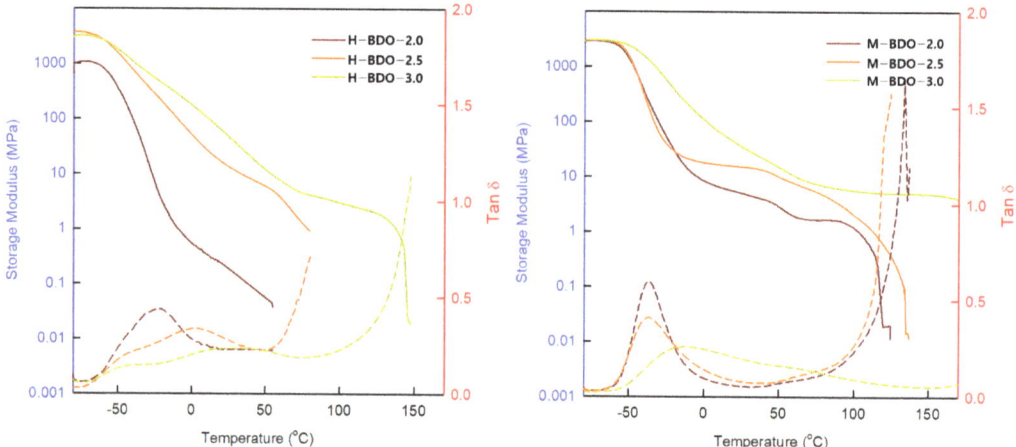

Figure 7. Storage modulus versus temperature and tan δ versus temperature for the obtained bio-based TPUs.

The G'_{25} storage modulus significantly increased with the HS content in Table 8. In the case of phase-separated HS, the modulus considerably increased with the higher HS content, making the materials stiffer. However, no melting transition was detected in the DSC thermogram, whereas the flow behavior of the TPU could be observed in DMA. The modulus above the T_g of the SS in the rubbery plateau region depends on the reinforcing effect of the HS on the soft matrix [66]. The flex temperature (T_{flex}) is defined as the temperature at the beginning of the rubber plateau region, that is, the intercept of the tangents. The flow temperature (T_{flow}) is defined as the temperature where the storage modulus G' reaches 1 MPa, and the storage modulus of the rubbery plateau is determined at room temperature. The modulus of the rubbery plateau, in the $T_{flex} - T_{flow}$ region is a function of the HS crosslink density and reinforcement by the separated HS. Usually, TPUs from segmented copolymers can be prepared with a range of rubbery moduli by changing the HS content [32]. With the [NCO]/[OH] molar ratio, the T_{flow} of the corresponding TPU is increased from 77.7 to 170 °C. These T_{flow} values are attributed to the melting of the phase-separated HS, and the increase in T_{flow} observed may result from an increase in the hydrogen bonding in the HS domain [30]. The T_{flow} decreased with the decreasing HS content, which is also in agreement with the results observed in other systems. The T_{flow} of the MDI-based TPU was considerably higher than that of the H_{12}MDI, which is also explained in the solvent effect theory by Flory [47]. In the MDI-based TPU, the rubbery plateau was extended and exhibited a higher rubbery elastic modulus and sharper peak compared to those of the H_{12}MDI-based TPU. The constant value of this storage modulus indicates that no phase transitions occurred within this temperature range, and phase separation is effective. This result is attributed to higher DPS, which exists in the samples of the M-BDO-series, as indicated by the FT-IR results.

Table 8. A summary of dynamic mechanical analysis results of the bio-based TPUs.

Sample	T_g (°C)	G'_{25} (MPa)	T_{flex} (°C)	T_{flow} (°C)
H-BDO-2.0	−54.8	0.65	-	-
H-BDO-2.5	−49.3	13.18	-	77.7
H-BDO-3.0	−40.6	46.34	80.1	138.0
M-BDO-2.0	−37.2	1.20	60.2	92.5
M-BDO-2.5	−37.2	5.05	94.2	102.0
M-BDO-3.0	−36.8	54.39	73.9	170.0

The MDI-based TPU demonstrated high values for T_g, G'_{25}, T_{flex}, and T_{flow} compared to those of H_{12}MDI-based TPU, suggesting a higher degree of stiffness of this sample. In addition, the regions of the rubbery plateau between T_{flex} and T_{flow} were extended. Typically, phase separation occurs in TPU materials because of the thermodynamic incompatibility between the SS and HS, resulting in elastomeric properties.

3.7. Thermogravimetric Analysis (TGA)

The TGA curves for TPUs prepared under different conditions are shown in Figure 8, and the corresponding data are summarized in Table 9. The temperature of 5% weight loss (T5) is typically considered the onset decomposition temperature [41]. The results showed differences at the beginning of thermal degradation (T5) depending on the [NCO]/[OH] molar ratio. Furthermore, the molecular weight and T5 increased with the [NCO]/[OH] molar ratio. The T5 of the M-BDO-series was 2–9 °C higher than that of the H-BDO-series. In particular, the M-BDO-3.0 sample was characterized by the best thermal stability in the initial stage of decomposition at 300–310 °C. Thus, in M-BDO-3.0, a higher temperature is required to obtain 5 and 10% weight loss compared to that in other materials. Therefore, it is advantageous to use MDI as isocyanate from the viewpoint of thermal properties. Despite the presence of a single-step drop, there are two peaks in the derivative curve (%/°C; DTG) owing to the change in the curve slope as it descends. The first step at 330–370 °C is related to HS decomposition (T_{HS}), and the second step around 390–426 °C is associated with SS decomposition (T_{SS}) as shown in Figure 8 [67]. All TPUs prepared in this study showed this characteristic, and the weight loss rate was dependent on the [NCO]/[OH] molar ratio. With an increase in the [NCO]/[OH] molar ratio, the maximum weight loss rate for the first and second decomposition steps decreased. All TPUs with a higher HS content also exhibited a higher weight loss rate for the first decomposition step and a lower weight loss rate for the second decomposition step. In conclusion, an increase in the [NCO]/[OH] molar ratio leads to a corresponding increase in thermal stability. This relationship is reversed in T_{HS} at a higher temperature, suggesting a more cross-linked structure of the polymer. In addition, the materials obtained from aromatic isocyanate showed higher thermal stability than those obtained from aliphatic isocyanate. Thus, the type of isocyanate was revealed to have more effect on thermal stability. According to the literature, the initial stage of decomposition temperature of bio-based TPUs was revealed at 310 °C with fatty acid dimer-based polyester polyols [19] and 315 °C [30] with bio-based polyether polyols-poly(trimethylene glycol). Additionally, TPU from Bio-based polyester polyols synthesized using esterification with azelaic acid, sebacic acid, and 1,3-propandiol showed 260 °C of the initial stage of decomposition temperature [68]. In the results of DTG peaks of bio-based TPU, Carmen et al. [69] using dimer acid-based polyol described the temperature range of around 311–360 °C and 390–440 °C, Paulina et al. [70] with the poly(propylene succinate)s showed the temperature of 5% mass loss at ca. 320 °C, and

T_{HS}/T_{SS} at 384.7/427.3 °C. These results represented that bio-based TPU in this study had similar thermal stability compared to the other bio-TPU.

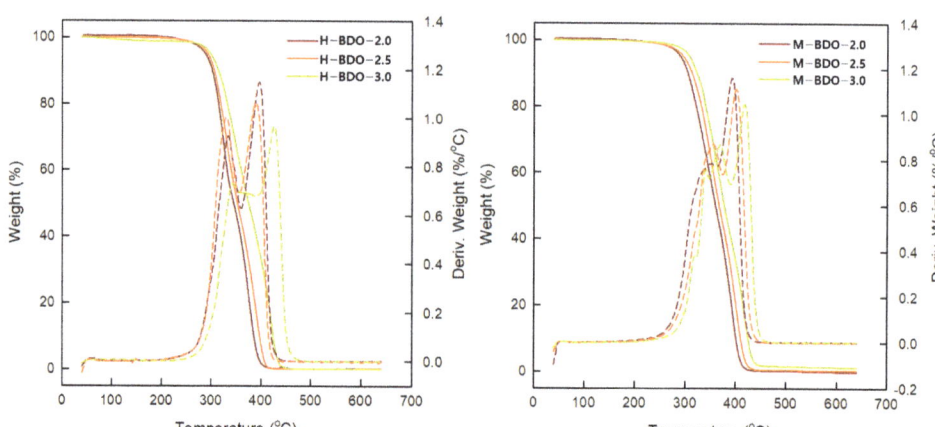

Figure 8. Thermogravimetric analysis curves and the respective derivative curves of bio-based TPUs.

Table 9. Thermal decomposition characteristics of the bio-based TPUs.

Sample	T5% (°C)	T10% (°C)	T50% (°C)	T_{HS} (°C)	T_{SS} (°C)
H-BDO-2.0	288.8	302.4	345.3	334.7	396.4
H-BDO-2.5	294.0	306.7	353.2	332.4	390.5
H-BDO-3.0	300.0	315.3	375.8	344.2	426.0
M-BDO-2.0	290.0	304.7	360.7	351.3	391.6
M-BDO-2.5	296.5	313.5	368.0	356.0	402.2
M-BDO-3.0	308.9	323.4	378.2	371.5	418.9

3.8. Mechanical Properties

Figure 9 and Table 10 illustrate the mechanical properties (initial modulus, tensile strength, and elongation at break) of all the samples at room temperature and hardness (Shore A). The mechanical performances of the TPUs are closely related to their compositions; the HS can act as physical cross-links and reinforcing units, while the SS is responsible for the material flexibility owing to the long linear polyol chain. From the data in Table 7, it can be concluded that the bio-based thermoplastic poly(ether-urethane)s prepared with the highest [NCO]/[OH] molar ratio has the highest HS content. The initial modulus values of the H-BDO- and M-BDO-series ranged from 12.20–42.40 MPa and 35.25–67.07 MPa, respectively. Therefore, the initial modulus values increased with the amount of HS content in the polymer, which was associated with higher stiffness and content of HS. Similar relationships were demonstrated for all prepared samples with respect to tensile strength and elongation at break. The increase in the HS content from 36.9 to 45.5 wt% resulted in a corresponding increase in the tensile strength from 4.61 to 31.61 MPa. The increase in the [NCO]/[OH] molar ratio also led to higher tensile strength. The HS content determines the tensile strength, while the SS content relates to the elongation at break values. The elongation at break is also related to the decrease in the chain mobility, and the permanent set after the break suggests the possibility of macromolecular chains returning to the states in which they were before the test. M-BDO-2.0 showed high elongation at break (1081%) and tensile strength values above 29 MPa. Moreover, M-BDO-3.0 showed higher hardness (Shore A), and the hardness value increased as the content of the HSs increased; hardness also increased with increasing [NCO]/[OH] molar ratio. The samples of M-BDO-2.0, M-

BDO-2.5, and M-BDO-3.0 showed elongation at break values of 1316.8, 629.9, and 635.7%, respectively. These differences in the elongation at break are related to the decreases in the chain mobility, which was confirmed for the H-BDO-series by shifting the T_g to higher values. Hardness is closely connected with the cross-link density in HS, which decreased the elasticity of the samples and led to rigid materials. The MDI-based samples had the highest hardness, with values ranging from 79 to 86 Shore A. The stiffness of the materials increased as the cross-link density increased.

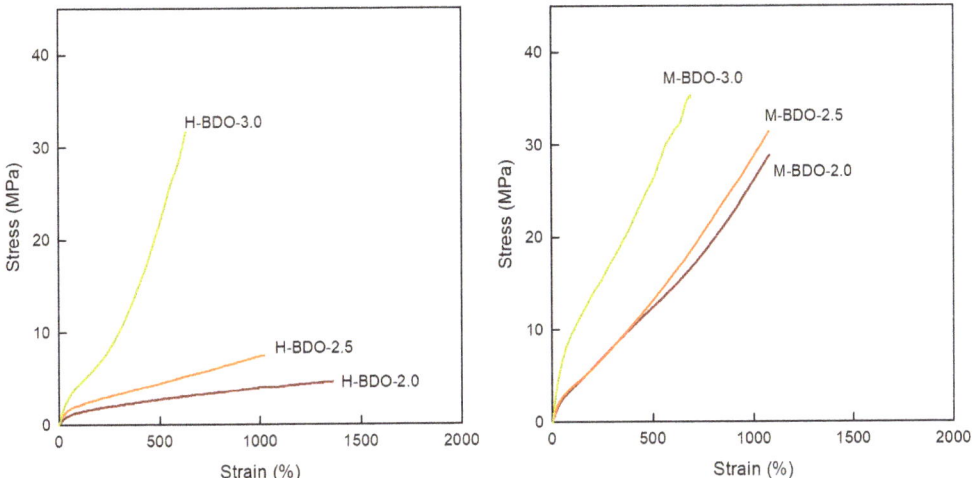

Figure 9. Tensile stress-strain curves of the bio-based TPUs.

Table 10. Mechanical properties and hardness of the bio-based TPUs.

Sample	Initial Modulus (MPa)	Tensile Strength(MPa)	Elongation (%)	Energy (J)	Hardness (Shore A)
H-BDO-2.0	12.20	4.61	1316.80	0.13	70
H-BDO-2.5	22.23	7.48	1024.50	0.14	79
H-BDO-3.0	42.40	31.61	635.70	1.01	85
M-BDO-2.0	35.25	28.84	1080.90	2.09	76
M-BDO-2.5	37.27	31.42	1076.90	3.55	77
M-BDO-3.0	67.07	35.04	693.90	3.24	86

3.9. Micro-Phase Separation Characteristics

Based on the results discussed above (FT-IR, AFM, DSC, etc.), four types of micro-phase separation in a two-phase system are suggested for the HS dispersed in the SS matrix, as shown in Figure 10. The figure shows (a) a disassociated structure of phase-mixed between HS and SS, (b) a hydrogen-bonded structure phase-separated between HS and SS that formed one-sided hard domains, (c) hydrogen-bonded structure of phase-mixed between the HS and SS, and (d) hydrogen-bonded structure of phase-separated between the HS and SS that formed dispersed hard domains. Due to this, H-BDO-2.0 is expected to exhibit the (a)-form in Figure 10 among the phase-separated forms. The (a)-form can be regarded as a phase-separated form when phase separation is not accomplished completely. The AFM image showed an exceptionally smooth surface with low roughness, lower DPS values, and tensile strength (4.61 MPa). In the case of H-BDO-2.5 and H-BDO-3.0, because no significant difference was observed in the DPS values, a phase-mixed system between HS and SS is expected to show the (c)-form. In particular, visible softening of the materials based on the tensile test was observed in the case of the H-BDO-series, which also suggested a less ordered phase among the structures. This result is different compared to other thermoplastic elastomers that often show a decrease of the rubbery plateau with increasing temperature due to incomplete phase separation or partial melting of the HS. In DSC, the slight increase in T_g observed for the MDI-based TPU suggests the presence of a relatively large amount of HS mixed within the soft domain, resulting in a higher degree of microphase separation for this TPU series caused by the increase in the molecular weight and hydrogen-bonded hard domain. These results are in agreement with the FT-IR spectra in the C=O region, where free and hydrogen-bonded carbonyl groups are involved. The formation of the hard domain and the T_m of the hard phase is drastically affected by the HS length distribution. Researchers [19,46] have reported that the properties of TPUs can be improved using uniform HS and have a more complete-phase separation. The disordering in the HS and the partial miscibility of the hard phase in the SS are also confirmed by the decrease in the hard domain. [47,48]. The difference in the behavior of T_m of the TPU materials suggests that the TPUs have different physical origins. The lower temperatures can be attributed to the melting of the less-ordered structures or suitable SSs, whereas the higher temperatures are associated with greater-order structures. The structural proposal in Figure 8 was supported by the segment structure of TPU, which was used as a guideline to elucidate the morphology of these polymers further. Based on this, the M-BDO-2.0 sample shows the (b)-form with low T_m of HS despite the high DPS value. The sample of M-BDO-2.5 and M-BDO-3.0 may demonstrate the (d)-form with the AFM images, which can be confirmed with high DPS values, T_m of HS (231.6 °C), rubbery plateau (73.9–170 °C), and tensile strength (35.0 MPa) from the FT-IR, DSC, and DMA results.

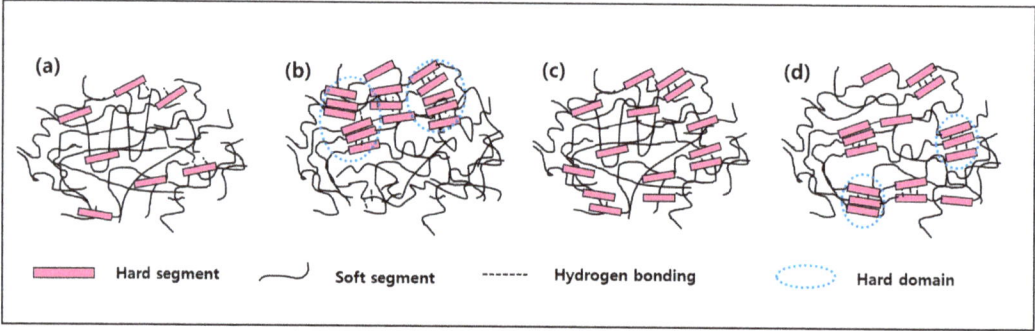

Figure 10. Schematic diagram of micro-phase separation in TPUs containing different conditions: (**a**) a disassociated structure of phase-mixed between HS and SS, (**b**) a hydrogen-bonded structure phase-separated between HS and SS that formed one-sided hard domains, (**c**) hydrogen-bonded structure of phase-mixed between the HS and SS, and (**d**) hydrogen-bonded structure of phase-separated between the HS and SS formed hard domains.

4. Conclusions

Recently, there has been increased interest in developing bio-based polyols and TPUs owing to environmental issues, which will lead to a growing demand for materials with soft and strong properties. In order to meet these demands of the times, this study aimed to synthesize a bio-based TPU. We used bio-based PO3G as a polyol, MDI, and H_{12}MDI as the isocyanate and bio-based BDO as the chain extender. Bio-based TPUs with [OH]/[NCO]/[OH] molar ratios ranging from 1:2:1 to 1:3:1 was successfully synthesized using a solvent-free one-shot process. This study investigated the effect of the equivalence ratio and isocyanate type on the compositions and properties of the HS and SS of the synthesized TPUs and then suggested four types of micro-phase separation of the HS and SS. The M_n and M_w of TPUs were in the ranges 37,723–112,117 and 70,953–236,689, respectively. The M_w of TPUs strongly increased as the [NCO]/[OH] molar ratio increased. The DPS also increased, and this result is expected to affect the thermal and physical properties and the micro-phase separation characteristics of the prepared film. When the HS content was increased by varying the [NCO]/[OH] molar ratio, the HS domains became larger; the roughness and the extent of phase separation increased, and the hard domain structures became more pronounced and visible in the sample morphology in the 3D AFM images. Considering the isocyanate type, the M-BDO-series showed higher average molecular weight and DPS than those of the H-BDO-series, despite having a lower content of HS. Consequently, the M-BDO-series had better-ordered, stronger, and more stable HS domains. The M-BDO-series with a high hard domain content showed a more rounded hilly break surface containing large globules and a high roughness value in the 3D AFM images. The SS T_g, SS T_m, HS T_g, and HS T_m of M-BDO_3.0 were observed at −44.3, 80.4, 157.3, and 231.57 °C, respectively, in the DSC analysis. The rubbery plateau was extended and exhibited a higher rubbery modulus and sharper peak, which was determined by DMA. According to TGA, the materials obtained from aromatic isocyanate showed higher thermal stability than those obtained from aliphatic isocyanate. In addition, the mechanical properties of the M-BDO-series were higher compared to those of the H-BDO-series. With respect to the micro-phase separation forms, H-BDO-2.0 is expected to show the (a)-form, indicating that phase separation is not accomplished completely because of the exceptionally smooth surface observed in AFM, with low roughness, lower DPS values, and tensile strength. H-BDO-2.5 and H-BDO-3.0 represented the (c)-form based on the tensile test and AFM, which also suggested a less-ordered phase among the micro-phase structures. The M-BDO-2.0 sample represented the (b)-form with low T_m of HS despite the high DPS value, while M-BDO-2.5 and M-BDO-3.0 can relate to the (d)-form with the AFM images, which can be confirmed with the high DPS values, T_m of HS, rubbery plateau, and tensile strength. The previous results showed that when an aromatic isocyanate (MDI) was used in the synthesis of TPU with a bio-based ether polyol, a soft sample with excellent thermal and physical properties could be obtained. In the future, we would like to investigate whether these TPU samples can be used as a shape memory polymer for 4D printing filaments.

Author Contributions: Conceptualization (lead), formal analysis (lead), writing—original draft (lead), Y.-S.J.; Conceptualization (supporting); formal analysis (supporting). writing—review and editing (supporting), E.-J.S.; Conceptualization, review, and editing (supporting), J.P.; Conceptualization (equal), funding acquisition (lead), S.L. All authors have read and agreed to the published version of the manuscript.

Funding: This research was supported by the Basic Science Research Program of the National Research Foundation of Korea (NRF) funded by the Ministry of Education (2021R1A4A1022059)

Institutional Review Board Statement: Not applicable.

Informed Consent Statement: Not applicable.

Data Availability Statement: The data presented in this study are available on request from the corresponding author.

Conflicts of Interest: The authors declare no conflict of interest.

References

1. Oh, J.; Kim, Y.K.; Hwang, S.H.; Kim, H.C.; Jung, J.H.; Jeon, C.H.; Kim, J.; Lim, S.K. Synthesis of Thermoplastic Polyurethanes Containing Bio-Based Polyester Polyol and Their Fiber Property. *Polymers* **2022**, *14*, 2033. [CrossRef] [PubMed]
2. Duval, A.; Sarbu, A.; Dalmas, F.; Albertini, D.; Averous, L. 2, 3-Butanediol as a Biobased Chain Extender for Thermoplastic Polyurethanes: Influence of Stereochemistry on Macromolecular Architectures and Properties. *Macromolecules* **2022**, *55*, 5371–5381. [CrossRef]
3. Gao, Z.; Wang, Z.; Liu, Z.; Fu, L.; Li, X.; Eling, B.; Pöselt, E.; Schander, E.; Wang, Z. Hard block length distribution of thermoplastic polyurethane determined by polymerization-induced phase separation. *Polymer* **2022**, *256*, 125236. [CrossRef]
4. Zhu, Y.; Romain, C.; Williams, C.K. Sustainable polymers from renewable resources. *Nature* **2016**, *540*, 354–362. [CrossRef]
5. Yilgör, I.; Yilgor, E.; Das, S.; Wilkes, G.L. Time-dependent morphology development in segmented polyetherurea copolymers based on aromatic diisocyanates. *J. Polym. Sci. Part B Polym. Phys.* **2009**, *47*, 471–483. [CrossRef]

6. Kasprzyk, P.; Głowińska, E.; Parcheta-Szwindowska, P.; Rohde, K.; Datta, J. Green TPUs from Prepolymer Mixtures Designed by Controlling the Chemical Structure of Flexible Segments. *Int. J. Mol. Sci.* **2021**, *22*, 7438. [CrossRef] [PubMed]
7. Zhang, C.; Garrison, T.F.; Madbouly, S.A.; Kessler, M.R. Recent advances in vegetable oil-based polymers and their composites. *Prog. Polym. Sci.* **2017**, *71*, 91–143. [CrossRef]
8. Donati, I.; Travan, A.; Pelillo, C.; Scarpa, T. Polyol Synthesis of Silver Nanoparticles: Mechanism of Reduction by Alditol Bearing Polysaccharides. *Biomacromolecules* **2009**, *10*, 210–221. [CrossRef]
9. Hakima, A.; Nassar, M.; Emam, A.; Sultan, M. Preparation and characterization of rigid polyurethane foam prepared from sugar-cane bagasse polyol. *Mater. Chem. Phys.* **2011**, *129*, 301–330. [CrossRef]
10. Lu, M.Y.; Surányi, A.; Viskolcz, B.; Fiser, B. Molecular design of sugar-based polyurethanes. *Croat. Chem. Acta* **1991**, *3*, 299–308. [CrossRef]
11. Li, Y.; Luo, X.; Hu, S. *Lignocellulosic Biomass-Based Polyols for Polyurethane Applications Bio-Based Polyols and Polyurethanes*; Springer Briefs in Molecular Science Book Series; Springer: Cham, Switzerland, 2015; pp. 45–64.
12. Zlatanić, A.; Lava, C.; Zhang, W.; Petrović, Z.S. Effect of structure on properties of polyols and polyurethanes based on different vegetable oils. *J. Polym. Sci. Part B Polym. Phys.* **2004**, *42*, 809–819. [CrossRef]
13. Sawpan, M. Polyurethanes from vegetable oils and applications: A review. *J. Polym. Res.* **2018**, *25*, 184. [CrossRef]
14. Petrovic, Z.; Yang, L.; Zlatanic, A.; Zhang, W.; Javni, I. Network structure and properties of polyurethanes from soybean oil. *J. Appl. Sci.* **2007**, *105*, 2717–2727. [CrossRef]
15. Gurunathan, T.; Mohanty, S.; Nayak, S.K. Isocyanate terminated castor oil-based polyurethane prepolymer: Synthesis and characterization. *Prog. Org. Coat.* **2015**, *80*, 39–48. [CrossRef]
16. Alagi, P.; Hong, S.C. Vegetable oil-based polyols for sustainable polyurethanes. *Macromol. Res.* **2015**, *23*, 1079–1086. [CrossRef]
17. Chuayjuljit, S.; Maungchareon, A.; Saravari, O. Preparation and properties of palm oil-based rigid polyurethane nanocomposite foams. *J. Reinf. Plast. Compos.* **2010**, *29*, 218–225. [CrossRef]
18. Shen, Y.; He, J.; Xie, Z.; Zhou, X.; Fang, C.; Zhang, C. Synthesis and characterization of vegetable oil based polyurethanes with tunable thermomechanical performance. *Ind. Crops Prod.* **2019**, *140*, 711–717. [CrossRef]
19. Ruan, M.; Luan, H.; Wang, G.; Shen, M. Bio-polyols synthesized from bio-based 1,3-propanediol and applications on polyurethane reactive hot melt adhesives. *Ind. Crops Prod.* **2019**, *128*, 436–444. [CrossRef]
20. Fakirov, S. *Handbook of Condensation Thermoplastic Elastomer*; Wiley-VHC Verlag: Weinheim, Germany, 2005.
21. Holden, G.; Kricheldorf, R.; Quirk, P. *Thermoplastic Elastomer*, 3rd ed.; Hanser: Munich, Germany, 2004.
22. Harrell, L.J. Segmented polyurethanes. Properties as function of segment size and distribution. *Macromolecules* **1969**, *2*, 607–612. [CrossRef]
23. Delebecq, E.; Pascault, J.P.; Boutevin, B.; Ganachaud, F. On the versatility of urethane/urea bonds: Reversibility, blocked isocyanate, and non-isocyanate polyurethane. *Chem. Rev.* **2013**, *113*, 80–118. [CrossRef] [PubMed]
24. Charlon, M.; Heinrich, B.; Matter, Y.; Couzigné, E.; Donnio, B.; Avérous, L. Synthesis, structure, and properties of fully biobased thermoplastic polyurethanes, obtained from a diisocyanate based on modified dimer fatty acids, and different renewable diols. *Eur. Polym. J.* **2014**, *61*, 197–205. [CrossRef]
25. Gaymans, R.J. Segmented copolymers with monodisperse crystallizable hard segments: Novel semi-crystalline materials. *Prog. Polym. Sci.* **2011**, *36*, 713–748. [CrossRef]
26. Wang, Y.; Ma, R.; Li, H.; Hu, S.; Gao, Y.; Liu, L.; Zhang, L. Effect of the content and strength of hard segment on the viscoelasticity of the polyurethane elastomer: Insights from molecular dynamics simulation. *Soft Matter* **2022**, *18*, 4090–4101. [CrossRef] [PubMed]
27. Kang, S.; Ji, Z.; Tseng, L.; Turner, S.; Villanueva, D.; Johnson, R.; Albano, A.; Langer, R. Design and Synthesis of Waterborne Polyurethanes. *Adv. Mater.* **2018**, *30*, 1706237. [CrossRef] [PubMed]
28. Biemond, G. Hydrogen Bonding in Segmented Block Copolymer. Ph.D. Thesis, University of Twentw, Enschede, The Netherlands, 2006.
29. Kasprzyk, P.; Benes, H.; Donato, R.K.; Datta, J. The role of hydrogen bonding on tuning hard-soft segments in bio-based thermoplastic poly(ether-urethane)s. *J. Clean. Prod.* **2020**, *274*, 122678. [CrossRef]
30. Niesten, M.; Feijen, J.; Gaymans, R.J. Synthesis and properties of segmented copolymers having aramid units of uniform length. *Polymer* **2000**, *41*, 8487–8500. [CrossRef]
31. Van der Schuur, M.; Gaymans, R.J. Segmented block copolymers based on poly(propylene oxide) and monodisperse polyamide-6,T segments. *J. Polym. Sci. Part A Polym. Chem.* **2006**, *44*, 4769–4781. [CrossRef]
32. Odian, G. *Step Polymerization, in Principles of Polymerization*; John Wiley & Sons: Hoboken, NJ, USA, 2004; pp. 39–197.
33. Kirpluks, M.; Cabulis, U.; Ivdre, A.; Kuranska, M.; Zieleniewska, M.; Auguscik, M. Mechanical and thermal properties of high-density rigid polyurethane foams from renewable resources. *J. Renew. Mater.* **2016**, *4*, 86–100. [CrossRef]
34. Li, X.X.; Sohn, M.H.; Cho, U.R. Synthesis and Properties of Bio-Thermoplastic Polyurethanes with Different Isocyanate Contents. *Elastomers Compos.* **2019**, *54*, 225–231.
35. Coleman, M.; Lee, K.; Skrovanek, D.; Painter, P. Hydrogen bonding in polymers. 4. Infrared temperature studies of a simple polyurethane. *Macromolecules* **1986**, *19*, 2149–2157. [CrossRef]
36. Lei, W.; Fang, C.; Zhou, X.; Cheng, Y.; Yang, R.; Liu, D. Morphology and thermal properties of polyurethane elastomer based on representative structural chain extenders. *Thermochim. Acta* **2017**, *653*, 116–125. [CrossRef]

37. Gorna, K.; Polowinski, S.; Gogolewski, S. Synthesis and characterization of biodegradable poly(e-caprolactone urethane)s. I. Effect of the polyol molecular weight, catalyst, and chain extender on the molecular and physical characteristics. *J. Polym. Sci. Part A Polym. Chem.* **2002**, *40*, 156–170. [CrossRef]
38. Xiaozhen, Y.; Decai, Y.; Hsu, S.L.; Meuse, C.W. Spectroscopic analysis of ordering and phase-separation behavior of model polyurethanes in a restricted geometry. *Macromolecules* **1992**, *25*, 925–932.
39. Elwell, M.J.; Ryan, A.J.; Grünbauer, H.C.; Lieshout, V. In-situ studies of structure development during the reactive processing of model flexible polyurethane foam systems using FT-IR spectroscopy, synchrotron SAXS, and rheology. *Macromolecules* **1996**, *29*, 2960–2968. [CrossRef]
40. Głowińska, E.; Datta, J. Bio polyetherurethane composites with high content of natural ingredients: Hydroxylated soybean oil based polyol, bio glycol and microcrystalline cellulose. *Cellulose* **2016**, *23*, 581–592. [CrossRef]
41. Ryan, A.J.; Willkomm, W.R.; Bergstrom, T.B.; Macosko, C.W.; Koberstein, J.T.; Yu, C.C.; Russell, T.P. Dynamics of (Micro)phase separation during fast, bulk copolymerization: Some synchrotron SAXS experiments. *Macromolecules* **1991**, *24*, 2883–2889. [CrossRef]
42. Król, P. Synthesis methods, chemical structures and phase structures of linear polyurethanes. Properties and applications of linear polyurethanes in polyurethane elastomers, copolymers and ionomers. *Prog. Mater. Sci.* **2007**, *52*, 915–1015. [CrossRef]
43. Saralegi, A.; Rueda, L.; Fernández-D'Arlas, B.; Mondragon, A.; Eceiza, C. Thermoplastic polyurethanes from renewable resources: Effect of soft segment chemical structure and molecular weight on morphology and final properties. *Polym. Int.* **2013**, *62*, 106–115. [CrossRef]
44. Suzuki, T.; Shibayama, M.; Hatano, K.; Ishii, M. [NCO]/[OH] and acryl-polyol concentration dependence of the gelation process and the microstructure analysis of polyurethane resin by dynamic light scattering. *Polymer* **2009**, *50*, 2503–2509. [CrossRef]
45. Kasprzyk, P.; Datta, J. Effect of Molar Ratio [NCO]/[OH] Groups during Prepolymer Chains Extending Step on the Morphology and Selected Mechanical Properties of Final Bio-Based Thermoplastic Poly (Ether-Urethane) Materials. *Polym. Eng. Sci.* **2018**, *58*, E119–E206. [CrossRef]
46. Versteegen, M.; Sijbesma, P.; Meijer, W. Synthesis and characterization of segmented copoly(ether urea)s with uniform hard segments. *Macromolecules* **2005**, *38*, 3176–3184. [CrossRef]
47. Flory, J. Thermodynamics of High Polymer Solutions. *J. Chem. Phys.* **1942**, *10*, 51. [CrossRef]
48. Verstraete, G.; Van Renterghem, J.; Van Bockstal, P.J.; Kasmi, S.; De Geest, B.G.; De Beer, T.; Vervaet, C. Hydrophilic thermoplastic polyurethanes for the manufacturing of highly dosed oral sustained release matrices via hot melt extrusion and injection molding. *Int. J. Pharm.* **2016**, *506*, 214–221. [CrossRef]
49. Kim, H.D.; Huh, J.H.; Kim, E.Y.; Park, C.C. Comparison of properties of thermoplastic polyurethane elastomers with two different soft segments. *J. Appl. Polym. Sci.* **1998**, *69*, 1349–1355. [CrossRef]
50. Tao, Y.; Hasan, A.; Deeb, G.; Hu, C.; Han, H. Rheological and mechanical behavior of silk fibroin reinforced waterborne polyurethane. *Polymers* **2016**, *8*, 94. [CrossRef]
51. Lluch, C.; Esteve-Zarzoso, B.; Bordons, A.; Lligadas, G.; Ronda, J.C.; Galia, M.; Cádiz, V. Antimicrobial polyurethane thermosets based on undecylenic acid: Synthesis and evaluation. *Macromol. Biosci.* **2014**, *14*, 1170–1180. [CrossRef]
52. Li, H.; Mahmood, N.; Ma, Z.; Zhu, M.; Wang, J.; Zheng, J.; Yuan, Z.; Wei, Q.; Xu, C. Preparation and characterization of bio-polyol and bio-based flexible polyurethane foams from fast pyrolysis of wheat straw. *Ind. Crops Prod.* **2017**, *103*, 64–72. [CrossRef]
53. Sheth, J.P.; Klinedinst, D.B.; Wilkes, G.L.; Yilgor, I.; Yilgor, E. Role of chain symmetry and hydrogen bonding in segmented copolymers with monodisperse hard segments. *Polymer* **2005**, *46*, 7317–7322. [CrossRef]
54. Kultys, A.; Rogulska, M.; Pikus, S.; Skrzypiec, K. The synthesis and characterization of new thermoplastic poly(carbonate-urethane) elastomers derived from HDI and aliphatic-aromatic chain extenders. *Eur. Polym. J.* **2009**, *45*, 2629–2643. [CrossRef]
55. Kojio, K.; Kugumiya, S.; Uchibaq, Y.; Nishino, Y.; Furukawa, M. The micro-separated structure of polyurethane bulk and thin films. *Polym. J.* **2009**, *41*, 118–124. [CrossRef]
56. Špírková, M.; Strachota, A.; Urbanová, M.; Baldrian, J.; Brus, J.; Šlouf, M.; Kuta, A.; Hrdlička, Z. Structural and surface properties of novel polyurethane films. *Mater. Manuf. Process.* **2009**, *24*, 1214–1216. [CrossRef]
57. Wang, Y.; Ma, C.; Mu, C.; Lin, W. Tailor-made zwitterionic polyurethane coatings: Microstructure, mechanical property and their antimicrobial performance. *RSC Adv.* **2017**, *7*, 27522–27529. [CrossRef]
58. Špírková, M.; Pavličević, J.; Strachota, A.; Poreba, R.; Bera, O.; Kaprálková, L.; Baldrian, J.; Šlouf, M.; Lazić, N.; Budinski-Simendić, J. Novel polycarbonate-based polyurethane elastomers: Composition–property relationship. *Eur. Polym. J.* **2011**, *47*, 959–972. [CrossRef]
59. Fuensanta, M.; Jofre-Reche, J.A.; Rodríguez-Llansola, F.; Costa, V.; Iglesias, J.I.; Martín-Martínez, J.M. Structural characterization of polyurethane ureas and waterborne polyurethane urea dispersions made with mixtures of polyester polyol and polycarbonate diol. *Prog. Org. Coat.* **2017**, *112*, 141–152. [CrossRef]
60. Jeffrey, T.K.; Adam, F. Galambos Multiple melting in segmented polyurethane block copolymers. *Macromolecules* **1992**, *25*, 5618–5624.
61. Fernández-d'Arlas, B.; Jens, B.; Peter, R.; Pöselt, E.; Raphael; Berend, T.; Müller, A. Tailoring the Morphology and Melting Points of Segmented Thermoplastic Polyurethanes by Self-Nucleation. *Macromolecules* **2016**, *49*, 7952–7964. [CrossRef]
62. Saiani, A.; Daunch, W.A.; Verbeke, H.; Leensclag, J.-W.; Higgins, J.S. Origin of Multiple Melting Endotherms in a High Hard Block Content Polyurethane. 1. Thermodynamic Investigation. *Macromolecules* **2001**, *34*, 9059–9068. [CrossRef]

63. Saiani, A.; Rochas, C.; Eeckhaut, G.; Daunch, W.A.; Leenslag, J.-W.; Higgins, J.S. Origin of Multiple Melting Endotherms in a High Hard Block Content Polyurethane. 2. Structural Investigation. *Macromolecules* **2004**, *37*, 1411–1421. [CrossRef]
64. Saiani, A.; Novak, A.; Rodier, L.; Eeckhaut, G.; Leenslag, J.-W. Higgins, J.S. Origin of Multiple Melting Endotherms in a High Hard Block Content Polyurethane: Effect of Annealing Temperature. *Macromolecules* **2007**, *40*, 7252–7262. [CrossRef]
65. Zhang, L.; Huang, M.R.; Yu, J.; Huang, X.; Dong, R.; Zhu, J. Bio-based shape memory polyurethanes (Bio-SMPUs) with short side chains in the soft segment. *J. Mater. Chem. A* **2014**, *2*, 11490. [CrossRef]
66. Parcheta, P.; Datta, J. Structure-rheology relationship of fully bio-based linear polyester polyols for polyurethane-Synthesis and investigation. *Polym. Test.* **2018**, *67*, 110–121. [CrossRef]
67. Chattopadhyay, D.K.; Webster, D.C. Thermal stability and flame retardancy of polyurethanes. *Prog. Polym. Sci.* **2009**, *34*, 1068–1113. [CrossRef]
68. Sohn, M.H.; Li, X.X.; Cho, U.R. Synthesis of Biomass-derived Polyurethane by Chain Extender Type. *Elastomers Compos.* **2019**, *54*, 279–285.
69. Bueno-Ferrer, C.; Hablot, E.; del Carmen Garrigós, M.; Bocchini, S.; Averous, L.; Jiménez, A. Relationship between morphology, properties and degradation parameters of novative biobased thermoplastic polyurethanes obtained from dimer fatty acids. *Polym. Degrad. Stab.* **2012**, *97*, 1964–1969. [CrossRef]
70. Paulina, P.; Ewa, G.; Janusz, D. Effect of bio-based components on the chemical structure, thermal stability and mechanical properties of green thermoplastic polyurethane elastomers. *Eur. Polym. J.* **2020**, *123*, 109422.

Article

Novel Rubber Composites Based on Copper Particles, Multi-Wall Carbon Nanotubes and Their Hybrid for Stretchable Devices

Vineet Kumar, Siraj Azam, Md. Najib Alam, Won-Beom Hong and Sang-Shin Park *

School of Mechanical Engineering, Yeungnam University, Gyeongsan 38541, Korea
* Correspondence: pss@ynu.ac.kr

Abstract: New technologies are constantly addressed in the scientific community for updating novel stretchable devices, such as flexible electronics, electronic packaging, and piezo-electric energy-harvesting devices. The device promoted in the present work was found to generate promising ~6V and durability of >0.4 million cycles. This stretchable device was based on rubber composites. These rubber composites were developed by solution mixing of room temperature silicone rubber (RTV-SR) and nanofiller, such as multi-wall carbon nanotube (MWCNT) and micron-sized copper particles and their hybrid. The hybrid composite consists of 50:50 of both fillers. The mechanical stretchability and compressive modulus of the composites were studied in detail. For example, the compressive modulus was 1.82 MPa (virgin) and increased at 3 per hundred parts of rubber (phr) to 3.75 MPa (MWCNT), 2.2 MPa (copper particles) and 2.75 MPa (hybrid). Similarly, the stretching ability for the composites used in fabricating devices was 148% (virgin) and changes at 3 phr to 144% (MWCNT), 230% (copper particles) and 199% (hybrid). Hence, the hybrid composite was found suitable with optimum stiffness and robust stretching ability to be useful for stretching electronic devices explored in this work. These improved properties were tested for a real-time stretchable device, such as a piezoelectric energy-harvesting device and their improved voltage output and durability were reported. In the end, a series of experiments conducted were summarized and a discussion on the best candidate with higher properties useful for prospective applications was reported.

Keywords: piezo-electric energy-harvesting device; stretchable devices; silicone rubber; multi-wall carbon nanotube; copper particles

Citation: Kumar, V.; Azam, S.; Alam, M.N.; Hong, W.-B.; Park, S.-S. Novel Rubber Composites Based on Copper Particles, Multi-Wall Carbon Nanotubes and Their Hybrid for Stretchable Devices. *Polymers* **2022**, *14*, 3744. https://doi.org/10.3390/polym14183744

Academic Editor: Shaojian He

Received: 4 August 2022
Accepted: 4 September 2022
Published: 7 September 2022

Copyright: © 2022 by the authors. Licensee MDPI, Basel, Switzerland. This article is an open access article distributed under the terms and conditions of the Creative Commons Attribution (CC BY) license (https://creativecommons.org/licenses/by/4.0/).

1. Introduction

The energy demands are increasing day by day and there is a sudden increase in this demand in the last few decades. With the limitation of resources, meeting these demands is becoming difficult to fulfill. Last century, most of the energy production was made through oil, petroleum and coal. However, these sources are limited and non-renewable and their use has created various environmental damages, such as global warming. These problems, such as global warming results in the melt of icebergs and the earth are slowly sinking into the oceans.

Thus, scientists around the globe are working on providing alternative routes of energy that are eco-friendly and mostly renewable. These green routes for producing and storing energy are super-capacitors [1], solar cells [2], batteries [3] and stretchable piezoelectric energy-harvesting devices [4]. Among them, batteries involve the use of acid that degrade the environment, while solar cells also have environmental issues.

However, capacitors and energy-harvesting devices are the cheapest, clean, and highly durable to produce or store energy [1,4]. Piezo-electric energy-harvesting devices or stretchable devices are promising sources of renewable energy sources and are thus explored in the present work. Piezo-electric materials include a family of materials ranging from crystalline materials, such as quartz-analogous crystal [5], ceramics materials, such as

PZT [6], lead-free ceramics, such as barium titanate [7], semiconductors [8], and dielectric polymers [9].

Among them, PZT was found to be the most promising piezo-electric material [10] but due to its poisoning effect, its use is limited [11]. Thus, new lead-free materials were used especially based on dielectric polymers as practiced in this work [12]. These polymer-based stretchable devices are composed of a flexible substrate and electrode [13]. Both electrodes and substrate are made up of elastomer matrix namely silicone rubber, which is soft and has the high stretchable ability as desired. The electrode is generally made up of conductive composites, which are generally composed of carbon nanotubes mixed with silicone rubber [14] while the substrate is made up of virgin or filled composites [15].

The piezo-electric effect based on dielectric polymer and the mechanism of energy harvesting from the piezo-electric device is shown in Scheme 1. The basic principle of piezo-electricity involves the development of electrical energy via mechanical deformation. When a mechanical strain is applied to piezo-electric material, it generates opposite charges on electrodes separated by the dielectric polymer substrate. The basic mechanism involves the generation of more and more opposite charges with more duration of the mechanical strain of the piezo-electric device and hence produces more voltage as described in Scheme 1.

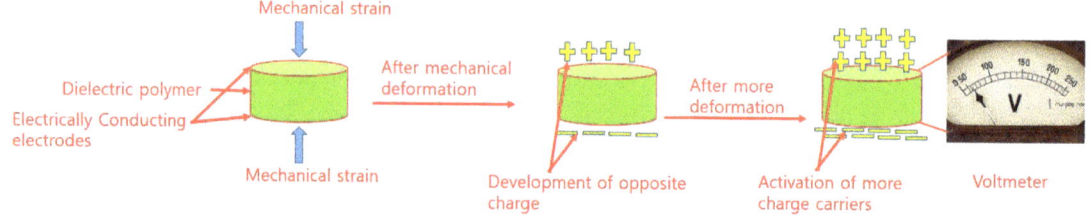

Scheme 1. Mechanism of piezo-electric effect in lead-free dielectric polymers.

The polymer composites are made up of different types of polymers, such as elastomers [16] or thermosets [17]. The elastomers are frequently used for stretchable devices. These elastomers are made of natural rubber [18] or synthesized rubber, such as diene rubbers [19] or silicone rubber [20]. Among them, silicone rubber is most promising due to its ease of process, ease of cure, and hardness of around 40 [21].

These properties make silicone rubber a promising candidate for soft and stretchable devices, such as actuators [22] and strain sensors [23]. Virgin rubber has poor properties and is sticky and not useful for any applications. Therefore, curing is performed to improve the composites' general properties (mechanical properties). Still, these mechanical properties are not enough to be suitable for practical applications. Thus, traditional fillers, such as carbon black were added while improving not only mechanical properties but also the thermal and electrical properties of rubber composites [24].

However, the carbon black used in these composites in high amounts (notably > 60 phr) and high loading alters the viscoelastic properties of rubber composites [25]. To fix this issue, nanofillers are employed in the last few decades that show robust improvement in properties at lower loading without altering the viscoelasticity of the composites. These nanofillers are clay minerals [26], silica [27], graphene [28] and carbon nanotube [29].

Among them, carbon nanotube was found utmost promise due to its favorable morphology and high aspect ratio [30]. Thus, the carbon nanotube is used in the present work as a nanofiller and high properties were reported at as small as 2 phr loading in the rubber matrix [31]. In addition to the carbon nanotube, copper particles are added to improve the overall properties of the composites [32]. Copper is considered a high-performance candidate for various applications due to its outstanding physio-mechanical properties. Due to different particle sizes of copper particles ranging from nanometer range to micron range. They can be used in rubber matrix as a reinforcing agent for improving various engineering applications.

Various studies demonstrate that copper is an outstanding material for improving electrical and thermal properties [33–37]. For example, Wang et al., report that copper can be used as a thermally conductive material in polybenzoxazine-based composites when added with boron nitride [33]. Their study further reports that copper can be successfully used to solve heat diffusion problems in electronic packaging [33]. Further studies by Yin et al., show that the copper nanowires introduced in boron nitride nanosheets can be useful for thermal management problems in flexible electronics [34].

Yang et al., studied the binary fillers, such as copper and tin in PVDF and improved the electrical conductivity by an order of magnitude as reported [35]. Moreover, Boudenne et al., studied the electrical and thermal behavior of the PP filled with two different types of copper particles and improved properties are reported [36]. They investigate the effect of particle size of copper particles and higher heat transport; an effective filler percolation threshold was attained from copper with small particle size [36].

In another study by Kumlutas et al., the effect of the shape of copper particles on thermal properties was noticed [37]. In their study, the copper particles with platelet or spherical or short fibers shapes were added to polyamide to study the orientation effect on the thermal properties of the composites [37]. The copper particles with shapes in form of fibers were found to be effective in enhancing thermal properties [37].

However, studies on copper particles in hybrid with MWCNT in silicone rubber for different types of mechanical properties, such as stretching ability, tribology, etc. are not yet reported and especially for piezoelectric energy harvesting applications. In our previous study, Mannikkavel et al., investigate the piezoelectric energy-harvesting device for MWCNT electrodes and HTV-RTV silicone rubber but only up to 1 V, and durability of 50,000 cycles was noticed [14].

One of the key limitations of our previous work was the poor voltage stability and durability of the device [14]. In this work, the substrate of the piezoelectric energy-harvesting device was filled with 1 phr of copper, MWCNT, or their hybrid, and their output voltage was monitored. The novelty in this work involves the use of copper, MWCNT or their hybrid as a reinforcing material in dielectric silicone rubber. The use of copper in single and hybrid forms is advantageous in piezoelectricity in the present work because it improves the electrical, mechanical, and thermal properties.

Mechanical properties, such as the stretching ability of piezo-electric devices are often ignored in studies but they are of great importance and studied in the present work. With the addition of MWCNT, the electrical properties are improved significantly but the stretching ability is greatly suppressed. It is because the addition of MWCNT promotes enhancing cross-linking density thereby making the composite stiff, and fragile and cracks are formed at an early stage of deformation thereby leading to the falling of voltage when used as electrodes in energy-harvesting devices [14].

Thus, we need a material that can improve stretching ability and maintain electrical conductivity with MWCNT to obtain robust performance in energy harvesting in stretching devices. For that, copper is ideal to achieve high performance by enhancing stretching ability and optimum crosslinking density without harming electrical properties as obtained by MWCNT in the electrode. From experiments, the output voltage was as high as ~6 V, and durability for MWCNT-based substrate was >0.4 million cycles. Thus, the present work was advantageous in terms of the amount of output voltage, stability of voltage and high durability.

2. Materials and Methods

2.1. Materials

The RTV silicone rubber with the commercial name "KE-441-KT" with a transparent appearance was obtained from Shin-Etsu Chemical Corp. Ltd. (Tokyo, Japan). It was used as an elastomeric matrix for the present work. The vulcanizing agent "CAT-RM" was obtained from Shin-Etsu Chemical Corp. Ltd. The multi-wall carbon nanotube (MWCNT) was used as a reinforcing nanofiller. The MWCNT has a diameter of <15 nm, lateral dimensions of 500 nm–1 μm and thus a high aspect ratio of >65. The chemical purity of the MWCNT was >95% and with a commercial name of "CM-100" and purchased by Hanwha Nanotech Corporation Ltd. (Seoul, Korea). The micron-size copper particles were obtained from Duskan Reagents (Ansan-si, Korea). The mold-releasing agent was obtained from Nabakem (Pyeongtaek-si, Korea).

2.2. Preparation of Composites

The fabrication of novel rubber composites was started following the previous work instructions [14]. The optimized route involves the preparation method of spraying the molds with a mold-releasing agent and keeping them drying at room temperature for 2–3 h. Then, the liquid state of RTV-SR (without the use of solvent) was poured into a beaker and a known amount of nanofillers was added (Table 1) to RTV-SR and mixed thoroughly for up to 10 min.

Table 1. Fabrication of the different rubber composites.

Formulation	RTV-SR (phr)	MWCNT (phr)	Copper (phr)	Vulcanizing Solution (phr)
Virgin	100	–	–	2
RTV-SR/MWCNT	100	1–3	–	2
RTV-SR/Hybrid	100	0.5–1.5	0.5–1.5	2
RTV-SR/Copper	100	–	1–3	2

After the filler-rubber mixing phase, the 2 phr of the vulcanizing agent were added to the composite and mixed for 1 min before pouring them into the sprayed molds. These molds were cylindrical (10 × 20 mm) for compressive properties or rectangular (2 × 60 × 60) for tensile properties. The molds were then manually pressed and kept at room temperature for curing for 24 h. Then, the samples were taken out of molds and tested for improved properties and novel applications, such as piezoelectric energy harvesting.

2.3. Characterization Technique

The morphological features and dispersion of filler were together studied through SEM microscopy (S-4800, Hitachi, Tokyo, Japan). The composite samples were sliced into 0.5 mm thick samples through a surgical blade and mounted onto an SEM stub before coating. Both powder samples and composite samples were sputtered with platinum coating for at least 2 min to make the surface electrically conductive for SEM examinations. The crosslink densities of rubber composites were calculated based on Flory–Rehner equation [38] as

$$V_c = -\frac{[\ln(1-V_r) + V_r + \chi V_r^2]}{V_s d_r \left(V_r^{\frac{1}{3}} - \frac{V_r}{2}\right)} \quad (1)$$

where V_c denotes the crosslink density, $\chi = 0.465$ is the interaction parameter for silicone rubber and toluene system [39], $V_s = 106.2$ is the molar volume of swelled solvent toluene, d_r is the density of the rubber, and V_r is the volume fraction of rubber in the swollen state. The volume fraction of rubber was calculated from equilibrium swelling data for 7 days in toluene. The V_r was calculated as—

$$V_r = \frac{(w_r/d_r)}{(w_r/d_r + w_s/d_s)} \quad (2)$$

where w_r and w_s are the weight of rubber and solvent, respectively; and d_r and d_s are the densities of rubber and solvent, respectively.

The compressive and tensile mechanical properties were examined through a universal testing machine (UTS, Lloyd Instruments, Bognor Regis, UK). The compressive mechanical properties were tested on cylindrical samples (20 × 10 mm) at a strain rate of 2 mm/min and up to 35% maximum strain. This strain value was selected because the sample fractured at a higher strain. The tensile mechanical properties were tested on dumbbell-shaped samples with a gauge length of 25 mm and thickness of 2 mm. The strain rate of tensile tests was maintained at 200 mm/min.

These mechanical measurements were obtained according to DIN 53 504 standards. The tribometer used in the present work was obtained from HM Hanmi Electronics, Korea. The model is "STM Smart", and the model name is "universal material testing machine". The experiments were performed at a load of 5 N, frequency of 3 Hz and distance of up to 50 m. The experimental set up and dimensions of pin used in experiments were described in Scheme 2. The sample dimensions are 25 cm × 10 cm, and the thickness is 0.8 cm.

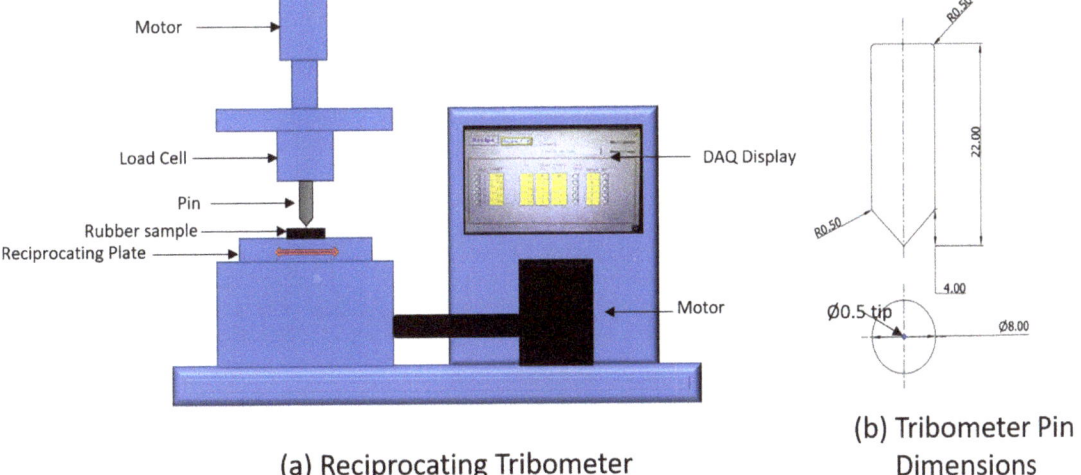

Scheme 2. Schematic of the (**a**) reciprocating tribometer; (**b**) Tribometer pin dimensions.

The output voltage and durability of the stretchable device were performed by a mechanical testing machine (Samick-THK, Daegu, Korea) under cyclic loading. The optical image, dimension of the sample, area of the electrode, dimension of substrate and electrode, and other useful information were described in the previous study [14]. The measurements were performed on an 8 mm thick substrate sandwiched by a 0.2 mm thick electrode in the stretchable device. The substrate was made up of different composites filled with 1 phr of copper, MWCNT or their hybrid and the electrode was made up of 2 phr MWCNT and 2 phr MoS_2. MoS_2 was added to improve the fracture strain of the MWCNT as optimized in a previous study [21].

3. Results and Discussions

3.1. Morphology of Filler Particles

The morphology of the filler particles in composites significantly influences the properties [40]. The filler with favorable morphology is dispersed easily and uniformly and leads to improved properties. Thus, it is important to study the morphology of the filler in the present study through SEM into a number of samples and their representative images are reported. Figure 1a shows the morphology examination of MWCNT. MWCNT has 1-dimensional (1-D) morphology with a tube shape appearance and is a nanofiller since its dimensions are in nm scale.

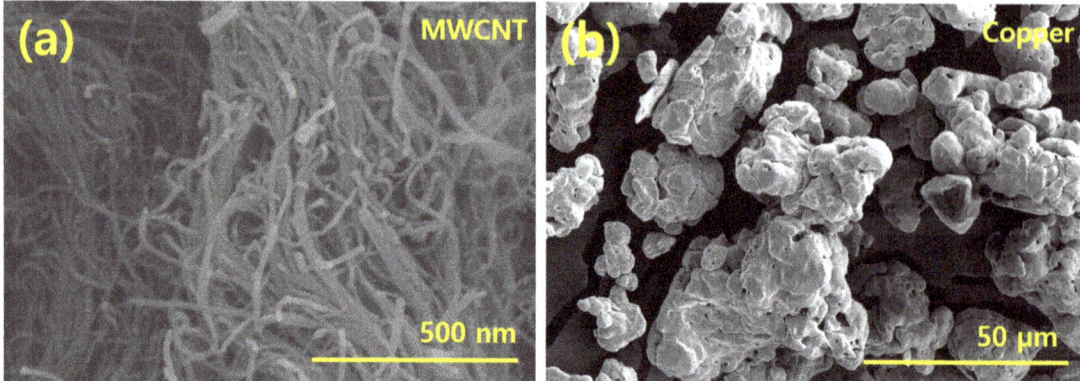

Figure 1. (**a**) SEM of filler particles: (**a**) MWCNT; (**b**) Copper particles.

Since its invention by S. Iijima in 1991 [41], it has become a promising candidate for filler, and various studies prove that it is a promising reinforcing filler and a candidate for improving electrical and thermal properties due to its high aspect ratio [42]. Moreover, its high surface area allows higher stress transfer from polymer chains to filler particles due to higher interfacial area and their interactions [43]. The copper particles are of irregular shape and 3-D in nature as shown in Figure 1b. The copper particles are micron-sized particles and usually have poor reinforcing properties as compared with MWCNT. Its small surface area is also responsible for its poor mechanical, electrical and thermal properties when used as filler in rubber composites. The dispersion of filler also influences the properties significantly and studied in Figure 2.

3.2. Filler Dispersion of Composites through SEM Microscope

In this work, filler dispersion is studied through an SEM microscope. Multiple images per sample were recorded and their representative images at different resolutions are presented in Figure 2. All the samples show uniform dispersion of filler with no signs of aggregations. However, the virgin samples show a neat rubber matrix with no presence of filler as excepted. The samples filled with MWCNT show that the MWCNT particles form long-range filler networks in the composite.

This is attributed to the higher aspect ratio of MWCNT (65) that allows continuous network formation in the composite. These features support the higher properties of the MWCNT-based composites [42,43]. Moreover, the interfacial interactions between MWCNT particles and polymer chains in rubber matrix are good, which also support higher properties of MWCNT composites.

The higher filler–polymer interfacial interactions are supported due to the higher interfacial area, which is provided by the higher surface area of MWCNT. In copper-based composites, since the particle size is in the micron range and the copper has a poor aspect ratio, the dispersion of copper is scarce in the rubber matrix due to the presence of a smaller number of copper particles in the rubber matrix. In some cases, the polymer chains are adsorbed on the copper particles, which expect higher reinforcement but due to the scarcity of several copper particles, the reinforcement excepted is low as compared to MWCNT composites.

However, in the case of hybrid composites, the SEM micrographs show that the polymer chains are not only adsorbed on micro copper particles but the MWCNT particles are also found in the vicinity of these copper particles. This behavior could result in synergism among the binary filler particles with improved reinforcing properties. In a few cases, the hollow structures are formed near copper particles while in some cases the polymer chains were found adsorbed on copper particles along with the MWCNT particles and this behavior affects the properties of the hybrid composites.

The features, such as the formation of the hollow structures around copper particles in the rubber matrix, mimic the sinking of solid copper particles in polymer solution with poor interfacial activity. These features support poor reinforcing effects, especially when compared with MWCNT as only fillers in composites. The XRD data of the hybrid composite to show that copper particles and MWCNTs are properly embedded in the rubber matrix is presented in Figure S1 (supporting information).

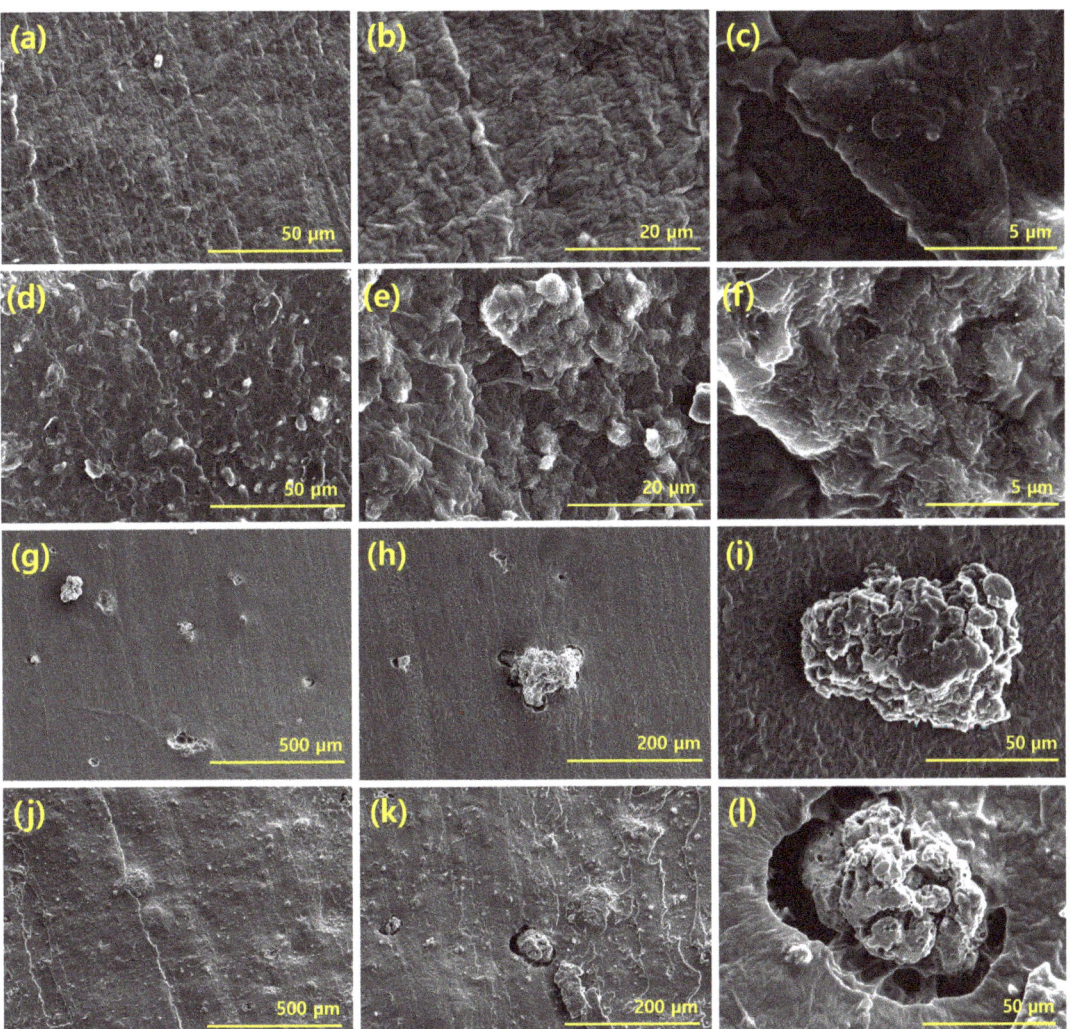

Figure 2. SEM of composites at 3 phr loading except virgin: (**a–c**) Virgin; (**d–f**) MWCNT; (**g–i**) Copper; (**j–l**) Hybrid.

3.3. Cross-Linking Density of Filled Composites

The cross-linking density of different composites at 3 phr is presented in Figure 3. We found from the experimental testing that the crosslink density was higher for hybrid composites and highest for MWCNT-based composites while lowest in copper composites.

The higher cross-linking density in MWCNT is due to its nanoscale dimensions, which help the even dispersion of the curatives in the rubber matrix.

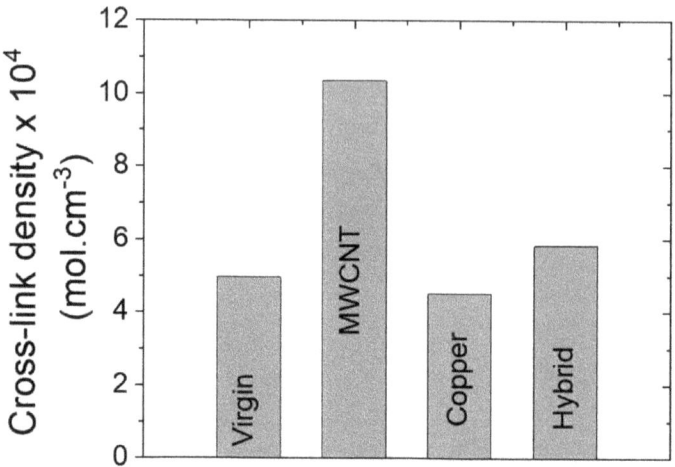

Figure 3. Cross-link density of different composites at 3 phr loading.

Another feature of MWCNT is its higher surface area, which also allows the better dispersion of curatives in rubber matrix [44]. The poor cross-linking density of copper is due to its micron range size and poor surface area, which leads to uneven dispersion of curatives and leads to a poor density of curatives. These measurements were correlated with the mechanical properties of composites in subsequent sections of the manuscript.

3.4. Mechanical Properties

The mechanical properties, such as mechanical stretching ability or stretch until fracture, of composites, are vital to be studied for stretchable devices [45] studied in this work. The mechanical stretchability of a device depends upon several parameters, such as filler dispersion [46], the morphology of filler [47], an aspect ratio of the filler [48], the type of rubber matrix used [49], or the type of mechanical deformation [14]. Here, compressive and tensile strain were applied, and the mechanical behavior of the stretchable device under different strains were explored.

3.4.1. Under Static Compressive Strain

The mechanical behavior of the composites under compressive strain was studied and presented in Figure 4a–c. We found that the compressive stress increases with an increase in compressive strain. It is attributed to the increase in packing density [50] of polymer chains and filler particles, which increases with an increase in compressive strain. Moreover, it was interesting to note that the compressive stress increases linearly up to 15% compressive strain and then increases exponentially up to 35%, especially in the case of MWNT composites in Figure 4a.

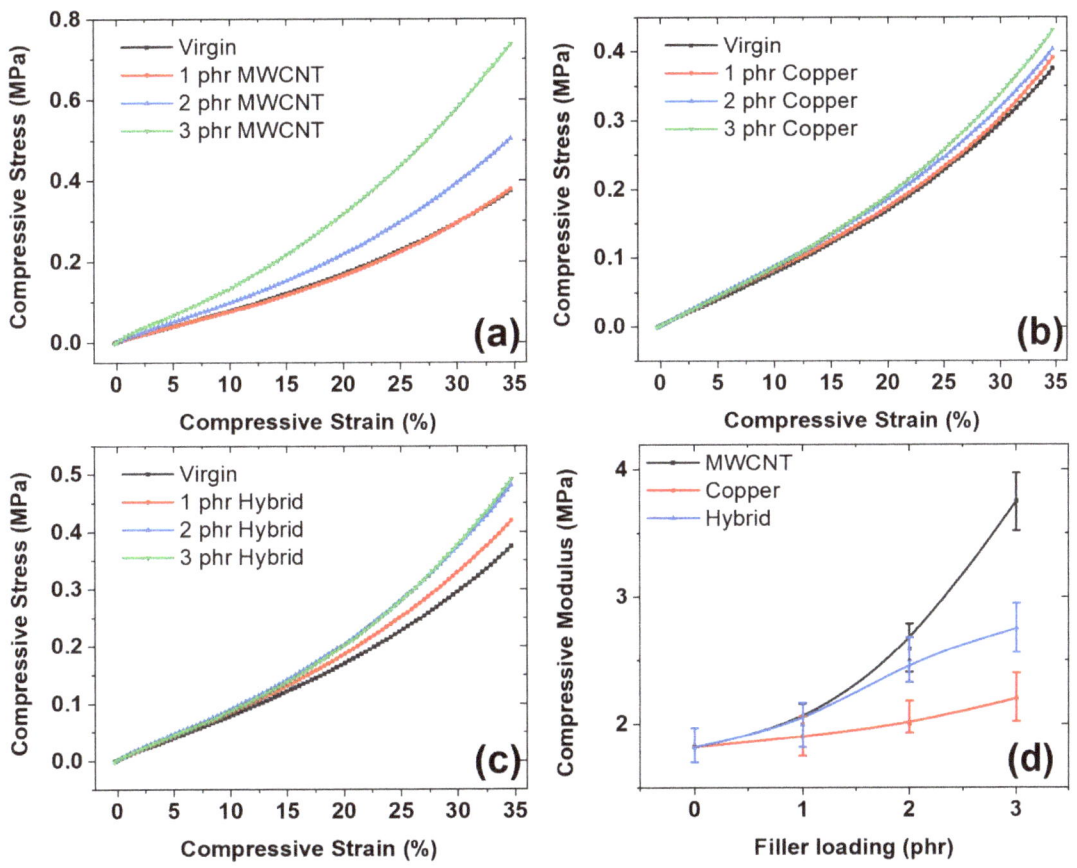

Figure 4. Compressive mechanical properties: (**a**) stress train of MWCNT; (**b**) stress–strain of Copper; (**c**) stress–strain of hybrid composites; (**d**) compressive modulus of different composites.

Such an increase is due to the higher anisotropic behavior of MWCNT and its higher aspect ratio [48] than copper particles. In addition to this, it was also interesting to note that with an increase in filler content from 1 to 3 phr, the compressive stress increases for all fillers. It is attributed to the formation of filler networks [51] and enhanced filler–polymer and filler–filler interactions [52] that lead to induced stiffness making the composite tough and highly reinforced.

In Figure 4d, the behavior of compressive modulus as a function of filler content is proposed. We found that the compressive modulus increases with an increase in filler content. It is due to improved interfacial interaction among filler and rubber matrix [52,53], which increases with the increase in filler volume fraction in the composite. In addition, it is noteworthy that up to 1 phr, the MWCNT, and hybrid composites show the highest compressive modulus while after 1 phr, MWCNT leads to higher values till 3 phr. It is attributed to higher filler anisotropy and aspect ratio of MWCNT, which make it a promising reinforcing agent for rubber matrix [48].

The higher modulus for hybrid composites at 1 phr is due to synergism among MWCNT and copper particles. After 1 phr, the modulus was lower than MWCNT while higher than copper particles. The lowest modulus of copper particles is due to its poor aspect ratio of nearly 1, poor reinforcing effect, and poor interfacial interaction due to small surface area and particle size in the micron range. The poor mechanical properties of copper are also due to a decrease in crosslinking density of the curatives with the addition of copper particles as justified in Figure 3.

3.4.2. Under Static Tensile Strain

The behavior of stress–strain profiles under tensile strain was reported in Figure 5a–c. Particularly, these measurements help us to determine properties, such as the modulus, tensile strength, and fracture strain of composites [54], as presented in Figure 5d–f. We found from stress–strain curves that the tensile stress increases with increasing tensile strain until fracture strain. Such a property can be stated as improved interactions in composites, such as cross-linking density [55] and other interfacial interactions [52,53].

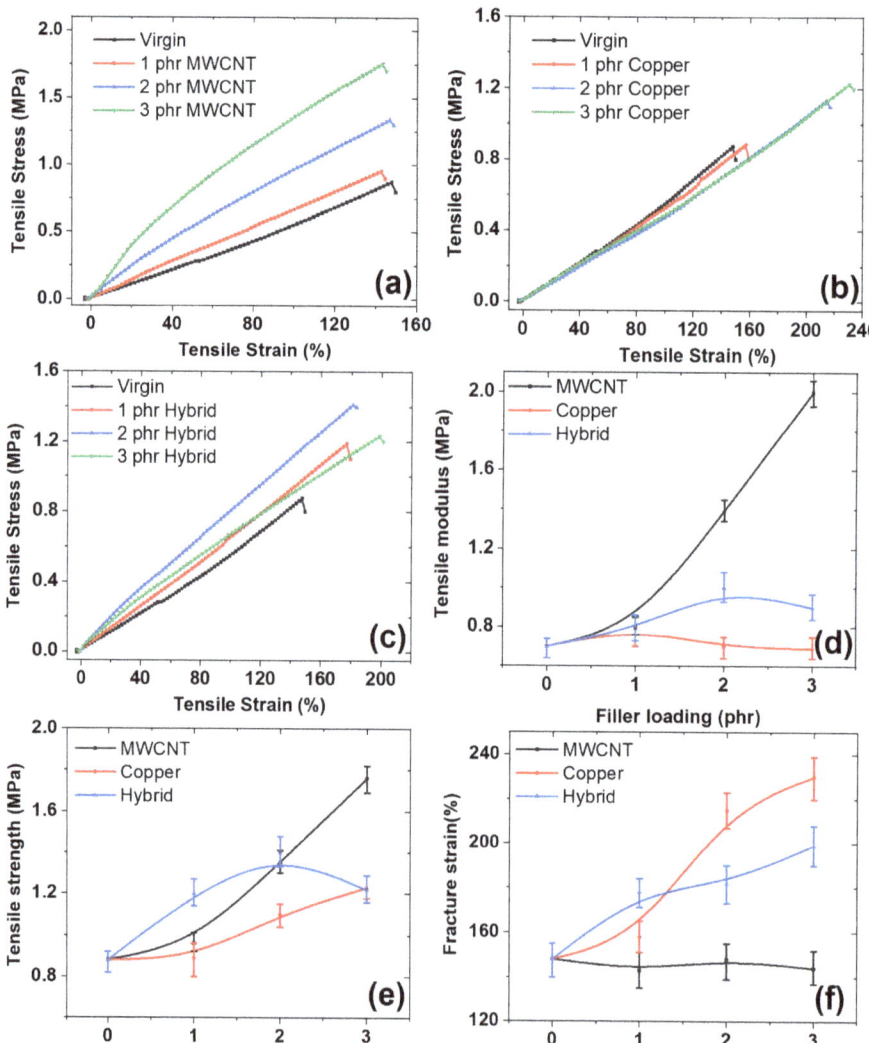

Figure 5. Stress–strain curves: (**a**) MWCNT composites. (**b**) Copper composites. (**c**) Hybrid composites. (**d**) Tensile modulus of different composites. (**e**) Tensile strength of composites. (**f**) Fracture strain of different composites.

With an addition of filler in the rubber matrix, the tensile stress increases with increasing filler content in the composite. This can be attributed to improved filler–filler and polymer-filler interactions in the composites [53]. It is also interesting to note that the mechanical properties except fracture strain were highest for MWCNT-based composites. It is attributed to the higher anisotropy of MWCNT that allows it to form percolative networks at lower filler loading of approx. 2 phr and higher mechanical properties, such as tensile modulus and tensile strength was reported.

The behavior of tensile modulus (Figure 5d), tensile strength (Figure 5e), and fracture strain (Figure 5f) of different composites [54] were studied and reported. We found that the tensile modulus increases with an increase in filler loading for MWCNT-based composites and was lower for hybrid and copper particles. Such an increase in modulus for MWCNT is attributed to the higher anisotropic effect of MWCNT particles [56] with improved interfacial interaction [52,53] and higher crosslink density resulting in a higher modulus.

However, the tensile modulus is lower for hybrid and copper-based particles, which is due to the poor reinforcing effect of copper particles due to the low anisotropic effect of filler, poor interfacial interaction, and lower crosslinking density. Similarly, the tensile strength was higher for hybrid composites up to 2 phr and higher for MWCNT at 3 phr. The higher tensile strength for hybrid composites is due to the synergistic effect and the higher tensile strength at 3 phr for MWCNT was due to the high anisotropic effect of MWCNT as a filler in composites [48].

In the end, the fracture strain was higher for copper and hybrid composites and lower for MWCNT-based composites. The higher fracture strain for a hybrid at 1 phr is due to the synergistic effect among the binary filler and the higher for copper composites is due to lower crosslinking density for copper-filled composites. The lower crosslinking density simulates the freer motion of polymer chains leading to higher fracture strain.

3.5. Theoretical Prediction of Hybrid Rubber Composites via Different Models

The theoretical prediction of modulus for one and two-component systems are well known in the literature [56]. These models generally depend on the aspect ratio of the filler, the volume fraction of the filler, and their interactive factors for a two-component system. Existing models, such as the Guth–Gold Smallwood model [56,57] and Halpin-Tsai theoretical equations [56,58] were used to study its deviation with the experimental results in the present study. After plotting the equations, it was found that the existing models fall within the experimental results, especially at lower volume fractions of the filler and the results are presented in Figure 6a,b.

The Guth–Gold Smallwood equation for two components system [56] can be written as—

$$E_{1+2} = E_o [(1 + 0.67 f_1 \phi_1) + (1 + 0.67 f_2 \phi_2)] \times i' \tag{3}$$

Here, the E_{1+2} is the combined predicted modulus of the hybrid system. Then, ϕ_1 and ϕ_2 are the volume fraction of the filler of two components. "i" is the interactive interaction parameters of the model. f_1 and f_2 are the aspect ratio of the fillers. The above model can be produced by setting the interactive species in a range of "0.4–0.5", which directly depends mainly on the volume fraction of the filler in the rubber matrix. The physical significance of the interactive factor is that the higher the value of the volume fraction of the filler particles and the higher the interaction among the filler particles and their interfacial interaction species. Similarly, the Halpin-Tsai theoretical equations for two components [56] can be written as—

$$E_{1+2} = E_o [(1 + 2 f_1 \phi_1)/(1 - \phi 1) + (1 + 2 f_2 \phi_2)/(1 - \phi_2)] \times i' \tag{4}$$

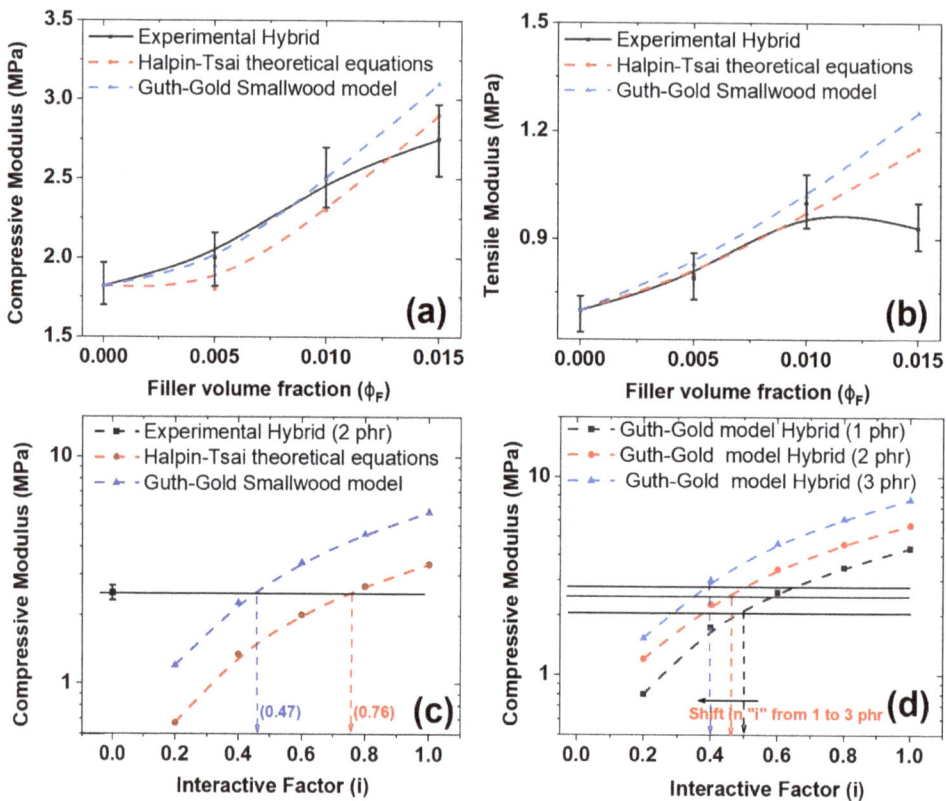

Figure 6. Theoretical model's vs. experimental values: (**a**) modeling for compressive modulus; (**b**) modeling for tensile modulus; (**c**) comparative of different theoretical models; (**d**) models' prediction via. different filler content.

Here, all the components are the same as the above models except "E_o", which means modulus of unfilled rubber. Both the models agree well with the experimental behavior of the composites up to 2 phr and then deviate. However, the Guth–Gold Smallwood model is closest to the experimental data. The interactive factor "i" for two components composite system was calculated and their corresponding values of modulus are reported in Figure 6c,d.

We found from the plots that the corresponding modulus increases with increasing interacting factors. It is also interesting to note that the interactive factor shifts as filler loading increases in the rubber matrix (Figure 6d). It is attributed to the synergistic effect of the two filler components in the composite. Moreover, the increases are due to filler–filler and interfacial interactions with increasing interactive factors in the composites. Here, the values of the interacting factor are in the range of 0.1 (poor interaction) to 1 (perfect interaction). However, experimentally, it is difficult to reach an interactive factor value of 1.

3.6. Tribology Properties of Rubber Composites

The tribological properties, such as the coefficient of friction of the rubber composites were studied and explored in the present study (Figure 7a,b). From the results, the properties were found to be correlated with the crosslink density (Figure 3) of the composites [44]. The copper with lower crosslink density was found to exhibit a higher coefficient of friction, which is higher than all composites including virgin samples. The experiments show that the best tribology properties, such as the lowest coefficient of friction were found in composites filled with MWCNT. The hybrid composite shows a coefficient of friction higher than MWCNT and lower than copper-based composites. Thus, a hybrid composite can be a candidate for optimum and medium tribological properties. The superior tribological properties of MWCNT are attributed due to its higher crosslinking density, high aspect ratio and higher polymer-filler interfacial interaction in the composite.

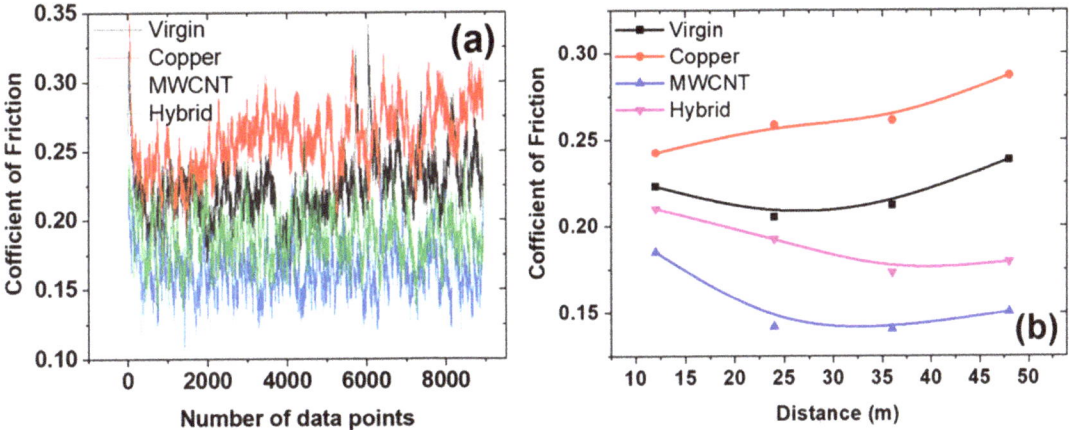

Figure 7. (a) coefficient of friction and data points; (b) coefficient of friction and distance.

4. Industrial Applications

4.1. Voltage Output of Different Types of Stretchable Devices

The specimen thickness is 8 mm in the compression loading, and the load is applied for 4 mm at 2 Hz. All three specimens are of the same dimensions. Electrode thickness is 0.2 mm painted on both sides of the substrate. Copper tape is attached to the electrode to connect it to the digital multimeter to record the voltage output, which is obtained from the flexible specimen. Load is applied for one hour, which applies 7200 cycles of loading to the specimen. The constant amplitude of loading is maintained during the testing. During the testing, the electrode material is kept constant. The only change is done in the substrate materials. CNT + MoS_2 binary filler reinforcement is chosen for preparing the electrode.

In the substrate, three different types of filler are reinforced in each specimen category. The performance of the substrate on reinforcing fillers with the polymer is studied in this experiment. When CNT is reinforced into the substrate, the voltage is around 1.25 V from the initial cycles. Due to the piezoelectric nature of the silicone polymer, the voltage is produced from the deformation. The geometry of the specimen is similar to the capacitor. Applying further loading during voltage production causes the activation of charge carriers from the substrate. The voltage production is kept constant until the end of 7200 repeated cycles for one hour of loading at 2 Hz. It represents the constant voltage production due to the activation of charge carriers in the steady state, which is represented in Figure 8a.

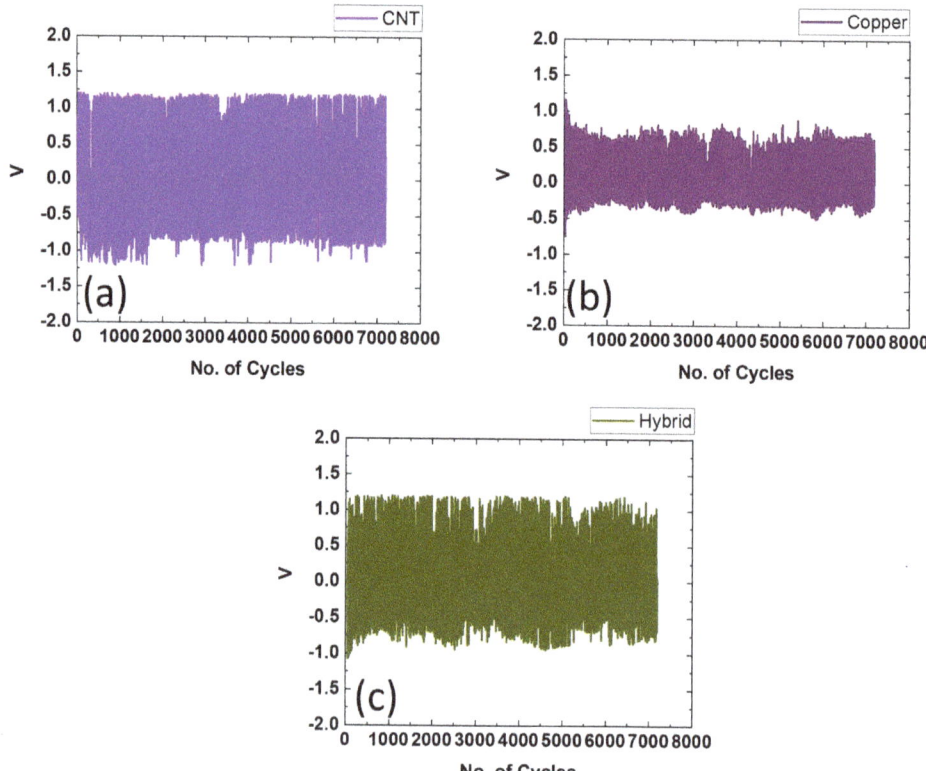

Figure 8. Piezoelectric energy-harvesting devices with different types of substrates: (**a**) MWCNT substrate; (**b**) copper substrate; and (**c**) hybrid substrate.

Figure 8b consists of a copper-reinforced substrate. Its voltage value in the initial loading is around 1 V. This value is suddenly decreasing due to the breakage of the electrode on initial repeated loading. Then, saturation occurs in the breakage, and a constant voltage is produced until the end of 7200 cycles. In comparison to the CNT-based substrate, the copper-based specimen can be able to produce less amount of voltage. Similarly, the hybrid specimen also produces a constant amount of voltage on repeated loading (Figure 8c). The voltage production from the hybrid specimen is slightly above 1 V. When compared with the CNT, CNT can produce more dense voltage than other materials. It is very clear from the obtained graphs of three different materials.

4.2. Durability of the Stretchable Device for the Best Candidate MWCNT Substrate

The specimen's substrate thickness is 8 mm over 0.2 mm electrodes painted on both sides. MWCNT is added to the substrate. The electrode is made of reinforcing MWCNT and MoS_2 nanoparticles. 21 mm hemispherical loader is used for applying load to the specimen. The flexible electrodes are connected to the multimeter to determine the produced voltage via the copper electrode. Over the 8 mm substrate, the deformation is applied for 50%, which is 4 mm. The load is applied at the frequency of 2 Hz.

On applying a deformation to the specimen, the piezoelectric specimen can be able to produce voltage output. The specimen can produce continuous voltage output during repeated loading cycles. During the initial cycles of loading, as increasing the number of cycles, the voltage is also getting increased. This is caused by charge carriers activating at an increasing rate [59]. After 8000 cycles, the voltage value is maintained constant, improving the number of cycles for up to 71,000. Then, the voltage value is slightly increasing due to the activation happening at the enhanced rate. The voltage value is slightly increasing for up to 135,000 cycles.

Again, the rise in voltage due to the activation occurs in the enhancing rate. After that, there is a sudden increase in voltage. It is mainly because of the electrodes' charging for a long time. Applying load further causes voltage enhancement in a more significant amount [59]. Then the voltage is gradually increasing. The voltage value around 100 thousand cycles is less than 2 V. During 200 thousand cycles, the voltage value is around 3 V. 50% voltage value is improved from 100 thousand to 200 thousand cycles.

Again, the voltage value improves as the number of cycles increases. The maximum voltage obtained around 300 thousand cycles is less than 5.5 V. Increasing the number of cycles further causes the improvement of voltage, which reaches the maximum voltage value of around less than 6 V at 350 thousand cycles. After that, the improvement of loading causes the decrement of voltage. This is due to the breakage of the electrode due to repeated cycles. The breakage in the electrode causes the conductivity path breakdown in the electrode. It causes a reduction in the voltage generated from the specimen. The electrode cannot transport the voltage generated from the specimen due to its reduced conductivity. The voltage value was reduced up to 380 thousand cycles. This is due to the gradual breakdown of the electrode, causing the regular voltage reduction. Then saturation occurs in the electrode breakdown causing the stabilization of voltage production on repeated cycles. The voltage variation for 0 to 400 thousand cycles is shown in Figure 9a,d.

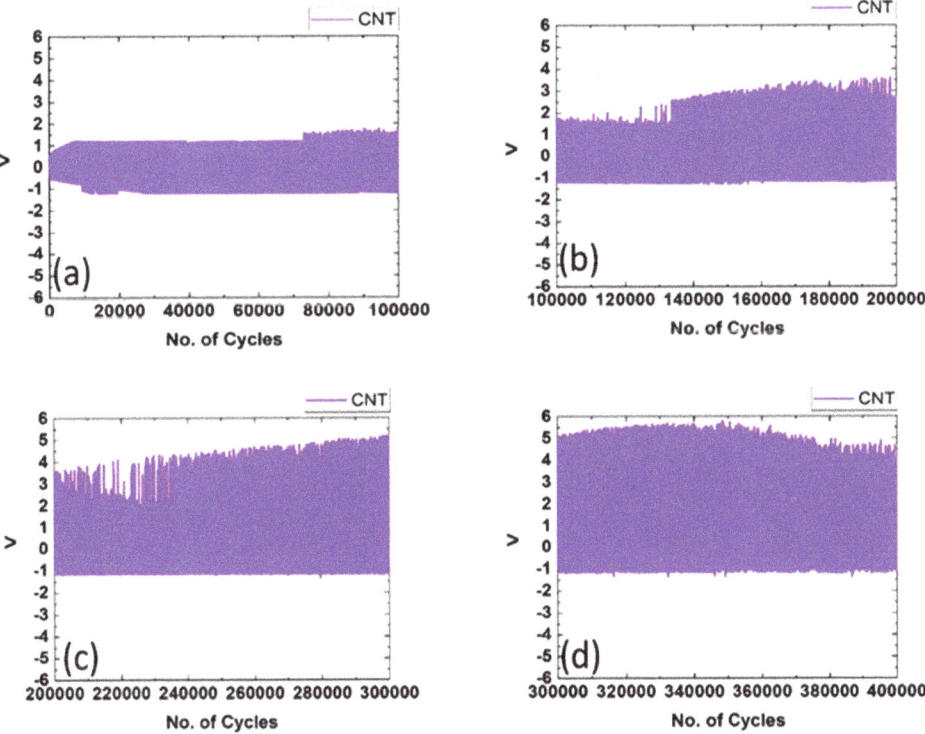

Figure 9. Piezoelectric energy-harvesting devices with different number of cycles for MWCNT substrate: (**a**) 0.1 million cycles; (**b**) 0.2 million cycles; (**c**) 0.3 million cycles; and (**d**) 0.4 million cycles.

5. Conclusions

The present work develops and studies impressive voltage generation of around ~6 V in stretchable devices based on rubber composites with high durability of >0.4 million cycles. These rubber composites were obtained by solution mixing of RTV-SR and different types of nanofiller, such as MWCNT or copper particles with micron size and their hybrids. The dispersion of these particles was obtained from SEM micrographs and uniform dispersion was noticed. MWCNT shows 1-D tube-shape morphology with an aspect ratio of around 65, while copper particles were 3-D, the irregular shape and a low aspect ratio of around 1.

The mechanical stretchability and compressive modulus were studied and correlated with the stretchable device performance. For example, the compressive modulus was 1.82 MPa (virgin) and increased at 3 per hundred parts of rubber (phr) to 3.75 MPa (MWCNT), 2.2 MPa (copper particles), and 2.75 MPa (hybrid). Similarly, the stretching ability for the composites used in fabricating devices was 148% (virgin) and changes at 3 phr to 144% (MWCNT), 230% (copper particles), and 199% (hybrid).

In the end, the results were summarized and concluded and the properties were correlated with the high performance of the stretchable devices. In conclusion, this work addresses the methods to obtain high-performance stretchable devices with novel properties and applications, such as flexible electronics or electronic packaging. The work also supports the use of RTV-SR suitability to obtain desired flexibility and stretchability to be useful for engineering applications. This work also recommends that hybrid fillers can also be useful to obtain optimum stiffness and stretchability.

Supplementary Materials: The following supporting information can be downloaded at: https://www.mdpi.com/article/10.3390/polym14183744/s1, Figure S1: XRD of the hybrid composite containing copper and MWCNT particles.

Author Contributions: Conceptualization, V.K., S.A., W.-B.H. and M.N.A.; methodology, V.K., S.A., W.-B.H. and M.N.A.; software, V.K.; validation, V.K., S.A., W.-B.H. and M.N.A.; formal analysis, V.K.; investigation, V.K., S.A., W.-B.H. and M.N.A.; resources, S.-S.P.; data curation, V.K.; writing—original draft preparation, V.K.; writing—review, and editing, V.K.; visualization, V.K., S.A. and M.N.A.; supervision, S.-S.P.; project administration, S.-S.P. All authors have read and agreed to the published version of the manuscript.

Funding: This research was supported by the Korea Institute for Advancement of Technology (KIAT) grant funded by the Korean Government (MOTIE) (P0002092, The Competency Development Program for Industry Specialist).

Institutional Review Board Statement: Not applicable.

Data Availability Statement: Not applicable.

Conflicts of Interest: The authors declare no conflict of interest.

References

1. González, A. Review on supercapacitors: Technologies and materials. *Renew. Sustain. Energy Rev.* **2016**, *58*, 1189–1206. [CrossRef]
2. Green, M.A. Thin-film solar cells: Review of materials, technologies and commercial status. *J. Mater. Sci. Mater. Electron.* **2007**, *18*, 15–19. [CrossRef]
3. Etacheri, V.; Marom, R.; Elazari, R.; Salitra, G.; Aurbach, D. Challenges in the development of advanced Li-ion batteries: A review. *Energy Environ. Sci.* **2011**, *4*, 3243–3262. [CrossRef]
4. Kim, H.S.; Kim, J.H.; Kim, J. A review of piezoelectric energy harvesting based on vibration. *Int. J. Precis. Eng. Manuf.* **2011**, *12*, 1129–1141. [CrossRef]
5. Jacobs, K.; Hofmann, P.; Reichow, J.; Görnert, P. Crystalline piezoelectric materials with special consideration of $GaPO_4$. *Funct. Mater.* **2000**, *13*, 391–396. [CrossRef]
6. Hayat, K.; Shah, S.S.; Ali, S.; Shah, S.K.; Iqbal, Y.; Aziz, M.A. Fabrication and characterization of $Pb(Zr_{0.5}Ti_{0.5})O_3$ nanofibers for nanogenerator applications. *J. Mater. Sci. Mater. Electron.* **2020**, *31*, 5859–15874.
7. Arlt, G.; Hennings, D.; De With, G. Dielectric properties of fine-grained barium titanate ceramics. *J. Appl. Phys.* **1985**, *58*, 1619–1625. [CrossRef]
8. Zhang, Y.; Jie, W.; Chen, P.; Liu, W.; Hao, J. Ferroelectric and piezoelectric effects on the optical process in advanced materials and devices. *Adv. Mater.* **2018**, *30*, e1707007. [CrossRef] [PubMed]

9. Wada, Y.; Hayakawa, R. Piezoelectricity and pyroelectricity of polymers. *Jpn. J. Appl. Phys.* **1976**, *15*, 2041–2057. [CrossRef]
10. Kang, M.-G.; Jung, W.-S.; Kang, C.-Y.; Yoon, S.-J. Recent progress on PZT based piezoelectric energy harvesting technologies. *Actuators* **2016**, *5*, 5. [CrossRef]
11. Panda, P.K.; Sahoo, B. PZT to lead free piezo ceramics: A review. *Ferroelectrics* **2015**, *474*, 128–143. [CrossRef]
12. Surmenev, R.A.; Orlova, T.; Chernozem, R.V.; Ivanova, A.A.; Bartasyte, A.; Mathur, S.; Surmeneva, M.A. Hybrid lead-free polymer-based nanocomposites with improved piezoelectric response for biomedical energy-harvesting applications: A review. *Nano Energy* **2019**, *62*, 475–506. [CrossRef]
13. Trung, T.Q.; Lee, N.-E. Recent progress on stretchable electronic devices with intrinsically stretchable components. *Adv. Mater.* **2017**, *29*, 1603167. [CrossRef]
14. Manikkavel, A. Investigation of high temperature vulcanized and room temperature vulcanized silicone rubber based on flexible piezo-electric energy harvesting applications with multi-walled carbon nanotube reinforced compo-sites. *Polym. Compos.* **2022**, *43*, 1305–1318. [CrossRef]
15. Kumar, V.; Kumar, A.; Wu, R.-R.; Lee, D.-J. Room-temperature vulcanized silicone rubber/barium titanate-based high-performance nanocomposite for energy harvesting. *Mater. Today Chem.* **2020**, *16*, 100232. [CrossRef]
16. Coran, A.Y. Chemistry of the vulcanization and protection of elastomers: A review of the achievements. *J. Appl. Polym. Sci.* **2003**, *87*, 24–30. [CrossRef]
17. Post, W.; Susa, A.; Blaauw, R.; Molenveld, K.; Knoop, R.J.I. A review on the potential and limitations of recyclable thermosets for structural applications. *Polym. Rev.* **2019**, *60*, 359–388. [CrossRef]
18. Bokobza, L. Natural rubber nanocomposites: A review. *Nanomaterials* **2019**, *9*, 12. [CrossRef]
19. Movahed, S.O.; Ansarifar, A.; Estagy, S. Review of the reclaiming of rubber waste and recent work on the recycling of ethylene-propylene-diene rubber waste. *Rubber Chem. Technol.* **2016**, *89*, 54–78. [CrossRef]
20. Warrick, E.L. Silicone elastomer developments 1967–1977. *Rubber Chem. Technol.* **1979**, *52*, 437–525. [CrossRef]
21. Kumar, V.; Manikkavel, A.; Kumar, A.; Alam, N.; Hwang, G.; Park, S. Stretchable piezo-electric energy harvesting device with high durability using carbon nanomaterials with different structure and their synergism with molybdenum disulfide. *J. Vinyl Addit. Technol* **2022**. [CrossRef]
22. Sun, Y.; Yun, S.S.; Jamie, P. Characterization of silicone rubber based soft pneumatic actuators. In Proceedings of the IEEE/RSJ International Conference on Intelligent Robots and Systems, Tokyo, Japan, 3–7 November 2013; IEEE: New York, NY, USA, 2013.
23. Giffney, T.; Bejanin, E.; Kurian, A.S.; Travas-Sejdic, J.; Aw, K. Highly stretchable printed strain sensors using multi-walled carbon nanotube/silicone rubber composites. *Sens. Actuators A Phys.* **2017**, *259*, 44–49. [CrossRef]
24. Balberg, I. A comprehensive picture of the electrical phenomena in carbon black-polymer composites. *Carbon* **2002**, *40*, 139–143. [CrossRef]
25. Sajjayanukul, T.; Saeoui, P.; Sirisinha, C. Experimental analysis of viscoelastic properties in carbon black-filled natural rubber compounds. *J. Appl. Polym. Sci.* **2005**, *97*, 2197–2203. [CrossRef]
26. Bergaya, F.; Lagaly, G. General introduction: Clays, clay minerals, and clay science. *Dev. Clay Sci.* **2006**, *1*, 1–18.
27. Jeelani, P.G.; Mulay, P.; Venkat, R.; Ramalingam, C. Multifaceted application of silica nanoparticles. A review. *Silicon* **2020**, *12*, 1337–1354. [CrossRef]
28. Neto, A.C.; Guinea, F.; Peres, N.M.; Novoselov, K.S.; Geim, A.K. The electronic properties of graphene. *Rev. Mod. Phys.* **2009**, *81*, 109. [CrossRef]
29. Coleman, J.N.; Khan, U.; Blau, W.J.; Gun'ko, Y.K. Small but strong: A review of the mechanical properties of carbon nanotube-polymer composites. *Carbon* **2006**, *44*, 1624–1652. [CrossRef]
30. Bokobza, L. Multiwall carbon nanotube elastomeric composites: A review. *Polymer* **2007**, *48*, 4907–4920. [CrossRef]
31. Kumar, V.; Lee, D.-J. Studies of nanocomposites based on carbon nanomaterials and RTV silicone rubber. *J. Appl. Polym. Sci.* **2016**, *134*. [CrossRef]
32. Chu, K.; Jia, C. Enhanced strength in bulk graphene-copper composites. *Phys. Status Solidi* **2013**, *211*, 184–190. [CrossRef]
33. Wang, Y.; Wu, W.; Drummer, D.; Liu, C.; Shen, W.; Tomiak, F.; Schneider, K.; Liu, X.; Chen, Q. Highly thermally conductive polybenzoxazine composites based on boron nitride flakes deposited with copper particles. *Mater. Des.* **2020**, *191*, 108698. [CrossRef]
34. Yin, C.-G.; Liu, Z.-J.; Mo, R.; Fan, J.-C.; Shi, P.-H.; Xu, Q.-J.; Min, Y.-L. Copper nanowires embedded in boron nitride nanosheet-polymer composites with enhanced thermal conductivities for thermal management. *Polymer* **2020**, *195*, 122455. [CrossRef]
35. Yang, Q.; Beers, M.H.; Mehta, V.; Gao, T.; Parkinson, D. Effect of thermal annealing on the electrical conductivity of copper-tin polymer composites. *ACS Appl. Mater. Interfaces* **2016**, *9*, 958–964. [CrossRef] [PubMed]
36. Boudenne, A.; Ibos, L.; Fois, M.; Majesté, J.; Géhin, E. Electrical and thermal behavior of polypropylene filled with copper particles. *Compos. Part A Appl. Sci. Manuf.* **2005**, *36*, 1545–1554. [CrossRef]
37. Tekce, H.S.; Kumlutas, D.; Tavman, I.H. Effect of particle shape on thermal conductivity of copper rein-forced polymer composites. *J. Reinf. Plast. Compos.* **2007**, *26*, 113–121. [CrossRef]
38. Flory, P.J.; Rehner Jr, J. Statistical mechanics of cross-linked polymer networks II swelling. *J. Chem. Phys.* **1943**, *11*, 521–526. [CrossRef]

39. Yang, X.; Li, Z.; Jiang, Z.; Wang, S.; Liu, H.; Xu, X.; Wang, D.; Miao, Y.; Shang, S.; Song, Z. Mechanical reinforcement of room-temperature-vulcanized silicone rubber using modified cellulose nanocrystals as cross-linker and nanofiller. *Carbohydr. Polym.* **2019**, *229*, 115509. [CrossRef] [PubMed]
40. Alig, I.; Pötschke, P.; Lellinger, D.; Skipa, T.; Pegel, S.; Kasaliwal, G.R.; Villmow, T. Establishment, morphology and properties of carbon nanotube networks in polymer melts. *Polymer* **2012**, *53*, 4–28. [CrossRef]
41. Iijima, S. Helical microtubules of graphitic carbon. *Nature* **1991**, *354*, 56–58. [CrossRef]
42. Li, J. Correlations between percolation threshold, dispersion state, and aspect ratio of carbon nanotubes. *Adv. Funct. Mater.* **2007**, *17*, 3207–3215. [CrossRef]
43. Zalamea, L.; Kim, H.; Pipes, R.B. Stress transfer in multi-walled carbon nanotubes. *Compos. Sci. Technol.* **2007**, *67*, 3425–3433. [CrossRef]
44. Kumar, V.; Kumar, A.; Alam, N.; Park, S. Effect of graphite nanoplatelets surface area on mechanical properties of room-temperature vulcanized silicone rubber nanocomposites. *J. Appl. Polym. Sci.* **2022**, *139*, e52503. [CrossRef]
45. Zhu, Y.; Xu, F. Buckling of aligned carbon nanotubes as stretchable conductors: A new manufacturing strategy. *Adv. Mater.* **2012**, *24*, 1073–1077. [CrossRef]
46. Choi, S. High-performance stretchable conductive nanocomposites: Materials, processes, and device applications. *Chem. Soc. Rev.* **2019**, *48*, 1566–1595. [CrossRef] [PubMed]
47. Ma, R.; Chou, S.-Y.; Xie, Y.; Pei, Q. Morphological/nanostructural control toward intrinsically stretchable organic electronics. *Chem. Soc. Rev.* **2019**, *48*, 1741–1786. [CrossRef]
48. Kato, Y.; Horibe, M.; Ata, S.; Yamada, T.; Hata, K. Stretchable electromagnetic-interference shielding materials made of a long single-walled carbon-nanotube–elastomer composite. *RSC Adv.* **2017**, *7*, 10841–10847. [CrossRef]
49. Khang, D.-Y.; Jiang, H.; Huang, Y.; Rogers, J.A. A stretchable form of single-crystal silicon for high-performance electronics on rubber substrates. *Science* **2006**, *311*, 208–212. [CrossRef]
50. Zhou, W.; Yu, D.; Wang, C.; An, Q.; Qi, S. Effect of filler size distribution on the mechanical and physical properties of alumina-filled silicone rubber. *Polym. Eng. Sci.* **2008**, *48*, 1381–1388. [CrossRef]
51. Galimberti, M. Filler networking of a nanographite with a high shape anisotropy and synergism with carbon black in poly (1,4-cis-isoprene)-based nanocomposites. *Rubber Chem. Technol.* **2014**, *87*, 197–218. [CrossRef]
52. Wang, M.-J. Effect of polymer-filler and filler-filler interactions on dynamic properties of filled vulcanizates. *Rubber Chem. Technol.* **1998**, *71*, 520–589. [CrossRef]
53. Ning, N.; Fu, S.; Zhang, W.; Chen, F.; Wang, K.; Deng, H.; Zhang, Q.; Fu, Q. Realizing the enhancement of interfacial interaction in semicrystalline polymer/filler composites via interfacial crystallization. *Prog. Polym. Sci.* **2012**, *37*, 1425–1455. [CrossRef]
54. Kumar, V. Investigation of silicone rubber composites reinforced with carbon nanotube, nanographite, their hybrid, and applications for flexible devices. *J. Vinyl Addit. Technol.* **2021**, *27*, 254–263. [CrossRef]
55. Urbaczewski-Espuche, E.; Galy, J.; Gerard, J.-F.; Pascault, J.-P.; Sautereau, H. Influence of chain flexibility and crosslink density on mechanical properties of epoxy/amine networks. *Polym. Eng. Sci.* **1991**, *31*, 1572–1580. [CrossRef]
56. Lee, J.-Y.; Kumar, V.; Tang, X.-W.; Lee, D.-J. Mechanical and electrical behavior of rubber nanocomposites under static and cyclic strain. *Compos. Sci. Technol.* **2017**, *142*, 1–9. [CrossRef]
57. Wolff, S.; Donnet, J.-B. Characterization of fillers in vulcanizates according to the einstein-guth-gold equation. *Rubber Chem. Technol.* **1990**, *63*, 32–45. [CrossRef]
58. Affdl, J.C.H.; Kardos, J.L. The halpin-tsai equations: A review. *Polym. Eng. Sci.* **1976**, *16*, 344–352. [CrossRef]
59. McKay, T.G.; Rosset, S.; Anderson, I.A.; Shea, H. An electroactive polymer energy harvester for wireless sensor networks. *J. Phys. Conf. Ser.* **2013**, *476*, 12117. [CrossRef]

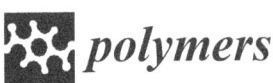

Article

Low-Velocity Impact Behavior of Sandwich Plates with FG-CNTRC Face Sheets and Negative Poisson's Ratio Auxetic Honeycombs Core

Chunhao Yang *, Wuning Ma, Zhendong Zhang and Jianlin Zhong

School of Mechanical Engineering, Nanjing University of Science and Technology, Nanjing 210094, China; kkmwn@163.com (W.M.); zzd1157@163.com (Z.Z.); zhongjianlin@njust.edu.cn (J.Z.)
* Correspondence: yang-ch@njust.edu.cn

Abstract: The combination of auxetic honeycomb and CNT reinforcement composite is expected to further improve the impact protection performance of sandwich structures. This paper studies the low-velocity impact response of sandwich plates with functionally graded carbon nanotubes reinforced composite (FG-CNTRC) face sheets and negative Poisson's ratio (NPR) auxetic honeycomb core. The material properties of FG-CNTRC were obtained by the rule of mixture theory. The auxetic honeycomb core is made of Ti-6Al-4V. The governing equations are derived based on the first-order shear deformation theory and Hamilton's principle. The nonlinear Hertz contact law is used to calculate the impact parameters. The Ritz method with Newmark's time integration schemes is used to solve the response of the sandwich plates. The (20/−20/20)s, (45/−45/45)s and (70/−70/70)s stacking sequences of FG-CNTRC are considered. The effects of the gradient forms of FG-CNTRC surfaces, volume fractions of CNTs, impact velocities, temperatures, ratio of plate length, width and thickness of surface layers on the value of the plate center displacement, the recovery time of deformation, contact force and contact time of low-velocity impact were analyzed in detail.

Keywords: FG-CNTRC; auxetic honeycomb core; negative Poisson's ratio; low-velocity impact

1. Introduction

As the "Nanometer" material science, typified by carbon nanotubes (CNTs), develops, the widespread use of CNTs reinforcement composite (CNTRC) has brought changes to the sensor, intelligent medical and shelter structure fields [1–3]. The CNTs could improve the mechanical properties of composite and are remarkable as an ideal reinforcement. Shen [4] introduced functionally graded properties into CNTRC by designing the volume fraction of CNTs along the thickness direction, which avoids material properties suffering degradation due to the high levels of CNTs. Then, Kwon et al. [5] successfully made FG-CNTRC using powder metallurgy technology. At this point, large numbers of studies on the buckling [6–17] and vibration [18–31] analyses of FG-CNTRC structures have been carried out. Because of the low-velocity impact during the manufacture, installation use and maintenance, the inside structure of composite could be damaged and the lifting capacity will decrease and even fail. Therefore, studies on the low-velocity impact of FG-CNTRC were also carried out [32–42].

Most natural materials have the properties of expanding (contracting) laterally when compressed (stretched) longitudinally, which can be defined as positive Poisson's ratio materials. In recent years, auxetic material has generated a lot of interest among researchers due to the negative Poisson's ratio (NPR) properties [43–45]. Re-entrant [46], chiral [47] and other various materials have been proposed. Due to the outstanding performance on energy absorption [48–50], crashworthiness [51,52], and low-velocity impact resistance [53,54], auxetic material has been increasingly applied in biological medicine, photonics, energy

Citation: Yang, C.; Ma, W.; Zhang, Z.; Zhong, J. Low-Velocity Impact Behavior of Sandwich Plates with FG-CNTRC Face Sheets and Negative Poisson's Ratio Auxetic Honeycombs Core. *Polymers* **2022**, *14*, 2938. https://doi.org/10.3390/polym14142938

Academic Editor: Md Najib Alam

Received: 29 June 2022
Accepted: 18 July 2022
Published: 20 July 2022

Copyright: © 2022 by the authors. Licensee MDPI, Basel, Switzerland. This article is an open access article distributed under the terms and conditions of the Creative Commons Attribution (CC BY) license (https://creativecommons.org/licenses/by/4.0/).

storage, thermal management, and acoustic areas [55]. As an ideal core of sandwich structures, auxetic material could be used in shield structures in aerospace and civil engineering. Therefore, the nonlinear mechanical response of the sandwich structure with an auxetic honeycomb core [56,57] was analyzed by Li, Shen, and Wang [58–64]. Wan et al. [65] analyzed the uniaxial compression or expanded properties of auxetic honeycombs. Grima et al. [66] proposed a hexagonal honeycomb with zero Poisson's ratios. Assidi and Ganghoffer [67] represented a composite with auxetic behavior and proved that the overall NPR could improve the mechanical properties. Grujicic et al. [68] focused on the sandwich structures with an auxetic hexagonal core and built the multi-physics model of fabrication and dynamic performance. Liu et al. [69] investigated the propagation of waves in a sandwich plate with a periodic composite core. Qiao and Chen [70] analyzed the impact response of auxetic double arrowhead honeycombs. Zhang et al. [71] analyzed the in-plane dynamic crushing behaviors and energy-absorbed characteristics of NPR honeycombs with cell microstructure. Zhang et al. [72] analyzed the dynamic mechanical and impact response on yarns with helical auxetic properties.

There are two main methods to propose auxetic structures: the first is using auxetic material as the core of sandwich plate [55]; and the second is changing the stacking sequence and orientation of laminate [73,74]. To realize a larger NPR value using the second method requires not only a specific stacking sequence but also a highly anisotropic properties of each ply [75]. Due to the mechanical properties of CNTs, the longitudinal elastic modulus E_{11} of CNTRC is much larger than the transverse elastic modulus E_{22} and large NPR properties can be proposed by designing the stacking sequence of CNTRC laminate. Then, Shen et al. [45,76] introduced the NPR property to the FG-CNTRC laminate and analyzed the nonlinear bending and free vibration response. Yang, Huang, and Shen [77,78], as well as Yu and Shen [79] analyzed the effects of an out-of-plane NPR property on large amplitude vibration and nonlinear bending of the FG-CNTRC laminated beam and plate. Fan, Wang [80] and Huang et al. [81,82] analyzed the dynamic response of the auxetic FG-CNTRC.

The combination of auxetic honeycomb and CNT reinforcement composite is expected to further improve the impact protection performance of sandwich structures. This paper studies the low-velocity impact response of the sandwich plates with functionally graded carbon nanotubes reinforced composite (FG-CNTRC) face sheets and a negative Poisson's ratio (NPR) auxetic honeycomb core. The rule of mixture theory was used to calculate the material properties of FG-CNTRC with the PmPV matrix and CNTs reinforcement, while the effective Poisson's ratio was obtained by laminate plate theory (Section 2.2). The NPR honeycomb core was made of Ti-6Al-4V (Section 2.3). The first-order shear deformation theory and Hamilton's principle were used to describe the governing equations of the plate (Section 3.1). The nonlinear Hertz contact law was used to calculate the impact parameters (Section 3.2). The Ritz method with Newmark's time integration schemes was used to solve the response of the sandwich plate (Section 3.3). After verifying the model, the (20/−20/20)s, (45/−45/45)s and (70/−70/70)s three kinds of stacking sequence of FG-CNTRC surfaces were considered. The effects of gradient forms of FG-CNTRC surfaces, volume fractions of CNTs, impact velocities, temperatures, ratio of plate length and the width and thickness of surface layers on low-velocity impact response were analyzed. The value of plate center displacement, recovery time of deformation, contact force and contact time were discussed in detail.

2. Modeling and Materials of Sandwich Plates

2.1. Modeling of Sandwich Plates

The sandwich plates with length a, width b and total thickness h are considered in this research, as shown in Figure 1. The face sheets with a thickness h_f are FG-CNTRC-laminated structures composed of CNTRC layers with various volume fractions of CNTs. The auxetic core with a thickness of h_c is the negative Poisson's ratio honeycomb structure using isotropic titanium alloy (Ti-6Al-4V). A coordinate system (x, y, z) with (x, y) plane in the middle surface of the plate and z in the thickness direction is considered.

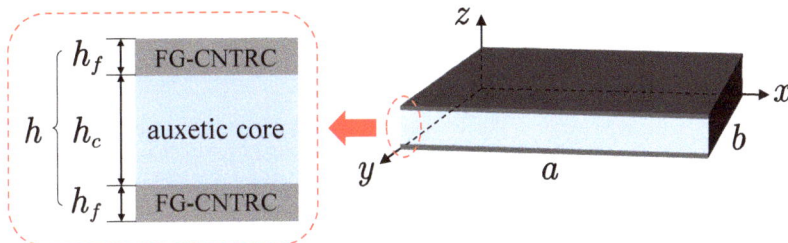

Figure 1. The sandwich plates with FG-CNTRC face sheets and auxetic honeycomb core.

2.2. Materials of FG-CNTRC Face Sheets

The CNTRC layers with the poly(m-phenylenevinylene)-co-((2,5-dioctoxy-p-phenylene) vinylene) (PmPV) matrix are considered in this research. The material properties of the face sheets can be obtained based on the rule of mixture theory [4].

$$E_{11} = \eta_1 V^c E_{11}^c + V^m E^m, \quad \rho = V^c \rho^c + V^m \rho^m,$$
$$\frac{\eta_2}{E_{22}} = \frac{V^c}{E_{22}^c} + \frac{V^m}{E^m}, \quad \frac{\eta_3}{G_{12}} = \frac{V^c}{G_{12}^c} + \frac{V^m}{G^m},$$
$$\alpha_{11} = \frac{V^c E_{11}^c \alpha_{11}^c + V^m E^m \alpha^m}{V^c E_{11}^c + V^m E^m}, \quad \nu_{12} = V^c \nu_{12}^c + V^m \nu^m$$
$$\alpha_{22} = (1 + \nu_{12}^c) V^c \alpha_{22}^c + (1 + \nu^m) V^m \alpha^m - \nu_{12} \alpha_{11}$$
(1)

where the superscript c and m represent the material properties of CNTs and the matrix, respectively. V is the volume fraction, in which $V^m + V^c = 1$. $\eta_j (j = 1, 2, 3)$ is the efficiency parameters of CNTs. The values are shown in Table 1. E, G, ν, ρ and α are the elastic module, shear module, Poisson's ratio, density and the thermal expansion of the materials, respectively. The (10, 10) SWCNTs are considered as the reinforcement in this research and the material properties are shown in Table 2. The material properties of the matrix PmPV are shown in Table 3.

Table 1. The efficiency parameter of CNTs [4].

V^{cnt}	η_1	η_2	η_3
0.11	0.149	0.934	0.934
0.14	0.150	0.941	0.941
0.17	0.149	1.381	1.381

Table 2. The material properties of (10, 10) SWCNTs (tube radius = 0.68 nm, thickness = 0.067 nm, length = 9.26 nm, $\nu_{12}^{cnt} = 0.175$) [10].

Temp (K)	E_{11}^{cnt} (TPa)	E_{22}^{cnt} (TPa)	G_{12}^{cnt} (TPa)	ν_{12}^{cnt}	α_{11}^{cnt} ($\times 10^{-6}$/K)	α_{22}^{cnt} ($\times 10^{-6}$/K)
300	5.6466	7.0800	1.9445	0.175	3.4584	5.1682
400	5.5308	6.9348	1.9643	0.175	4.1496	5.0905
500	5.4744	6.8641	1.9644	0.175	4.5361	5.0189

Table 3. The material properties of PmPV [76].

Temp (K)	E^{pm} (GPa)	ν^{pm}	α^{pm} ($\times 10^{-6}$/K)
300	2.10	0.34	45.00
400	1.63	0.34	47.25
500	1.16	0.34	49.50

The functionally graded properties of the CNTRC laminated structure are established according to the arrangement of CNTRC layers with the CNTs' volume fractions of 0.11, 0.14 and 0.17. As shown in Figure 2, four types of FG-CNTRC, namely FG-V, FG-A, FG-O, FG-X and a uniformly distributed CNTRC with CNTs' volume fractions of 0.14, namely UD, can be obtained. The laminated arrangement of FG-CNTRC can be expressed as

$$\begin{aligned} &\text{FG} - \text{V}: [(0.17)_2/(0.14)_2/(0.11)_2] \\ &\text{FG} - \text{A}: [(0.11)_2/(0.14)_2/(0.17)_2] \\ &\text{FG} - \text{O}: [0.11/0.14/0.17]_s \\ &\text{FG} - \text{X}: [0.17/0.14/0.11]_s \end{aligned} \quad (2)$$

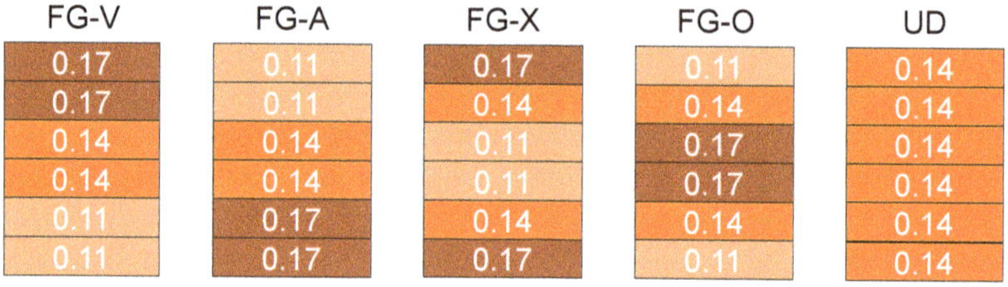

Figure 2. The CNTs' volume fractions arrangement of five types of CNTRC laminate.

For an anisotropic laminated plate, the effective Poisson's ratios ν_{13}^e and ν_{23}^e can be expressed as [44]

$$\nu_{13}^e = -\frac{\begin{vmatrix} \mathbf{A}_{13} & \mathbf{B}_{6-1} \\ \mathbf{B}_{5-3} & \mathbf{D} \end{vmatrix}}{\begin{vmatrix} \mathbf{A}_{5-1} & \mathbf{B}_{6-1} \\ \mathbf{B}_{5-1} & \mathbf{D} \end{vmatrix}}, \quad \nu_{23}^e = \frac{\begin{vmatrix} \mathbf{A}_{23} & \mathbf{B}_{6-2} \\ \mathbf{B}_{5-3} & \mathbf{D} \end{vmatrix}}{\begin{vmatrix} \mathbf{A}_{5-2} & \mathbf{B}_{6-2} \\ \mathbf{B}_{5-2} & \mathbf{D} \end{vmatrix}} \quad (3)$$

where **A**, **B** and **D** are the stiffness matrix of the FG-CNTRC laminated surface. The aforementioned elements of the matrix are presented in Appendix A.

Combining the gradient forms of FG-CNTRC, the effective Poisson's ratios could be calculated as shown in Figure 3. Three typical stacking sequences including $(20/-20/20)s$, $(45/-45/45)s$ and $(70/-70/70)s$ are considered to analyze the low-velocity impact response under various effective Poisson's ratios.

2.3. Materials of Auxetic Honeycomb Core

The honeycomb core made of Ti-6Al-4V with negative Poisson's ratio properties is considered in this research. The unit cell of the honeycomb is shown in Figure 4 and the material properties of the honeycomb core can be obtained by [56]

$$E_1^h = E_{Ti}\left(\frac{t_h}{l_h}\right)^3 \frac{\cos\theta_h}{(h_h/l_h + \sin\theta_h)\sin^2\theta_h}, \quad E_2^h = E_{Ti}\left(\frac{t_h}{l_h}\right)^3 \frac{(h_h/l_h + \sin\theta_h)}{\cos^3\theta_h}$$

$$\nu_{12}^h = \frac{\cos^2\theta_h}{(h_h/l_h + \sin\theta_h)\sin\theta_h}, \quad G_{12}^h = E_{Ti}\left(\frac{t_h}{l_h}\right)^3 \frac{(h_h/l_h + \sin\theta_h)}{(h_h/l_h)^2(1 + 2h_h/l_h)\cos\theta_h} \quad (4)$$

$$G_{13}^h = G_{Ti}\frac{t_h}{l_h}\frac{\cos\theta_h}{h_h/l_h + \sin\theta_h}, \quad G_{23}^h = G_{Ti}\frac{t_h}{l_h}\frac{1 + 2\sin^2\theta_h}{2\cos\theta_h(h_h/l_h + \sin\theta_h)},$$

$$\rho^h = \rho_{Ti}\frac{t_h/l_h(h_h/l_h + 2)}{2\cos\theta_h(h_h/l_h + \sin\theta_h)}$$

where the superscript h and subscript Ti represent the material properties of honeycomb and Ti-6Al-4V, respectively. l_h represents the length of the inclined cell rib; t_h represents the thickness of the cell rib; h_h represents the length of the vertical cell rib; and θ_h represents the inclined angle. The original properties of the honeycomb can be controlled by the parameters above. The material properties of the Ti-6Al-4V are mentioned in Table 4.

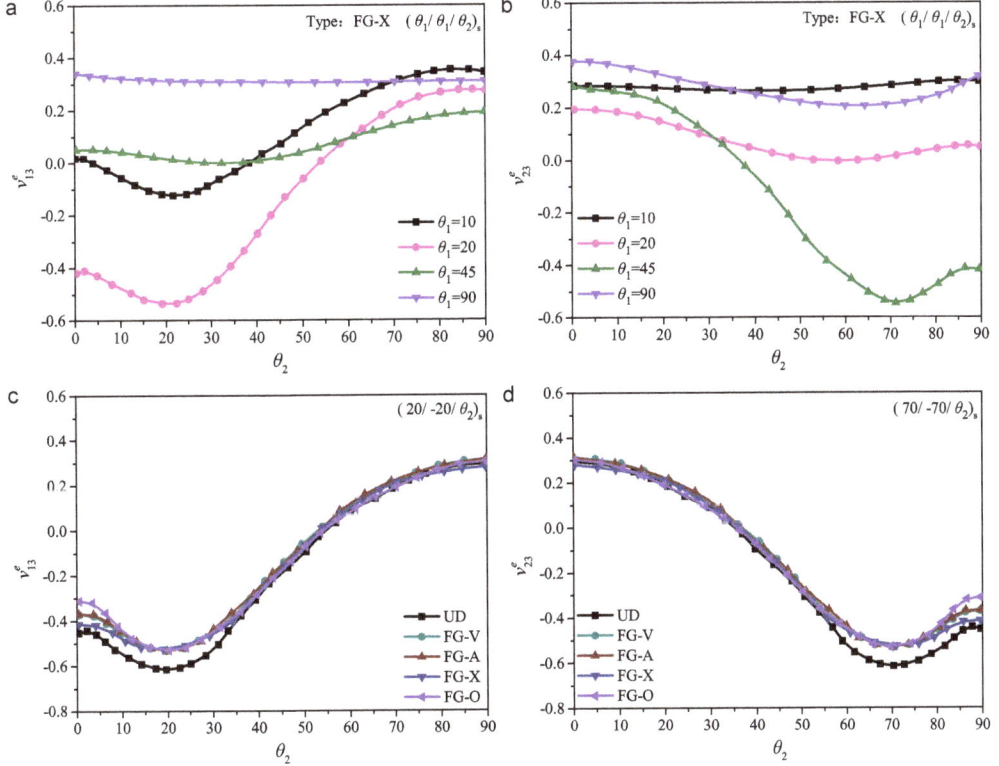

Figure 3. The effective Poisson's ratios of FG-CNTRC laminated plates: (**a**) ν_{13}^e for $(\theta_1/\theta_1/\theta_2)_s$ laminates of type FG-X; (**b**) ν_{23}^e for $(\theta_1/\theta_1/\theta_2)_s$ laminates of type FG-X; (**c**) ν_{13}^e for $(20/-20/\theta_2)_s$ laminates; and (**d**) ν_{23}^e for $(20/-20/\theta_2)_s$ laminates.

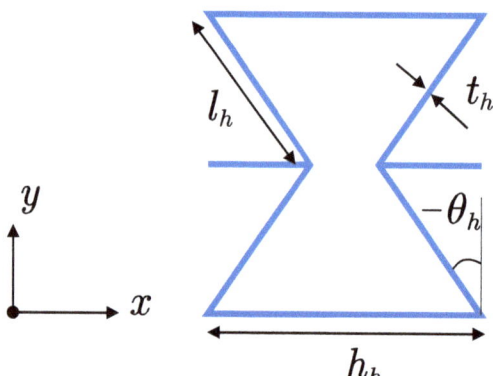

Figure 4. The structure of the auxetic honeycomb core.

Table 4. The material properties of Ti-6Al-4V.

Material Properties	E_{Ti} (GPa)	v_{12}	ρ_{Ti} (g/cm^3)
Ti-6Al-4V	$122.56 \times (1\text{--}4.586 \times 10^{-4}T)$	0.342	4.43

3. Computational Methods

3.1. Governing Equations

The first-order shear deformation theory is used to describe the sandwich plate with length a, width b and thickness h, as shown in Figure 1. The displacement field $(\bar{u}, \bar{v}, \bar{w})$ can be expressed as

$$\bar{u}(x,y,z,t) = u(x,y,t) + z\phi_x(x,y,t)$$
$$\bar{v}(x,y,z,t) = v(x,y,t) + z\phi_y(x,y,t) \quad (5)$$
$$\bar{w}(x,y,z,t) = w(x,y,t)$$

where u, v and w are the translation displacement components at the mid-plane in the x, y and z directions, respectively. ϕ_x and ϕ_y denote the rotation of the normal to the mid-plane along the y axis and x axis, respectively. The relationship between strain and displacement can be expressed as

$$\varepsilon = \varepsilon^0 + z\kappa^0$$
$$\gamma = \gamma^0 \quad (6)$$

where

$$\varepsilon = \begin{bmatrix} \varepsilon_{xx} \\ \varepsilon_{yy} \\ \gamma_{xy} \end{bmatrix}, \; \varepsilon^0 = \begin{bmatrix} \dfrac{\partial u}{\partial x} \\ \dfrac{\partial v}{\partial y} \\ \dfrac{\partial v}{\partial x} + \dfrac{\partial u}{\partial y} \end{bmatrix}, \; \kappa^0 = \begin{bmatrix} \dfrac{\partial \phi_x}{\partial x} \\ \dfrac{\partial \phi_y}{\partial y} \\ \dfrac{\partial \phi_y}{\partial x} + \dfrac{\partial \phi_x}{\partial y} \end{bmatrix},$$

$$\gamma = \begin{bmatrix} \gamma_{yz} \\ \gamma_{xz} \end{bmatrix}, \; \gamma^0 = \begin{bmatrix} \phi_y + \dfrac{\partial w}{\partial y} \\ \phi_x + \dfrac{\partial w}{\partial x} \end{bmatrix}. \quad (7)$$

Considering the temperature effect, the stress component based on a linear constitutive relationship can be written as

$$\left\{\begin{array}{c}\sigma_{xx}\\ \sigma_{yy}\\ \tau_{xy}\\ \tau_{yz}\\ \tau_{xz}\end{array}\right\}=\left[\begin{array}{ccccc}\bar{Q}_{11}&\bar{Q}_{12}&0&0&0\\ \bar{Q}_{21}&\bar{Q}_{22}&0&0&0\\ 0&0&\bar{Q}_{66}&0&0\\ 0&0&0&\bar{Q}_{44}&0\\ 0&0&0&0&\bar{Q}_{55}\end{array}\right]\left(\left\{\begin{array}{c}\varepsilon_{xx}\\ \varepsilon_{yy}\\ \gamma_{xy}\\ \gamma_{yz}\\ \gamma_{xz}\end{array}\right\}-\left\{\begin{array}{c}\alpha_{11}\\ \alpha_{22}\\ 0\\ 0\\ 0\end{array}\right\}\Delta T\right) \qquad (8)$$

where ΔT is the temperature change and the transformed stiffness $\bar{Q}$ can be calculated by

$$\left[\begin{array}{c}\bar{Q}_{11}\\ \bar{Q}_{12}\\ \bar{Q}_{22}\\ \bar{Q}_{16}\\ \bar{Q}_{26}\\ \bar{Q}_{66}\end{array}\right]=\left[\begin{array}{cccc}c^4&2c^2s^2&s^4&4c^2s^2\\ c^2s^2&c^4+s^4&c^2s^2&-4c^2s^2\\ s^4&2c^2s^2&c^4&4c^2s^2\\ c^3s&(cs^3-c^3s)&-cs^3&-2cs(c^2-s^2)\\ cs^3&(c^3s-cs^3)&-c^3s&2cs(c^2-s^2)\\ c^2s^2&-2c^2s^2&c^2s^2&(c^2-s^2)^2\end{array}\right]\left[\begin{array}{c}Q_{11}\\ Q_{12}\\ Q_{22}\\ Q_{66}\end{array}\right] \qquad (9)$$

$$\left[\begin{array}{c}\bar{Q}_{44}\\ \bar{Q}_{45}\\ \bar{Q}_{55}\end{array}\right]=\left[\begin{array}{cc}c^2&s^2\\ -cs&cs\\ s^2&c^2\end{array}\right]\left[\begin{array}{c}Q_{44}\\ Q_{55}\end{array}\right]$$

where s and c are the sin and cos of the lamination angle against the x axis of the plate. Furthermore, the stiffness parameters can be given as

$$Q_{11}=\frac{E_{11}}{1-v_{12}v_{21}},\ Q_{22}=\frac{E_{22}}{1-v_{12}v_{21}},\ Q_{12}=\frac{v_{21}E_{11}}{1-v_{12}v_{21}} \qquad (10)$$

$$Q_{44}=G_{23},\ Q_{55}=G_{13},\ Q_{66}=G_{12}$$

The strain energy of the sandwich plate U_p can be expressed as

$$U_p=\frac{1}{2}\int_\Omega \bar{\varepsilon}^T S\bar{\varepsilon}d\Omega \qquad (11)$$

where $\bar{\varepsilon}=(\varepsilon^0,\kappa^0,\gamma^0)^T$ is the strain matrix, S is the material constant matrix and

$$S=\left[\begin{array}{ccc}A&B&0\\ B&D&0\\ 0&0&A_s\end{array}\right]=\left[\begin{array}{cccccccc}A_{11}&A_{12}&A_{16}&B_{11}&B_{12}&B_{16}&0&0\\ A_{12}&A_{22}&A_{26}&B_{12}&B_{22}&B_{26}&0&0\\ A_{16}&A_{26}&A_{66}&B_{16}&B_{26}&B_{66}&0&0\\ B_{11}&B_{12}&B_{16}&D_{11}&D_{12}&D_{16}&0&0\\ B_{12}&B_{22}&B_{26}&D_{12}&D_{22}&D_{26}&0&0\\ B_{16}&B_{26}&B_{66}&D_{16}&D_{26}&D_{66}&0&0\\ 0&0&0&0&0&0&A^s_{44}&A^s_{45}\\ 0&0&0&0&0&0&A^s_{45}&A^s_{55}\end{array}\right] \qquad (12)$$

where A, B, D, A_s are the matrices of the plate stiffness, which can be calculated by

$$(A,B,D)=\sum_{k=1}^{N}\int_{h_{k-1}}^{h_k}(\bar{Q})_k\left(1,z,z^2\right)dz,\ A_s=K_s\sum_{k=1}^{N}\int_{h_{k-1}}^{h_k}(\bar{Q})_k dz \qquad (13)$$

where the transverse shear correction coefficient K_s can be calculated by

$$K_s=\begin{cases}\dfrac{5}{6}, & \text{isotropic material}\\ \dfrac{5}{6-v_1V_1-v_2V_2}, & \text{functionally graded material}\end{cases} \qquad (14)$$

where v and V are the Poisson's ratios and volume fraction of each material in the entire cross-section. The kinetic energy of the sandwich plate T can be obtained by

$$T = \frac{1}{2}\int_\Omega \int_{-h/2}^{h/2} \rho(z)\left(\dot{u}^2 + \dot{v}^2 + \dot{w}^2\right)\mathrm{d}z\,\mathrm{d}\Omega \tag{15}$$

The external virtual work δW can be obtained by

$$\delta W = F_c(t)\delta\mu \tag{16}$$

where $F_c(t)$ is the contact force between the plate and the impactor, and μ is the deflection of the sandwich plate. Then, the total energy function based on Hamilton's principle can be expressed as

$$\int_0^t (\delta U_p - \delta T - \delta W)\mathrm{d}t = 0 \tag{17}$$

The boundary conditions for the clamped of the plate edge can be expressed as

$$u = 0, v = 0, w = 0, \phi_x = 0, \phi_y = 0 \tag{18}$$

3.2. Low-Velocity Impact Response

Based on the nonlinear Hertz contact law, the contact force $F_c(t)$ between the sandwich plate and a steel ball can be obtained by [83]

$$F_c(t) = \begin{cases} K_c \mu^{\frac{3}{2}}(t) & \text{loading} \\ F_{cm}\left(\dfrac{\mu}{\mu_m}\right)^{\frac{5}{2}} & \text{unloading} \end{cases} \tag{19}$$

where $\mu = w_i - w_p$ is the deflection of the sandwich plate, and w_i, w_p refers to the displacement of the impactor and plate center, respectively. The subscript m refers to the maximum value of the variables. K_c is the contact coefficient, which can be expressed as [83],

$$K_c = \frac{4}{3}\left(\frac{1-v_i^2}{E_i} + \frac{1}{E_2}\right)^{-1}\sqrt{r_i} \tag{20}$$

where E_i, v_i, r_i are the elasticity modulus, Poisson's ratios and the radius of the impactor, respectively. E_2 is the transverse elasticity modulus of the sandwich plate. The displacement of the impactor w_i can be calculated by

$$w_i = v_i t - \frac{1}{m_i}\int_0^t F_c(\tau)(t-\tau)\mathrm{d}\tau \tag{21}$$

where v_i and m_i are the velocity and mass of the impactor, respectively. Then, the Equation (19) can be obtained by

$$\left(\frac{F_c(t)}{K_c}\right)^{2/3} = v_i t - \frac{1}{m_i}\int_0^t F_c(t-\tau)\mathrm{d}\tau - w_p \tag{22}$$

3.3. Solution Procedure

The Ritz method is considered to deduce the governing equations of motion from the total energy function in the spatial domain, and the functions of the displacement field can be expressed as

$$u = \sum_{n=1}^{N} p_n^u(x,y) U_n(t)$$
$$v = \sum_{n=1}^{N} p_n^v(x,y) V_n(t)$$
$$w = \sum_{n=1}^{N} p_n^w(x,y) W_n(t) \quad (23)$$
$$\phi_x = \sum_{n=1}^{N} p_n^{\phi_x}(x,y) \Phi_{xn}(t)$$
$$\phi_y = \sum_{n=1}^{N} p_n^{\phi_y}(x,y) \Phi_{yn}(t)$$

where $p_n(x,y)$ are the shape functions. $n = 1, 2, \cdots, N$ and N is the number of terms in the basis. $U_n(t), V_n(t), W_n(t), \Phi_{xn}(t), \Phi_{yn}(t)$ are the unknown coefficients chosen according to the boundary conditions. The shape functions of the polynomial are considered in this research [84,85].

The equations of motion of the sandwich plate and impactor can be obtained by

$$\begin{aligned} \mathbf{M}\ddot{\mathbf{q}} + \mathbf{K}\mathbf{q} &= \mathbf{F} \\ m_i \ddot{w}_i + F_c &= 0 \end{aligned} \quad (24)$$

where $\mathbf{q}, \mathbf{M}, \mathbf{K}, \mathbf{F}$ are the degrees of the freedom vector, mass matrix, stiffness matrix and impact load vector, respectively. Furthermore, the components of the mass matrix and the stiffness matrix are presented in Appendix B. The dot over the variable refers to the differentiation of that variable with respect to time. The Newmark's time integration schemes is considered to solve the time-dependent equations after assembling the process and implementing boundary conditions. By using Taylor series expansions, the $\mathbf{q}_{t+\Delta t}$, $\dot{\mathbf{q}}_{t+\Delta t}$ and $\ddot{\mathbf{q}}_{t+\Delta t}$ can be transformed into

$$\begin{aligned} \mathbf{q}_{t+\Delta t} &= q(t) + \Delta t \dot{q}_t + \frac{1}{2}\Delta t^2 \ddot{q}_t - \frac{1}{2}\beta_2 \Delta t^2 \ddot{q}_t + \frac{1}{2}\beta_2 \Delta t^2 \ddot{q}_{t+\Delta t} \\ \dot{\mathbf{q}}_{t+\Delta t} &= \dot{q}_t + \Delta t \ddot{q}_t - \beta_1 \Delta t \ddot{q}_t + \beta_1 \Delta t \ddot{q}_{t+\Delta t} \\ \ddot{\mathbf{q}}_{t+\Delta t} &= \frac{2}{\beta_2 \Delta t^2}(q_{t+\Delta t} - q_t) - \frac{2}{\beta_2 \Delta t}\dot{q}_t - \frac{1}{\beta_2}\ddot{q}_t + \ddot{q}_t \end{aligned} \quad (25)$$

Substituting Equation (25) into Equation (24):

$$\left(\frac{2}{\beta_2 \Delta t^2}\mathbf{M} + \mathbf{K}\right)\mathbf{q}_{t+\Delta t} = \mathbf{F}_{t+\Delta t} + \mathbf{M}\left(\frac{2}{\beta_2 \Delta t^2}\mathbf{q}_t + \frac{2}{\beta_2 \Delta t}\dot{\mathbf{q}}_t + \left(\frac{1}{\beta_2} - 1\right)\ddot{\mathbf{q}}_t\right) \quad (26)$$

where the Newmark's parameters $\beta_1 = 0.5$ and $\beta_2 = 0.5$ are considered in this research according to the Newmark β-method.

4. Results and Discussion

4.1. Validation Studies

To validate the calculation method, the relative examples of Refs. [38,86] are considered by contrast. The parameters of the plate are set to 1 m in length, 1 m in width and 0.01 m in thickness. The gradient form is UD while the V^c is 0.28. The parameters of the impactor are set as a mass of 0.5 kg and a radius of 0.25 m. The working conditions are a temperature of 300 K and an initial impact velocity of 3 m/s. The displacement–time curve comparative result is shown in Figure 5. It can be inferred that the results are in good agreement. The maximum displacement and contact time error could be accepted for analysis.

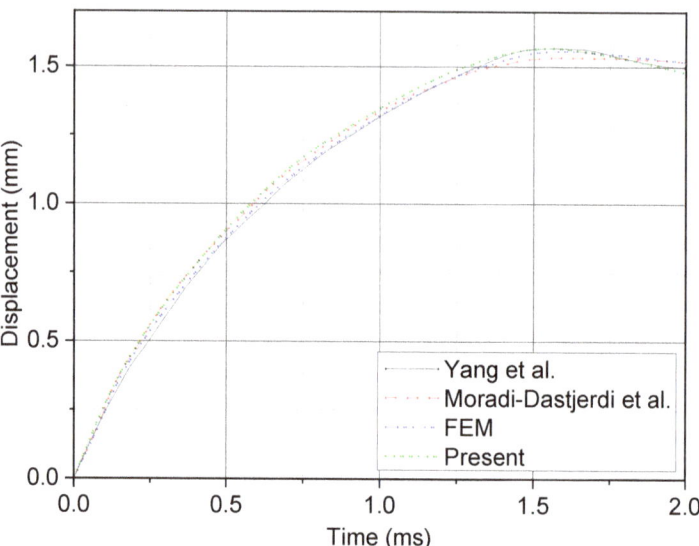

Figure 5. Comparison of the plate center displacement with the results obtained from the Ref. [38,86] and FEM method.

In order to validate the equivalent layer model for the relative soft honeycomb core, a full-scale finite element simulation with an auxetic honeycomb core model was performed in contrast using the ABAQUS software, as shown in Figure 6. The sandwich structure with 0.5 mm thickness Ti-6Al-4V face sheets and auxetic honeycomb core was considered. The parameters of honeycomb core were set as: thickness h_c = 23 mm; length of inclined cell rib l_h = 5 mm; length of the vertical cell rib h_h = 10 mm; and inclined angle θ_h = $-40°$. The second-order accuracy S4R elements were used to mesh the structure. Moreover, the meshes of face sheets are designed to share nodes with cores along the two interfaces, indicating the perfectly adhered to assumption. The impactor was set as an analytically rigid body ball with radius 10 mm. Furthermore, the mass was calculated according to the density 7.8 g/cm^3. The general contact method with frictionless property was used to define the contact behavior. The initial impact velocity was 3 m/s, using predefined fields. All six degrees of freedoms of the boundary nodes were constrained to simulate clamped boundary conditions. The displacement–time curve comparative result is shown in Figure 7. It can be inferred that the results are in good agreement and the equivalent layer model could be used for the present research.

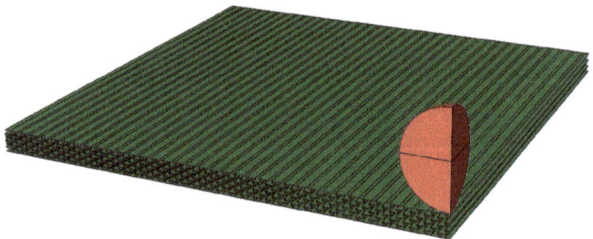

Figure 6. Low-velocity impact simulation in ABAQUS software.

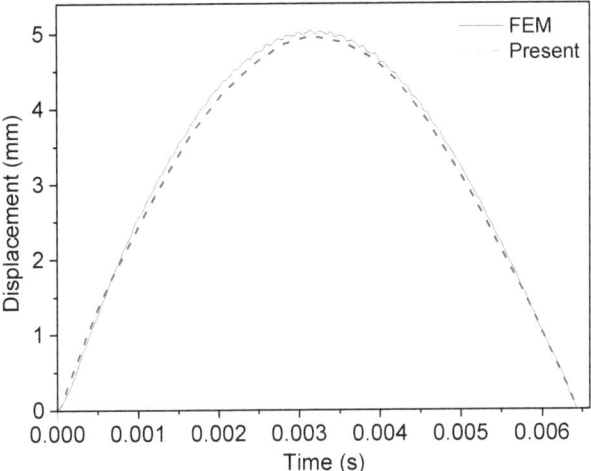

Figure 7. Comparison of the plate center displacement with the results obtained from FEM and present method.

To be sure, the modeling method based on continuum mechanics theory in this paper was verified. The molecular dynamic theories or nano-scale continuum modeling is a more accurate simulation method for nanomaterials such as SCNT. However, this research focuses on the qualitative study of each parameter on the structural impact response, and the continuum mechanics theory can be used to show the trend of response after verification.

4.2. Parameter Studies

After verifying the model and computing method of this research, we focus on the $(20/-20/20)s$, $(45/-45/45)s$ and $(70/-70/70)s$ stacking sequences of the FG-CNTRC surface, the function gradient, volume fraction of CNTs, impact velocity, temperature, length/width ratio and FG-CNTRC surface thickness effects on the low-velocity impact response of the sandwich plate with FG-CNTRC face sheets and NPR auxetic honeycomb core are analyzed. The plate center displacement w_p, recovery time of deformation t_r, contact force F_c and contact time t_c are considered in detail. The initial parameters of the sandwich plate structure and boundary conditions are set as:

- Sandwich plate—length/width ratio $a/b = 1$, total thickness $h = 25.4$ mm;
- FG-CNTRC surface—thickness $h_s = 1.2$ mm, gradient form FG-V;
- Honeycomb core—thickness $h_c = 23$ mm, length of inclined cell rib $l_h = 5$ mm, length of the vertical cell rib $h_h = 10$ mm, inclined angle $\theta_h = -40°$;
- Calculate conditions—temperature $T = 300$ K, impact velocity $v_i = 2$ m/s, boundary conditions clamped.

4.2.1. Gradient Forms of FG-CNTRC Surfaces

The low-velocity impact of gradient forms FG-V, FG-A, FG-X, FG-O and UD are considered. The plate center displacement of the three stacking sequences are shown in Figure 8. The $(20/-20/20)s$ ply has the largest plate center displacement w_p, reaches the maximum value first and has the shortest recovery time of deformation t_r. The $(45/-45/45)s$ ply has the smallest plate center displacement w_p. The $(70/-70/70)s$ ply has the longest recovery time of deformation t_r. The value of the plate center displacement w_p, recovery time of deformation t_r, contact force F_c and contact time t_c are shown in Table 5 in detail. The UD form of $(20/-20/20)s$ ply and $(70/-70/70)s$ ply has the largest w_p, smallest F_c and longest t_r. The FG-O form of $(20/-20/20)s$ ply has the smallest w_p, largest F_c and shortest t_r. While the

FG-X form of (70/−70/70)s ply has the smallest w_p, largest F_c and shortest t_r. The response of the (45/−45/45)s ply is more complicated. The UD form has the largest w_p and longest t_r. The FG-X form has the largest F_c and shortest t_r. The FG-O form has the smallest w_p. The FG-V form has the smallest F_c. The contact time t_c of each gradient forms are nearly the same.

Table 5. Low-velocity impact response of the sandwich structure with various gradient forms.

Type	Gradient Forms	w_p (mm)	F_c (N)	t_r (ms)	t_c (ms)
(20/−20/20)s	FG-A	2.522	1155.943	4.970	5.250
	FG-O	2.518	1156.313	4.966	5.250
	UD	2.534	1149.806	4.982	5.250
	FG-V	2.522	1154.605	4.972	5.250
	FG-X	2.521	1155.906	4.969	5.250
(45/−45/45)s	FG-A	2.436	1164.017	5.304	5.650
	FG-O	2.433	1164.171	5.303	5.650
	UD	2.444	1163.188	5.311	5.650
	FG-V	2.439	1162.674	5.306	5.650
	FG-X	2.434	1164.324	5.300	5.650
(70/−70/70)s	FG-A	2.491	1106.108	5.876	5.750
	FG-O	2.490	1106.045	5.875	5.750
	UD	2.498	1104.619	5.888	5.750
	FG-V	2.494	1105.591	5.879	5.750
	FG-X	2.488	1107.330	5.874	5.750

It is observed that the (45/−45/45)s ply with nearly zero Poisson's ratio has the smallest w_p, and the (70/−70/70)s ply with the native v^e_{23} has the smallest F_c. Within three stacking sequences and five gradient forms, (45/−45/45)s ply with FG-O type has the smallest w_p, while (70/−70/70)s ply with UD type has the smallest F_c. The percentage decrease is approximately 5% by changing the stacking sequence and gradient form of the surface sheets.

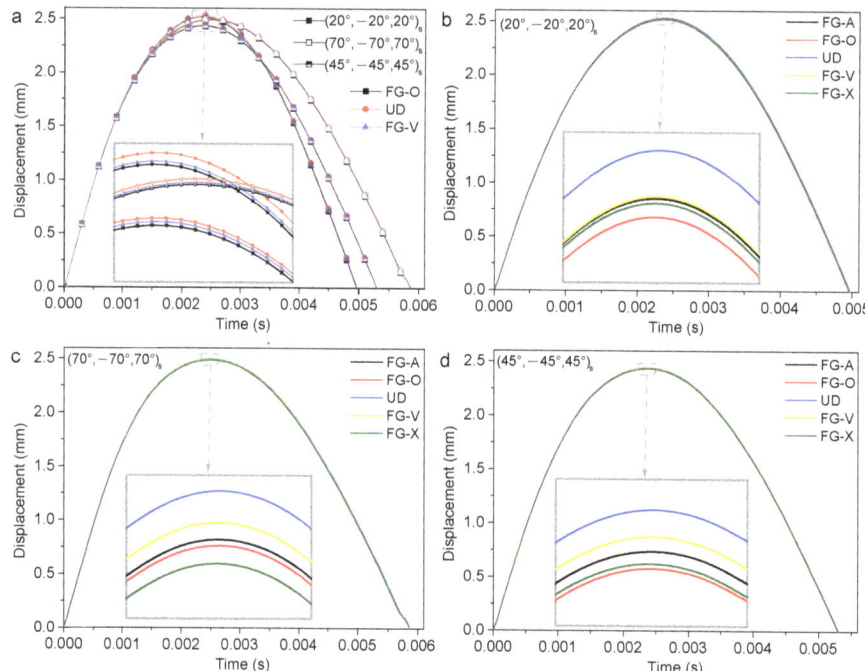

Figure 8. Plate center displacement response of a sandwich structure with various gradient forms: (**a**) FG-O, UD and FG-V face sheets plate; (**b**) (20/−20/20)s plate; (**c**) (70/−70/70)s plate; and (**d**) (45/−45/45)s plate.

4.2.2. Volume Fractions of CNTs

The 0.11, 0.14 and 0.17 volume fractions of CNTs are considered. The surface layer of this part of the research is set as uniform distribution. The plate center displacement are shown in Figure 9. The $(20/-20/20)s$ ply has the largest plate center displacement w_p and shortest recovery time of deformation t_r. The $(45/-45/45)s$ ply has the smallest plate center displacement w_p and the $(70/-70/70)s$ ply has the longest recovery time of deformation t_r. According to Table 6, the response of three stacking sequences is similar. With the volume fractions of CNTs increasing, the plate center displacement w_p, recovery time of deformation t_r and contact time t_c decreases, while the contact force F_c increases. It can be inferred that the contact stiffness increases with the volume fractions of CNTs increasing.

It is observed that increasing the stiffness of the sandwich structure by increasing the volume fraction of CNTs can lead to a reduction in the w_p and an increase of the F_c. Furthermore, this phenomenon is more sensitive to $(20/-20/20)s$ ply with a reduction in w_p by approximately 6.4%.

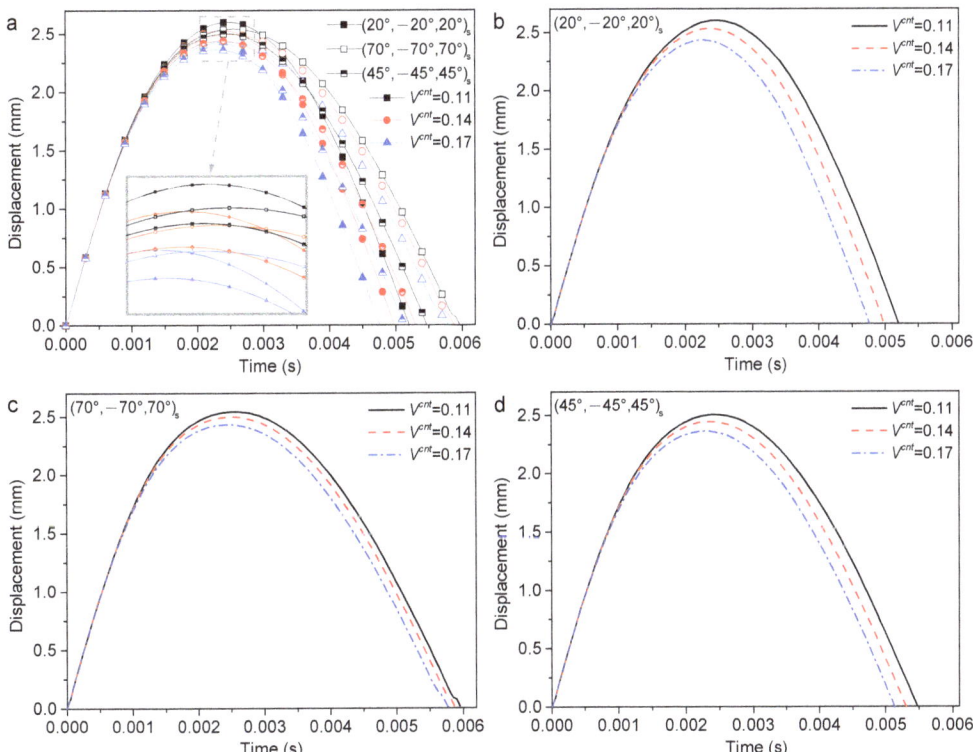

Figure 9. Plate center displacement response of the sandwich structure with various volume fractions of CNTs: (**a**) FG-O, UD and FG-V face sheets plate; (**b**) $(20/-20/20)s$ plate; (**c**) $(70/-70/70)s$ plate; (**d**) $(45/-45/45)s$ plate.

Table 6. Low-velocity impact response of a sandwich structure with various volume fraction of CNTs.

Type	Volume Fraction	w_p (mm)	F_c (N)	t_r (ms)	t_c (ms)
(20/−20/20)s	0.11	2.602	1120.218	5.202	5.500
	0.14	2.534	1149.936	4.978	5.250
	0.17	2.436	1196.062	4.762	5.000
(45/−45/45)s	0.11	2.503	1134.566	5.478	5.800
	0.14	2.444	1163.188	5.311	5.650
	0.17	2.365	1194.500	5.139	5.450
(70/−70/70)s	0.11	2.436	1092.485	5.969	5.850
	0.14	2.498	1104.619	5.880	5.750
	0.17	2.433	1121.323	5.783	5.600

4.2.3. Impact Velocity

The impact velocity plays an important role in the impact response. Considering 1 m/s, 2 m/s and 3 m/s impact velocity, the plate center displacements of three stacking sequences are shown in Figure 10. The (20/−20/20)s ply has the largest plate center displacement w_p and has the shortest recovery time of deformation t_r. The (45/−45/45)s ply has the smallest plate center displacement w_p. The (70/−70/70)s ply has the longest recovery time of deformation t_r. According to Table 7, with the increased impact velocity, the plate center displacement w_p and the contact force F_c increased, while the recovery time of deformation t_r and contact time t_c decreased.

It is observed that the three stacking sequences have a slight impact on the variable ratio of w_p and F_c. Increasing the impact velocity from 1 m/s to 3 m/s can lead to an increase in the w_p and F_c by approximately 62.5% and 68%, respectively.

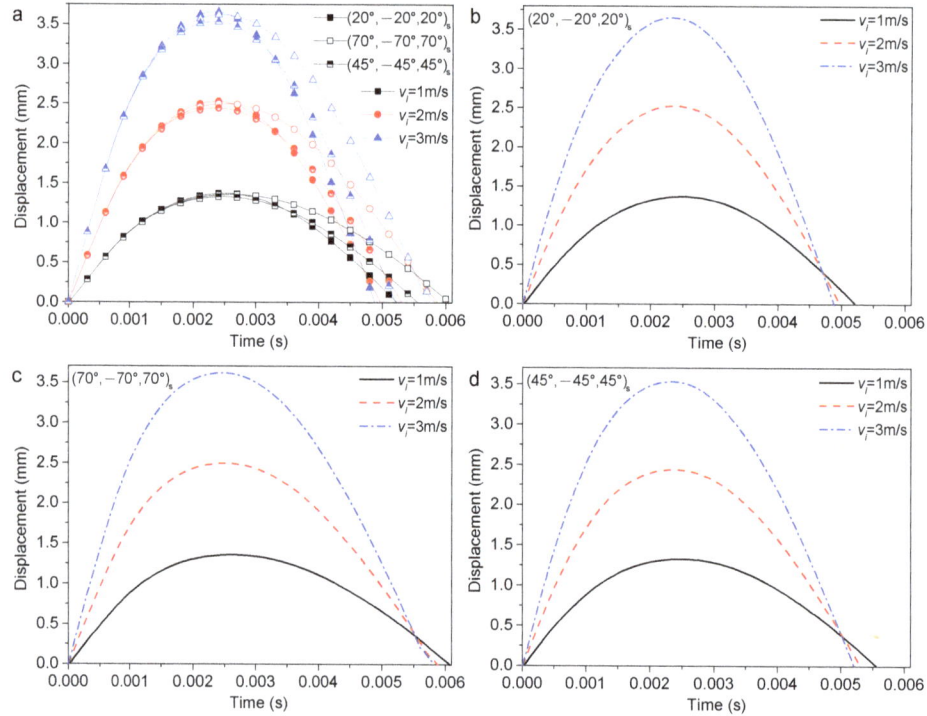

Figure 10. Plate center displacement response of the sandwich structure with various impact velocities: (a) FG-O, UD and FG-V face sheets plate; (b) (20/−20/20)s plate; (c) (70/−70/70)s plate; and (d) (45/−45/45)s plate.

Table 7. Low-velocity impact response of the sandwich structure with various impact velocities.

Type	Impact Velocity (m/s)	w_p (mm)	F_c (N)	t_r (ms)	t_c (ms)
(20/−20/20)s	1	1.372	563.495	5.223	5.500
	2	2.522	1154.605	4.972	5.250
	3	3.654	1784.333	4.881	5.150
(45/−45/45)s	1	1.329	575.804	5.558	5.850
	2	2.439	1162.674	5.306	5.650
	3	3.532	1777.370	5.201	5.350
(70/−70/70)s	1	1.357	551.596	6.074	7.300
	2	2.494	1105.591	5.879	5.750
	3	3.616	1706.855	5.848	7.200

4.2.4. Temperature

The low-velocity impact response of FG-CNTRC plates under various temperatures is the hotspot of its application under extreme conditions. The temperatures of 300 K, 400 K and 500 K are considered, as shown in Figure 11. Similarly to the result of various impact velocities, the (20/−20/20)s ply has the largest plate center displacement w_p and has the shortest recovery time of deformation t_r. The (45/−45/45)s ply has the smallest plate center displacement w_p. The (70/−70/70)s ply has the longest recovery time of deformation t_r. According to Table 8, with the increased temperature, the plate center displacement w_p, recovery time of deformation t_r and contact time t_c increased, while the contact force F_c decreased.

It is observed that the stiffness of the sandwich structure will reduce by increasing the temperature. From 300 K to 500 K, the w_p will increase by approximately 8.4%.

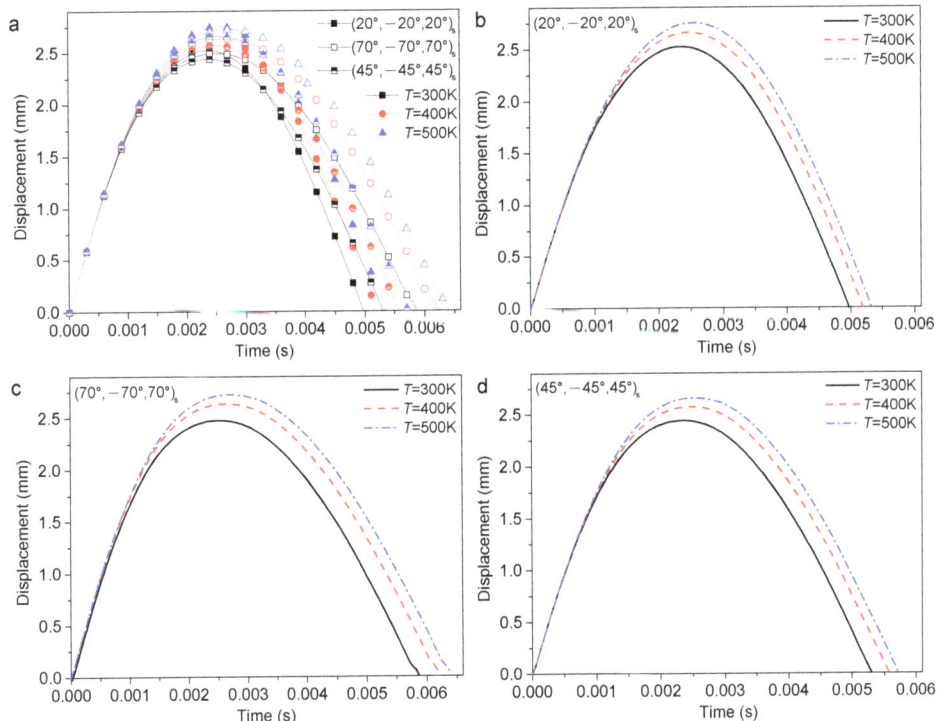

Figure 11. Plate center displacement response of the sandwich structure with various temperatures: (**a**) FG-O, UD and FG-V face sheets plate; (**b**) (20/−20/20)s plate; (**c**) (70/−70/70)s plate; and (**d**) (45/−45/45)s plate.

Table 8. Low-velocity impact response of the sandwich structure with various temperatures.

Type	Temperature (K)	w_p (mm)	F_c (N)	t_r (ms)	t_c (ms)
(20/−20/20)s	300	2.522	1154.605	4.972	5.250
	400	2.659	1092.760	5.193	5.550
	500	2.753	1119.235	5.332	5.560
(45/−45/45)s	300	2.439	1162.674	5.306	5.650
	400	2.571	1104.613	5.570	5.950
	500	2.659	1098.925	5.714	6.100
(70/−70/70)s	300	2.494	1105.591	5.879	5.750
	400	2.635	1044.606	6.221	6.050
	500	2.723	1011.190	6.405	6.250

4.2.5. Ratio of Plate Length and Width

The length/width ratio a/b = 0.5, 1.0 and 2.0 are considered, as shown in Figure 12. The coupling between stacking sequence and a/b makes the low-velocity impact response complicated. The a/b = 2.0 has the largest plate center displacement w_p, while a/b = 0.5 is the smallest of all three stacking sequences. The responses are shown in Table 9 in detail. When a/b = 0.5, the (70/−70/70)s ply has the largest w_p and smallest F_c, the (45/−45/45)s ply has the smallest w_p and largest F_c. When a/b = 2.0, whilst the (45/−45/45)s ply has the largest w_p and smallest F_c, the (20/−20/20)s ply has the smallest w_p and largest F_c. However, the t_r decreases at first and then increases with the increase in a/b. The t_c increases with the increase in a/b. The results inferred that the ratio of plate length and width has a large influence on the low-velocity impact, which causes the nonlinear change phenomenon.

It is observed that the geometry scale has more influence on the impact response, due to the anisotropic honeycomb core. Using the honeycomb section as the long side of the structure can reduce the F_c.

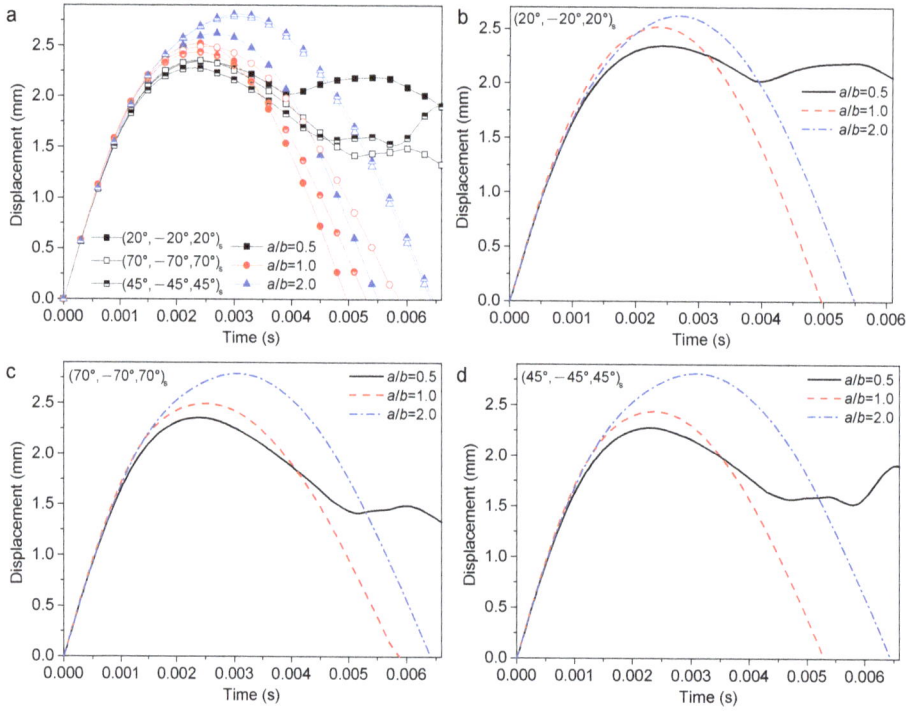

Figure 12. Plate center displacement response of the sandwich structure with various a/b: (**a**) FG-O, UD, and FG-V face sheets plate; (**b**) (20/−20/20)s plate; (**c**) (70/−70/70)s plate; and (**d**) (45/−45/45)s plate.

4.2.6. Thickness of Surface Layer

The thickness of the FG-CNTRC surface layer h_s = 0.6 mm, 1.2 mm and 2.4 mm are considered, and the low-velocity impact response is shown in Figure 13. When h_s = 1.2 mm and 2.4 mm, the stacking sequence has a large influence on the plate displacement w_p. According to Table 10, when h_s = 0.6 mm, the (20/−20/20)s ply has the smallest w_p, largest F_c and shortest t_r and t_c. The (45/−45/45)s ply has the largest w_p, longest t_r and t_c. The (70/−70/70)s ply has the smallest F_c. When h_s = 2.4 mm, the (20/−20/20)s ply has the smallest w_p, largest F_c and shortest t_r and t_c. The (45/−45/45)s ply has the largest w_p, smallest F_c and longest t_r and t_c.

Table 9. Low-velocity impact response of a sandwich structure with various a/b.

Type	a/b	w_p (mm)	F_c (N)	t_r (ms)	t_c (ms)
(20/−20/20)s	0.5	2.342	1147.677	9.380	3.900
	1.0	2.522	1154.605	4.972	5.250
	2.0	2.624	1044.604	5.498	5.600
(45/−45/45)s	0.5	2.275	1165.578	8.125	4.350
	1.0	2.439	1162.674	5.306	5.650
	2.0	2.813	955.473	6.454	6.550
(70/−70/70)s	0.5	2.354	1111.640	8.070	4.750
	1.0	2.494	1105.591	5.879	5.750
	2.0	2.791	965.506	6.417	6.500

It is observed that increasing h_s can lead to a reduction in the w_p and an increase in the F_c by increasing the stiffness of the structure.

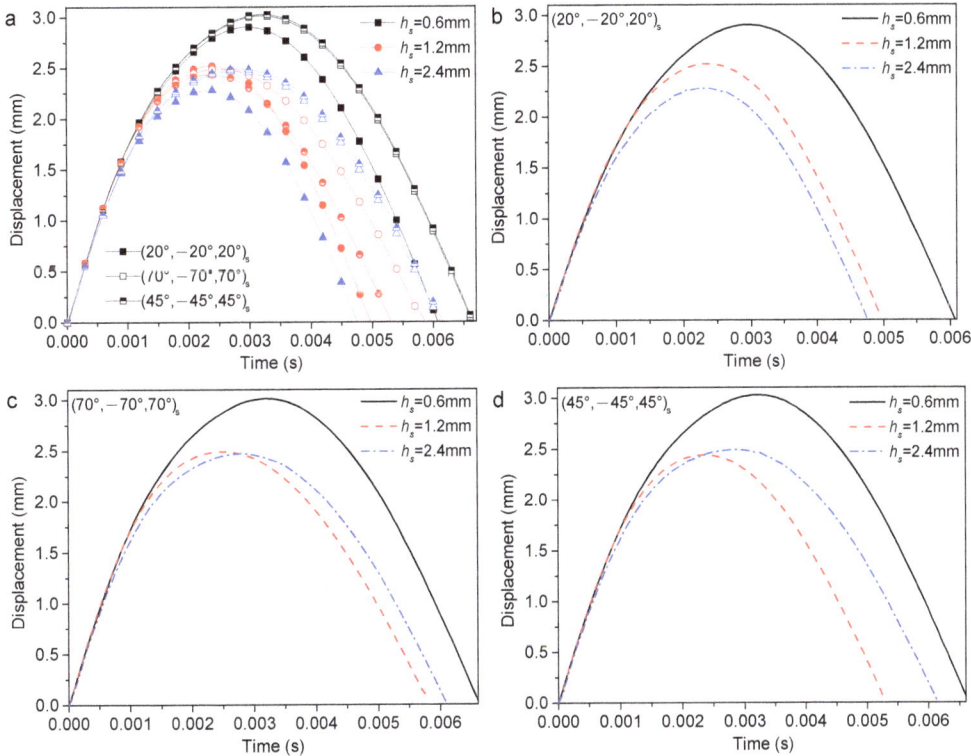

Figure 13. Plate center displacement response of the sandwich structure with various h_s: (**a**) FG-O, UD and FG-V face sheets plate; (**b**) (20/−20/20)s plate; (**c**) (70/−70/70)s plate; and (**d**) (45/−45/45)s plate.

Table 10. Low-velocity impact response of the sandwich structure with various h_s.

Type	h (mm)	w_p (mm)	F_c (N)	t_r (ms)	t_c (ms)
(20/−20/20)s	0.6	2.903	946.210	6.072	6.250
	1.2	2.522	1154.605	4.970	5.250
	2.4	2.287	1746.733	4.746	4.850
(45/−45/45)s	0.6	3.027	904.395	6.650	6.900
	1.2	2.439	1162.674	5.302	5.650
	2.4	2.494	1209.172	6.149	6.350
(70/−70/70)s	0.6	3.013	902.615	6.625	6.850
	1.2	2.494	1105.591	5.877	5.750
	2.4	2.476	1469.982	6.102	6.300

5. Conclusions

In this research, a numerical method on the low-velocity impact response of the sandwich plate with an FG-CNTRC surface and NPR honeycomb core was proposed and verified. Three kinds of stacking sequences of FG-CNTRC, namely (20/−20/20)s, (45/−45/45)s and (70/−70/70)s, were considered. The effects of gradient forms of FG-CNTRC surfaces, volume fractions of CNTs, impact velocities, temperatures, the ratio of the plate length and the width and thickness of surface layers on the low-velocity impact response were analyzed. The results of the plate center displacement w_p, recovery time of deformation t_r, contact force F_c and contact time t_c show that:

- Gradient forms of FG-CNTRC surfaces:

  (20/−20/20)s ply—the UD form has the largest w_p, smallest F_c and longest t_r; and the FG-O form has the smallest w_p, largest F_c and shortest t_r;

  (45/−45/45)s ply—the UD form has the largest w_p and longest t_r; the FG-X form has the largest F_c and shortest t_r; the FG-O form has the smallest w_p; and the FG-V form has the smallest F_c;

  (70/−70/70)s ply—the UD form has the largest w_p, smallest F_c and longest t_r; the FG-X form has the smallest w_p, largest F_c and shortest t_r.

  Within three stacking sequences and five gradient forms, the (45/−45/45)s ply with FG-O type has the smallest w_p, while the (70/−70/70)s ply with the UD type has the smallest F_c. The percentage decrease is approximately 5% by changing the stacking sequence and gradient form of the surface sheets.

- Volume fractions of CNTs:

  The (20/−20/20)s ply has the largest w_p and shortest t_r. The (45/−45/45)s ply has the smallest w_p and the (70/−70/70)s ply has the longest t_r;

  The plate center displacement w_p, recovery time of deformation t_r and contact time t_c decreased, while the contact force F_c increased with the increased volume fractions of CNTs.

  Increasing the volume fraction of CNTs from 0.11 to 0.17 can lead to a reduction in the w_p and an increase in the F_c. Furthermore, this phenomenon is more sensitive to (20/−20/20)s ply with a reduction in w_p by approximately 6.4%.

- Impact velocities:

  The (20/−20/20)s ply has the largest w_p and has the shortest t_r. The (45/−45/45)s ply has the smallest w_p. The (70/−70/70)s ply has the longest t_r.

  The plate center displacement w_p and contact force F_c increased, while the recovery time of deformation t_r and contact time t_c decreased as the impact velocity increased. The three stacking sequences have a slight impact on the variable ratio of w_p and F_c. Increasing the impact velocity from 1 m/s to 3 m/s can lead to an increase in the w_p and F_c of approximately 62.5% and 68%, respectively.

- Temperatures:

  The (20/−20/20)s ply has the largest w_p and the shortest t_r. The (45/−45/45)s ply has the smallest w_p. The (70/−70/70)s ply has the longest t_r.

The plate center displacement w_p, recovery time of deformation t_r and contact time t_c increased, while the contact force F_c decreased as the temperature increased.
The stiffness of the structure will reduce by increasing the temperature. From 300 K to 500 K, the w_p will increase by approximately 8.4%.

- Ratio of plate length and width:
  $(20/-20/20)$s ply: $a/b = 2.0$ has the smallest w_p and largest F_c.
  $(45/-45/45)$s ply: $a/b = 0.5$ has the smallest w_p and largest F_c; $a/b = 2.0$ has the largest w_p and smallest F_c.
  $(70/-70/70)$s ply: $a/b = 0.5$ has the largest w_p and smallest F_c.
  The t_r decreased at first and then increased as a/b increased.
  The t_c increased as a/b increased.
  Due to the anisotropic honeycomb core, the geometry scale has more influence on the impact response. Using the honeycomb section as the long side of the structure can reduce the F_c.

- Thickness of surface layers:
  $(20/-20/20)$s ply: $h_s = 0.6$ mm has the smallest w_p, largest F_c and shortest t_r and t_c; $h_s = 2.4$ mm has the smallest w_p, largest F_c and shortest t_r and t_c.
  $(45/-45/45)$s ply: $h_s = 0.6$ mm has the largest w_p, longest t_r and t_c; $h_s = 2.4$ mm has the largest w_p, smallest F_c and longest t_r and t_c.
  $(70/-70/70)$s ply: $h_s = 0.6$ mm has the smallest F_c.
  Increasing h_s can lead to a reduction in the w_p and an increase in the F_c by increasing the stiffness of the structure.

Author Contributions: Conceptualization, C.Y.; data curation, J.Z.; funding acquisition, W.M.; investigation, Z.Z.; methodology, C.Y. and W.M.; project administration, C.Y.; resources, Z.Z.; software, W.M.; supervision, J.Z.; validation, C.Y.; visualization, W.M.; writing—original draft, C.Y.; writing—review and editing, C.Y. All authors have read and agreed to the published version of the manuscript.

Funding: This research was funded by the National Natural Science Foundation of China, grant number 12002169, 11902160. Jiangsu Province Postdoctoral Science Foundation, grant number 2020Z226. The fundamental research funds for the central universities, grant number 309181B8807.

Institutional Review Board Statement: Not applicable.

Informed Consent Statement: Not applicable.

Data Availability Statement: Not applicable.

Conflicts of Interest: The authors declare that the work described has not been published before; they have no conflict of interest regarding the publication of this article.

Appendix A

$$\mathbf{A}_{13} = \begin{vmatrix} A_{21} & A_{22} & 0 & 0 & A_{26} \\ A_{31} & A_{32} & 0 & 0 & A_{36} \\ 0 & 0 & A_{44} & A_{45} & 0 \\ 0 & 0 & A_{45} & A_{55} & 0 \\ A_{61} & A_{62} & 0 & 0 & A_{66} \end{vmatrix}, \quad \mathbf{A}_{23} = \begin{vmatrix} A_{11} & A_{12} & 0 & 0 & A_{16} \\ A_{31} & A_{32} & 0 & 0 & A_{36} \\ 0 & 0 & A_{44} & A_{45} & 0 \\ 0 & 0 & A_{45} & A_{55} & 0 \\ A_{61} & A_{62} & 0 & 0 & A_{66} \end{vmatrix},$$

$$\mathbf{A}_{5-1} = \begin{vmatrix} A_{22} & A_{23} & 0 & 0 & A_{26} \\ A_{32} & A_{33} & 0 & 0 & A_{36} \\ 0 & 0 & A_{44} & A_{45} & 0 \\ 0 & 0 & A_{45} & A_{55} & 0 \\ A_{62} & A_{63} & 0 & 0 & A_{66} \end{vmatrix}, \quad \mathbf{A}_{5-2} = \begin{vmatrix} A_{11} & A_{13} & 0 & 0 & A_{16} \\ A_{31} & A_{33} & 0 & 0 & A_{36} \\ 0 & 0 & A_{44} & A_{45} & 0 \\ 0 & 0 & A_{45} & A_{55} & 0 \\ A_{61} & A_{63} & 0 & 0 & A_{66} \end{vmatrix},$$

$$\mathbf{B}_{5-1} = \begin{vmatrix} B_{12} & B_{13} & 0 & 0 & B_{16} \\ B_{22} & B_{23} & 0 & 0 & B_{26} \\ B_{32} & B_{33} & 0 & 0 & B_{36} \\ 0 & 0 & B_{44} & B_{45} & 0 \\ 0 & 0 & B_{45} & B_{55} & 0 \\ B_{62} & B_{63} & 0 & 0 & B_{66} \end{vmatrix}, \quad \mathbf{B}_{5-2} = \begin{vmatrix} B_{11} & B_{13} & 0 & 0 & B_{16} \\ B_{21} & B_{23} & 0 & 0 & B_{26} \\ B_{31} & B_{33} & 0 & 0 & B_{36} \\ 0 & 0 & B_{44} & B_{45} & 0 \\ 0 & 0 & B_{45} & B_{55} & 0 \\ B_{61} & B_{63} & 0 & 0 & B_{66} \end{vmatrix},$$

$$\mathbf{B}_{5-3} = \begin{vmatrix} B_{11} & B_{12} & 0 & 0 & B_{16} \\ B_{21} & B_{22} & 0 & 0 & B_{26} \\ B_{31} & B_{32} & 0 & 0 & B_{36} \\ 0 & 0 & B_{44} & B_{45} & 0 \\ 0 & 0 & B_{45} & B_{55} & 0 \\ B_{61} & B_{62} & 0 & 0 & B_{66} \end{vmatrix}, \quad \mathbf{B}_{6-1} = \begin{vmatrix} B_{21} & B_{22} & B_{23} & 0 & 0 & B_{26} \\ B_{31} & B_{32} & B_{33} & 0 & 0 & B_{36} \\ 0 & 0 & 0 & B_{44} & B_{45} & 0 \\ 0 & 0 & 0 & B_{45} & B_{55} & 0 \\ B_{61} & B_{62} & B_{63} & 0 & 0 & B_{66} \end{vmatrix},$$

$$\mathbf{B}_{6-2} = \begin{vmatrix} B_{11} & B_{12} & B_{13} & 0 & 0 & B_{16} \\ B_{31} & B_{32} & B_{33} & 0 & 0 & B_{36} \\ 0 & 0 & 0 & B_{44} & B_{45} & 0 \\ 0 & 0 & 0 & B_{45} & B_{55} & 0 \\ B_{61} & B_{62} & B_{63} & 0 & 0 & B_{66} \end{vmatrix}, \quad \mathbf{D} = \begin{vmatrix} D_{11} & D_{12} & D_{13} & 0 & 0 & D_{16} \\ D_{21} & D_{22} & D_{23} & 0 & 0 & D_{26} \\ D_{31} & D_{32} & D_{33} & 0 & 0 & D_{36} \\ 0 & 0 & 0 & D_{44} & D_{45} & 0 \\ 0 & 0 & 0 & D_{45} & D_{55} & 0 \\ D_{61} & D_{62} & D_{63} & 0 & 0 & D_{66} \end{vmatrix}.$$

Appendix B

$$\mathbf{M} = \int_\Omega \left(\mathbf{P}_n^T \mathbf{m} \mathbf{P}_n^{-1}\right) d\Omega, \quad \mathbf{K} = \int_\Omega \left\{ \begin{Bmatrix} \mathbf{B}_m^T \\ \mathbf{B}_b^T \\ \mathbf{B}_s^T \end{Bmatrix}^T \mathbf{S} \begin{Bmatrix} \mathbf{B}_m \\ \mathbf{B}_b \\ \mathbf{B}_s \end{Bmatrix} \right\} d\Omega$$

where

$$\mathbf{P}_n = \begin{bmatrix} p_n & 0 & 0 & 0 & 0 \\ 0 & p_n & 0 & 0 & 0 \\ 0 & 0 & p_n & 0 & 0 \\ 0 & 0 & 0 & p_n & 0 \\ 0 & 0 & 0 & 0 & p_n \end{bmatrix}, \quad \mathbf{m} = \begin{bmatrix} I_0 & 0 & 0 & I_1 & 0 \\ 0 & I_0 & 0 & 0 & I_1 \\ 0 & 0 & I_0 & 0 & 0 \\ I_1 & 0 & 0 & I_2 & 0 \\ 0 & I_1 & 0 & 0 & I_2 \end{bmatrix},$$

$$(I_0, I_1, I_2) = \int_{-h/2}^{h/2} \rho\left(1, z, z^2\right) dz$$

$$\mathbf{B}_m = \begin{bmatrix} \frac{\partial p_n}{\partial x} & 0 & 0 & 0 & 0 \\ 0 & \frac{\partial p_n}{\partial y} & 0 & 0 & 0 \\ \frac{\partial p_n}{\partial y} & \frac{\partial p_n}{\partial x} & 0 & 0 & 0 \end{bmatrix}, \quad \mathbf{B}_b = \begin{bmatrix} 0 & 0 & 0 & \frac{\partial p_n}{\partial x} & 0 \\ 0 & 0 & 0 & 0 & \frac{\partial p_n}{\partial y} \\ 0 & 0 & 0 & \frac{\partial p_n}{\partial y} & \frac{\partial p_n}{\partial x} \end{bmatrix}, \quad \mathbf{B}_s = \begin{bmatrix} 0 & 0 & \frac{\partial p_n}{\partial y} & 0 & p_n \\ 0 & 0 & \frac{\partial p_n}{\partial x} & p_n & 0 \end{bmatrix}$$

References

1. Iijima, S. Carbon nanotubes: Past, present, and future. *Phys. B Condens. Matter* **2002**, *323*, 1–5. [CrossRef]
2. Esawi, A.M.; Farag, M.M. Carbon nanotube reinforced composites: Potential and current challenges. *Mater. Des.* **2007**, *28*, 2394–2401. [CrossRef]
3. Hu, K.; Kulkarni, D.D.; Choi, I.; Tsukruk, V.V. Graphene-polymer nanocomposites for structural and functional applications. *Prog. Polym. Sci.* **2014**, *39*, 1934–1972. [CrossRef]
4. Shen, H.S. Nonlinear bending of functionally graded carbon nanotube-reinforced composite plates in thermal environments. *Compos. Struct.* **2009**, *91*, 9–19. [CrossRef]
5. Kwon, H.; Bradbury, C.R.; Leparoux, M. Fabrication of Functionally Graded Carbon Nanotube-Reinforced Aluminum Matrix Composite. *Adv. Eng. Mater.* **2011**, *13*, 325–329. [CrossRef]
6. Shen, H.S. Postbuckling of nanotube-reinforced composite cylindrical shells in thermal environments, Part II: Pressure-loaded shells. *Compos. Struct.* **2011**, *93*, 2496–2503. [CrossRef]

7. Jafari Mehrabadi, S.; Sobhani Aragh, B.; Khoshkhahesh, V.; Taherpour, A. Mechanical buckling of nanocomposite rectangular plate reinforced by aligned and straight single-walled carbon nanotubes. *Compos. Part Eng.* **2012**, *43*, 2031–2040. [CrossRef]
8. Zhang, L.; Lei, Z.; Liew, K. An element-free IMLS-Ritz framework for buckling analysis of FG–CNT reinforced composite thick plates resting on Winkler foundations. *Eng. Anal. Bound. Elem.* **2015**, *58*, 7–17. [CrossRef]
9. Fan, Y.; Wang, H. Nonlinear bending and postbuckling analysis of matrix cracked hybrid laminated plates containing carbon nanotube reinforced composite layers in thermal environments. *Compos. Part B Eng.* **2016**, *86*, 1–16. [CrossRef]
10. Zhang, L.; Liew, K. Postbuckling analysis of axially compressed CNT reinforced functionally graded composite plates resting on Pasternak foundations using an element-free approach. *Compos. Struct.* **2016**, *138*, 40–51. [CrossRef]
11. Zhang, L.; Liew, K.; Reddy, J. Postbuckling of carbon nanotube reinforced functionally graded plates with edges elastically restrained against translation and rotation under axial compression. *Comput. Methods Appl. Mech. Eng.* **2016**, *298*, 1–28. [CrossRef]
12. Kiani, Y. Buckling of FG-CNT-reinforced composite plates subjected to parabolic loading. *Acta Mech.* **2017**, *228*, 1303–1319. [CrossRef]
13. Kiani, Y. Thermal buckling of temperature-dependent FG-CNT-reinforced composite skew plates. *J. Therm. Stress.* **2017**, *40*, 1442–1460. [CrossRef]
14. Kiani, Y. Thermal post-buckling of temperature dependent sandwich plates with FG-CNTRC face sheets. *J. Therm. Stress.* **2018**, *41*, 866–882. [CrossRef]
15. Kiani, Y.; Mirzaei, M. Rectangular and skew shear buckling of FG-CNT reinforced composite skew plates using Ritz method. *Aerosp. Sci. Technol.* **2018**, *77*, 388–398. [CrossRef]
16. Ansari, R.; Hassani, R.; Gholami, R.; Rouhi, H. Thermal postbuckling analysis of FG-CNTRC plates with various shapes and temperature-dependent properties using the VDQ-FEM technique. *Aerosp. Sci. Technol.* **2020**, *106*, 106078. [CrossRef]
17. Hieu, P.T.; Van Tung, H. Thermomechanical postbuckling of pressure loaded CNT reinforced composite cylindrical shells under tangential edge constraints and various temperature conditions. *Polym. Compos.* **2020**, *41*, 244–257. [CrossRef]
18. Wang, Z.X.; Shen, H.S. Nonlinear vibration of nanotube-reinforced composite plates in thermal environments. *Comput. Mater. Sci.* **2011**, *50*, 2319–2330. [CrossRef]
19. Zhu, P.; Lei, Z.; Liew, K. Static and free vibration analyses of carbon nanotube-reinforced composite plates using finite element method with first order shear deformation plate theory. *Compos. Struct.* **2012**, *94*, 1450–1460. [CrossRef]
20. Lei, Z.; Liew, K.; Yu, J. Free vibration analysis of functionally graded carbon nanotube-reinforced composite plates using the element-free kp-Ritz method in thermal environment. *Compos. Struct.* **2013**, *106*, 128–138. [CrossRef]
21. Abdollahzadeh Shahrbabaki, E.; Alibeigloo, A. Three-dimensional free vibration of carbon nanotube-reinforced composite plates with various boundary conditions using Ritz method. *Compos. Struct.* **2014**, *111*, 362–370. [CrossRef]
22. Kamarian, S.; Shakeri, M.; Yas, M.; Bodaghi, M.; Pourasghar, A. Free vibration analysis of functionally graded nanocomposite sandwich beams resting on Pasternak foundation by considering the agglomeration effect of CNTs. *J. Sandw. Struct. Mater.* **2015**, *17*, 632–665. [CrossRef]
23. Mehar, K.; Panda, S.K. Geometrical nonlinear free vibration analysis of FG-CNT reinforced composite flat panel under uniform thermal field. *Compos. Struct.* **2016**, *143*, 336–346. [CrossRef]
24. Wu, C.P.; Li, H.Y. Three-dimensional free vibration analysis of functionally graded carbon nanotube-reinforced composite plates with various boundary conditions. *J. Vib. Control* **2016**, *22*, 89–107. [CrossRef]
25. Wang, Q.; Qin, B.; Shi, D.; Liang, Q. A semi-analytical method for vibration analysis of functionally graded carbon nanotube reinforced composite doubly-curved panels and shells of revolution. *Compos. Struct.* **2017**, *174*, 87–109. [CrossRef]
26. Thomas, B.; Roy, T. Vibration and damping analysis of functionally graded carbon nanotubes reinforced hybrid composite shell structures. *J. Vib.Control* **2017**, *23*, 1711–1738. [CrossRef]
27. Hasrati, E.; Ansari, R.; Torabi, J. Nonlinear Forced Vibration Analysis of FG-CNTRC Cylindrical Shells Under Thermal Loading Using a Numerical Strategy. *Int. J. Appl. Mech.* **2017**, *9*, 1750108. [CrossRef]
28. Ansari, R.; Hasrati, E.; Torabi, J. Nonlinear vibration response of higher-order shear deformable FG-CNTRC conical shells. *Compos. Struct.* **2019**, *222*, 110906. [CrossRef]
29. Sofiyev, A.; Hui, D. On the vibration and stability of FGM cylindrical shells under external pressures with mixed boundary conditions by using FOSDT. *Thin-Walled Struct.* **2019**, *134*, 419–427. [CrossRef]
30. Maji, P.; Rout, M.; Karmakar, A. The free vibration response of temperature-dependent carbon nanotube-reinforced composite stiffened plate. *Mech. Adv. Mater. Struct.* **2021**, *29*, 2555–2569. [CrossRef]
31. Quoc, T.H.; Van Tham, V.; Tu, T.M. Active vibration control of a piezoelectric functionally graded carbon nanotube-reinforced spherical shell panel. *Acta Mech.* **2021**, *232*, 1005–1023. [CrossRef]
32. Wang, Z.X.; Xu, J.; Qiao, P. Nonlinear low-velocity impact analysis of temperature-dependent nanotube-reinforced composite plates. *Compos. Struct.* **2014**, *108*, 423–434. [CrossRef]
33. Jam, J.; Kiani, Y. Low velocity impact response of functionally graded carbon nanotube reinforced composite beams in thermal environment. *Compos. Struct.* **2015**, *132*, 35–43. [CrossRef]
34. Song, Z.G.; Zhang, L.W.; Liew, K.M. Dynamic responses of CNT reinforced composite plates subjected to impact loading. *Compos. Part B Eng.* **2016**, *99*, 154–161. [CrossRef]
35. Malekzadeh, P.; Dehbozorgi, M. Low velocity impact analysis of functionally graded carbon nanotubes reinforced composite skew plates. *Compos. Struct.* **2016**, *140*, 728–748. [CrossRef]

36. Ebrahimi, F.; Habibi, S. Low-velocity impact response of laminated FG-CNT reinforced composite plates in thermal environment. *Adv. Nano Res.* **2017**, *5*, 69–97.
37. Yang, C.H.; Ma, W.N.; Ma, D.W.; He, Q.; Zhong, J.L. Analysis of the low velocity impact response of functionally graded carbon nanotubes reinforced composite spherical shells. *J. Mech. Sci. Technol.* **2018**, *32*, 2681–2691. [CrossRef]
38. Yang, C.H.; Ma, W.N.; Ma, D.W. Low-velocity impact analysis of carbon nanotube reinforced composite laminates. *J. Mater. Sci.* **2018**, *53*, 637–656. [CrossRef]
39. Fallah, M.; Daneshmehr, A.R.; Zarei, H.; Bisadi, H.; Minak, G. Low velocity impact modeling of functionally graded carbon nanotube reinforced composite (FG-CNTRC) plates with arbitrary geometry and general boundary conditions. *Compos. Struct.* **2018**, *187*, 554–565. [CrossRef]
40. Bayat, M.R.; Rahmani, O.; Mosavi Mashhadi, M. Nonlinear low-velocity impact analysis of functionally graded nanotube-reinforced composite cylindrical shells in thermal environments. *Polym. Compos.* **2018**, *39*, 730–745. [CrossRef]
41. Ma, W.; Yang, C.; Ma, D.; Zhong, J. Low-velocity impact response of nanotube-reinforced composite sandwich curved panels. *Sādhanā* **2019**, *44*, 227. [CrossRef]
42. Khalkhali, A.; Geran Malek, N.; Bozorgi Nejad, M. Effects of the impactor geometrical shape on the non-linear low-velocity impact response of sandwich plate with CNTRC face sheets. *J. Sandw. Struct. Mater.* **2020**, *22*, 962–990. [CrossRef]
43. Evans, K.E.; Nkansah, M.A.; Hutchinson, I.J.; Rogers, S.C. Molecular network design. *Nature* **1991**, *353*, 124. [CrossRef]
44. Shen, H.S.; Li, C.; Reddy, J. Large amplitude vibration of FG-CNTRC laminated cylindrical shells with negative Poisson's ratio. *Comput. Methods Appl. Mech. Eng.* **2020**, *360*, 112727. [CrossRef]
45. Yang, J.; Huang, X.H.; Shen, H.S. Nonlinear vibration of temperature-dependent FG-CNTRC laminated plates with negative Poisson's ratio. *Thin-Walled Struct.* **2020**, *148*, 106514. [CrossRef]
46. Whitty, J.; Alderson, A.; Myler, P.; Kandola, B. Towards the design of sandwich panel composites with enhanced mechanical and thermal properties by variation of the in-plane Poisson's ratios. *Compos. Part A Appl. Sci. Manuf.* **2003**, *34*, 525–534. [CrossRef]
47. Scarpa, F.; Blain, S.; Lew, T.; Perrott, D.; Ruzzene, M.; Yates, J. Elastic buckling of hexagonal chiral cell honeycombs. *Compos. Part A Appl. Sci. Manuf.* **2007**, *38*, 280–289. [CrossRef]
48. Evans, K.E. Auxetic polymers: A new range of materials. *Endeavour* **1991**, *15*, 170–174. [CrossRef]
49. Scarpa, F.; Yates, J.R.; Ciffo, L.G.; Patsias, S. Dynamic crushing of auxetic open-cell polyurethane foam. *Proc. Inst. Mech. Eng. Part C J. Mech. Eng. Sci.* **2002**, *216*, 1153–1156. [CrossRef]
50. Liu, W.; Wang, N.; Luo, T.; Lin, Z. In-plane dynamic crushing of re-entrant auxetic cellular structure. *Mater. Des.* **2016**, *100*, 84–91. [CrossRef]
51. Mohsenizadeh, S.; Alipour, R.; Shokri Rad, M.; Farokhi Nejad, A.; Ahmad, Z. Crashworthiness assessment of auxetic foam-filled tube under quasi-static axial loading. *Mater. Des.* **2015**, *88*, 258–268. [CrossRef]
52. Hou, S.; Liu, T.; Zhang, Z.; Han, X.; Li, Q. How does negative Poisson's ratio of foam filler affect crashworthiness? *Mater. Des.* **2015**, *82*, 247–259. [CrossRef]
53. Lakes, R. Foam Structures with a Negative Poisson's Ratio. *Science* **1987**, *235*, 1038–1040. [CrossRef] [PubMed]
54. Evans, K.E.; Alderson, A. Auxetic Materials: Functional Materials and Structures from Lateral Thinking! *Adv. Mater.* **2000**, *12*, 617–628. [CrossRef]
55. Surjadi, J.U.; Gao, L.; Du, H.; Li, X.; Xiong, X.; Fang, N.X.; Lu, Y. Mechanical Metamaterials and Their Engineering Applications. *Adv. Eng. Mater.* **2019**, *21*, 1800864. [CrossRef]
56. Duc, N.D.; Seung-Eock, K.; Tuan, N.D.; Tran, P.; Khoa, N.D. New approach to study nonlinear dynamic response and vibration of sandwich composite cylindrical panels with auxetic honeycomb core layer. *Aerosp. Sci. Technol.* **2017**, *70*, 396–404. [CrossRef]
57. Cong, P.H.; Khanh, N.D.; Khoa, N.D.; Duc, N.D. New approach to investigate nonlinear dynamic response of sandwich auxetic double curves shallow shells using TSDT. *Compos. Struct.* **2018**, *185*, 455–465. [CrossRef]
58. Li, C.; Shen, H.S.; Wang, H. Nonlinear Vibration of Sandwich Beams with Functionally Graded Negative Poisson's Ratio Honeycomb Core. *Int. J. Struct. Stab. Dyn.* **2019**, *19*, 1950034. [CrossRef]
59. Li, C.; Shen, H.S.; Wang, H. Nonlinear bending of sandwich beams with functionally graded negative Poisson's ratio honeycomb core. *Compos. Struct.* **2019**, *212*, 317–325. [CrossRef]
60. Li, C.; Shen, H.S.; Wang, H. Thermal post-buckling of sandwich beams with functionally graded negative Poisson's ratio honeycomb core. *Int. J. Mech. Sci.* **2019**, *152*, 289–297. [CrossRef]
61. Li, C.; Shen, H.S.; Wang, H.; Yu, Z. Large amplitude vibration of sandwich plates with functionally graded auxetic 3D lattice core. *Int. J. Mech. Sci.* **2020**, *174*, 105472. [CrossRef]
62. Li, C.; Shen, H.S.; Wang, H. Postbuckling behavior of sandwich plates with functionally graded auxetic 3D lattice core. *Compos. Struct.* **2020**, *237*, 111894. [CrossRef]
63. Li, C.; Shen, H.S.; Wang, H. Nonlinear dynamic response of sandwich plates with functionally graded auxetic 3D lattice core. *Nonlinear Dyn.* **2020**, *100*, 3235–3252. [CrossRef]
64. Li, C.; Shen, H.S.; Wang, H. Full-scale finite element modeling and nonlinear bending analysis of sandwich plates with functionally graded auxetic 3D lattice core. *J. Sandw. Struct. Mater.* **2020**, *23*, 109963622092465. [CrossRef]
65. Wan, H.; Ohtaki, H.; Kotosaka, S.; Hu, G. A study of negative Poisson's ratios in auxetic honeycombs based on a large deflection model. *Eur. J. Mech.-A/Solids* **2004**, *23*, 95–106. [CrossRef]

66. Grima, J.N.; Oliveri, L.; Attard, D.; Ellul, B.; Gatt, R.; Cicala, G.; Recca, G. Hexagonal Honeycombs with Zero Poisson's Ratios and Enhanced Stiffness. *Adv. Eng. Mater.* **2010**, *12*, 855–862. [CrossRef]
67. Assidi, M.; Ganghoffer, J.F. Composites with auxetic inclusions showing both an auxetic behavior and enhancement of their mechanical properties. *Compos. Struct.* **2012**, *94*, 2373–2382. [CrossRef]
68. Grujicic, M.; Galgalikar, R.; Snipes, J.; Yavari, R.; Ramaswami, S. Multi-physics modeling of the fabrication and dynamic performance of all-metal auxetic-hexagonal sandwich-structures. *Mater. Des.* **2013**, *51*, 113–130. [CrossRef]
69. Liu, X.F.; Wang, Y.F.; Wang, Y.S.; Zhang, C. Wave propagation in a sandwich plate with a periodic composite core. *J. Sandw. Struct. Mater.* **2014**, *16*, 319–338. [CrossRef]
70. Qiao, J.; Chen, C. Impact resistance of uniform and functionally graded auxetic double arrowhead honeycombs. *Int. J. Impact Eng.* **2015**, *83*, 47–58. [CrossRef]
71. Zhang, X.C.; An, L.Q.; Ding, H.M.; Zhu, X.Y.; El-Rich, M. The influence of cell micro-structure on the in-plane dynamic crushing of honeycombs with negative Poisson's ratio. *J. Sandw. Struct. Mater.* **2015**, *17*, 26–55. [CrossRef]
72. Zhang, G.; Ghita, O.R.; Evans, K.E. Dynamic thermo-mechanical and impact properties of helical auxetic yarns. *Compos. Part B Eng.* **2016**, *99*, 494–505. [CrossRef]
73. Hadi Harkati, E.; Bezazi, A.; Scarpa, F.; Alderson, K.; Alderson, A. Modelling the influence of the orientation and fibre reinforcement on the Negative Poisson's ratio in composite laminates. *Phys. Status Solidi* **2007**, *244*, 883–892. [CrossRef]
74. Lim, T.C. Vibration of thick auxetic plates. *Mech. Res. Commun.* **2014**, *61*, 60–66. [CrossRef]
75. Zhang, R.; Yeh, H.L.; Yeh, H.Y. A Preliminary Study of Negative Poisson's Ratio of Laminated Fiber Reinforced Composites. *J. Reinf. Plast. Compos.* **1998**, *17*, 1651–1664. [CrossRef]
76. Shen, H.S.; Huang, X.H.; Yang, J. Nonlinear bending of temperature-dependent FG-CNTRC laminated plates with negative Poisson's ratio. *Mech. Adv. Mater. Struct.* **2020**, *27*, 1141–1153. [CrossRef]
77. Yang, J.; Huang, X.H.; Shen, H.S. Nonlinear flexural behavior of temperature-dependent FG-CNTRC laminated beams with negative Poisson's ratio resting on the Pasternak foundation. *Eng. Struct.* **2020**, *207*, 110250. [CrossRef]
78. Yang, J.; Huang, X.H.; Shen, H.S. Nonlinear Vibration of Temperature-Dependent FG-CNTRC Laminated Beams with Negative Poisson's Ratio. *Int. J. Struct. Stab. Dyn.* **2020**, *20*, 2050043. [CrossRef]
79. Yu, Y.; Shen, H.S. A comparison of nonlinear vibration and bending of hybrid CNTRC/metal laminated plates with positive and negative Poisson's ratios. *Int. J. Mech. Sci.* **2020**, *183*, 105790. [CrossRef]
80. Fan, Y.; Wang, Y. The effect of negative Poisson's ratio on the low-velocity impact response of an auxetic nanocomposite laminate beam. *Int. J. Mech. Mater. Des.* **2021**, *17*, 153–169. [CrossRef]
81. Huang, X.h.; Yang, J.; Bai, L.; Wang, X.e.; Ren, X. Theoretical solutions for auxetic laminated beam subjected to a sudden load. *Structures* **2020**, *28*, 57–68. [CrossRef]
82. Huang, X.h.; Yang, J.; Wang, X.e.; Azim, I. Combined analytical and numerical approach for auxetic FG-CNTRC plate subjected to a sudden load. *Eng. Comput.* **2020**. [CrossRef]
83. Abrate, S. Hull Slamming. *Appl. Mech. Rev.* **2013**, *64*, 060803. [CrossRef]
84. Lin, F.; Xiang, Y. Numerical Analysis on Nonlinear Free Vibration of Carbon Nanotube Reinforced Composite Beams. *Int. J. Struct. Stab. Dyn.* **2014**, *14*, 1350056. [CrossRef]
85. Lin, F.; Xiang, Y. Vibration of carbon nanotube reinforced composite beams based on the first and third order beam theories. *Appl. Math. Model.* **2014**, *38*, 3741–3754. [CrossRef]
86. Moradi-Dastjerdi, R.; Payganeh, G.; Tajdari, M. Resonance in functionally graded nanocomposite cylinders reinforced by wavy carbon nanotube. *Polym. Compos.* **2017**, *38*, E542–E552. [CrossRef]

MDPI AG
Grosspeteranlage 5
4052 Basel
Switzerland
Tel.: +41 61 683 77 34

Polymers Editorial Office
E-mail: polymers@mdpi.com
www.mdpi.com/journal/polymers

Disclaimer/Publisher's Note: The statements, opinions and data contained in all publications are solely those of the individual author(s) and contributor(s) and not of MDPI and/or the editor(s). MDPI and/or the editor(s) disclaim responsibility for any injury to people or property resulting from any ideas, methods, instructions or products referred to in the content.

www.ingramcontent.com/pod-product-compliance
Lightning Source LLC
LaVergne TN
LVHW070204100526
838202LV00015B/1993